CONSERVATION AND
THE GENETICS OF POPULATIONS

Fred W. Allendorf
University of Montana
and
Victoria University of Wellington

and

Gordon Luikart
Université Joseph Fourier, CNRS
and
University of Montana

With illustrations by Agostinho Antunes

Blackwell
Publishing

BLACKWELL PUBLISHING   1004918714
350 Main Street, Malden, MA 02148-5020, USA
9600 Garsington Road, Oxford OX4 2DQ, UK
550 Swanston Street, Carlton, Victoria 3053, Australia

The right of Fred W. Allendorf and Gordon Luikart to be identified as the Authors of this Work
has been asserted in accordance with the UK Copyright, Designs, and Patents Act 1988.

First published 2007 by Blackwell Publishing Ltd

1   2007

Library of Congress Cataloging-in-Publication Data

Allendorf, Frederick William.
    Conservation and the genetics of populations / Fred W. Allendorf and
Gordon Luikart.
        p.   cm.
    ISBN-13: 978-1-4051-2145-3 (pbk. : alk. paper)
    ISBN-10: 1-4051-2145-9 (pbk. : alk. paper)
    1. Biological diversity conservation.   2. Population genetics.   3.
Evolutionary genetics.   I. Luikart, Gordon.   II. Title.
QH75.A42 2006
576.5'8—dc22

                                                2006001707

A catalogue record for this title is available from the British Library.

Set in 10.5 / 12.5pt Dante
by Graphicraft Limited, Hong Kong
Printed and bound in Singapore
by Markono Print Media Pte Ltd

The publisher's policy is to use permanent paper from mills that operate a sustainable forestry
policy, and which has been manufactured from pulp processed using acid-free and elementary
chlorine-free practices. Furthermore, the publisher ensures that the text paper and cover board
used have met acceptable environmental accreditation standards.

For further information on
Blackwell Publishing, visit our website:
www.blackwellpublishing.com

Chapter opening images remain copyright of Agostinho Antunes

Contents

PART I: INTRODUCTION

Authors of Guest Boxes

C. Scott Baker, School of Biological Sciences, University of Auckland, Auckland, New Zealand (cs.baker@auckland.ac.nz). Chapter 9.

Louis Bernatchez, Département de biologie, Université Laval, Quebec QC G1K 7P4, Canada (Louis.Bernatchez@bio.ulaval.ca). Chapter 5.

James V. Briskie, School of Biological Sciences, University of Canterbury, Christchurch, New Zealand (jim.briskie@canterbury.ac.nz). Chapter 18.

Vincent Castric, Laboratoire de génétique et évolution des populations végétales, Villeneuve d'Ascq Cedex, France (Vincent.Castric@univ-lille1.fr). Chapter 5.

Michael L. Collyer, Department of Statistics, Department of Ecology, Evolution, and Organismal Biology, Iowa State University, Ames, IA 50011, USA (collyer@iastate.edu). Chapter 8.

David W. Coltman, Department of Biological Sciences, University of Alberta, Edmonton, Alberta, Canada (dcoltman@ualberta.ca). Chapter 11.

James F. Crow, Laboratory of Genetics, University of Wisconsin, Madison, WI 53706 USA (jfcrow@wisc.edu). Appendix.

Nancy N. FitzSimmons, School of Resource, Environmental and Heritage Sciences, University of Canberra, Canberra, ACT 2601, Australia (Nancy.Fitzsimmons@canberra.edu.au), Chapter 4.

Chris J. Foote, Fisheries and Aquaculture Department, Malaspina University College, Nainamo, British Columbia, Canada (footec@mala.bc.ca). Chapter 2.

Stephen H. Forbes, Division of Medical Genetics, Department of Medicine, University of Washington, Seattle, WA 98195, USA (stephenforbes@earthlink.net). Chapter 10.

Robert C. Lacy, Department of Conservation Science, Chicago Zoological Society, Brookfield, IL 60513, USA (rlacy@ix.netcom.com). Chapter 13.

Paul L. Leberg, Department of Biology, University of Louisiana-Lafayette, Lafayette, LA 70504, USA (pll6743@louisiana.edu). Chapter 6.

John L. Maron, Division of Biological Sciences, University of Montana, Missoula, MT 59812, USA (john.maron@mso.umt.edu). Chapter 19.

Craig R. Miller, Department of Biology, Eastern Washington University, Cheney, WA 99004, USA (cmiller2@mail.ewu.edu). Chapter 7.

L. Scott Mills, Wildlife Biology Program, University of Montana, Missoula, MT 59812, USA (smills@forestry.umt.edu). Chapter 1.

Brian G. Murray, School of Biological Sciences, University of Auckland, Auckland, New Zealand (b.murray@auckland.ac.nz). Chapter 3.

Michael W. Nachman, Department of Ecology and Evolutionary Biology, University of Arizona, Tucson, AZ 85721, USA (nachman@u.arizona.edu). Chapter 12.

Franz B. Pichler, School of Biological Sciences, University of Auckland, Auckland, New Zealand (f.pichler@auckland.ac.nz). Chapter 9.

Loren H. Rieseberg, Indiana University, Department of Biology, Bloomington, IN 47405, USA (lriesebe@bio.indiana.edu). Chapter 17.

David L. Rogowski, School of Biological Sciences, Queen's University Belfast, Belfast BT9 7BL, Northern Ireland, UK (d.rogowski@qub.ac.uk). Chapter 6.

Michael E. Soulé, P.O. Box 1808, Paonia, CO 81428, USA (rewild@tds.net). Chapter 1.

Craig A. Stockwell, Department of Biological Sciences, North Dakota State University, Fargo, ND 58105, USA (Craig.Stockwell@ndsu.nodak.edu). Chapter 8.

Andrea C. Taylor, School of Biological Sciences, Monash University, Victoria 3800, Australia (Andrea.Taylor@sci.monash.edu.au). Chapter 14.

Robert C. Vrijenhoek, Monterey Bay Aquarium Research Institute, Moss Landing, CA 95039, USA (vrijen@mbari.org). Chapter 15.

Lisette P. Waits, Fish and Wildlife Resources, University of Idaho, Moscow, ID 83844, USA (lwaits@uidaho.edu). Chapters 7 and 20.

Robin S. Waples, Northwest Fisheries Science Center, Seattle, WA 98112, USA (Robin.Waples@noaa.gov). Chapter 16.

Andrew Young, Centre for Plant Biodiversity Research, CSIRO Plant Industry, Canberra, ACT 2601, Australia (andrew.young@csiro.au). Chapter 3.

Preface

The many beings are numberless; I vow to save them all.

Traditional Zen vow

The one process now going on that will take millions of years to correct is the loss of genetic and species diversity by the destruction of natural habitats. This is the folly our descendants are least likely to forgive us.

Edward O. Wilson, 1984

This book is about applying the concepts and tools of genetics to problems in conservation. Our guiding principle in writing has been to provide the conceptual basis for understanding the genetics of biological problems in conservation. We have not attempted to review the extensive and ever growing literature in this area. Rather we have tried to explain the underlying concepts and to provide enough clear examples and key citations for further consideration. We also have strived to provide enough background so that students can read and understand the primary literature.

Our primary intended audience is broadly trained biologists who are interested in understanding the principles of conservation genetics and applying them to a wide range of particular issues in conservation. This includes advanced undergraduate and graduate students in biological sciences or resource management, as well as biologists working in conservation biology for management agencies. The treatment is intermediate and requires a basic understanding of ecology and genetics.

This book is not an argument for the importance of genetics in conservation. Rather, it is designed to provide the reader with the appropriate background to determine how genetic information may be useful in any specific case. The primary current causes of extinction are anthropogenic changes that affect ecological characteristics of populations (habitat loss, fragmentation, introduced species, etc.). However, genetic information and principles can be invaluable in developing conservation plans for species threatened with such effects.

The usefulness of genetic tools and concepts in the conservation of biological diversity is continually expanding as new molecular technologies, statistical methods, and computer programs are being developed at an increasing rate. Conservation genetics and molecular ecology are under explosive growth, and this growth is likely to continue for the foreseeable future. Indeed we have recently entered the age of genomics. New laboratory and computational technologies for generating and analyzing molecular genetic data are emerging at a rapid pace.

There are several excellent texts in population genetics available (e.g., Hartl and Clark 1997; Halliburton 2004; Hedrick 2005). These texts concentrate on questions related to the central focus of population and evolutionary genetics, which is to understand the processes and mechanisms by which evolutionary changes occur. There is substantial overlap between these texts and this book. However, the theme underlying this book is the application of an understanding of the genetics of natural populations to conservation.

We have endeavored to present a balanced view of theory and data. The first four chapters (Part I) provide an overview of the study of genetic variation in natural populations of plants and animals. The middle eight chapters (Part II) provide the basic principles of population genetics theory with an emphasis on concepts especially relevant for problems in conservation. The final eight chapters (Part III) synthesis these principles and apply them to a variety of topics in conservation.

We emphasize the interpretation and understanding of genetic data to answer biological questions in conservation. Discussion questions and problems are included at the end of each chapter to engage the reader in understanding the material. We believe well written problems and questions are an invaluable tool in learning the information presented in the book. These problems feature analysis of real data from populations, conceptual theoretical questions, and the use of computer simulations. A website contains example data sets and software programs for illustrating population genetic processes and for teaching methods for data analysis.

We have also included a comprehensive glossary. Words included in the glossary are bolded the first time that they are used in the text. Many of the disagreements and long-standing controversies in population and conservation genetics result from people using the same words to mean different things. It is important to define and use words precisely.

We have asked many of our colleagues to write guest boxes that present their own work in conservation genetics. Each chapter contains a guest box that provides further consideration of the topics of that chapter. These boxes provide the reader with a broader voice in conservation genetics, as well as familiarity with recent case study examples, and some of the major contributors to the literature in conservation genetics.

The contents of this book have been influenced by lecture notes and courses in population genetics taken by the senior author from Bob Costantino and Joe Felsenstein. We also thank Fred Utter for his contagious passion to uncover and describe genetic variation in natural populations. This book began as a series of notes for a course in conservation genetics that the senior author began while on sabbatical at the University of Oregon in 1993. About one-quarter of the chapters were completed within those first 6 months. However, as the demands of other obligations took over, progress slowed to a near standstill. The majority of this book has been written by the authors in close collaboration over the last 2 years.

Acknowledgments

We are grateful to the students in the Conservation Genetics course at the University of Montana over many years who have made writing this book enjoyable by their enthusiasm and comments. Earlier versions of this text were used in courses at the University of Oregon in 1993 and at the University of Minnesota in 1997; we are also grateful to those students for their encouragement and comments. We also gratefully acknowledge the University of Montana and Victoria University of Wellington for their support. Much appreciated support also was provided by Pierre Taberlet and the CNRS Laboratoire d'Ecologie Alpine. Part of this was written while FWA was supported as a Fulbright Senior Scholar. This book could not have been completed without the loving support of our wives Michel and Shannon.

We thank the many people from Blackwell Publishing for their encouragement, support, and help throughout the process of completing this book. We are grateful to the authors of the Guest Boxes who quickly replied to our many inquiries. We are indebted to Sally Aitken for her extremely thorough and helpful review of the entire book, John Powell for his excellent help with the glossary, and to Kea Allendorf for her help with the literature cited. We also thank many colleagues who have helped us by providing comments, information, unpublished data, and answers to questions: Teri Allendorf, Jon Ballou, Mark Beaumont, Albano Beja-Pereira, Steve Beissinger, Pierre Berthier, Giorgio Bertorelle, Matt Boyer, Brian Bowen, Ron Burton, Chris Cole, Charlie Daugherty, Sandie Degnan, Dawson Dunning, Norm Ellstrand, Dick Frankham, Chris Funk, Oscar Gaggiotti, Neil Gemmell, John Gilliespie, Mary Jo Godt, Dave Goulson, Ed Guerrant, Bengt Hansson, Sue Haig, Kim Hastings, Phil Hedrick, Kelly Hildner, Rod Hitchmough, J. T. Hogg, Denver Holt, Jeff Hutchings, Mike Ivie, Mike Johnson, Bill Jordan, Carrie Kappel, Joshua Kohn, Peter Lesica, Laura Lundquist, Shujin Luo, Lisa Meffert, Don Merton, Scott Mills, Andy Overall, Per Palsbøll, Jim Patton, Rod Peakall, Robert Pitman, Kristina Ramstad, Reg Reisenbichler, Bruce Rieman, Pete Ritchie, Bruce Rittenhouse, Bruce Robertson, Rob Robichaux, Nils Ryman, Mike Schwartz, Jim Seeb, Brad Shaffer, Pedro Silva, Paul Spruell, Paul Sunnucks, David Tallmon, Kathy Traylor-Holzer, Randal Voss, Hartmut Walter, Robin Waples, and Andrew Whiteley.

Fred W. Allendorf is a Regents Professor at the University of Montana and a Professorial Research Fellow at Victoria University of Wellington in New Zealand. His primary research interests are conservation and population genetics. He has published over 200 articles on the population genetics and conservation of fish, amphibians, mammals, invertebrates, and plants. He is a past President of the American Genetic Association, served as Director of the Population Biology Program of the National Science Foundation, and has served on the editorial boards of *Conservation Biology, Molecular Ecology, Evolution, Conservation Genetics, Molecular Biology and Evolution,* and the *Journal of Heredity.* He has taught conservation genetics at the University of Montana, University of Oregon, University of Minnesota, and Victoria University of Wellington.

Gordon Luikart is a Research Associate Professor at the University of Montana and a Visiting Professor in the Center for Investigation of Biodiversity and Genetic Resources at the University of Porto, Portugal. He was a Research Scientist with the Centre National de la Recherche Scientifique (CNRS) at the University Joseph Fourier in Grenoble, France, where he was awarded the CNRS bronze medal for research excellence. His research focuses on the conservation and genetics of wild and domestic animals, and includes nearly 50 publications in the field. He was a Fulbright Scholar at La Trobe University, Melbourne, Australia, is a member of the IUCN specialists group for Caprinae (mountain ungulate) conservation, and has served on the editorial boards of *Conservation Biology* and *Molecular Ecology Notes.*

List of Symbols

This list includes mathematical symbols with definitions and references to the primary chapters in which they are used. There is some duplication of usage which reflects the general usage in the literature. However, the specific meaning should be apparent from the context and chapter.

Symbol	Definition	Chapter
Latin symbols		
A	number of alleles at a locus	3, 4, 5, 6
B	number of lethal equivalents per gamete	13
c	probability of recolonization	15
d^2	squared difference in number of repeat units between the two microsatellite alleles in an individual	10
D	Nei's genetic distance	9
D	coefficient of gametic disequilibrium	10, 17
D'	standardized measure of gametic disequilibrium	10
D_B	gametic disequilibrium caused by population subdivision	10
D_C	composite measure of gametic disequilibrium	10
D_{ST}	proportion of total heterozygosity due to genetic divergence between subpopulations	19
e	probability of extinction of a subpopulation	15
E	probability of an event	A
f	inbreeding coefficient	6
F	pedigree inbreeding coefficient	6, 10, 13, A
F	proportion by which heterozygosity is reduced relative to heterozygosity in a random mating population	9

Symbol	Definition	Chapter
F_{IS}	departure from Hardy–Weinberg proportions within local demes or subpopulations	5, 9, 13
F_{IT}	overall departure from Hardy–Weinberg proportions	9
F_{SR}	proportion of total differentiation due to differences between subpopulations within regions	9
F_{ST}	proportion of genetic variation due to differences among populations	3, 9, 12, 13, 16
g	generation interval	7, 18
G_i	frequency of gamete i	10
G_{ST}	F_{ST} extended for three or more alleles	9
h	gene diversity, computationally equivalent to H_e, and especially useful for haploid marker systems	4, 7
h	degree of dominance of an allele	12
h	heterozygosity	6
H_0	null hypothesis	A
H_A	alternative hypothesis	A
H_B	broad sense heritability	11
H_e	expected proportion of heterozygotes	3, 4, 5, 9, A
H_0	observed heterozygosity	3, 9, A
H_N	narrow sense heritability	11
H_S	mean expected heterozygosity	3, 9, 12, 15
H_T	total genetic variation	3, 9, 15
k	number of gametes contributed by an individual to the next generation	7
k	number of populations	16
K	equilibrium size of a population	6
m	proportion of migrants	9, 15, 16
MP	match probability	20
n	sample size	
n	haploid chromosome number	3
N	population size	6, 7, 8, 19, A
N_c	census population size	7, 14, 18, A
N_e	effective population size	7, 8, 11, 12, 14, 18, A
N_{el}	inbreeding effective population size	7
N_{eV}	variance effective population size	7
N_f	number of females in a population	6, 7, 9
N_m	number of males in a population	6, 7, 9
p	frequency of allele A_1 (or A)	5, 6, 8
p	proportion of patches occupied in a metapopulation	15
P	proportion of loci that are polymorphic	3, 5
P	probability of an event	5
PE	probability of paternity exclusion; average probability of excluding (as father) a randomly sampled nonfather	20
PI_{av}	average probability of identity	20
q	frequency of allele A_2 (or a)	5, 8

Symbol	Definition	Chapter
q	proportion of an individual's genome that comes from one parental population in an admixed population	17
Q	probability of an individual's genotype originating from each population	16
Q_{ST}	proportion of total genetic variation for a phenotypic trait due to genetic differentiation among populations (analogous to F_{ST})	11
r	frequency of allele A_3	5
r	correlation coefficient	A
r	rate of recombination	10
r	intrinsic population growth rate	6, 14
r_A	correlation between two traits	10
R	correlation coefficient between alleles at two loci	10
R	response to selection	11, 18
R	number of recaptured individuals	14
$R(g)$	allelic richness in a sample of g genes	5
R_{ST}	analog to F_{ST} that accounts for differences length of microsatellite alleles	9, 16
s	selection coefficient (intensity of selection)	8, 9, 12
s_x	standard deviation	A
s_x^2	sample variance	A
S	selection differential	11, 18
S	effects of inbreeding on probability of survival	13
S	selfing rate	9
t	number of generations	6
V_A	proportion of phenotypic variability due to additive genetic differences between individuals	11, 14
V_D	proportion of phenotypic variability due to dominance effects (interactions between alleles)	11
V_E	proportion of phenotypic variability due to environmental differences between individuals	2, 11, 14
V_G	proportion of phenotypic variability due to genetic differences between individuals	2, 11
V_I	proportion of phenotypic variability due to epistatic effects	11
V_m	increase in additive genetic variation per generation due to mutation	12, 14
V_P	total phenotypic variability for a trait	2, 11
w	relative fitness	8
W	absolute fitness	8
x	number of observations or times that an event occurs	A
$\bar{x}$	sample mean	A

Greek symbols

Symbol	Definition	Chapter
α	probability of a Type I error	A
β	probability of a Type II error	A

Symbol	Definition	Chapter
δ	proportional reduction in fitness due to selfing	13
Δ	change in value from one generation to the next	6, 7, 8
λ	factor by which population size increases in each time unit	14
μ	population mean	A
μ	neutral mutation rate	12
π	probability of an event	A
σ_x^2	population variance	A
Φ_{ST}	analogous to F_{ST} but incorporates genealogical relationships among alleles	9
X^2	chi-square statistic	5, A

Other symbols

$\hat{x}$	estimate of parameter x
x^*	equilibrium value of parameter x
$\bar{x}$	mean value of parameter x
x'	value of parameter x in the next generation

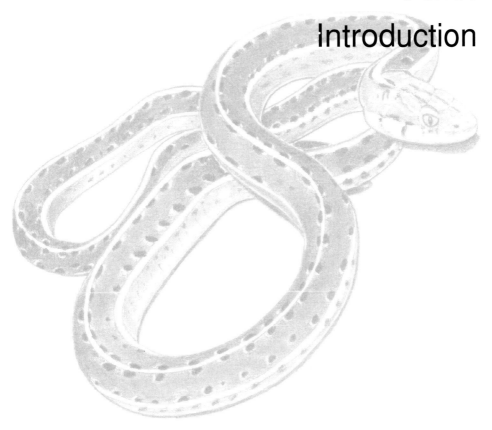

Part I

Introduction

Introduction

Whitebark pine, Section 1.1

We are at a critical juncture for the conservation and study of biological diversity: such an opportunity will never occur again. Understanding and maintaining that diversity is the key to humanity's continued prosperous and stable existence on Earth.

US National Science Board Committee on International Science's Task Force on Global Biodiversity (1989)

The extinction of species, each one a pilgrim of four billion years of evolution, is an irreversible loss. The ending of the lines of so many creatures with whom we have traveled this far is an occasion of profound sorrow and grief. Death can be accepted and to some degree transformed. But the loss of lineages and all their future young is not something to accept. It must be rigorously and intelligently resisted.

Gary Snyder (1990)

We are living in a time of unprecedented extinctions (Myers and Knoll 2001). Current extinction rates have been estimated to be 50–500 times background rates and are increasing; an estimated 3,000–30,000 species go extinct annually (Woodruff 2001). Projected extinction rates vary from 5 to 25% of the world's species by 2015 or 2020. Approximately 25% of mammals, 11% of birds, 20% of reptiles, 34% of fish, and 9–34% of major plant taxa are threatened with extinction over the next few decades (IUCN 2001). Over 50% of animal species are considered to be either critically endangered, endangered, or vulnerable to extinction (IUCN 2001).

Conservation biology provides perhaps the most difficult and important questions ever faced by science (Pimm et al. 2001). The problems are difficult because they are so complex and cannot be approached by the reductionist methods that have worked so well in other areas of science. Moreover, solutions to these problems require a major readjustment of our social and political systems. There are no more important scientific challenges because these problems threaten the continued existence of our species and the future of the biosphere itself.

Aldo Leopold inspired a generation of biologists to recognize that the actions of humans are imbedded into an ecological network that should not be ignored (Meine 1998). The organized actions of humans are controlled by sociopolitical systems that operate into the future on a time scale of a few years at most. All too often our systems of conservation are based on the economic interest of humans in the immediate future. We tend to disregard, and often mistreat, elements that lack economic value but that are essential to the stability of the ecosystems upon which our lives and the future of our species depend.

1.1 Genetics and conservation

Genetics has a long history of application to human concerns. The domestication of animals is thought to have been perhaps the key step in the development of civilization (Diamond 1997). Early peoples directed genetic change in domestic and agricultural species to suit their needs. It has been estimated that the dog was domesticated over 15,000 years ago, followed by goats and sheep around 10,000 years ago (Darlington 1969; Zeder and Hess 2000). Wheat and barley were the first crops to be domesticated in the Old World approximately 10,000 years ago; beans, squash, and maize were domesticated in the New World at about the same time (Darlington 1969).

The initial genetic changes brought about by cultivation were not due to intentional selection but apparently were inherent in cultivation itself. Genetic change under domestication was later accelerated by thousands of years of purposeful selection as animals and crops were chosen to be more productive or to be used for other purposes. This process became formalized in the discipline of agricultural genetics after the rediscovery of Mendel's principles at the beginning of the 20th century.

The "success" of these efforts can be seen everywhere. Humans have transformed much of the landscape of our planet into croplands and pasture to support the over 6 billion humans alive today. Recently, however, we have begun to understand the cost at which this success has been achieved. The replacement of wilderness by human exploited environments is causing the rapidly increasing loss of species and ecosystems throughout the world. The continued growth of the human population imperils a large proportion of the wild species that now remain.

In 1974, Otto Frankel published a landmark paper entitled "Genetic conservation: our evolutionary responsibility" that set out conservation priorities from a genetic perspective:

> First, . . . we should get to know much more about the structure and dynamics of natural populations and communities. . . . Second, even now the geneticist can play a part in injecting genetic considerations into the planning of reserves of any kind. . . . Finally, reinforcing the grounds for nature conservation with an evolutionary perspective may help to give conservation a permanence which a utilitarian, and even an ecological grounding, fail to provide in men's minds.

Frankel, an agricultural plant geneticist, came to the same conclusions as Leopold, a wildlife biologist, by a very different path. In Frankel's view, we cannot anticipate the future world in which humans will live in a century or two. Therefore, it is our responsibility to "keep evolutionary options open" for future humans. It is time to apply our understanding of genetics to conserving the natural ecosystems that are threatened by human civilization.

Recent advances in molecular genetics, including sequencing of the entire genomes of many species, have revolutionized applications of genetics (e.g., medicine and agriculture). For example, it recently has been suggested that genetic engineering should be considered as a conservation genetics technique (Adams et al. 2002). Many native trees in the northern temperate zone have been devastated by introduced diseases (e.g., European and North American elms, and the North American chestnut). Adams et al. (2002) have suggested that transfer of resistance genes by genetic engineering is perhaps the only available method for preventing the loss of important tree species.

The loss of key tree species is likely to affect many other species as well. For example, whitebark pine is currently one of the two most import food resources for grizzly bears in the Yellowstone National Park ecosystem (Mattson and Merrill 2002). However, virtually all of the whitebark pine in this region is projected to be **extirpated** because of an exotic pathogen (Mattson et al. 2001). The use of genetic engineering to improve crop plants has been very controversial. There no doubt will be a lively debate in the near future about the use of these procedures to prevent the extinction of natural populations.

As in other areas of genetics, model organisms have played an important research role in conservation genetics (Frankham 1999). Many important theoretical issues in conservation biology cannot be answered by research on threatened species (e.g., how much gene flow is required to prevent the inbreeding effects of small population size?). Such empirical questions are often best resolved in species that can be raised in captivity in large numbers with a rapid generation interval (e.g., the fruit fly *Drosophila*, the guppy, deer mouse, and the mustard plant). Such laboratory investigations can also provide excellent training opportunities for students. We have tried to provide a balance of examples from model and threatened species. Nevertheless, where possible we have chosen examples from threatened species even though many of the principles were first demonstrated with model species.

1.2 What should we conserve?

Conservation can be viewed as an attempt to protect the genetic diversity that has been produced by evolution over the previous 3.5 billion years on our planet (Eisner et al. 1995).

Genetic diversity is one of three forms of biodiversity recognized by the IUCN as deserving conservation, along with species and ecosystem diversity (McNeely et al. 1990). We can consider the implications of this relationship between genetic diversity and conservation at many levels: genes, individuals, populations, species, genera, etc.

1.2.1 Phylogenetic diversity

The amount of genetic divergence based upon **phylogenetic** relationships is often considered when setting conservation priorities for different species (Mace et al. 2003). For example, the United States Fish and Wildlife Service (USFWS) assigns priority for listing under the United States Endangered Species Act (ESA) on the basis of "taxonomic distinctiveness" (USFWS 1983). Species of a **monotypic** genus receive the highest priority. The tuatara raises several important issues about assigning conservation value and allocating our conservation efforts based upon taxonomic distinctiveness (Example 1.1).

Example 1.1 The tuatara: A living fossil

The tuatara is a lizard-like reptile that is the remnant of a taxonomic group that flourished over 200 million years ago during the Triassic Period (Figure 1.1). Tuatara are now confined to some 30 small islands off the coast of New Zealand (Daugherty et al. 1990). Three species of tuatara were recognized in the 19th century. One of these species is now extinct. A second species, *Sphenodon guntheri*, has been ignored by legislation designed to protect the tuatara, which "lumped" all extant tuatara into a single species, *S. punctatus*. The legislatively defined monotypy of tuatara has led to the belief that the species is relatively widespread and that the extinction of 10 of the total of 40 populations over the past century is not of serious concern. The failure to recognize *S. guntheri* has brought this species to the brink of extinction.

Figure 1.1 Adult male tuatara.

Daugherty et al. (1990) reported allozyme and morphological differences from 24 of the 30 islands on which tuatara are thought to remain. These studies support the status of *S. guntheri* as a distinct species and indicate that fewer than 300 individuals of this species remain on a single island, North Brother Island in Cook Strait. Another population of *S. guntheri* became extinct earlier in this century. Daugherty et al. (1990) argue that not all tuatara populations are of equal conservation value. As the last remaining population of a separate species, the tuatara on North Brother Island represent a greater proportion of the genetic diversity remaining in the genus *Sphenodon* and deserve special recognition and protection.

On a larger taxonomic scale, how should we value the tuatara relative to other species of reptiles? The two tuatara species are the last remaining representatives of the Sphenodontida, one of four extant orders of reptiles (tuatara, snakes and lizards, alligators and crocodiles, and tortoises and turtles). In contrast, there are approximately 5,000 species in the Squamata, the speciose order that contains lizards and snakes.

One position is that conservation priorities should regard all species as equally valuable. This position would equate the two tuatara species with any two species of reptiles. Another position is that we should take phylogenetic diversity into account in assigning conservation priorities. The extreme phylogenetic position is that we should assign equal conservation value to each major sister group in a phylogeny. According to this position, the two tuatara species would be weighed equally with the over 5,000 species of other snakes and lizards. Some intermediate between these two positions seems most reasonable.

Vane-Wright et al. (1991) have presented a method for assigning conservation value on the basis of phylogenetic relationships. This system is based upon the information content of the topology of a particular phylogenetic hierarchy. Each extant species is assigned an index of taxonomic distinctness that is inversely proportional to the number of branching points to other extant lineages. May (1990) has estimated that the tuatara represents between 0.3 and 7% of the taxonomic distinctness, or perhaps we could say genetic information, among reptiles. This is equivalent to saying that each of the two tuatara species are equivalent to approximately 10–200 of the "average" reptile species. Crozier and Kusmierski (1994) have developed an approach to set conservation priorities that is based upon phylogenetic relationships and genetic divergence among taxa. Faith (2002) recently has presented a method for quantifying biodiversity for the purpose of identifying conservation priorities that considers phylogenetic diversity both between and within species.

There is great appeal to placing conservation emphasis on distinct evolutionary lineages with few living relatives. Living fossils, such as the tuatara or the coelacanth (Thomson 1991), represent important pieces in the jigsaw puzzle of evolution. Such species are relics that are representatives of taxonomic groups that once flourished. Study of the primitive morphology, physiology, and behavior of living fossils can be extremely important in understanding evolution. For example, tuatara morphology has hardly changed in nearly 150 million years. Among the many primitive features of the tuatara is a rudimentary third, or pineal, eye on the top of the head. Tuatara also represent an important ancestral outgroup for understanding vertebrate evolution.

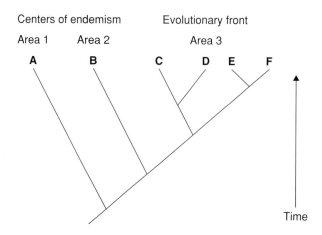

Figure 1.2 Hypothetical phylogeny of seven species. Redrawn from Erwin (1991).

In contrast, others have argued that our conservation strategies and priorities should be based primarily upon conserving the evolutionary process rather than particular pieces of the evolutionary puzzle that are of interest to humans (Erwin 1991). Those species that will be valued most highly under the schemes that weigh phylogenetic distinctness are those that may be considered evolutionary failures. Evolution occurs by changes within a single evolutionary lineage (**anagenesis**) and the branching of a single evolutionary lineage into multiple lineages (**cladogenesis**). Conservation of primitive, nonradiating taxa is not likely to be beneficial to the protection of the evolutionary process and the environmental systems that are likely to generate future evolutionary diversity (Erwin 1991).

Figure 1.2 illustrates the phylogenetic relations among seven hypothetical species (from Erwin 1991). Species A and B are phylogenetically distinct taxa that are endemic to small geographic areas (e.g., tuataras in New Zealand). Such lineages carry information about past evolutionary events, but they are relatively unlikely to be sources of future evolution. In contrast, the stem resulting in species C, D, E, and F is relatively likely to be a source of future anagenesis and cladogenesis. In addition, species such as C, D, E, and F may be widespread, and, therefore, are not likely to be the object of conservation efforts.

The problem is more complex than just identifying species with high conservation value; we must take a broader view and consider the habitats and environments where our conservation efforts could be concentrated. Conservation emphasis on A and B type species will lead to protection of environments that are not likely to contribute to future evolution (e.g., small islands on the coast of New Zealand). In contrast, geographic areas that are the center of evolutionary activity for diverse taxonomic groups could be identified and targeted for long-term protection.

Recovery from our current extinction crisis should be a central concern of conservation (Myers et al. 2000). It is important to maintain the potential for the generation of future biodiversity. We should identify and protect contemporary hotspots of evolutionary radiation and the functional taxonomic groups from which tomorrow's biodiversity is likely to originate. In addition, we should protect those phylogenetically distinct species that are of special value for our understanding of biological diversity and the evolutionary process. These species are also potentially valuable for future evolution of biodiversity because of their combination of unusual phenotypic characteristics that may give rise to a future evolutionary radiation.

1.2.2 Populations, species, or ecosystems

A related, and sometimes impassioned, dichotomy between protecting centers of biodiversity or phylogenetically distinct species is the dichotomy between emphasis on species conservation or on the conservation of habitat or ecosystems (Soulé and Mills 1992). Conservation efforts to date have emphasized the concerns of individual species. For example, the **ESA** has been the legal engine behind much of the conservation efforts in the United States. However, it is frustrating to see enormous resources being spent on a few high profile species when little is spent on less charismatic taxa or in preventing environmental deterioration that would benefit many species. It is clear that a more comprehensive and proactive conservation strategy emphasizing protection of habitat and ecosystems, rather than species, is needed. Some have advocated a shift from saving things, the products of evolution (species, communities, or ecosystems), to saving the underlying processes of evolution "that underlie a dynamic biodiversity at all levels" (Templeton et al. 2001).

It has been argued that more concern about extinction should be focused on the extinction of genetically distinct populations, and less on the extinction of species (Hughes et al. 1997; Hobbs and Mooney 1998). The conservation of many distinct populations is required to maximize evolutionary potential of a species and to minimize the long-term extinction risks of a species. In addition, a population focus would also help prevent costly and desperate "last-minute" conservation programs that occur when only one or two small populations of a species remain. The first attempt to estimate the rate of population extinction worldwide was published by Hughes et al. (1997). They estimate that tens of millions of local populations that are genetically distinct go extinct each year. Approximately 16 million of the world's three billion genetically distinct natural populations go extinct each year in tropical forests alone.

Luck et al. (2003) have considered the effect of population diversity on the functioning of ecosystems and so-called **ecosystem services**. They argue that the relationship between biodiversity and human well being is primarily a function of the diversity of populations within species. They have also proposed a new approach for describing population diversity that considers the value of groups of individuals to the services that they provide.

Ceballos and Ehrlich (2002) have recently compared the historical and current distributions of 173 declining mammal species from throughout the world. Their data included all of the terrestrial mammals of Australia and subsets of terrestrial mammals from other continents. Nearly 75% of all species they included have lost over 50% of their total geographic range. Approximately 22% of all Australian species are declining, and they estimated that over 10% of all Australian terrestrial mammal populations have been extirpated since the 19th century. These estimates, however, all assume that population extirpation is proportional to loss of range area.

The amount of genetic variation within a population may also play an important ecosystem role in the relationships among species. Bangert et al. (2005) found in experiments that greater genetic diversity in cottonwood trees resulted in greater species richness in the arthropod community dependent upon these trees. The generality of this result needs to be tested. This study used interspecific hybrids rather than just differing amounts of genetic variation within intraspecific populations. Nevertheless, this is an interesting concept that deserves to be explored.

Conservation requires a balanced approach that is based upon habitat protection that also takes into account the natural history and viability of individual species. Consider chinook salmon in the Snake River basin of Idaho that are listed under the ESA. These fish spend their first 2 years of life in small mountain rivers and streams far from the ocean. They then migrate over 1,500 km downstream through the Snake and Columbia Rivers and enter the Pacific Ocean. There they spend two or more years ranging as far north as the coast of Alaska before they return to spawn in their natal freshwater streams. There is no single ecosystem that encompasses these fish, other than the biosphere itself. Protection of this species requires a combination of habitat measures and management actions that take into account the complex life history of these fish.

1.3 How should we conserve biodiversity?

Extinction is a **demographic** process: the failure of one generation to replace itself with a subsequent generation. Demography is of primary importance in managing populations for conservation (Lacy 1988; Lande 1988). Populations are subject to uncontrollable **stochastic** demographic factors as they become smaller. It is possible to estimate the expected mean and variance of a population's time to extinction if one has an understanding of a population's demography and environment (Belovsky 1987; Goodman 1987; Lande 1988).

There are two main types of threats causing extinction: deterministic and stochastic threats (Caughley 1994). **Deterministic** threats are habitat destruction, pollution, overexploitation, species translocation, and global climate change. Stochastic threats are random changes in genetic, demographic, or environmental factors. Genetic stochasticity is random genetic change (drift) and increased inbreeding (Shaffer 1981). Genetic stochasticity leads to loss of genetic variation (including beneficial alleles) and increase in frequency of harmful alleles. An example of demographic stochasticity is random variation in sex ratios, e.g., producing only male offspring. Environmental stochasticity is simply random environmental variation, such as the occasional occurrence of several harsh winters in a row.

Under some conditions, extinction is likely to be influenced by genetic factors. Small populations are also subject to genetic stochasticity that can lead to loss of genetic variation through genetic drift. The inbreeding effect of small populations is likely to lead to a reduction in the fecundity and viability of individuals in small populations. For example, Frankel and Soulé (1981, p. 68) have suggested that a 10% decrease in genetic variation due to the inbreeding effect of small populations is likely to cause a 10–25% reduction in reproductive performance of a population. This, in turn, is likely to cause a further reduction in population size, and thereby reduce a population's ability to persist (Gilpin and Soulé 1986).

Some have argued that genetic concerns can be ignored when projecting the viability of small populations because they are in much greater danger of extinction by purely demographic stochastic effects (Lande 1988; Pimm et al. 1988; Frankham 2003). It has been argued that such small populations are not likely to persist long enough to be affected by inbreeding depression and that efforts to reduce demographic stochasticity will also reduce the loss of genetic variation. The disagreement over whether or not genetics should be considered in demographic predictions of population persistence has been unfortunate and misleading. Extinction is a demographic process that is likely to be influenced by genetic effects **under some circumstances**. The important issue is to determine under what conditions genetic concerns are likely to influence population persistence (Nunney and Campbell 1993).

Perhaps most importantly, we need to recognize when management recommendations based upon demographic and genetic considerations may be in conflict with each other. For example, small populations face a variety of genetic and demographic effects that threaten their existence. Management plans aim to increase the population size as soon as possible to avoid the problems associated with small populations. However, efforts to maximize growth rate may actually increase the rate of loss of genetic variation by relying on the reproductive success of a few individuals (Caughley 1994).

Ryman and Laikre (1991) considered supportive breeding in which a portion of wild parents are brought into captivity for reproduction and their offspring are released back into the natural habitat where they mix with wild conspecifics. Programs similar to this are carried out in a number of species to increase population size and thereby temper stochastic demographic effects. Under some circumstances, supportive breeding may reduce effective population size and cause a drastic reduction in genic heterozygosity.

Genetic information also can provide valuable insight into the demographic structure and history of a population (Escudero et al. 2003). Examination of the number of unique genotypes in populations that are difficult to census are currently being used to estimate total population size (Schwartz et al. 1998). Many demographic models assume a single random mating population. Examination of the distribution of genetic variation over the distribution of a species can identify what geographic units can be considered as separate demographic units. Consider the simple example of a population of trout found within a single small lake, for which it would seem appropriate to consider these fish a single demographic unit. However, under some circumstances the trout in a single small lake may actually represent two or more separate reproductive (and demographic) groups with little or no exchange between them (e.g., Ryman et al. 1979).

The issue of population persistence is a multidisciplinary problem that involves many aspects of the biology of the populations involved. A similar statement can be made about most of the issues we are faced with in conservation biology. We can only resolve these problems by an integrated approach that incorporates demography and genetics, as well as other biological considerations that are likely to be critical for a particular problem (e.g., behavior, physiology, or interspecific interactions).

1.4 Applications of genetics to conservation

Darwin (1896) was the first to consider the importance of genetics in the persistence of natural populations. He expressed concern that deer in British nature parks may be subject to loss of vigor because of their small population size and isolation. Voipio (1950) presented the first comprehensive consideration of the application of population genetics to the management of natural populations. He was primarily concerned with the effects of genetic drift in game populations that were reduced in size by trapping or hunting and fragmented by habitat loss.

The modern concern for genetics in conservation began around 1970 when Sir Otto Frankel (1970) began to raise the alarm about the loss of primitive crop varieties and their replacement by genetically uniform cultivars (Guest Box 1). It is not surprising that these initial considerations of conservation genetics dealt with species that were used directly as resources by humans. Conserving the genetic resources of wild relatives of agricultural species remains an important area of conservation genetics (Maxted 2003).

The application of genetics to conservation in a more general context did not blossom until around 1980, when three publications established the foundation for applying the principles of genetics to conservation of biodiversity (Soulé and Wilcox 1980; Frankel and Soulé 1981; Schonewald-Cox et al. 1983).

Perpetuation of biodiversity primarily depends upon the protection of the environment and maintenance of habitat. Nevertheless, genetics has played an important and diverse role in conservation biology in the last few years. Nearly 10% of the articles published in the journal *Conservation Biology* since its inception in 1988 have "genetic" or "genetics" in their title. Probably at least as many other articles deal with largely genetic concerns but do not have the term in their title. Thus, approximately 15% of the articles published in *Conservation Biology* have genetics as a major focus.

The subject matter of papers published on conservation genetics is extremely broad. However, many of articles dealing with conservation and genetics fit into one of the following five broad categories:

1 Management and reintroduction of captive populations, and the restoration of biological communities.
2 Description and identification of individuals, genetic population structure, kin relationships, and taxonomic relationships.
3 Detection and prediction of the effects of habitat loss, fragmentation, and isolation.
4 Detection and prediction of the effects of hybridization and introgression.
5 Understanding the relationships between adaptation or fitness and the genetic characters of individuals or populations.

These topics are listed in order of increasing complexity and decreasing uniformity of agreement among conservation geneticists. Although the appropriateness of captive breeding in conservation has been controversial (Snyder et al. 1996), procedures for genetic management of captive populations are well developed with relatively little controversy. However, the relationship between specific genetic types and fitness or adaptation has been a particularly vexing issue in evolutionary and conservation genetics. Nevertheless, recent studies have shown that natural selection can bring about rapid genetic changes in populations that may have important implications for conservation (Stockwell et al. 2003) (see Guest Box 8).

Genetics is likely to play an even greater role in conservation biology in the future (Hedrick 2001; Frankham 2003). Invasive species are currently recognized as one of the top two threats to global biodiversity (Walker and Steffen 1997). Studies of genetic diversity and the potential for rapid evolution of invasive species may provide useful insights into what causes species to become invasive (Lee 2002). More information about the genetics and evolution of invasive species or native species in invaded communities, as well as their interactions, may lead to predictions of the relative susceptibility of ecosystems to invasion, identification of key alien species, and predictions of the subsequent effects of removal. In addition, genetic engineering and cloning of endangered species may also come to play a role in conservation (Ryder and Benirschke 1997; Ryder et al. 2000; Loi et al. 2001; Adams et al. 2002).

This is an exciting time to be interested in the genetics of natural populations. Molecular techniques make it possible to detect genetic variation in any species of interest, not just those that can be bred and studied in the laboratory (Wayne and Morin 2004). This work is meant to provide a thorough examination of our understanding of the genetic variation in natural populations. Based upon that foundation, we will consider the application of this understanding to the many problems faced by conservation biologists with the hope that our more informed actions can make a difference.

Guest Box 1 The role of genetics in conservation
L. Scott Mills and Michael E. Soulé

Although most conservationists have ignored genetics and most geneticists have ignored the biodiversity catastrophe, by 1970 some agricultural geneticists, led by Sir Otto Frankel (1974) had begun to sound an alarm about the disappearance of thousands of land races – crop varieties coaxed over thousands of years to adapt to local soils, climates, and pests. Frankel challenged geneticists to help promote an "evolutionary ethic" focused on maintaining evolutionary potential and food security in a rapidly changing world.

Frankel's pioneering thought inspired the first international conference on conservation biology in 1978. It brought together ecologists and evolutionary geneticists to consider how their fields could help slow the extinction crisis. Some of the chapters in the proceedings (Soulé and Wilcox 1980) foreshadowed population viability analysis and the interactions of demography and genetics in small populations (the extinction vortex). Several subsequent books (Frankel and Soulé 1981; Schonewald-Cox et al. 1983; Soulé 1987) consolidated the role of genetic thinking in nature conservation.

Thus, topics such as inbreeding depression and loss of heterozygosity were prominent since the beginning of the modern discipline of conservation biology, but like inbred relatives, they were conveniently forgotten at the end of the 20th century. Why? Fashion. Following the human proclivity to champion simple, singular solutions to complex problems, a series of papers on population viability in the late 1980s and early 1990s argued that – compared to demographic and environmental accidents – inbreeding and loss of genetic variation were trivial contributors to extinction risk in small populations. Eventually, however, this swing in scientific fashion was arrested by the friction of real world complexity.

Thanks to the work of F_1 and F_2 conservation geneticists, it is now clear that inbreeding depression can increase population vulnerability by interacting with random environmental variation, not to mention deterministic factors including habitat degradation, new diseases, and invasive exotics. Like virtually all dualisms, the genetic versus nongenetic battles abated in the face of the overwhelming evidence for the relevance of both.

Genetic approaches have become prominent in other areas of conservation biology as well. These include: (1) the use of genetic markers in forensic investigations concerned with wildlife and endangered species; (2) genetic analyses of hybridization and invasive species; (3) noninvasive genetic estimation of population size; and (4) studies of taxonomic affiliation and distance. And, of course, the specter of global climate change has renewed interest in the genetic basis for adaptation as presaged by Frankel 30 years ago. Nowadays, genetics is an equal partner with ecology, systematics, physiology, epidemiology, and behavior in conservation, and both conservation and genetics are enriched by this pluralism.

Problem 1.1

An extensive bibliography that includes thousands of papers considering the role of genetics in conservation is available online at http://www.lib.umt.edu/guide/allendorf.htm. Search this bibliography for the papers with the words genetics and conservation in the title or keywords to get an idea of the variety of papers published on this topic. You should find several hundred papers in this category. Note which types of problems are of greatest concern and which taxonomic groups are included and which are poorly represented. Use this bibliography to answer questions 2 and 3 below.

Problem 1.2

What are the top five journals that publish papers in conservation and genetics as determined by the number of published papers in the bibliography?

Problem 1.3

Pick a species of conservation that you are interested in. Search for all of the papers published on this species in the bibliography. Remember to search for both the common name (e.g., silversword) and the scientific name (*Argyroxiphium*). Based upon these titles, what are the major conservation issues of concern for the species that you have chosen?

Problem 1.4

A paper was published in May 2005 proclaiming that the ivory-billed woodpecker is not extinct (Fitzpatrick et al. 2005). This species had not been seen since 1944, and was thought to be long extinct. This paper reported evidence of at least a handful of individuals persisting in the Big Woods region of Arkansas. Bird lovers and conservationist have been celebrating the exciting news for days. However, the recovery of this species is uncertain. Why?

2

Phenotypic Variation in Natural Populations

Western terrestrial garter snake, Section 2.3

Few persons consider how largely and universally all animals are varying. We know however, that in every generation, if we would examine all the individuals of any common species, we should find considerable differences, not only in size and color, but in the form and proportions of all the parts and organs of the body.
Alfred Russel Wallace (1892, p. 57)

It would be of great interest to determine the critical factors controlling the variability of each species, and to know why some species are so much more variable than others.
David Lack (1947)

Genetics has been defined as the study of differences among individuals (Sturtevant and Beadle 1939). If all of the individuals within a species were identical, we could still study and describe their morphology, physiology, behavior, etc. However, geneticists would be out of work. Genetics and the study of inheritance are based upon comparing the similarity of parents and their progeny relative to the similarity among unrelated individuals within populations or species.

Variability among individuals is also essential for adaptive evolutionary change. **Natural selection** cannot operate unless there are **phenotypic** differences between individuals. Transformation of individual variation within populations to differences between populations or species by the process of natural selection is the basis for adaptive evolutionary change described by Darwin almost 150 years ago (Darwin 1859). Nevertheless, there is surprisingly little in Darwin's extensive writings about the extent and pattern of differences between individuals in natural populations. Rather Darwin relied heavily on examples from animal breeding and the success of artificial selection to argue for the potential of evolutionary change by natural selection (Ghiselin 1969).

Alfred Russel Wallace (the co-founder of the principle of natural selection) was perhaps the first biologist to emphasize the extent and importance of variability within natural populations (Figure 2.1). Wallace felt that "Mr. Darwin himself did not fully recognise the enormous amount of variability that actually exists" (Wallace 1923, p. 82). Wallace concluded that for morphological measurements, individuals commonly varied by up to 25% of the mean value; that is, from 5 to 10% of the individuals within a population differ from the population mean by 10–25% (Wallace 1923, p. 81). This was in opposition to the commonly held view of naturalists in the 19th century that individual variation was comparatively rare in nature.

Mendel's classic work was an attempt to understand the similarity of parents and progeny for traits that varied in natural populations. The original motivation for Mendel's

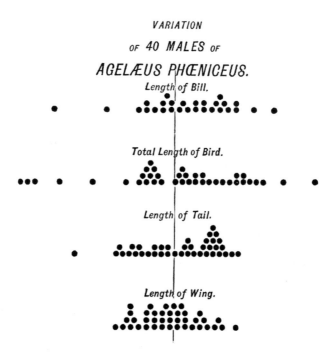

Figure 2.1 Diagram from Alfred Russel Wallace of variation in body dimensions of 40 red-winged blackbirds in the United States. From Wallace (1923, p. 64).

work was to test a theory of evolution proposed by his botany professor (Unger 1852) that proposed that "variants arise in natural populations which in turn give rise to varieties and subspecies until finally the most distinct of them reach species level" (Mayr 1982, p. 711). The importance of this inspiration can be seen in the following quote from Mendel's original paper (1865): ". . . this appears, however, to be the only right way by which we can finally reach the solution of a question the importance of which cannot be overestimated in connection with the history of the evolution of organic forms".

Population genetics was limited to the study of species that could be studied experimentally in the laboratory for most of the 20th century. Experimental population genetics was dominated by studies that dealt with *Drosophila* fruit flies until the mid-1960s because of the difficulty in determining the genetic basis of phenotypic differences between individuals (Lewontin 1974). *Drosophila* that differed phenotypically in natural populations could be brought into the laboratory for detailed analysis of the genetic differences underlying the phenotypic differences. Similar studies were not possible for species with long generation times that could not be raised in captivity in large numbers. However, population genetics underwent an upheaval in the 1960s when biochemical techniques allowed genetic variation to be studied directly in natural populations of any organism.

Molecular techniques today make it possible to study differences in the DNA sequence itself in any species. Projects are currently underway to sequence the entire genome of several species. However, even this level of detail will not provide sufficient information to understand the significance genetic variation in natural populations. Adaptive evolutionary change within populations consists of gradual changes in morphology, life history, physiology, and behavior. Such traits are usually affected by a combination of many genes and the environment so that it is difficult to identify single genes that contribute to the genetic differences between individuals for many of the phenotypic traits that are of interest.

This difficulty has been described as a paradox in the study of the genetics of natural populations (Lewontin 1974). We are interested in the phenotype of those characters for which genetic differences at individual loci have only a slight phenotypic effect relative to the contributions of other loci and the environment. "What we can measure is by definition uninteresting and what we are interested in is by definition unmeasurable" (Lewontin 1974, p. 23). This paradox can only be resolved by a multidisciplinary approach that combines molecular biology, developmental biology, and population genetics so that we can understand the developmental processes that connect the genotype and the phenotype (e.g. Lewontin 1999; Clegg and Durbin 2000).

The more complex and distant the connection between the genotype and the phenotype, the more difficult it is to determine the genetic basis of observed phenotypic differences. For example, human behavior is influenced by many genes and an extremely complex developmental process that continues to be influenced by the environment throughout the lifetime of an individual (Figure 2.2). It is relatively easy to identify genetic differences in the structural proteins, enzymes, and hormones involved in behavior because they are direct expressions of DNA sequences in the genotype. We also know that these genetic differences result in differences in behavior between individuals; for example, a mutation in the enzyme monoamine oxidase *A* apparently results in a tendency for aggressive and violent behavior in humans (Morell 1993). However, it is extremely difficult in general to identify the specific differences in the genotype that are responsible for such differences in behavior.

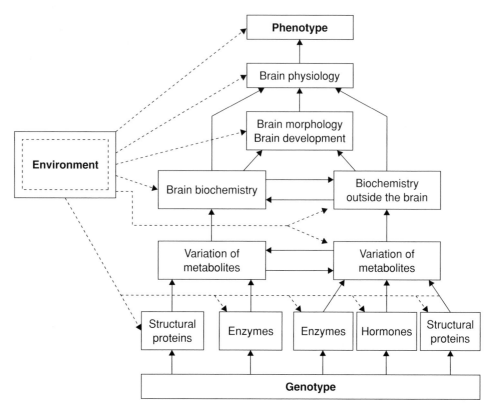

Figure 2.2 Diagrammatic representation of interconnections between genotype and environment resulting in expression of the phenotype of a human. Redrawn from Vogel and Motulsky (1986).

We are faced with a dilemma in our study of the genetic basis of phenotypic variation in natural populations. We can start with the genotype (bottom of Figure 2.2) and find genetic differences between individuals at specific loci; however, it is difficult to relate those differences to the phenotypic differences between individuals that are of interest. The alternative is to start with the phenotype of interest (top of Figure 2.2), and determine if the phenotypic differences have a genetic basis; however, it is usually extremely difficult to identify which specific genes contribute to those phenotypic differences (Mackay 2001).

In this chapter, we consider the amount and pattern of phenotypic variation in natural populations. We introduce approaches and methodology used to understand the genetic basis of phenotypic variation. In the next chapter, we will examine genetic variation directly in chromosomes and molecules and consider how it relates to evolution and conservation.

2.1 Color pattern

Mendel chose to study the inheritance of seven characters that had clearly distinguishable forms without intermediates: tall versus dwarf plants, violet versus white flowers, green versus yellow pods, etc. His landmark success depended upon the selection of qualitative traits in which the variation could be classified into discrete categories rather than quantitative traits in which individuals vary continuously (e.g., weight, height, etc.).

The presence of such discrete polymorphisms in species has sometimes been problematic for naturalists and taxonomists. For example, the king coat color pattern in cheetahs was first described as a cheetah–leopard hybrid (van Aarde and van Dyk 1986). Later, animals with this pattern were recognized as a new species of cheetah (*Acinonyx rex*). It was then suggested that this coat pattern was a genetic polymorphism within cheetahs. Inheritance results with captive cheetahs eventually confirmed that this phenotype results from a recessive allele at an **autosomal** locus (van Aarde and van Dyk 1986).

Rare color phenotypes in some species sometimes attract wide interest from the public. For example, a photojournalist published a picture of a white-phase black bear near Juneau, Alaska, in the summer of 2002. In response to public concerns, the Alaska Board of Game ordered an emergency closure of hunting on all "white-phase" black bears in the Juneau area during the 2002 hunting season. The Kermode or "spirit" bear is a white phase of the black bear that occurs at low to moderate frequencies in coastal British Columbia and Alaska (Marshall and Ritland 2002). The white color is caused by a recessive allele with single base pair substitution. White phases of species like the tiger are often maintained in zoos because of public interest. A single, recessive autosomal allele is responsible for the white color. However, this allele also causes abnormal vision in tigers (Thornton 1978).

Discrete color polymorphisms are widespread in plants and animals. For example, a recent review of color and pattern polymorphisms in anurans (frogs and toads) cites polymorphisms in 225 species (Hoffman and Blouin 2000). However, surprisingly little work has been done that describes the genetic basis of these polymorphisms or their adaptive significance. Hoffman and Blouin (2000) report that the mode of inheritance has been described in only 26 species, but conclusively demonstrated in only two! Nevertheless, available results suggest that in general color pattern polymorphisms are highly heritable in anurans.

Color pattern polymorphisms have been described in many bird species (Hoekstra and Price 2004). Polymorphism in this context can be considered to be the occurrence of two or more discrete, genetically based phenotypes in a population in which the frequency of the rarest type is greater than 1% (Hoffman and Blouin 2000). For example, red and gray morphs of the eastern screech owl occur throughout its range (VanCamp and Henny 1975). This polymorphism has been recognized since 1874 when it was realized that red and gray birds were conspecific and that the types were independent of age, sex, or season of the year.

The genetic basis of this phenotypic polymorphism has been studied by observing progeny produced by different mating types in a population in northern Ohio (VanCamp and Henny 1975). Matings between gray owls produced all gray progeny. The simplest explanation of this observation is a single locus with two alleles, where the red allele (*R*) is dominant to the gray allele (*r*). Under this model, gray owls are homozygous *rr*, and red owls are either homozygous *RR* or heterozygous *Rr*. Homozygous *RR* red owls are expected to produce all red progeny regardless of the genotype of their mate. One-half of the progeny between heterozygous *Rr* owls and gray bird (*rr*) are expected to be red and one-half are expected to be gray. We cannot predict the expected progeny from matings involving red birds without knowing the frequency of homozygous *RR* and heterozygous *Rr* birds in the population. Progeny frequencies from red parents in Table 2.1 are compatible

Table 2.1 Inheritance of color polymorphism in eastern screech owls from northern Ohio (VanCamp and Henny 1975).

Mating	Number of families	Progeny	
		Red	Gray
Red × red	8	23	5
Red × gray	46	68	63
Gray × gray	135	0	439

with most red birds being heterozygous *Rr*; this is expected because the red morph is relatively rare in northern Ohio based upon the number of families with red parents in Table 2.1. We will take another look at these results in Chapter 5 after we have considered estimating genotypic frequencies in natural populations.

A series of papers of flower color polymorphism in the morning glory provides a promising model system for connecting adaptation with the developmental and molecular basis of phenotypic variation (reviewed in Clegg and Durbin 2000). Flower color variation in this species is determined primarily by allelic variation at four loci that affect flux through the flavonoid biosynthetic pathway. Perhaps the most surprising finding is that almost all of the mutations that determine the color polymorphism are the result of the insertion of mobile elements called **transposons**. In addition, the gene that is most clearly subject to natural selection is not a structural gene that encodes a protein, but is rather a regulatory gene that determines the floral distribution of pigmentation.

We have entered an exciting new era where for the first time it has become possible to identify the genes responsible for color polymorphisms. A recent series of papers have shown that a single gene, melanocortin-1 receptor (*Mc1r*), is responsible for color polymorphism in a variety of birds and mammals (Majerus and Mundy 2003; Mundy et al. 2004) (see Guest Box 12). Field studies of natural selection, combined with study of genetic variation in *Mc1r*, will eventually lead to understanding of the roles of selection and mutation in generating similarities and differences between populations and species.

2.2 Morphology

Morphological variation is everywhere. Plants and animals within the same population differ in size, shape, and numbers of body parts. However, there are serious difficulties with using morphological traits to understand patterns of genetic variation. The biggest problem is that variations in morphological traits are caused by both genetic and environmental differences among individuals. Therefore, variability in morphological traits cannot be used to estimate the amount of genetic variation within populations or the amount of genetic divergence between populations. In fact, we will see in Section 2.4 that morphological differences between individuals in different populations may actually be misleading in terms of genetic differences between populations.

Morphological differences are sometimes observed over time in natural populations. For example, pink salmon on the west coast of North America have tended to become

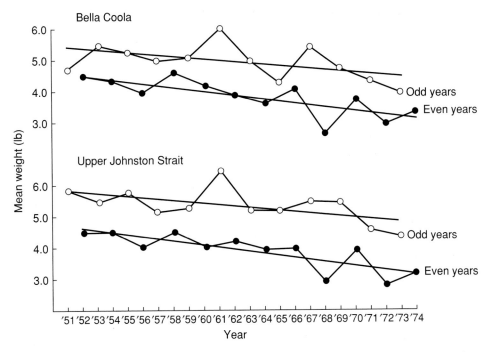

Figure 2.3 Decrease in size of pink salmon caught in two rivers in British Columbia, Canada, between 1950 and 1974. Two lines are drawn for each river: one for the salmon caught in odd-numbered years, the other for even years. Redrawn from Ricker (1981).

smaller at sexual maturity between 1950 and 1974 (Figure 2.3). Pink salmon have an unusual life history in that all individuals become sexually mature and return from the ocean to spawn in fresh water at 2 years of age, and all fish die after spawning (Heard 1991). Therefore, pink salmon within a particular stream comprise separate odd- and even-year populations that are reproductively isolated from each other (Aspinwall 1974). Both the odd- and even-year populations of pink salmon have become smaller over this time period. This effect is thought to be largely due to the effects of selective fishing for larger individuals.

Most phenotypic differences between individuals within populations have **both** genetic and environmental causes. Geneticists often represent this distinction by partitioning the total phenotypic variability for a trait (V_P) within a populations into two components:

$$V_P = V_G + V_E \tag{2.1}$$

where V_G is the proportion of phenotypic variability due to genetic differences between individuals and V_E is the proportion due to environmental differences. The heritability of a trait is defined as the proportion of the total phenotypic variation that has a genetic basis (V_G/V_P).

One of the first attempts to tease apart genetic and environmental influences on morphological variation in a natural population was by Punnett (of Punnett square fame) in 1904. He obtained a number of velvet belly sharks from the coast of Norway to study the

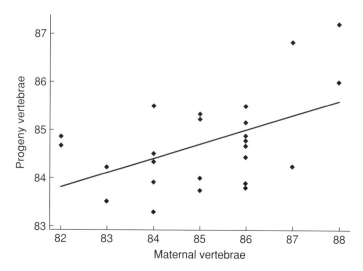

Figure 2.4 Regression of the mean number of total vertebrae in sharks before birth on the total number of vertebrae in their mothers ($P < 0.01$). Data from Punnett (1904).

development of the limbs in vertebrates. The velvet belly is a small, round-bodied, viviparous shark that is common along the European continental shelf. Punnett counted the total number of vertebrae in 25 adult females that each carried from two to 14 fully developed young. He estimated the correlation between vertebrae number in females and their young to test the inheritance of this morphological character. He assumed that the similarity between females and their progeny would be due to inheritance since the females and their young developed in completely different environments (Figure 2.4).

Punnett (1904) concluded "the values of these correlations are sufficiently large to prove that the number of units in a primary linear meristic series is not solely due to the individual environment but is a characteristic transmitted from generation to generation". In fact, approximately 25% of the total variation in progeny vertebrae number can be attributed to the effect of their mothers ($r = 0.504$; $P < 0.01$). We will take another look at these data in Chapter 11 when we consider the genetic basis of morphological variation in more detail (also see Tave 1984).

Some phenotypic variation can be attributed to neither genetic nor environmental differences among individuals. Bilateral characters of an organism may differ in size, shape, or number. Take, for example, the number of gill rakers in fish species. The left and right branchial arches of the same individual usually have the same number of gill rakers (Figure 2.5). However, some individuals are asymmetric; that is, they have different number of gill rakers on the left and right sides. **Fluctuating asymmetry** of such bilateral traits occurs when most individuals are symmetric and there is no tendency for the left or right side to be greater in asymmetric individuals (Palmer and Strobeck 1986).

What is the source of such fluctuating asymmetry? The cells on the left and right sides are genetically identical, and it seems unreasonable to attribute such variability to environmental differences between the left and right side of the developing embryo. Fluctuating asymmetry is thought to be the result of the inability of individuals to control and integrate development so that random physiological differences occur during development and

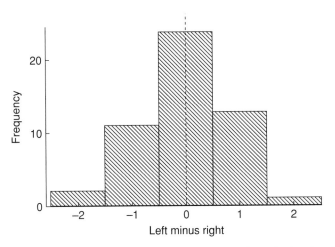

Figure 2.5 Fluctuating asymmetry for gill rakers on lower first branchial arches from a randomly mating population of rainbow trout. Redrawn from Leary and Allendorf (1989).

result in asymmetry (Palmer and Strobeck 1997). That is, fluctuating asymmetry is a measure of developmental noise – random molecular events (Lewontin 2000). Mather (1953) called the regulation or suppression of these chance physiological differences to be the genotypic stabilization of development, and proposed that developmental stability could be measured by fluctuating asymmetry. Thus, increased "noise" or accidents during development (i.e., decreased developmental stability) will result in greater fluctuating bilateral asymmetry. The amount of fluctuating asymmetry in populations may be a useful measure of stress resulting from either genetic or environmental causes in natural populations (Leary and Allendorf 1989; Clarke 1993; Zakharov 2001).

2.3 Behavior

Behavior is another aspect of the phenotype and thus will be affected by natural selection and other evolutionary processes, just as any phenotypic characteristic will be. Human behavioral genetic research has been controversial over the years because of concerns that the results from behavioral genetic studies might be used to stigmatize individuals or groups of people. Genetically based differences in behavior are of special interest in conservation because many behavioral differences are of importance for local adaptation and because captive breeding programs often result in changes in behavior as a result of adaptation to captivity.

Most research in behavioral genetics has used laboratory species such as mice and *Drosophila*. These studies have focused on determining the genetic, neurological, and molecular basis of differences in behavior among individuals. *Drosophila* behavioral geneticists are especially creative in naming genes affecting behavior; they have recently identified a gene known as *couch potato* that is associated with reduced activity in adults (Bellen et al. 1992).

The extent to which genetic factors are involved in differences in bird migratory behavior has been studied systematically over the last 20 years in the blackcap, a common

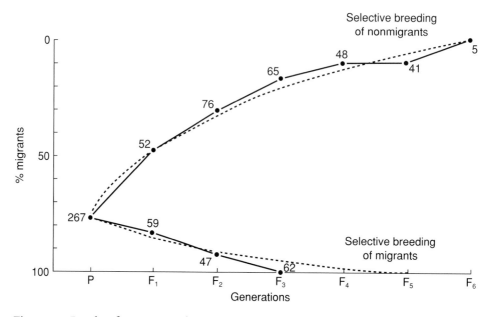

Figure 2.6 Results of a two-way selective breeding experiment for migratory behavior with blackcaps from a partially migratory Mediterranean population. P, parental. From Berthold and Helbig (1992).

warbler of Western Europe (Berthold 1991; Berthold and Helbig 1992). Selection experiments have shown that the tendency to migrate itself is inherited and is based upon a multilocus system with a threshold for expression (Figure 2.6). In addition, differences between geographic populations in the direction of migration is also genetically influenced (Figure 2.7).

The importance of genetically based differences in behavior for adaptation to local conditions has been shown by an elegant series of experiments with the western terrestrial garter snake (Arnold 1981). This garter snake occurs in a wide variety of habitats throughout the west of North America from Baja to British Columbia, and occurs as far east as South Dakota. Arnold has compared the diets of snakes living in the foggy and wet coastal climate of California and the drier, high elevation, inland areas of that state. As hard as it may be to believe, the major prey of coastal snakes is the banana slug; in contrast, banana slugs do not occur at the inland sites.

Arnold captured pregnant females from both locations and raised the young snakes in isolation away from their littermates and mother to remove this possible environmental influence on behavior. The young snakes were offered a small chunk of freshly thawed banana slug. Naive coastal snakes usually ate the slugs; inland snakes did not (Figure 2.8). Hybrid snakes between the coastal and inland sites were intermediate in slug-eating proclivity. These results confirm that the difference between populations in slug-eating behavior has a strong genetic component.

Studies with several salmon and trout species have demonstrated innate differences in migratory behavior that correspond to specializations in movement from spawning and incubation habitat in streams to lakes favorable for feeding and growth (reviewed in Allendorf and Waples 1996). Fry emerging from lake outlet streams typically migrate

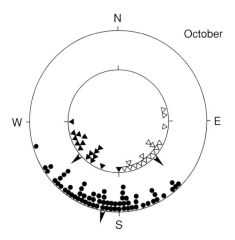

Figure 2.7 Results of tests in orientation cages allowing blackcaps to choose direction of migration. Solid and open triangles represent birds from Germany and Austria, respectively. The difference in direction of migration corresponds to geographic differences between these populations. The solid circles show the orientation of hybrids experimentally produced between the two parental groups. From Berthold and Helbig (1992).

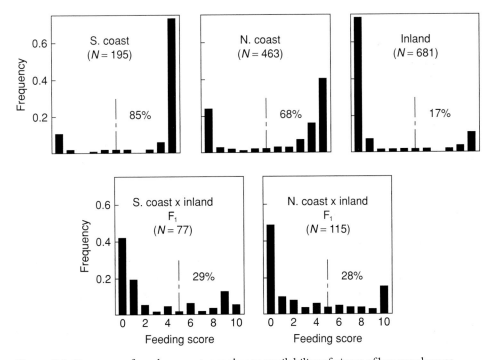

Figure 2.8 Response of newborn garter snakes to availability of pieces of banana slugs as food. Snakes from coastal populations tend to have a high slug feeding score. A score of 10 indicates that a snake ate a piece of slug on each of the 10 days of the experiment. Inland snakes rarely ate a piece of slug on even one day. From Arnold (1981).

upstream upon emergence, and fry from inlet streams typically migrate downstream. Differences in the compass orientation behavior of newly emerged sockeye salmon correspond to movements to feeding areas.

Kaya (1991) has shown that behavioral adaptations have evolved in Arctic grayling in just a few generations under strong selection. Arctic grayling from the Big Hole River have been planted in lakes throughout Montana, USA, over the last 50 years. Arctic grayling from mountain lakes emerge as fry from gravel in streams and immediately migrate into a nearby lake. Fry from adults that spawn upstream from the lake innately swim downstream after emerging. Conversely, fry from adults that spawn downstream innately swim upstream after emerging. Fry that swim in the wrong direction upon emergence would be expected to have greatly reduced survival. Such differences between upstream and downstream spawning populations have apparently arisen by natural selection since the introduction of these populations.

2.4 Differences among populations

Populations from different geographic areas are detectably different for many phenotypic attributes in almost all species. Gradual changes across geographic or environment gradients are found in many species. However, there is no simple way to determine if such a **cline** for a particular phenotype results from genetic or environmental differences between populations. One way to test for genetic differences between populations is to eliminate environmental differences by raising individuals under identical environmental differences in a so-called common-garden. That is, by making V_E in expression 2.1 equal to zero, any remaining phenotypic differences must be due to genetic differences between individuals.

The classic common-garden experiments were conducted with altitudinal forms of yarrow plants along an altitudinal gradient from the coast of central California to over 10,000 ft (over 3,000 m) in the Sierra Nevada Mountains (Clausen et al. 1948). Individual plants were cloned into genetically identical individuals by cutting them into pieces and rooting the cuttings. The clones were then raised at three different altitudes (Figure 2.9). Phenotypic differences among plants from different altitudes persisted when the plants were grown in common locations at each of the altitudes (Figure 2.9). Coastal plants had poor survival at high altitude, but grew much faster than high altitude plants when grown at sea level.

Transplant and common-garden experiments are much more difficult with animals than with plants for several obvious reasons. However, James and her colleagues have partitioned clinal variation in size and shape of the red-winged blackbird into genetic and environmental components by conducting transplantation experiments (reviewed in James 1991). Eggs were transplanted between nests in northern and southern Florida, and between nests in Colorado and Minnesota. A surprisingly high proportion of the regional differences in morphology were explained by the locality in which the eggs developed (James 1983, 1991).

James (1991) has reviewed experimental studies of geographic variation in bird species. She found a remarkably consistent pattern of intraspecific variation in size in breeding populations of North American bird species. Individuals from warm humid climates tend to be smaller than birds from increasingly cooler and drier regions. In addition, birds from

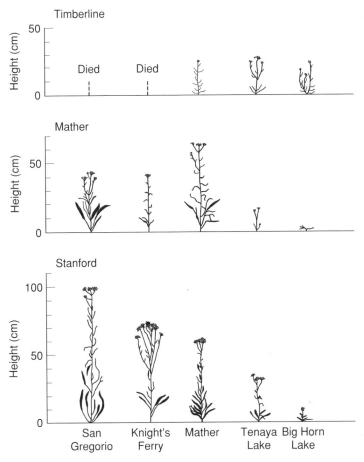

Figure 2.9 Representative clones of yarrow plants originating from five different altitudinal locations grown at three altitudes: 30 m above sea level at Stanford, 1,200 m above sea level at Mather, and 3,000 m above sea level at Timberline. The San Gregorio clone was from a coastal population, and the Big Horn Lake clone was from the highest altitude site (over 3,000 m); the other three clones were from an altitudinal gradient between these two extremes. From Clausen et al. (1948); redrawn from Strickberger (2000).

regions with greater humidity tend to have more darkly colored feathers. The consistent patterns in body size and coloration among many species suggest that these patterns are adaptations that have evolved by natural selection in response to differential selection in different environments.

For example, there is some evidence that the color polymorphism that we considered earlier in eastern screech owls affects the survival and reproductive success of individuals. The frequency of red owls increases from north (less than 20% red) to south (approximately 80% red) (Pyle 1997). VanCamp and Henny (1975) found evidence that red owls suffered relatively greater mortality than gray owls during severe winter conditions in Ohio, and suggested that this may be due to metabolic differences between red and gray birds (Mosher and Henny 1976). A similar north–south clinal pattern of red and gray morphs has also been reported in the ruffed grouse; Gullion and Marshall (1968) reported that the red morph has lower survival during extreme winter conditions than the gray morph.

2.4.1 Countergradient variation

Countergradient variation is a pattern in which genetic influences counteract environmental influences so that phenotypic change along an environmental gradient is minimized (Conover and Schultz 1995) (see Guest Box 2). For example, Berven et al. (1979) used transplant and common-garden experiments in the laboratory to examine the genetic basis in life history traits of green frogs. In the wild, montane tadpoles experience lower temperatures; they grow and develop slowly and are larger at metamorphosis than are lowland tadpoles that develop at higher temperatures. Egg masses collected from high and low altitude populations were cultured side by side in the laboratory at temperatures that mimic developmental conditions at high and low altitude (18 and 28°C). The differences observed between low and high altitude frogs raised under common conditions in the laboratory for some traits were **opposite** in direction to the differences observed in nature. That is, at low (montane like) temperatures, lowland tadpoles grew even slower than, took longer to complete metamorphosis, and were larger than montane tadpoles.

A reversal of naturally occurring phenotypic differences under common environments may occur when natural selection favors development of a similar phenotype in different environments. Consider the developmental rate in a frog or fish species and assume that there is some optimal developmental rate. Individuals from populations occurring naturally at colder temperatures will be selected for a relatively fast developmental rate to compensate for the reduction in developmental rate caused by lower temperatures. Individuals in the lower temperature environment may still develop more slowly in nature. However, if grown at the same temperature, the individuals from the colder environment will develop more quickly. This will result in countergradient variation (Figure 2.10).

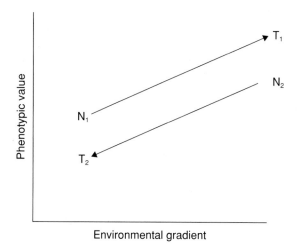

Figure 2.10 Diagram of countergradient variation. The end points of the lines represent outcomes of a reciprocal transplant experiment: N_1 and N_2 are the native phenotypes of each population in its home environment. T_1 and T_2 are the phenotypes when transplanted to the other environment. Redrawn from Conover and Schultz (1995).

Therefore, phenotypic differences between populations are not a reliable indicator of genetic differences between populations without additional information. In some cases, all of the phenotypic differences between populations may result from environmental conditions. And, even if genetic differences do exist, they actually may be in the opposite direction of the observed phenotypic differences between populations.

Differences among populations in the amount of total phenotypic variation within populations can also be misleading. Using the relationship represented by expression 2.1, we would expect a positive association between V_P and V_G. That is, if V_E is constant, then greater total phenotypic variability (V_P) in a population would be indicative of greater genetic variability (V_G). However, assuming that V_E is constant is a very poor assumption because different populations are subject to different environmental conditions. In addition, the reduction in genetic variation associated with small population size sometimes can decrease developmental stability and thereby increase total phenotypic variability in populations (Leary and Allendorf 1989).

Thus, it is not appropriate to use the amount of total phenotypic variability (V_P) in separate populations to detect differences in the amount of genetic variation between populations. The relationship between V_P and V_G is not straightforward. It differs for different traits within a single population and also depends on the history of the population. The genetic analysis of polygenic phenotypic variation is considered in more detail in Chapter 12.

Guest Box 2 Looks can be deceiving: countergradient variation in secondary sexual color in sympatric morphs of sockeye salmon
Chris J. Foote

Sockeye salmon and kokanee are respectively the anadromous (sea-going, physically large) and nonanadromous (lake-dwelling, small) morphs of sockeye salmon (Figure 2.11). Both morphs occur throughout the native range of the species in North Pacific drainages of North America and Asia. The morphs are polyphyletic, with one morph, likely sockeye, having given rise to the other on numerous independent occasions throughout their range (Taylor et al. 1996). The morphs occur together or separately in lakes (where sockeye typically spend their first year of life), but wherever they occur sympatrically, even where they spawn together, they are always genetically distinct in a wide array of molecular and physical traits (Taylor et al. 1996; Wood and Foote 1996). This reproductive isolation appears to result from significant, size-related, prezygotic isolation coupled with the large selective differences between marine and lacustrine environments (Wood and Foote 1996).

On viewing sockeye and kokanee on the breeding grounds, one is struck by the large size difference between them (sockeye can be 2–3 times the length and 20–30 times the weight of kokanee), and by their shared, striking, bright red breeding color (Craig and Foote 2001) (Figure 2.11). However, with respect to genetic differentiation, looks can be deceiving. The size difference between the morphs results largely from differences in food availability in the marine versus lacustrine environments, with only a slight genetic difference in growth evident between the morphs when grown in a common environment (Wood and Foote 1990).

Figure 2.11 Photograph of sockeye salmon (large) and kokanee (small) on the breeding grounds in Dust Creek – a tributary of Takla Lake, British Columbia. Both kokanee and sockeye salmon have bright red bodies and olive green heads. Photo by Chris Foote.

In contrast, their similarity in breeding color masks substantial polygenic differentiation in the mechanism by which they produce their red body color. The progeny of sockeye that rear in fresh water throughout their life cannot turn red at maturity like kokanee; rather they turn green (Craig and Foote 2001). Therefore, as sockeye repeatedly gave rise to kokanee over the last 10,000 years, they did so by at first producing a green freshwater morph that over time changed genetically to converge on the ancestral red breeding color.

The convergence in breeding color in sockeye and kokanee is an example of countergradient variation. Kokanee, which live in carotenoid-poor lake environments are three times more efficient in utilizing carotenoids than sockeye, which live in a carotenoid-rich marine environment (Craig and Foote 2001). Interestingly, the selective force for the re-emergence of red in kokanee appears to be inherited from ancestral sockeye. Sockeye possess a very strong, and apparently innate, preference for red mates (Foote et al. 2004), a preference that is shared by kokanee. This pre-existing bias appears to have independently driven the evolution of red breeding colour in kokanee throughout their distribution. This contrasts with other examples of countergradient variation, where natural selection, not **sexual selection**, is thought to be the driving selective force (Conover and Shultz 1995).

Problem 2.1

Several state fish and game departments from northern states in the United States (e.g., Illinois) have stocked largemouth bass from Florida into lakes because they are well known for their large size. However, the introduced fish have been found to grow slower than native largemouth bass (Philipp 1991). Write a short letter to the Director of the Illinois Department of Fish and Game that explains the likely basis for these observations.

Problem 2.2

Cooke (1987) observed the following patterns of inheritance of a color polymorphism in the lesser snow goose:

	Progeny	
Mating	**White**	**Blue**
White × white	35,104 (98%)	809 (2%)
White × blue	3,873 (40%)	5,691 (60%)
Blue × blue	1,064 (10%)	9,707 (90%)

Assume that this trait is determined by a single locus. Propose a model that would explain these results (see Table 2.1). Are there any progeny that do not conform to the expectations of your model? How might these progeny be explained?

Problem 2.3

Select some plant or animal species that you have observed that exhibits phenotypic variation for some trait. Record your observations in some systematic way. That is, either measure a number of individuals or classify a number of individuals into two or more discrete categories for the trait that you have selected. Do you think that the phenotypic differences you have observed have any genetic basis? Propose an experimental procedure for testing for some genetic influence on the phenotypic variation that you have observed.

Problem 2.4

The plots below show the regression of mean progeny values in 14 families of rainbow trout for two meristic traits on the mid-parent value (the mid-parent value is the mean of the maternal and paternal value; data from Leary et al. 1985). The regression of the mean number of dorsal rays on the mid-parent values for these families is highly significant ($P < 0.001$); the regression of the mean number of

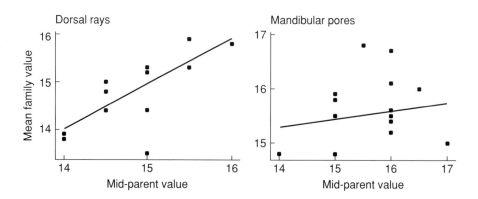

mandibular pores on the mid-parent values for these families is not significant. Which of these two traits do you think has the greater heritability (that is, is most influenced by genetic differences between individuals in this population)? Why?

Problem 2.5

The white-phase Kermode bear has been recognized as a **subspecies** of the black bear (Nagorsen 1990). However, as discussed in Section 2.1, this color phase is caused by a recessive allele at a single locus, and both white- and black-phase bears occur within the same breeding population. Do you think that the white-phase Kermode bears should be recognized as a separate subspecies or group for purposes of conservation? How should the white-phase Kermode bears best be protected?

Genetic Variation in Natural Populations: Chromosomes and Proteins

Sumatran orangutan, Section 3.1.7

The empirical study of population genetics has always begun with and centered around the characterization of the genetic variation in populations.

Richard C. Lewontin (1974)

Geographic chromosomal variation, which in many instances does not correlate with variation in phenotype, is increasingly being detected within both large and small species of mammals. We argue that this cryptic chromosome variation can pose a significant threat to translocation practices involving the admixture of specimens between geographically distant populations.

Terrence J. Robinson and Frederick F. B. Elder (1993)

Genetic variation is the raw material of evolution. Change in the genetic composition of populations and species is the primary mechanism of evolutionary change within species.

In Chapter 2, we examined phenotypic variation in natural populations. In this chapter, we will examine the genetic basis of this phenotypic variation by examining genetic differences between individuals in their chromosomes and proteins. In Chapter 4, we will examine variation in DNA sequences. This order, from the chromosomes that are visible under a light microscope down to the study of molecules, reflects the historical sequence of study of natural populations.

In our consideration of conservation, we are concerned with genetic variation at two fundamentally different hierarchical levels:

1   Genetic differences among individuals within local populations.
2   Genetic differences among populations within the same species.

The amount of genetic variation within a population provides insight into the demographic structure and evolutionary history of a population. For example, lack of genetic variation may indicate that a population has gone through a recent dramatic reduction in population size. Genetic divergence among populations is indicative of the amount of genetic exchange that has occurred among populations, and can play an important role in the conservation and management of species. For example, genetic analysis of North Pacific minke whales has shown that some 20–40% of the whale meat sold in Korean and Japanese markets comes from a protected and genetically isolated population of minke whales in the Sea of Japan (Baker et al. 2000).

Population geneticists struggled throughout most of the 20th century to measure genetic variation in natural populations (Table 3.1). Before the advent of biochemical and molecular techniques, genetic variation could only be examined by bringing individuals into the laboratory and using experimental matings. The fruit fly (*Drosophila* spp) was the workhorse of empirical population genetics during this time because of its short generation time and ease of laboratory culture. For example, 41% of the papers (9 of 22) in the first volume of the journal *Evolution* published in 1947 used *Drosophila*; while just 10% of the papers (21 of 212) in the volume of *Evolution* published in the year 2000 were on *Drosophila*.

The tools used to examine chromosomal and allozyme variation have been used for many years. Their utility has been eclipsed by powerful new techniques that we will consider in the next chapter, which allow direct examination of genetic variation in entire genomes! Nevertheless, these "old" tools and the information that they have provided continue to be valuable. The technique of allozyme **electrophoresis** will continue to fade

Table 3.1 Historical overview of primary methods used to study genetic variation in natural populations.

Time period	Primary techniques
1900–1970	Laboratory matings and chromosomes
1970s	Protein electrophoresis (allozymes)
1980s	Mitochondrial DNA
1990s	Nuclear DNA
2000s	Genomics

away in the next decade or so as it is replaced by techniques that examine genetic variation in the DNA that encodes the proteins studied by allozyme electrophoresis (Utter 2005). Study of DNA sequences, however, cannot replace examination of chromosomes. We expect there will be a rejuvenation of chromosomal studies in evolutionary and conservation genetics when new technologies are developed that allow rapid examination of chromosomal differences between individuals (deJong 2003).

3.1 Chromosomes

Surprisingly little emphasis has been placed on chromosomal variability in conservation genetics (Benirschke and Kumamoto 1991; Robinson and Elder 1993). This is unfortunate because heterozygosity for chromosomal differences often causes reduction in fertility (Nachman and Searle 1995; Rieseberg 2001). For example, the common cross between a female horse with 64 chromosomes and a female mule with 62 chromosomes produces a sterile mule that has 63 chromosomes. Some captive breeding programs unwittingly have hybridized individuals that are morphologically similar but have distinct chromosomal complements: orangutans (Ryder and Chemnick 1993), gazelles (Ryder 1987), and dik-diks (Ryder et al. 1989). Similarly, translocation or reintroduction programs may cause problems if individuals are translocated among chromosomally distinct groups (for example see Guest Box 3).

The possible occurrence of hybridization in captivity of individuals from chromosomally distinct populations is much more common than would generally be expected because small, isolated populations have a greater rate of chromosomal evolution than common widespread taxa (Lande 1979). Thus, the very demographic characteristics that make a species a likely candidate for captive breeding are the same characteristics that favor the evolution of chromosomal differences between groups. For example, extensive chromosomal variability has been reported in South American primates (Matayoshi et al. 1987) and some plants (Example 3.1).

The direct examination of genetic variation in natural population began with the description of differences in chromosomes between individuals. One of the first reports of differences in the chromosomes of individuals within populations was by Stevens (1908), who described different numbers of **supernumerary chromosomes** in beetles (White 1973). For many years, study of chromosomal variation in natural populations was dominated by the work of Theodosius Dobzhansky and his colleagues on *Drosophila* (Dobzhansky 1970) because of the presence of giant polytene chromosomes in the salivary glands (Painter 1933). However, the study of chromosomes in other species lagged far behind. For example, until 1956 it was thought that humans had 48, rather than 46 chromosomes in each cell. It is amazing that the complete human genome was sequenced within 50 years of the development of the technical ability to even count the number of human chromosomes.

3.1.1 Karyotypes

A **karyotype** is the characteristic chromosome complement of a cell, individual, or species. Chromosomes in the karyotype of a species are usually arranged beginning with the largest chromosome (Figure 3.1). The large number of **microchromosomes** in the

---

Example 3.1 Cryptic chromosomal species in the graceful tarplant

The graceful tarplant (*Holocarpha virgata*) is a classic example of the importance of chromosomal differentiation between populations for conservation and management. Clausen (1951) described the karyotype of plants from four populations of this species ranging from Alder Springs in northern California to a population near San Diego, California. These populations can hardly be distinguished morphologically and live in similar habitats. Plants from all of these populations had a haploid set of four chromosomes ($n = 4$). However, the size and shape of these four chromosomes differed among populations. Experimental crossings revealed that matings between individuals in different populations either failed to produce F_1 individuals or the F_1 individuals were sterile.

Clausen (1951) concluded that these populations were distinct species because of their chromosomal characteristics and infertility. Nevertheless, he felt that it would be "impractical" to classify them as taxonomic species because of their morphological similarity and lack of ecological distinctness. These populations are classified as the same species today. Nevertheless, for purposes of conservation, each of the chromosomally distinct populations should be treated as separate species because of their reproductive isolation. Translocations of individuals among populations could have serious harmful effects because of reduction in fertility or the production of sterile hybrids.

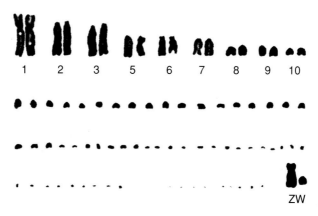

Figure 3.1 Karyotype ($2n = 84$) of a female cardinal. From Bass (1979).

shown karyotype is typical for many bird species (Shields 1982). Evidence suggests that bird microchromosomes are essential, unlike the supernumerary chromosomes discussed later in this section (Shields 1982).

Chromosomes of eukaryotic cells consist of DNA and associated proteins. Each chromosome consists of a single highly folded and condensed molecule of DNA. Some large chromosomes would be several centimeters long if they were stretched out – thousands of times longer than a cell nucleus. The DNA in a chromosome is coiled again and again and

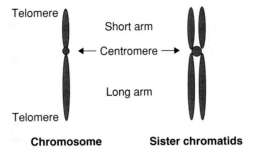

Figure 3.2 Diagram of an unreplicated chromosome and a chromosome that has replicated into two identical sister chromatids that are joined at the centromere.

is tightly packed around histone proteins. Chromosomes are generally thin and difficult to observe even with a microscope. Before cell division (mitosis and meiosis), however, they condense into thick structures that are readily seen with a light microscope. This is the stage when we usually observe chromosomes (Figure 3.1). The chromosomes right before cell division have already replicated so that each chromosome consists of two identical sister chromatids (Figure 3.2).

Chromosomes function as the vehicles of inheritance during the processes of mitosis and meiosis. Mitosis is the separation of the sister chromatids of replicated chromosomes during somatic cell division to produce two genetically identical cells. Meiosis is the pairing of and separation of homologous replicated chromosomes during the division of sex cells to produce gametes.

Certain physical characteristics and landmarks are used to describe and differentiate chromosomes. The first is size; the chromosomes are numbered from the largest to the smallest. The **centromere** appears as a constricted region and serves as the attachment point for spindle microtubules that are the filaments responsible for chromosomal movement during cell division (Figure 3.2). The centromere divides a chromosome into two arms. Chromosomes in which the centromere occurs approximately in the middle are called **metacentric**. In **acrocentric** chromosomes, the centromere occurs near one end of the chromosome. Staining techniques have been developed that differentially stain different regions of a chromosome to help distinguish chromosomes that have similar size and centromere location (Figure 3.3).

3.1.2 Sex chromosomes

Many groups of plants and animals have evolved sex-specific chromosomes that are involved in the process of sex determination (see Rice 1996 for an excellent review). In mammals, females are homogametic XX and males are heterogametic XY (Figure 3.3). The chromosomes that do not differ between the sexes are called **autosomes**. For example, there are 28 pairs of chromosomes in the karyotype of African elephants ($2n = 56$) (Houck et al. 2001); thus, each African elephant has 54 autosomes and two **sex chromosomes**. The heterogametic sex is reversed in birds: males are homogametic ZZ and females are heterogametic ZW (see Figure 3.1). Note that the XY and ZW notations are strictly arbitrary and are used to indicate which sex is homogametic. For example,

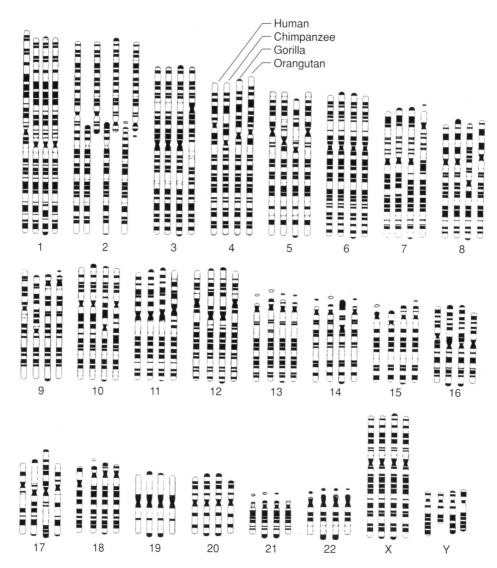

Figure 3.3 Karyotypes and chromosomal banding patterns of humans, chimpanzees, gorillas, and orangutans. From Strickberger (2000); from Yunis and Prakash (1982).

Lepidopterans (butterflies and moths) are ZZ/ZW; this indicates that the males are the homogametic sex.

The heterogametic sex often differs between species in some taxa (Charlesworth 1991). Some fish species are XX/XY, some are ZZ/ZW, some do not have detectable sex chromosomes, and a few species even have more than two sex chromosomes (Devlin and Nagahama 2002). Many plant species do not have separate sexes, and, therefore, they do not have sex chromosomes (Charlesworth 2002). However, both XX/XY and ZZ/ZW sex-determination systems occur in dioecious plant species with separate male and female individuals.

Heteromorphic sex chromosomes can provide useful markers for conservation, and the sex of individuals can be determined by karyotypic examination. However, many other far easier procedures can be used to sex individuals by their sex chromosomes complement. For example, one of the two X chromosomes in most mammal species is inactivated and forms a darkly coiling structure (a Barr body) that can be readily detected with a light microscope in epithelial cells scraped from the inside of the mouth of females but not males (White 1973). We will see in the next chapter that DNA sequences specific to one of the sex chromosomes can be used in many taxa to identify the sex of individuals.

3.1.3 Polyploidy

Most animal species contain two sets of chromosomes and therefore are diploid ($2n$) for most of their life cycles. The eggs and sperm of animals are haploid and contain only one set of chromosomes ($1n$). Some species, however, are polyploid because they possess more than two sets of chromosomes: triploids ($3n$), tetraploids ($4n$), pentaploids ($5n$), hexaploids ($6n$), and even greater number of chromosome sets. Polyploidy is relatively rare in animals, but it does occur in invertebrates, fishes, amphibians, and lizards (White 1973). Polyploid is common in plants and is a major mechanism of speciation (Stebbins 1950; Soltis and Soltis 1999). Approximately 40% of all flowering plant species are polyploids and nearly 75% of all grasses are polyploid.

Perhaps the most interesting cases of polyploidy occur when diploid and tetraploid forms of the same taxon exist in sympatry. For example, both diploid (*Hyla chrysoscelis*) and tetraploid (*H. versicolor*) forms of gray tree frogs occur throughout the central United States (Ptacek et al. 1994). Reproductive isolation between diploids and tetraploids is maintained by call recognition; the larger cells of the tetraploid males result in a lower calling frequency, which is recognized by the females (Bogart 1980). Hybridization between diploids and tetraploids does occur and results in triploid progeny that are not fertile (Gerhardt et al. 1994).

A thorough treatment of polyploidy is beyond the scope of this chapter. Nevertheless, examination of ploidy levels is an important taxonomic tool when describing units of conservation in some plant taxa (see Guest Box 3; Chapter 16).

3.1.4 Numbers of chromosomes

Many closely related species have different numbers of chromosomes, as the horse and mule discussed in Section 3.1. For example, a haploid set of human chromosomes has $n = 23$ chromosomes, while the extant species that are most closely related to humans all have $n = 24$ chromosomes (see Figure 3.3). This difference is due to a fusion of two chromosomes to form a single chromosome (human chromosome 2) that occurred sometime in the human evolutionary lineage following its separation from the ancestor of the other species. This is an example of a **Robertsonian fusion** which we will consider in Section 3.1.8.

Chromosome numbers have been found to evolve very slowly in some taxa. For example, the approximately or so 100 cetaceans have either $2n = 42$ or 44 chromosomes (Benirschke and Kumamoto 1991). In contrast, chromosome numbers have diverged rather rapidly in other taxa. For example, the $2n$ number in horses (genus *Equus*) varies

Table 3.2 Characteristic chromosome numbers of some living members of the horse family (White 1973).

Species		2n
Przewalski's horse	*Equus przewalski*	66
Domestic horse	*E. cabbalus*	64
Donkey	*E. asinus*	62
Kulan	*E. hemionus*	54
Grevy's zebra	*E. grevyi*	46
Burchell's zebra	*E. burchelli*	44
Mountain zebra	*E. zebra*	34

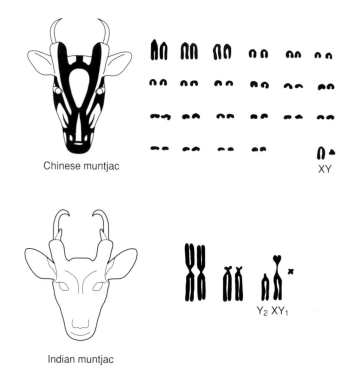

Chinese muntjac

XY

Indian muntjac

Y_2 XY_1

Figure 3.4 The Chinese and Indian muntjac and their karyotypes. The Indian muntjac has the lowest known chromosome number of any mammal. First-generation hybrids ($2n = 27$) created in captivity between these species are viable (Wang and Lan 2000). The two Y chromosomes in this species have resulted from a centric fusion between an autosome and the sex chromosomes (White 1973). From Strickberger (2000).

from $2n = 34$ to 66 (Table 3.2). The Indian muntjac has a karyotype that is extremely divergent from other species in the same genus (Figure 3.4). In the next few sections we will consider the types of chromosome rearrangements that bring about karyotypic changes among species.

3.1.5 Supernumerary chromosomes

Supernumerary chromosomes (also called B chromosomes) are not needed for normal development and vary in number in many plant and animal species. They are usually small, lack functional genes (**heterochromatic**), and do not pair and segregate during meiosis (White 1973; Jones 1991). In general, the presence or absence of B chromosomes does not affect the phenotype or the fitness of individuals (Battaglia 1964). It is thought that B chromosomes are "parasitic" genetic elements that do not play a role in adaptation (Jones 1991). B chromosomes have been reported in many species of higher plants (Müntzing 1966). In animals, B chromosomes have been described in many invertebrates, but they are rarer in vertebrates. However, Green (1991) has described extensive polymorphism for B chromosomes in populations of the Pacific giant salamander along the west coast of North America.

3.1.6 Chromosomal size

In many species differences in size between homologous chromosomes have been detected. In most of the cases it appears that the "extra" region is due to a heterochromatic segment that does not contain functional genes (White 1973, p. 306). These extra heterochromatic regions resulting in size differences between homologous chromosomes are analogous to supernumerary chromosomes, except that they are inherited in a Mendelian manner. Heterochromatic differences in chromosomal size seem to be extremely common in several species of South American primates (Matayoshi et al. 1987).

3.1.7 Inversions

Inversions are segments of chromosomes that have been turned around so that the gene sequence has been reversed. Inversions are produced by two chromosomal breaks and a rejoining, with the internal piece inverted (Figure 3.5). An inversion is **paracentric** if both breaks are situated on the same side of the centromere (Figure 3.6), and **pericentric** if the two breaks are on opposite side of the centromere (Figure 3.7).

Heterozygosity for inversions is often associated with reduced fertility. **Recombination** (crossing over) within inversions produces aneuploid gametes that form inviable zygotes (Figures 3.6 and 3.7). The allelic combinations at different loci within inversion loops will tend to stay together because of the low rate of successful recombination within inversions. In situations where several loci within an inversion affect the same trait, the allelic combinations at the loci are referred to as a **supergene**). Examples of phenotypes controlled by supergenes include shell color and pattern in snails (Ford 1971), mimicry in butterflies (Turner 1985), and flower structure loci in *Primula* (Kurian and Richards 1997).

Inversions are exceptionally common in some taxonomic groups because they do not have the usual harmful effects of producing inviable zygotes (White 1973, p. 241). For example, crossing over and recombination does not occur in male dipteran (two-winged) flies (e.g., *Drosophila*). Therefore, lethal chromatids will only be produced in females in these species. However, there seems to be a meiotic mechanism in these species so that chromatids without a centromere or with two centromeres pass into the polar body rather than into the egg nucleus so that fertility is not reduced. Chromosomes with a single

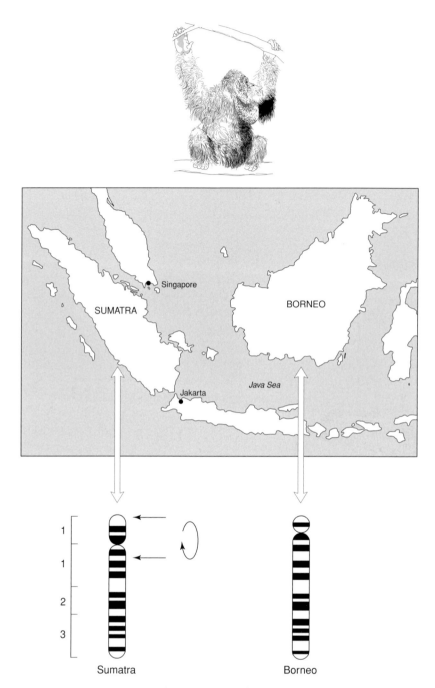

Figure 3.5 Pericentric inversion in chromosome 2 of two subspecies of orangutans from Sumatra and Borneo (chromosomes from Seuanez 1986).

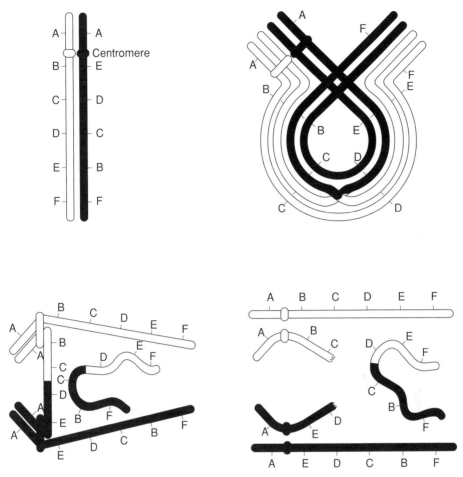

Figure 3.6 Crossing over in a heterozygote for a paracentric inversion. The two chromosomes are shown in the upper left. The pairing configuration and crossing over between two nonsister chromatids is shown in the upper right. Separation of chromosomes during the first meiotic division is shown in the lower left. The resulting chromosomal products of meiosis are shown in the lower right. Only two chromosomes have complete sets of genes; these are noncrossover chromosomes that have the same sequences as the two original chromosomes. From Dobzhansky et al. (1977).

centromere having deficiencies or duplications that are produced in heterozygotes for pericentric inversions (Figure 3.7) are just as likely to be passed into the egg nucleus as normal chromosomes.

Paracentric inversions are difficult to detect because they do not change the relative position of the centromere on the chromosome. They can only be detected by examination of meiotic pairing or by using some technique that allows visualization of the genic sequence on the chromosome, such as in polytenic chromosomes of *Drosophila* and other Dipterans. Several chromosome-staining techniques that reveal banding patterns were discovered in the early 1970s (see Figure 3.3; Comings 1978). These techniques have

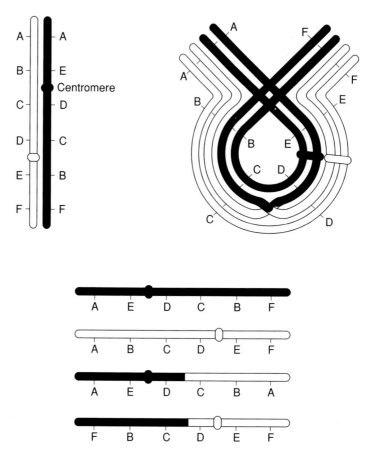

Figure 3.7 Crossing over in a heterozygote for a pericentric inversion. The two chromosomes are shown in the upper left. The pairing configuration and crossing over between two nonsister chromatids is shown in the upper right. The resulting products of meiosis are shown below. Only the two top chromosomes have complete sets of genes; these are noncrossover chromosomes that have the same sequences as the two original chromosomes. From Dobzhansky et al. (1977).

been extremely helpful in identifying homologous chromosomes in karyotypes and for detecting chromosomal rearrangements such as paracentric inversions (e.g., Figure 3.6). However, relatively few species have been studied with these techniques, so that we know little about the frequency of paracentric inversions in natural populations.

Pericentric inversions are more readily detected than paracentric inversions because they change their relative position on the centromere. Two frequent pericentric inversions have been described in orangutans (*Pongo pygmaeus*). Orangs from Borneo (*P. p. pygmaeus*) and Sumatra (*P. p. abelii*) are fixed for different forms of an inversion at chromosome 2 (Seuanez 1986; Ryder and Chemnick 1993) (see Figure 3.5). All wild captured orangs have been homozygous for these two chromosomal types, while over a third of all captive-born orangs have been heterozygous (Table 3.3). A pericentric inversion of chromosome 9 is polymorphic in both subspecies (Table 3.3). The persistence of the polymorphism in chromosome 9 for this period of time is surprising.

Table 3.3 Chromosomal inversion polymorphisms in the orangutan (Ryder and Chemnick 1993). The inversion in chromosome 2 distinguishes the Sumatran (S) and Bornean (B) subspecies. The two inversion types in chromosome 9 (C and R) are polymorphic in both subspecies.

	Chromosome 2			Chromosome 9		
	BB	**SB**	**SS**	**CC**	**CR**	**RR**
Wild born	51	0	41	67	22	3
Zoo born	90	44	82	71	34	3

Divergence in proteins and mitochondrial DNA sequences between Bornean and Sumatran orangutans support the chromosomal evidence that these two subspecies have been isolated for over a million years (Ryder and Chemnick 1993). Two groups of authors proposed in 1996 that the orangs of Borneo and Sumatra should be recognized as separate species on the basis of these chromosomal differences and molecular genetic divergence (Xu and Arnason 1996; Zhi et al. 1996). Muir et al. (1998) strongly disagreed with this recommendation based on a variety of arguments. They conclude in a more detailed study of mtDNA that the pattern of genetic divergence among orangutans is complex and is not adequately described by a simple Sumatra–Borneo split (Muir et al. 2000).

Chromosomal polymorphisms seem to be unusually common in some bird species (Shields 1982). Figure 3.1 shows the karyotype of a cardinal that is heterozygous for a pericentric inversion of chromosome 5 (Bass 1979). Figure 3.8 shows all three karyotypic combinations of these two inversions. There is evidence that suggests that such

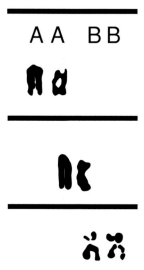

Figure 3.8 Three possible karyotypes for the chromosome 5 polymorphism for a pericentric inversion in cardinals. Chromosome 5A is acrocentric, while chromosome 5B is metacentric. Modified from Bass (1979).

chromosomal polymorphisms may be associated with important differences in morphology and behavior among individuals. For example, Rising and Shields (1980) have described pericentric inversion polymorphisms in slate-colored juncos that are associated with morphological differences in bill size and appendage size. They suggest that this polymorphism is asso-ciated with habitat partitioning during the winter when they live and forage in flocks.

Throneycroft (1975) has described an extremely interesting pericentric inversion in chromosome 2 of white-throated sparrows. Individuals with the inversion have a white median stripe on their head while individuals lacking the inversion have a tan stripe. There is strong **disassortative mating** for this polymorphism; that is, birds tend to select mates with the opposite color pattern. Tuttle (2003) has described a variety of behavioral differences between these two morphs: white males are more aggressive, spend less time guarding their mates, and provide less parental care.

3.1.8 Translocations

A translocation is a chromosomal rearrangement in which part of a chromosome becomes attached to a different chromosome. Reciprocal or mutual translocations result from a break in each of two nonhomologous chromosomes and an exchange of chromosomal sections. In general, polymorphisms for translocations are rare in natural populations because of infertility problems in heterozygotes.

Robertsonian translocations are a special case of translocations in which the break points occur very close to either the centromere or the **telomere** (Figure 3.9). A Robertsonian fusion occurs when a break occurs in each of two acrocentric chromosomes near their centromeres, and the two chromosomes join to form a single metacentric chromosome. Robertsonian fissions also occur where this process is reversed. Robertsonian polymorphisms can be relatively frequent in natural populations because the translocation involves the entire chromosome arm and balanced gametes are usually produced by heterozygotes (Searle 1986; Nachman and Searle 1995). Lamborot (1991) has presented an elegant example of several Robertsonian polymorphisms in a lizard species from central Chile (Figure 3.10).

The Western European house mouse (*Mus musculus domesticus*) has an exceptionally variable karyotype (Nachman and Searle 1995). The rate of evolution of Robertsonian changes in this species is nearly 100 times greater than in most other mammals (Nachman and Searle 1995). Over 40 chromosomal races of this species have been described in Europe and North Africa on the basis of Roberstonian translocations (Hauffe and Searle

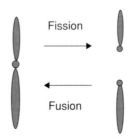

Figure 3.9 Diagram of Robertsonian fusion and fission.

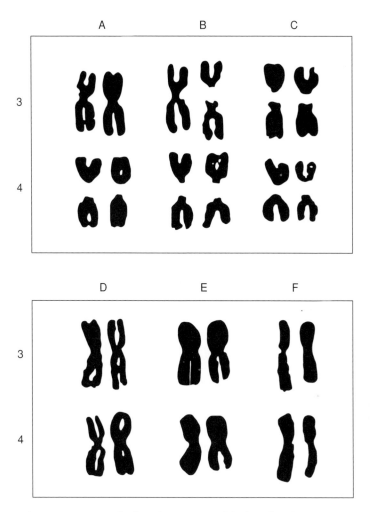

Figure 3.10 Chromosomes 3 and 4 from karyotypes of the lizard *Liolaemus monticola* separated by the Maipo and Yeso Rivers in central Chile. A through C are the northern chromosomal type ($2n = 38$, 39, and 40); D through F are the southern chromosomal type ($2n = 34$). The southern and northern types are distinguished by a Robertsonian rearrangement in chromosome 4. In addition, the northern type is polymorphic from a Robertsonian rearrangement of chromosome 3. From Lamborot (1991).

1998). As expected, hybrids between three of these races have reduced fertility. On average, the litter size of crosses with one hybrid parent is 44% less than crosses between two parents from the AA race (Table 3.4).

3.2 Protein electrophoresis

The first major advance in our understanding of genetic variation in natural populations began in the mid-1960s with the advent of protein electrophoresis (Powell 1994). The

Table 3.4 Litter sizes produced by mice heterozygous for Robertsonian translocations characteristic of three different chromosomal races (AA, POS, and UV). From Hauffe and Searle (1998).

Female	Male	No. of litters	Litter size
AA	AA (control)	17	6.7 ± 0.8
AA	(AA × POS)	16	4.1 ± 0.4
AA	(AA × UV)	18	2.6 ± 0.3
AA	(UV × POS)	19	3.8 ± 0.3
AA (control)	AA	18	6.8 ± 0.4
(AA × POS)	AA	7	1.0 ± 0
(AA × UV)	AA	10	3.1 ± 0.6
(POS × UV)	AA	11	4.0 ± 0.5

study of variation in amino acid sequences of proteins by electrophoresis allowed an immediate assessment of genetic variation in a wide variety of species (Lewontin 1974). There is a direct relationship between genes (DNA base pair sequences) and proteins (amino acid sequences). Proteins have an electric charge and migrate in an electric field at different rates depending upon their charge, size, and shape. A single amino acid substitution can affect migration rate and thus can be detected by electrophoresis. Moreover, the genomes of all animals (from elephants to *Drosophila*), all plants (from sequoias to the Furbish lousewort), and all prokaryotes (from *Escherichia coli* to HIV) encode proteins. Empirical population genetics has become universal, as genetic variation has been described in natural populations of thousands of species in the last 30 years.

Figure 3.11 outlines the procedures in the gel electrophoresis of enzymes (also see May 1998). There are two fundamental steps to electrophoresis. The first is to separate proteins with different mobilities in some kind of a supporting medium (usually a gel of starch or polyacrylamide). However, most tissues contain proteins encoded by hundreds of different genes. The second step of the process, therefore, is to locate the presence of specific proteins. This step is usually accomplished by taking advantage of the specific catalytic activity of different enzymes. Specific enzymes can be located by using a chemical solution containing the substrate specific for the enzyme to be assayed, and a salt that reacts with the product of the reaction catalyzed by the enzyme, producing a visible product.

For example, the enzyme lactate dehydrogenase (LDH) catalyzes the reversible interconversion of lactic acid and pyruvic acid and causes the release of hydrogen ions (H^+):

$$\text{Lactic acid} + (NAD^+) \xrightleftharpoons{\text{LDH}} \text{Pyruvic acid} + NADH + (H^+)$$

We can "see" LDH on a gel by using a dye that is soluble and colorless but that becomes nonsoluble and purple when coupled with the hydrogen ions produced wherever LDH is present in the gel. Figure 3.12 shows the variation at an LDH gene (*LDH-B2*) that is expressed in the liver tissue of rainbow trout. All of the individuals in this figure are progeny from a laboratory mating between a single female and a single male that were both heterozygous (*100/69*) at this locus. Alleles are generally identified by their relative

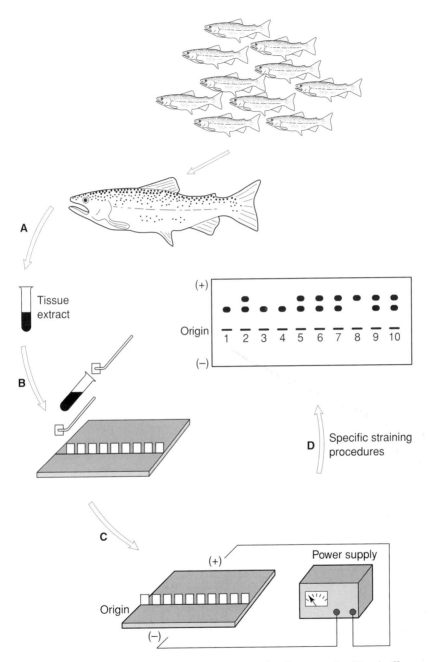

Figure 3.11 Allozyme electrophoresis. (A) A tissue sample is homogenized in a buffer solution and centrifuged. (B) The supernatant liquid is placed in the gel with filter paper inserts. (C) Proteins migrate at different rates in the gel because of differences in their charge, size, or shape. (D) Specific enzymes are visualized in the gel by biochemical staining procedures. From Utter et al. (1987).

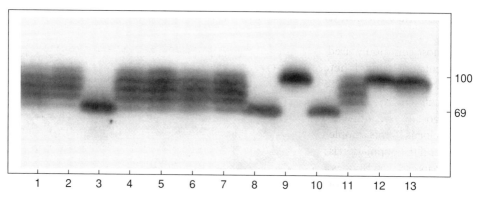

Figure 3.12 Inheritance of a polymorphism for a lactate dehydrogenase (LDH) locus (*LDH-B2*) expressed in liver tissue of rainbow trout. Heterozygotes for LDH are five banded because four protein subunits combine to form a single functional enzyme. All 13 samples on this gel are liver samples from progeny produced by a mating between two heterozygotes (*100/69*). Genotypes are: 1 and 2 (*100/69*), 3 (*69/69*), 4–7 (*100/69*), 8 (*69/69*), 9 (*100/100*), 10 (*69/69*), 11 (*100/69*), and 12–13 (*100/100*). From Utter et al. (1987).

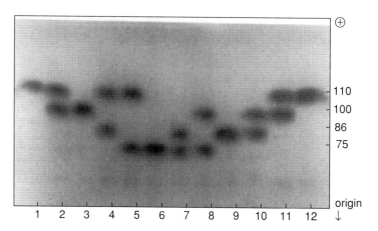

Figure 3.13 Gel electrophoresis of the enzyme aconitate dehydrogenase in the livers of 12 Chinook salmon. The relative mobilities of the allozymes encoded by four alleles at this locus are on the right. The genotypes of all 12 individuals are: 1 (*110/110*), 2 (*100/110*), 3 (*100/100*), 4 (*110/86*), 5 (*110/75*), 6 (*75/75*), 7 (*86/75*), 8 (*100/75*), 9 (*86/86*), 10 (*100/86*), 11 (*110/100*), and 12 (*110/110*). From Utter et al. (1987).

migration distance in the gel. Thus, the *69* allele migrates approximately 69% as far as the common allele, *100*. The observed frequencies (3, 7, and 3) of the three genotypes *100/100*, *100/69*, and *69/69* are close to the expected 1 : 2 : 1 Mendelian proportions from a cross between two heterozygotes.

We can stain different gels for many enzymes and thus examine genetic variation at many protein loci. Figure 3.13 shows variation at the enzyme aconitate dehydrogenase in Chinook salmon from a sample from the Columbia River of North America.

3.3 Genetic variation within natural populations

The most commonly used measure to compare the amount of genetic variation within different populations is **heterozygosity**. At a single locus, heterozygosity is the proportion of individuals that are heterozygous; heterozygosity (H) ranges between zero and one. Two different measures of heterozygosity are used. H_0 is the observed proportion of heterozygotes. For example, seven of the 12 individuals in Figure 3.13 are heterozygotes so H_0 at this locus in this sample is 0.58 ($7/12 = 0.58$). H_e is the expected proportion of heterozygotes if the population is mating at random; the estimation of H_e is discussed in detail in the Chapter 5. H_e provides a better standard to compare the relative amount of variation in different populations as long as the populations are mating at random (Nei 1977). Protein electrophoresis is commonly used to estimate heterozygosity at many loci in a population by taking the mean heterozygosities over all loci.

Another measure often used is **polymorphism** or the proportion of loci that are genetically variable (P). The likelihood of detecting genetic variation at a locus increases as more individuals are sampled from a population. This dependence on sample size is partially avoided by setting an arbitrary limit for the frequency of the most common allele. We use the criterion that the most common allele must have a frequency of 0.99 or less.

Data from natural populations

Harris (1966) was one of the first to describe protein heterozygosity at multiple loci (Table 3.5). He described genetic variation in the human population in England and found three of 10 loci to be polymorphic ($P = 0.30$).

Nevo et al. (1984) have summarized the results of protein electrophoresis surveys of some 1,111 species! Average heterozygosities for major taxonomic groups are shown in Figure 3.14. Different species sometimes have enormous differences in the amount of genetic variation they possess (Table 3.6). Differences between species in amounts of genetic variation can have important significance. Remember, evolutionary change cannot occur unless there is genetic variation present and this can have significant implications for conservation.

Table 3.5 Genetic variation at 10 protein loci in humans (Harris 1966).

	Allele frequency				
Locus	**1**	**2**	**3**	H_e	H_0
AP	0.600	0.360	0.040	0.509	0.510
PGM-1	0.760	0.240	–	0.365	0.360
AK	0.950	0.050	–	0.095	0.100
7 loci	1.000	–	–	0.000	0.000
Total				0.097	0.097

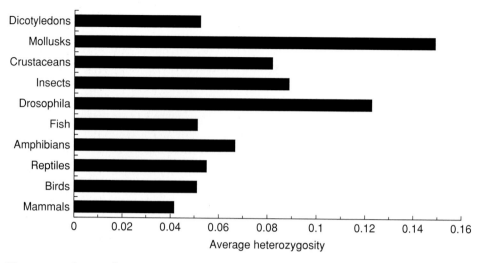

Figure 3.14 Average heterozygosities (H_S) from major taxa. From Gillespie (1992); data from Nevo et al. (1984).

Table 3.6 Summary of genetic variation demonstrating range of genetic variation found in different species of animals and plants (Nevo et al. 1984). H_S is the mean expected heterozygosity (H_e) over all loci for all populations examined.

Species	No. of loci	P (%)	H_S
Roundworm	21	29	0.027
American toad	14	34	0.116
Polar bear	29	2	0.000
Moose	23	9	0.018
Humans	107	47	0.125
Apache trout	30	0	0.000
Alligator	44	7	0.016
Red pine	35	3	0.007
Ponderosa pine	35	83	0.180
Yellow evening primrose	20	25	0.028
Salsify	21	9	0.026
Gilia	13	52	0.106

3.4 Genetic divergence among populations

The total amount of genetic variation within a species (H_T) can be partitioned into genetic differences among individuals within a single population (H_S) and genetic differences among different populations (Nei 1977). The proportion of total genetic variation within a species that is due to differences among populations is generally

Example 3.2 Do rare plants have less genetic variation?

There are several reasons to expect that rare species with restricted geographic distributions will have less genetic variation. First, loss of genetic variation caused by chance events (e.g., genetics and the founder effect, Chapter 6) will be greater in smaller populations. In addition, species with restricted geographic distributions will occur in a limited number of environments and will therefore be less affected by natural selection to exist under different environmental conditions.

Karron (1991) provided a very interesting test of this expectation by comparing the amount of genetic variation at allozyme loci in closely related species. He compared congeneric species from 10 genera in which both locally **endemic** and widespread species were present. One to four species of each type (rare and widespread) were used in each genus. In nine of 10 cases, the widespread species had a greater number of average alleles per locus (Figure 3.15).

These data support the prediction that rare species contain less genetic variation than widespread species. On the average, widespread species tend to have greater molecular genetic variation than rare endemic species with a limited distribution in nine of 10 genera. However, this relationship is not so simple. The amount of genetic variation in a species will be profoundly affected by the history of a species, as well as by its current condition. For example, some very common and widespread species have little genetic variation because they may have gone through a recent population bottleneck (see the red pine, Example 11.2). Rare endemic species should have relatively little genetic variation unless their current rareness is recent and they historically were more widespread. Thus, "rareness" should be used cautiously as a predictor of the amount of genetic variation within individual species.

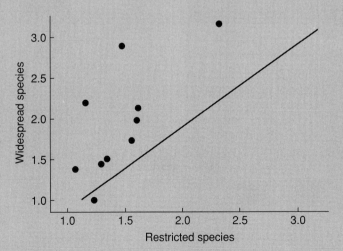

Figure 3.15 Comparison of the mean number of alleles per allozyme locus in widespread and restricted species of plants in 10 genera (Karron 1991). One to four species of each type were compared within each genus. The solid line shows equal mean values. In nine of 10 cases the mean of the widespread species was greater than the mean of the restricted species.

Table 3.7 Comparison of H_T, H_S, and F_{ST} for different major taxa of animals (Ward et al. 1992) and plants classified by their geographic range (Hamrick and Godt 1990, 1996).

Taxa	H_T	H_S	F_{ST}	No. of species
Amphibians	0.136	0.094	0.315	33
Birds	0.059	0.054	0.076	16
Fish	0.067	0.054	0.135	79
Mammals	0.078	0.054	0.242	57
Reptiles	0.124	0.090	0.258	22
Crustaceans	0.088	0.063	0.169	19
Insects	0.138	0.122	0.097	46
Mollusks	0.157	0.121	0.263	44
Endemic plants	0.096	0.063	0.248	100
Regional plants	0.150	0.118	0.216	180
Widespread plants	0.202	0.159	0.210	85

represented by F_{ST}:

$$F_{ST} = 1 - \frac{H_S}{H_T}$$

The estimation of F_{ST} is considered in detail in Chapter 9. Ward et al. (1992) have summarized the estimated values of these parameters in animal species (Table 3.7).

Some interesting patterns emerge from Table 3.7. First, invertebrates tend to have greater variation within populations (H_S) than vertebrates. This reflects the tendency for local populations of invertebrates to be larger than vertebrates because of the relationship where species with smaller body size tend to have larger population size (Cotgreave 1993). An analogous pattern is seen for plants. Species with a wider range, and therefore greater total population size, have greater total heterozygosity (H_T) than endemic plants, which have a more narrow range (Example 3.2). We will examine the relationship between population size and genetic variation in Chapter 6.

In addition, those taxa that we would expect to have greater ability for movement and exchange among populations have less genetic divergence among populations. For example, bird species have the same mean amount of genetic variation within populations, but they have much less genetic divergence among populations than fish or mammals. This difference reflects the greater ability of birds for exchange among geographically isolated populations because of flight. We will consider the relationship between exchange among populations and genetic divergence in Chapter 9.

3.5 Strengths and limitations of protein electrophoresis

Protein markers have been the workhorse for describing the genetic structure of natural populations over the last 30 years (Lewontin 1991). The strengths of protein electrophoresis are many. First, genetic variation at a large number of nuclear loci can be studied with relative ease, speed, and low cost. In addition, the genetic basis for variation of protein loci can often be inferred directly from electrophoretic patterns because of the codominant expression of isozyme loci, the constant number of subunits for the same enzyme in different species, and consistent patterns of tissue-specific expression of different loci. Third, it is relatively easy for different laboratories to examine the same loci and use identical allelic designations so that data sets from different laboratories can be combined (White and Shaklee 1991).

Protein electrophoresis also has several weaknesses. First, it can examine only a specific set of genes within the total genome – those that code for water-soluble enzymes. In addition, this technique cannot detect genetic changes that do not affect the amino acid sequence of a protein subunit. Thus, silent substitutions within codons or genetic changes in noncoding regions within genes cannot be detected with protein electrophoresis. Finally, this technique usually requires that multiple tissues be taken for analysis and stored in ultra-cold freezers. Thus, individuals to be analyzed often must be sacrificed and the samples must be treated with care and stored under proper refrigeration. Techniques that examine DNA directly using the **polymerase chain reaction** (PCR) do not require lethal sampling and can often use old (or even ancient) specimens that have not been carefully stored.

Protein electrophoresis is still the best tool for certain questions. For example, so-called cryptic species occur in many groups of invertebrates. Protein electrophoresis is the quickest and best initial method for detecting cryptic species in a sample of individuals from an unknown taxonomic group. Individuals from different species will generally be fixed for different alleles at some loci. The absence of heterozygotes at these loci would suggest the presence of two reproductively isolated, genetically divergent groups (Ayala and Powell 1972). As we will see in the next chapter, PCR-based DNA techniques generally either require prior genetic information or they rely upon techniques in which heterozygotes cannot be distinguished from some homozygotes.

Guest Box 3 Management implications of polyploidy in a cytologically complex self-incompatible herb
Andrew Young and Brian G. Murray

The button wrinklewort (Asteraceae) is a perennial herb that occurs in the temperate grasslands of Australia's southeast (Figure 3.16). This ecosystem has been substantially reduced in extent and condition over the last 150 years due to pasture improvement for sheep grazing. The fate of the button wrinklewort populations has paralleled that of its habitat, with the species now persisting in only 27 populations ranging in size from as few as seven to approximately 90,000 plants, but over half of the populations have less than 200 reproductive individuals.

Many smaller populations are declining (Young et al. 2000a), especially in the southern part of the species range, with reductions in fruit production of up to 90%. Like many of the Asteraceae, the button wrinklewort has a genetically controlled sporophytic self-incompatibility system (Young et al. 2000a, 2000b). Crossing experiments show that the most likely cause of this reduced fecundity

A

1 2 3 4 5 6 7 8 9 10 11 12 13 14 15 16 17 18 19 20 21 22

B

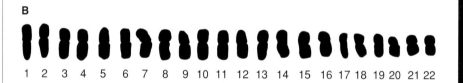

1 2 3 4 5 6 7 8 9 10 11 12 13 14 15 16 17 18 19 20 21 22

23 24 25 26 27 28 29 30 31 32 33 34 35 36 37 38 39 40 41 42 43 44

Figure 3.16 Sketch and karyoptyes of diploid (A, $2n = 22$), and tetraploid (B, $2n = 44$) button wrinkleworts. From Murray and Young (2001).

in small populations is mate limitation resulting from low allelic richness at the self-incompatibility (S) locus (see Section 8.4). Pollen grains can only fertilize plants that do not have the same S allele as carried by the pollen. Genetic and demographic rescue of small button wrinklewort populations could be achieved by increasing S-allele richness through translocation of plants among populations – especially from large, genetically diverse, and demographically viable northern populations to small, declining southern ones. However, this approach is complicated by the substantial cytological variation exhibited by the species.

Cytogenetic analysis reveals that button wrinklewort is chromosomally variable (Murray and Young 2001) (Figure 3.16). In the northern populations the dominant cytotype is diploid $2n = 22$, with occasional individuals exhibiting a second stable haploid chromosome number of $x = 13$. In this case $2n = 26$ individuals have an additional pair of each of the small and large metacentric chromosomes. In the south, the majority of the populations are $2n = 44$ autotetraploids for the common $x = 11$ cytotype, though $2n = 52$ individuals based on $x = 13$ are observed at very low frequencies.

Despite maintaining higher allelic richness on average than equivalent-sized diploid populations (Brown and Young 2000), polyploid populations are more mate-limited than diploids owing to the greater likelihood of matching S alleles among tetraploid genotypes (Young et al. 2000b). Note that while this makes the inclusion of divergent genetic material even more of an imperative, importing S alleles from northern diploid populations presents a range of genetic problems: (1) diploid × tetraploid crosses, though viable, produce substantially fewer fruits than crosses within ploidy level; (2) the triploid progeny of diploid × tetraploid crosses have reduced pollen fertility due to the production of unbalanced gametes during meiosis; and (3) backcrossing of triploids to either diploids or tetraploids produces a range of aneuploids with low fertility (Young and Murray 2000).

Taken together these limitations argue against mixing diploid and tetraploid plants despite the potential advantages of increased mate availability through the introduction of novel S alleles. However, analysis of the tetraploid populations shows that they appear to tolerate a good deal of chromosomal variation, with the presence of aneuploids ranging from $2n = 43$ to $2n = 46$ at frequencies of up to 0.23 (Murray and Young 2001). The presence of such individuals suggests the possibility of recent natural gene flow between diploid and tetraploid races indicating that fertility and fitness barriers to S-allele transfer across ploidy levels may not be insurmountable. Nevertheless, currently, a conservative approach of separate management of chromosome races is advisable, with translocation of plants between populations with the same chromosome number being the best management method to affect genetic rescue and restore demographic viability.

Problem 3.1

Herzog et al. (1992) have described a polymorphism in chromosome 3 of the black-handed spider monkey. The centromere is in the middle of the chromosome in one type (type A) of chromosome 3, while in the other type (type B) the centromere is near the end of the chromosome; both chromosomal types are of equal total length. There are no observable differences in any of the other chromosomes.

(a) What type of chromosomal rearrangement is likely to be responsible for this polymorphism? Be as specific as possible.

(b)　Assume that a zoo that is interested in developing a captive breeding program for this species contacts you as a consulting geneticist. Would you recommend that individuals brought into captivity be screened for this chromosomal polymorphism? If so, how should this information be used in the breeding program? Write a letter to the director of the zoo that explains and justifies your answer. Make sure that you address the potential effect of this rearrangement both in the captive population and in the wild following reintroduction.

Problems 3.2–3.7

The diagrams below are hypothetical results of protein electrophoresis of a sample of 20 individuals from a population of sea otters. Note that samples from each individual were electrophoresed and stained for three different enzymes that each are each encoded by a single locus (A–C). This population is diploid and mates at random with respect to these loci.

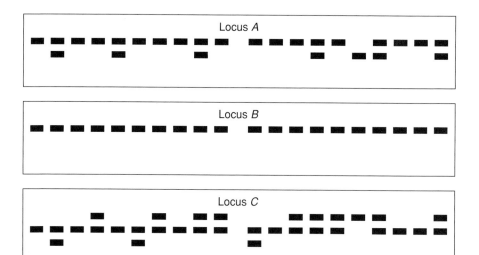

Problem 3.2

Designate the most common allele at each locus as 1, the second most common allele as 2, and so on. Write down the genotype (e.g., 11, 12, etc.) of each individual in the table opposite. For example, the first individual is homozygous for the common allele at locus A (11); the second individual is heterozygous (12) at A, and individual 16 is homozygous for the rare allele at this locus (22).

Indiv.	Locus *A*	Locus *B*	Locus *C*
1	*11*		
2	*12*		
3			
4			
5			
6			
7			
8			
9			
10			
11			
12			
13			
14			
15			
16	*22*		
17			
18			
19			
20			

Problem 3.3

What proportion of the three loci in this sample are polymorphic?

Problem 3.4

What is the observed heterozygosity for each locus?

Problem 3.5

What is the average observed heterozygosity over all three loci?

Problem 3.6

What are the expected genotypic distributions for locus *A* in progeny from a mating between individuals 1 and 2?

Problem 3.7

What are the expected genotypic distributions for locus *C* in progeny from a mating between individuals 1 and 16? How about between individuals 11 and 13?

Problem 3.8

Ling-Ling was a female giant panda at the National Zoo in Washington, DC (O'Brien et al. 1984). In March of 1983, Ling-Ling copulated with her male companion Hsing-Hsing. Ling-Ling was also artificially inseminated with sperm from a male giant panda at the London Zoo, Chia-Chia. On 21 July 1983 she gave birth to a baby that died shortly after birth. The following genotypes were detected at six protein loci by protein electrophoresis. Which male was the father of Ling-Ling's baby?

Locus	Ling-Ling	Baby	Hsing-Hsing	Chia-Chia
1	*AA*	*Aa*	*Aa*	*AA*
2	*Bb*	*bb*	*bb*	*bb*
3	*Cc*	*Cc*	*cc*	*cc*
4	*dd*	*Dd*	*Dd*	*dd*
5	*Ee*	*Ee*	*EE*	*Ee*
6	*FF*	*FF*	*FF*	*Ff*

Problem 3.9

Molecular genetic markers have allowed us to study reproductive behavior in wild populations that was not possible previously. For example, evidence of extra-pair copulations is accumulating in many bird species. Price et al. (1989) used a single allozyme locus (lactate dehydrogenase, *LDH*) to study parentage in house wrens. The authors were interested in detecting both extra-pair copulations (when the female mates with a male other that her "mate") and egg dumping (when a female lays an egg in the nest of another female).

They presented the following data shown:

Parental genotypes (female × male)	No. of pairs	Chicks		
		FF	*FS*	*SS*
SS × *SS*	6	0	4	29
SS × *FS*	6	0	13	18
FS × *SS*	4	0	10	11
FF × *SS*	2	0	8	2

Do these data provide evidence for either extra-pair copulations or egg dumping in this population?

Problem 3.10

The eastern reef egret in Australia usually has solid gray plumage. There is also a white morph that is fairly common in some populations. Assume that the white phenotype is due to a recessive allele (*g*) and that the dominant allele (*G*) produces gray plumage.

(a) What are the expected phenotypic ratios for male and female progeny from a cross between a white female and a gray male that had a white mother? Assume that this locus is on an autosome.

(b) Assume instead that the *G* locus occurs on the Z chromosome. How would this affect the expected phenotypic ratios for male and female progeny from a cross between a white female and a gray male that had a white mother?

Remember that in birds males are ZZ, and females are ZW. In addition, the W chromosome does not contain functional gene copies for many of the genes that are found on the Z. Therefore,

Males: *GG* = gray   Females: *GW* = gray
        Gg = gray   *gW* = white
        gg = white

Problem 3.11

Black and Johnson (1979) reported a highly unusual pattern of inheritance of allozyme polymorphisms in the intertidal anemone *Actina tenebrosa* from Rottnest Island in Western Australia. This species is viviparous, and up to five young are brooded by adults at a time until they are released as relatively large juveniles. The following parental and progeny genotypes were found at three allozyme loci:

Locus	Parental genotype	No. of broods	Progeny genotypes		
			FF	*FS*	*SS*
MDH	*FF*	25	68	0	0
	FS	53	0	158	0
	SS	11	0	0	35
PGM	*FF*	44	145	0	0
	FS	9	0	33	0
SOD	*FF*	71	225	0	0
	FS	18	0	50	0
	SS	1	0	0	2

How would you explain these results? That is, what system of mating and reproduction would explain the observed parent–progeny combinations?

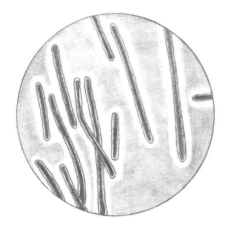

Genetic Variation in Natural Populations: DNA

Thermus aquaticus *("Taq"), Box 4.1*

The number of genes prescribing a eukaryotic life form such as a Douglas fir or human being runs into the tens of thousands. The nucleotide pairs composing them vary among species from one billion to ten billion. If the DNA helices in one cell of a mouse, a typical animal species, were placed end on end and magically enlarged to have the same width as wrapping string, they would extend for over nine hundred kilometers, with about four thousand nucleotide pairs packed into every meter. Measured in bits of pure information, the genome of a cell is comparable to all editions of the Encyclopedia Britannica *published since its inception in 1768.*

E. O. Wilson (2002)

My mitochondria comprise a very large proportion of me. I cannot do the calculation, but I suppose there is almost as much of them in sheer dry bulk as there is the rest of me. Looked at in this way, I could be taken for a very large, motile colony of respiring bacteria, operating a complex system of nuclei, microtubules, and neurons, and, at the moment, running a typewriter.

Lewis Thomas (1974)

A constantly expanding number of methods are used for detecting variation in DNA sequences in natural populations. We will discuss only some of the primary approaches that are used to study variation of DNA in natural populations. Sunnucks (2000) has reviewed the principle methods for DNA analysis and their advantages and disadvantages (also see Schlötterer 2004). It is important to remember that there is no universal "best" technique. The best technique to examine genetic variation depends upon the question being asked. The tool-kit of a molecular geneticist is analogous to the tool-box of a carpenter. Whether a hammer or power screwdriver is the appropriate tool depends on whether you are trying to drive in a nail or set a screw.

This chapter provides a conceptual overview of the primary techniques employed to study genetic variation in natural populations. Our emphasis is on the nature of the genetic information produced by each technique and how it can be used in conservation genetics. Detailed descriptions of the techniques and procedures can be found in the original papers. Most of these techniques are reviewed in Smith and Wayne (1996) and Hoelzel (1998).

4.1 Mitochondrial and chloroplast DNA

The first studies of DNA variation in natural populations examined animal mitochondrial DNA (mtDNA) because it is a relatively small circular molecule (approximately 16,000 bases in vertebrates and many other animals) that is relatively easy to isolate from genomic DNA and occurs in thousands of copies per cell. These characteristics allowed investigators to isolate thousands of copies of mtDNA molecules by ultracentrifugation.

In 1979, two independent groups published the first reports of genetic variation in DNA from natural populations. Avise et al. (1979a, 1979b) used **restriction enzyme** analysis of mtDNA to describe sequence variation and the genetic population structure of mice and pocket gophers. Avise (1986) provides an overview of the early work by Avise and his colleagues. Brown and Wright (1979) examined mtDNA to determine the sex of lizard species that hybridized to produce parthenogenetic species. A paper by Brown et al. (1979) compared the rate of evolution of mtDNA and the nuclear DNA in primates. This latter work was done in collaboration with Allan C. Wilson whose laboratory became a center for the study of the evolution of mtDNA (Wilson et al. 1985).

Several characteristics of animal mtDNA make it especially valuable for certain applications in understanding patterns of genetic variation. First, it is haploid and maternally inherited in most species. That is, a progeny generally inherits a single mtDNA genotype from its mother (Figure 4.1). There are thousands of mtDNA molecules in an egg, but relatively few in sperm. In addition, mitochondria from the sperm are actively destroyed once they are inside the egg. There are many exceptions to strict maternal inheritance. For example, there is evidence of some incorporation of male mitochondria ("paternal leakage") in species that generally show maternal inheritance, e.g., mice (Gyllensten et al.

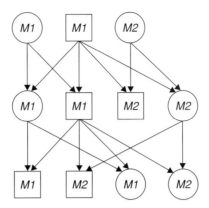

Figure 4.1 Pedigree showing maternal inheritance of two mtDNA genotypes: *M1* and *M2*. By convention in pedigrees, males are represented by squares and females are represented by circles. Each progeny inherits the mtDNA of its mother.

1991) and humans (Awadalla et al. 1999). In addition, some species show double uniparental inheritance of mtDNA in which there are separate maternally and paternally inherited mtDNA molecules (Sutherland et al. 1998). Paternal leakage may lead to **heteroplasmy** (the presence of more than one mitochondrial genotype within an individual).

Mitochondrial DNA molecules are especially valuable for reconstructing **phylogenetic** trees because there is generally no recombination between mtDNA molecules. Unlike nuclear DNA, the historical genealogical record of descent is not "shuffled" by recombination between different mtDNA lineages during gamete production, as occurs in nuclear DNA during meiosis. Recombination between lineages is not likely because mtDNA generally occurs in only one lineage per individual (one haploid genome) because the male gamete does not contribute mtDNA to the zygote. Thus, the mtDNA of a species can be considered a single nonrecombining genealogical unit with multiple alleles or haplotypes (Avise 2004).

Plant mitochondrial and chloroplast DNA have been somewhat less useful than animal mtDNA for genetic studies of natural populations primarily because of low variation (Clegg 1990; Powell 1994). However, the discovery of microsatellite sequences in chloroplast DNA has provided a very useful marker (Provan et al. 2001) (see Section 4.2.1). The pattern of inheritance of these cytoplasmic molecules is variable in plants. For example, mtDNA is generally maternally inherited in plants but is paternally inherited in many conifers. There is also no recombination for mitochondrial and chloroplast molecules in many plant species. McCauley et al. (2005) found primarily maternal inheritance (9% of all individuals) in *Silene vulgaris*, a gynodioecious plant, and that heteroplasmy occurred in over 20% of all individuals.

4.1.1 Restriction endonucleases and RFLPs

The discovery of restriction **endonucleases** (restriction enzymes) in 1968 (Meselson and Yuan 1968) marked the beginning of the era of genetic engineering (i.e., the cutting and splicing together of DNA fragments from different chromosomes or organisms). Restriction endonucleases are enzymes in bacteria that cleave foreign DNA, such as DNA from intracellular viral pathogens (bacteriophages) harmful to the bacteria. The bacterial DNA is protected from cleavage because it is methylated. The most commonly used restriction endonuclease is *Eco*RI from the bacterium *Escherichia coli*. *Eco*RI cleaves a specific six base sequence: GAATTC (and the reverse compliment CTTAAG). The cleavage is uneven, such that each strand is left with an overhang of AATT as follows:

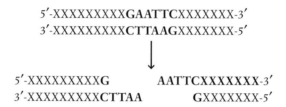

5′-XXXXXXXXX**GAATTC**XXXXXXX-3′
3′-XXXXXXXXX**CTTAAG**XXXXXXX-5′

5′-XXXXXXXXX**G**   **AATTC**XXXXXXX-3′
3′-XXXXXXXXX**CTTAA**   **G**XXXXXXX-5′

where the overhang is in bold and the Xs represent the sequence flanking the restriction site sequence.

How can we use restriction enzymes to detect DNA sequence polymorphisms? If we cut a DNA sequence with a restriction enzyme, some individuals will have only one restriction site, while others might have two or three. A circular DNA molecule (such as mtDNA) with one restriction site will yield one linear DNA fragment after cleavage (Figure 4.2). If two cleavage sites exist, then two linear DNA fragments are produced from the cleavage. We can visualize the number of fragments using gel electrophoresis to separate them by length; short fragments migrate faster than long ones (Figure 4.2).

This is the basis of the **restriction fragment length polymorphism** (RFLP) technique for detecting DNA polymorphisms. Restriction site polymorphisms are usually generated by a single nucleotide substitution in the restriction site (e.g., from GA*A*TTC to

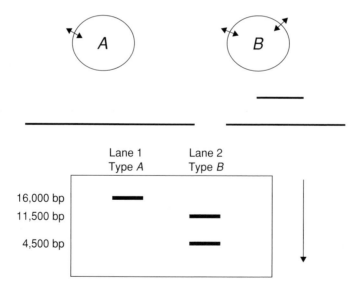

Figure 4.2 Hypothetical examination of sequence differences in mtDNA revealed by restriction enzyme analysis. Type *A* has only one cleavage site, which produces a single linear fragment of 16,000 base pairs; type *B* has two cleavage sites, which produce two linear fragments of 11,500 and 4,500 base pairs. Electrophoresis of the digested products results in the pattern shown. The DNA fragments move in the direction indicated by the arrow, and the smaller fragments migrate faster.

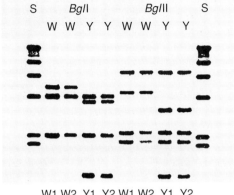

Figure 4.3 RFLP of mtDNA in cutthroat trout digested by two restriction enzymes (*Bgl*I and *Bgl*II). The W lanes are two (W1 and W2) westslope cutthroat trout and the Y lanes are two (Y1 and Y2) Yellowstone cutthroat trout. The S lanes are size standards. From Forbes (1990).

GA*T*TTC). This causes the loss of the restriction site in the individual because the enzyme will no longer cleave the individual's DNA. Thus a restriction site polymorphism is detectable as an RFLP following digestion of the molecule with a restriction enzyme and gel electrophoresis.

Each restriction enzyme cuts a different specific DNA sequence (usually four or six base pairs in length). For example, *Taq*1 cuts at TCGA. More than 400 different enzymes are commercially available. Thus we can easily study polymorphism across an mtDNA molecule by using a large number of different restriction enzymes.

Figure 4.3 shows RFLP variation in the mtDNA molecule of two subspecies of cutthroat trout digested by two restriction enzymes (*Bgl*I and *Bgl*II) that each recognize six base pair sequences. There are three cut sites for *Bgl*I in the W (westslope cutthroat trout) haplotype; there is an additional cut site in the Y (Yellowstone cutthroat trout) haplotype so that the largest fragment in the W haplotype is cut into two smaller pieces. The Y haplotype also has an additional cut site for *Bgl*II resulting in one more fragment than in the W haplotype. RFLP analysis is also useful for studies of nuclear genes following PCR amplification of a gene fragment (see below).

4.1.2 Polymerase chain reaction

Detection and screening of mtDNA polymorphism is most often conducted using polymerase chain reaction (PCR) (Box 4.1), followed by restriction enzyme analysis or by directly sequencing of the PCR product. For conducting PCR, "universal" primers are available for both mtDNA (Kocher et al. 1989) and chloroplast DNA (Taberlet et al. 1991). These primers will amplify a specific sequence (e.g., the cytochrome *b* gene) across a wide range of taxa. This universality has facilitated the accumulation of many DNA studies since the 1980s. The restriction enzyme approach involves cutting the PCR fragment into smaller pieces and visualizing the fragments by gel electrophoresis (see above). Sequencing is becoming more common as the process becomes less expensive and automated (see Section 4.5). For example, the complete mtDNA sequences for 53 humans from diverse origins have been published (e.g., Ingman et al. 2000).

Box 4.1 Polymerase chain reaction

The polymerase chain reaction (PCR) can generate millions of copies of a specific target DNA sequence in about 3 hours, even when starting from small DNA quantities (e.g., one target DNA molecule!). Millions of copies are necessary to facilitate analysis of DNA sequence variation. PCR involves the following three steps conducted in a small (0.5 ml) plastic tube in a thermocycling machine: (1) denature (make single-stranded) a DNA sample from an individual

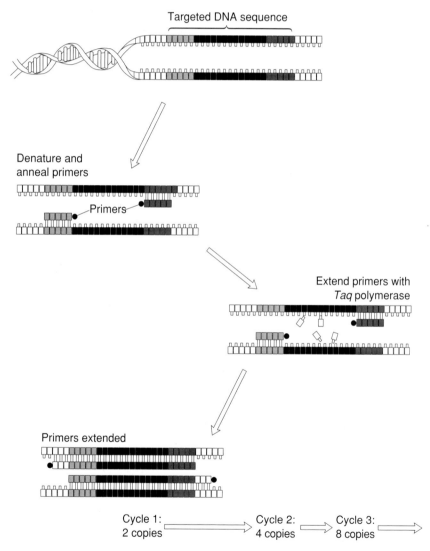

Figure 4.4 The main steps of the polymerase chain reaction (PCR): denaturing of the double-stranded template DNA, annealing of primers flanking the target sequence, and extension from each primer by *Taq* polymerase to add nucleotides across the target sequence and generate a double-stranded DNA molecule.

by heating the DNA to 95°C; (2) cool the sample to about 60°C to allow hybridization (i.e., annealing) of a primer (i.e., a DNA fragment of approximately 20 bp)
to each flanking region of the target sequence; and (3) reheat slightly (72°C)
to facilitate extension of the single strand into a double-stranded DNA by the
enzyme *Taq* polymerase (Figure 4.4). These three steps are repeated 30–40
times until millions of copies result.

PCR was invented by Mullis in 1988. He used a DNA polymerase enzyme
from the heat-stable organism *Thermus aquaticus* (thus the name "*Taq*"). This
bacterium was originally obtained from hot springs of Yellowstone National Park
in Montana and Wyoming. PCR has revolutionized modern biology and has
widespread applications in the areas of genomics, population genetics, forensics,
medical diagnostics, and gene expression analysis. Mullis was awarded the Nobel
Prize in Chemistry in 1993 for his contributions to the development of PCR.

4.2 Single copy nuclear loci

4.2.1 Microsatellites

Microsatellites have become the most widely used DNA marker in population genetics for
genome mapping, molecular ecology, and conservation studies. Microsatellite DNA markers
were first discovered in the 1980s (Schlötterer 1998). They are also called VNTRs (variable
number of tandem repeats) or SSRs (simple sequence repeats) and consist of tandem repeats
of a short sequence motif of one to six nucleotides (e.g., cgtcgtcgtcgtcgt, which can be represented by $(cgt)^n$ where $n = 5$). The number of repeats at a polymorphic locus ranges
from approximately five to 100. PCR primers are designed to hybridize to the conserved
DNA sequences flanking the variable repeat units (Example 4.1). Microsatellite PCR products are generally between 75 and 300 base pairs (bp) long, depending on the locus.
Microsatellites are usually analyzed using PCR followed by gel electrophoresis (Figure 4.5).

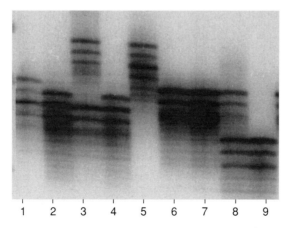

1 2 3 4 5 6 7 8 9

Figure 4.5 Microsatellite variation in the hairy-nosed wombat. The sub-bands or stutter
bands occur as a result of "slippage" during PCR amplification. Allele sizes for each individual
are: 1 (*187/191*), 2 (*187/189*), 3 (*187/199*), 4 (*187/189*), 5 (*195/199*), 6 (*191/191*), 7 (*191/191*),
8 (*183/191*), and 9 (*183/183*). From Taylor et al. (1994).

Example 4.1

Modified GenBank sequence data base entry for the *Lla71CA* locus in the hairy-nosed wombat (Figure 4.5). The primers in the sequence at the bottom have been capitalized and the dinucelotide repeat region (CA) is shown in bold. The "**n**"s in the sequence are base pairs that could not be resolved in the sequencing process.

1: AF185107. *Lasiorhinus latif*

LOCUS	AF185107 310 bp DNA linear MAM 01-JAN-2000
DEFINITION	Lasiorhinus latifrons microsatellite Lla71CA sequence.
AUTHORS	Beheregeray, L. B., Sunnucks, P., Alpers, D. L. and Taylor, A. C.
TITLE	Microsatellite loci for the hairy-nosed wombats (*Lasiorhinus krefftii* and *Lasiorhinus latifrons*)
JOURNAL	Unpublished
AUTHORS	Taylor, A. C.
JOURNAL	Submitted (31-AUG-1999) Biological Sciences, Monash University, Wellington Rd., Clayton, VIC 3168, Australia
FEATURES	Location/Qualifiers
source	1..310
<u>repeat_region</u>	109..154
	/rpt_type=tandem
	/rpt_unit=ca
BASE COUNT	99 a 94 c 42 g 68 t 7 others
ORIGIN	

```
  1 gngctcggnn cccctggatc acagaatcta aatctgagca tctcagAATG AGAAGGTATC
 61 TCCAGGataa ccannnccct ctacctaaac aagaattcca ctcccctaca cacacacaca
121 cacacacaca cacacacaca cacacacaca cacactcaat agacccaaca agtggaatgt
181 cacacagcct ttggggnagg tgggggatat acttCCTATG ACATAGCCTA TACCacttct
241 gaatagtaac tttcctatcc ataaatctaa aacctacttc ccactctttt ctgctagttc
301 tataatctgg
```

The main advantage of these markers is that they are usually highly polymorphic, even in small populations and endangered species (e.g., polar bears or cheetahs). This high polymorphism results from a high mutation rate (see Chapter 12). A microsatellite mutation usually results in a change in the number of repeats (usually an increase or decrease of one repeat unit). The rate of mutation is typically around one mutation in every 1,000 or 10,000 meioses (10^{-3} or 10^{-4} per generation).

Primer pairs developed in one species can often be used in closely related species because primer sites are generally highly conserved. For example, about 50% of primers designed from cattle will work in wild sheep and goats that diverged approximately 20 million years ago (Maudet et al. 2001). This is an enormous advantage because over 3,500 microsatellites have been mapped in cattle; thus cattle primers can be tested to find polymorphic markers across the genome of any ungulate, without the time and cost of

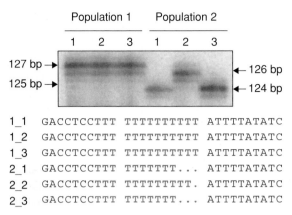

Figure 4.6 Chloroplast microsatellite polymorphism in six individuals of the leguminous tree *Caesalpinia echinata* from two populations (from Provan et al. 2001). The three individuals from Population 1 all have the *127* allele; the sequence of these individuals, shown below the gel, indicates that they have 13 copies of the (T) mononucleotide repeat. Two different alleles are present in the three individuals from Population 2.

cloning and mapping for each new species. Similar transfer of makers is possible for wild canids, felids, primates, salmonids, and galliform birds because genome maps are available with microsatellites.

Microsatellite primer sets for thousands of species can be found at several websites. *Molecular Ecology Notes* has a website that contains all primers published in that journal plus many others. The United States National Center for Biotechnology Information (NCBI) maintains one of the primary websites for sequence information: http://www.ncbi.nih.gov/. This resource was established in 1988 (Wheeler et al. 2000). For example, Example 4.1 shows the GenBank sequence database entry for the microsatellite locus shown in Figure 4.5.

Chloroplast microsatellites

An unusual type of microsatellite has been found to occur in the genome of chloroplasts (Provan et al. 2001). Chloroplast microsatellites are usually mononucleotide repeats and often have less than 15 repeats (Figure 4.6). These markers have proved to be exceptionally useful in the study of a wide variety of plants. Furthermore, the uniparental inheritance of chloroplasts (usually maternal in angiosperms and paternal in gymnosperms) make them useful for distinguishing the relative contributions of seed and pollen flow to the genetic structure of natural populations by comparing nuclear and chloroplast markers.

4.2.2 PCRs of protein coding loci

PCR primers can be designed to detect genetic variation at protein coding loci that include protein coding regions (**exons**) which often contain noncoding regions (**introns**). Coding regions tend to be much less variable than noncoding regions. Therefore, PCR primers can be designed using exon sequences that will produce a PCR product that consists

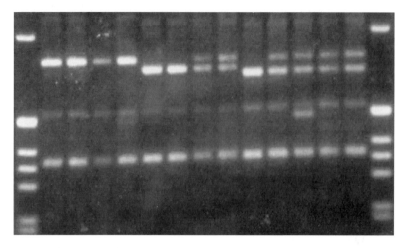

Figure 4.7 Length polymorphism in an intron for a growth hormone gene (*GH-1*) in coho salmon (*Oncorhynchus kisutch*). The lanes at the ends are size standards. There are three alleles at this locus that differ by the number of copies of a 31-base pair repeat. The repeat occurs 11 times in the *a allele, 9 times in *b, and 8 in *c. The genotypes from left to right are *a/a, a/a, a/a, a/a, b/b, b/b, a/b, a/b, c/c, a/c, a/c, a/c,* and *a/c.* From S. H. Forbes (unpublished data).

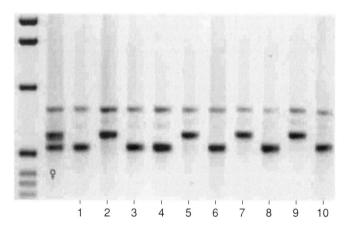

Figure 4.8 Inheritance of a length polymorphism in an intron for a growth hormone gene (*GH-2*) in pink salmon. The first lane on the left is a size standard and the next is the mother of a family of gynogenetic haploid progeny from a single female produced by fertilizing the eggs with sperm that had been irradiated so that the male genome was not incorporated into the developing embryo (Spruell et al. 1999a). Progeny 1, 3, 4, 6, 8, and 10 inherited the smaller allele (*GH-2*C446*) and progeny 2, 5, 7, and 9 inherited the *GH-2*C527* allele.

primarily of the more variable introns. Figure 4.7 shows a polymorphism in an intron from a growth hormone gene (*GH-1*) in coho salmon. There are three alleles at this locus that differ by the number of copies of a 31 bp repeat (Forbes et al. 1994). The repeat occurs 11 times in the *a allele, nine times in *b, and eight times in *c.

Comparison of progeny and parental genotypes is the only direct way to confirm the genetic basis of polymorphisms such as seen in Figure 4.7. For example, Figure 4.8 shows the of inheritance of a polymorphism in intron C of the growth hormone gene (*GH-2*) in pink salmon. All of the individuals on this gel are gynogenetic haploid progeny produced by fertilizing the eggs from a single female with sperm that had been irradiated so that the male genome was not incorporated into the developing embryo (Spruell et al. 1999a). The mother of this family was a heterozygote for two alleles (*GH-2*C446* and *GH-2*C527*) that differ in size by 81 bp. Four of the offspring inherited the **C527* allele and six inherited the alternative smaller allele.

4.2.3 Single nucleotide polymorphisms

Single base polymorphisms (SNPs) are the most abundant polymorphism in the genome, with one occurring about every 500 bp in many wild animal populations (Brumfield et al. 2003; Morin et al. 2004). For example, a G and a C might exist in different individuals at a particular nucleotide position within a population (or within a heterozygous individual). Because the mutation rate at a single base pair is low (about 10^{-8} changes per nucleotide per generation), SNPs usually consist of only two alleles. Thus SNPs are usually biallelic markers. Transitions are a replacement of a purine with a purine (G $\leftrightarrow$ A) or a pyrimidine with a pyrimidine (C $\leftrightarrow$ T). Transversions are a replacement of a purine with a pyrimidine (A or G to C or T) or vice versa (C or T to A or G). Even though there are twice as many possible transversions as transitions, SNPs in most species tend to be transitions. This is because of the nature of the mutation process (transitions are more common than transversions) and that transversions in coding regions are more likely to cause an amino acid substitution than transitions.

   SNPs have great potential for many applications in describing genetic variation in natural populations. For example, Akey et al. (2002) described allele frequencies at over 26,000 SNPs in three human populations! Two randomly chosen humans will differ at up to several million single nucleotide sites over their entire genomes. SNPs may be even more common in other species because humans arose recently in evolutionary terms from relatively few founders and thus have somewhat limited genome variation.

   SNPs could are likely to replace microsatellites as the marker of choice for many applications in conservation genetics. They should be especially useful for studies involving partially degraded DNA (from noninvasive and ancient DNA samples) because they are short and thus can be PCR amplified from DNA fragments of less than 50 bases (PCR primers flanking SNPs must each be about 20 bases long). Theoretical population geneticists have begun developing statistical approaches for analyzing the masses of SNP data that are expected to emerge in the next few years (e.g., Kuhner et al. 2000). The time it takes until SNPs become popular in conservation depends on the speed with which new technologies become available to permit rapid and inexpensive screening of SNPs in many species. A recent paper (Smith et al. 2005a) describes genetic variation at 10 SNP loci in over 1,000 Chinook salmon from throughout the rim of the north Pacific Ocean (Figure 4.9).

   Ascertainment bias is a crucial issue in many applications of SNPs (Morin et al. 2004). Ascertainment bias results from the selection of loci from an unrepresentative sample of individuals, or using a particular method, which yields loci that are not representative of the spectrum of allele frequencies in a population. For example, if few individuals are used for SNP discovery (e.g., via DNA sequencing), then SNP loci with rare alleles will be underrepresented, and future genotyping studies using those SNPs will reveal a (false) deficit of rare alleles (e.g., false bottleneck signature). Ascertainment bias has the potential

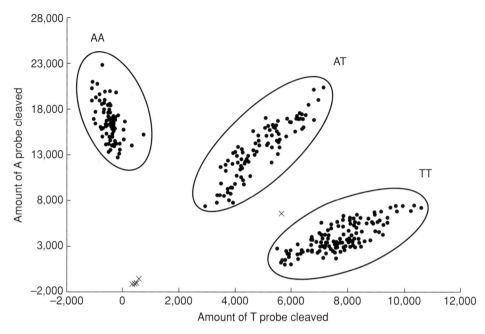

Figure 4.9 SNP genotyping assay in Chinook salmon (locus *OtspP450*). Each dot represents an individual fish whose genotype is determined by its position with respect to the two axes, which indicate the amount of each allele-specific probe (adenine [A] or thymine [T]) cleaved during the course of the assay. The ellipses indicate clusters of single genotypes. The Xs represent unreadable samples (due to air bubbles, failed PCR, etc.). From Smith et al. (2005).

to introduce a systematic bias in estimates of variation within and among populations. The protocol used to identify SNPs for a study must be recorded in detail, including the number and origin of individuals screened, to enable ascertainment bias to be assessed and potentially corrected.

4.3 Multilocus techniques

Multilocus techniques assay many genome locations simultaneously with a single PCR reaction (Bruford et al. 1998). The advantage of these techniques is that many loci can be examined readily with little or no information about sequences from the genome. The major disadvantages are that it is generally difficult to associate individual bands with particular loci, and heterozygotes can often not be distinguished from one of the homozygous classes. Thus we usually cannot resolve between a heterozygote (with one band) and the homozygote "dominant" type (also with one band, but two copies of it). Consequently, we cannot compute individual (observed) heterozygosity to test for Hardy–Weinberg proportions. In addition, these markers are biallelic and thus provide less information per locus than the more polymorphic microsatellites. It can require 5–10 times more of these loci to provide the same information as multiple allelic microsatellite loci (e.g., Waits et al. 2001).

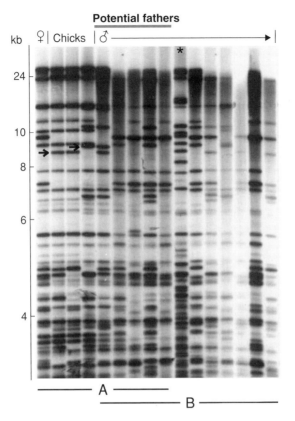

Figure 4.10 Minisatellite gel of kakapo from New Zealand. Two comparisons are shown on this gel: (A) paternity analysis showing the mother, her chicks, and all potential fathers; examples of bands not present in the mother are marked with an arrow; (B) comparison of minisatellite profiles for unrelated males from Stewart Island and a single bird, Richard Henry (asterisked), captured on the South Island. These profiles demonstrate substantial genetic divergence between the Fiordland and Stewart Island populations. Males from Stewart Island share an average of 69% of their bands while the Fiordland bird shares an average of 48% of its bands with males from Stewart Island. From Miller et al. (2003).

However, more loci (10–25) can be analyzed per PCR and per gel lane using some of these techniques, compared to microsatellites (5–10 loci per lane using fluorescent labels).

4.3.1 Minisatellites

Minisatellites are tandem repeats of a sequence motif that is approximately 20 to several hundred nucleotides long – much longer than *micro*satellite motifs. Minisatellites were first discovered by Jefferys et al. in 1985, and used for DNA "fingerprinting" in human forensics cases. They were soon after used in wildlife populations, for example, to study paternity and detect extra-pair copulations in birds though to be monogamous. For example, Figure 4.10 shows a minisatellite analysis of the endangered kakapo, a highly endangered flightless parrot (see Example 18.1) that is endemic to New Zealand.

Minisatellites are generally highly polymorphic and thus are most useful for interindividual studies such as parentage analysis and individual identification (e.g., DNA "fingerprinting"). Alleles are identified by the number of tandem repeats of the sequence motif. Disadvantages include a difficulty determining allelic relationships (identifying alleles that belong to one locus). This can be difficult because most minisatellite typing systems reveal bands (alleles) from many loci that are all visualized together in one gel lane. Also, the repeat motifs are long, so they cannot be studied in samples of partially degraded DNA generally containing small fragments of only 100–300 bp.

4.3.2 RAPDs

The RAPD (randomly amplified polymorphic DNA) method is a fingerprinting technique involving PCR amplification using an arbitrary primer sequence. Each PCR contains only one primer of about 10 bp long. Thus the primer acts as both the forward and reverse primer. Individual primers hybridize to hundreds of sites in the genome. However amplification occurs only between sites located less than about 2 kb apart (i.e., the maximum size of a PCR product). PCR conditions (e.g., annealing temperature) are usually chosen such that 10–20 fragments are amplified per primer. Thus relatively few primers can produce many fragments and screen many loci throughout the genome.

RAPDs were first described in late 1990 by two independent teams (Welsh and McClelland 1990; Williams et al. 1990). One team coined a different name (AP-PCR or arbitrary primed PCR), which has not been used nearly as much as the simpler acronym RAPD. RAPD kits are available from commercial companies. They can be used to find polymorphic markers for any species, even if no sequence information exists. The main drawback is it is often difficult to achieve reproducible results. Some journals will no longer publish studies using RAPDs because of these problems with repeatability.

4.3.3 AFLPs

The AFLP technique uses PCR to generate DNA fingerprints (i.e., multilocus band profiles). These "fingerprints" are generated by selective PCR amplification of DNA fragments produced by cleaving genomic DNA. This technique was named AFLP because it resembles the RFLP technique (Vos et al. 1995). However, these authors said that AFLP is not an acronym for amplified fragment length polymorphism because it does not detect length polymorphisms. The main advantage of AFLP is that many polymorphic markers can be developed quickly for most species even if no sequence information exists for the species. In addition, the markers generally provide a broad sampling of the genome. AFLP is faster, less labor intensive, and provides more information than other commonly used techniques. Furthermore, AFLPs are more reproducible than RAPDs. The AFLP technique can be used to develop diagnostic markers for different animal breeds, ecotypes, or sexes (Griffiths and Orr 1999).

4.3.4 PINEs and ISSRs

There are a variety of small nuclear elements that occur thousands of times throughout the genomes of most species (Avise 2004). For example, a single 300 bp repetitive element (*Alu*) occurs more than 500,000 times in the human genome, and constitutes an amazing 5–6% of the human genome! It is possible to design PCR primers using the sequences of

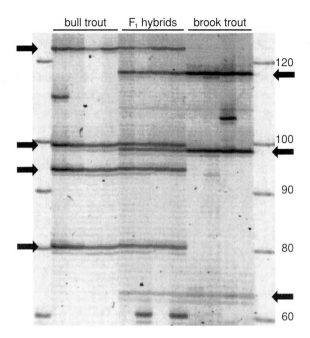

Figure 4.11 PINE fragments used to identify hybrids between bull and brook trout. Fragments found in bull trout but not brook trout are indicated by arrows along the left side of the gel. Fragments found in brook trout but not bull trout are indicated by arrows along the right side of the gel. Sizes of fragments of known length are given in base pairs on the right. From Spruell et al. (2001).

repetitive elements to generate multiple DNA fragments from a single PCR (Nelson et al. 1989; Buntjer and Lenstra 1998). For example, PINEs (paired interspersed nuclear elements) use primers identical to one end of the element that are designed so that they initiate DNA replication away from the end of the element, progressing into the surrounding genomic DNA (Spruell et al. 1999a). This procedure results in a number of fragments from a single PCR reaction that can be used to detect genetic variation (Figure 4.11). PINE fragments are more reproducible and reliable than RAPD fragments because longer primers can be used, and they are easier to use than AFLPs because they require only a single PCR without any preparatory steps. These markers have been especially valuable at detecting hybridization between species (Spruell et al. 2001) and for identifying species (Buntjer and Lenstra 1998) (see Chapter 20).

Inter-simple sequence repeat (ISSR) markers use a similar procedure to generate a large number of DNA fragments from a single PCR. However, rather than using the sequences of repetitive elements for PCR primers, ISSR primers are based upon the simple sequence repeats found in microsatellites.

4.4 Sex-linked markers

Genetic markers in sex determining regions can be especially valuable in understanding genetic variation in natural populations. For example, markers that are specific to sex determining the Y or W chromosomes can be used to identify the gender of individuals in species in which it is difficult to identify gender phenotypically, as in many bird species (Ellegren 2000b). In addition, Y chromosome markers, like mtDNA, are especially useful for phylogenetic reconstruction because Y chromosome DNA is haploid and nonrecombining in mammals.

Mammals can be sexed using PCR amplification of Y chromosome fragments, or by co-amplification of a homologous sequence on both the Y and X that are subsequently discriminated by either size, restriction enzyme cleavage of diagnostic sites, or by sequencing (Fernando and Melnick 2001). Similar molecular sexing techniques exist for birds and other taxa (e.g., amphibians). The W chromosome of birds has conserved sequences not found on the Z, allowing nearly universal avian sexing PCR techniques (e.g., Huynen et al. 2002). Some plants also have sex-linked sequences (Korpelainen 2002).

4.5 DNA sequences

The first application of DNA sequencing to the study of genetic variation in natural populations was by Kreitman (1983) who published the DNA sequences of 11 alleles at the alcohol dehyrogenase locus from *Drosophila melanogaster*. Initial studies of DNA variation were technically involved and time consuming so that it was expensive and difficult to sample large numbers of individuals from natural populations. However, the advent of the polymerase chain reaction in the mid-1980s removed these obstacles.

DNA sequencing methods were first developed independently by Walter Gilbert and Frederick Sanger. Gilbert and Sanger, along with Paul Berg, were awarded the Nobel Prize in Chemistry in 1980. Sanger and his colleagues used their own sequencing method to determine the complete nucleotide sequence of the bacteriophage *fX174*, the first genome ever completely sequenced. The worldwide genome sequencing capacity in the year 2001 was 2,000 bp per second.

4.6 Additional techniques and the future

There is expanding potential for genetics to contribute to conservation, thanks to technological advances in molecular biology allowing automated screening of huge numbers of DNA markers more quickly and cheaply in a growing number of species. Along with new approaches for DNA typing (e.g., DNA arrays) new disciplines are emerging such as **genomics** and population genomics. We briefly discuss below aspects of genomics and recent developments likely to aid conservation genetics in the very near future.

4.6.1 Genomics

Genomics can be defined as investigations into the structure and function of very large numbers of genes. Structural genomics includes the genetic mapping, physical mapping, and sequencing of entire genomes or chromosomes. Functional genomics investigates the function of DNA sequences – for example, the effect of a nucleotide substitution on fitness or disease susceptibility. Comparative genomics assesses the nature and significance of differences between genomes by determining the relationship between genotype and phenotype through comparing different genomes and morphological and physiological attributes (O'Brien et al. 1999). Comparative genomics allows reconstruction of relationships among taxa and an understanding of molecular and genome evolution. It is also a powerful way to build new genome maps in previously unstudied (but related) species (Lyons et al. 1997).

"Population genomics" refers to the study of many DNA markers (e.g., mapped markers and coding genes) in many individuals from different populations (Luikart et al. 2003). The two main advantages of studying many loci are increased power for most statistical analyses and improved genome coverage, allowing less biased inference and the identification of certain loci (or nucleotides or chromosome regions) that behave differently and that thus could be under selection and may be important for fitness, mate choice, local adaptation, or speciation (see Section 9.6.3).

Informative population genomic studies require: (1) the development of genome resources such as numerous DNA markers (preferably mapped) or sequences from gene fragments (e.g., ESTs or **expressed sequence tags**); and (2) large-scale genotyping capabilities facilitated by PCR multiplexing or rapid genotyping assays.

4.6.2 More informative molecular markers

The availability of molecular genetic markers has clearly been a major driving force of the expanding role that genetics plays in conservation. The number and kind of available molecular markers influences our ability to address questions crucial to conservation. It is difficult to predict which kinds of markers and analysis systems will be adopted in the near future to improve studies in conservation genetics. But trends will likely follow the trends in human genetics and model species, with perhaps a 5–10-year lag time (Schlötterer 2004). Thus, markers in or near genes, linked markers, SNPs, and DNA sequences will be increasing used.

We predict that major advances in the "quality" of marker data over the next few years will include: (1) markers in or near genes of known and important function; and (2) sets of physically linked markers with known interlocus map distances. Markers in genes will allow us to address questions concerning the genetic basis of adaptation and speciation, as well as to understand the importance of locus-specific versus genome-wide effects. The availability of sets of linked markers will allow the use of linkage disequilibrium information to improve estimates of population parameters such as the magnitude and date of population growth events, the amount and date of admixture or hybridization events, and to conduct individual-based assignment tests, for example. The development of markers in genes and linked markers will be facilitated by the discovery and sequencing of expressed sequence tags (ESTs).

Expressed sequence tags are useful for developing informative DNA markers for conservation genetic studies, because they are in coding genes. ESTs are unique DNA sequences derived from a **cDNA library** (and thus from a transcribed sequence). ESTs are discovered by extracting mRNA from a tissue and using reverse transcriptase PCR. The PCR products (cDNAs) are then sequenced. The sequence can be used in a BLAST search to determine if the gene (or a similar gene) from which the sequence originates is known and **annotated**. Once ESTs are available, primers can be designed to sequence them and identify polymorphic markers (e.g., SNPs or microsatellites; Vigouroux et al. 2002). Conserved primers (in exons) can be developed to amplify genes among divergent taxa.

Huge EST data bases are being generated for species of economic and ecological import-ance (e.g., galliform birds, salmonid fish, many mammals, and crop plants). EST libraries and data bases also can be developed for nonmodel organisms.

4.6.3 Improved genotyping technologies

One can predict that in the future data sets will become larger with more loci, individuals, and populations. The principal benefits from larger data sets will be more accurate and precise estimates of population parameters. Larger data sets will become more common thanks to advances such as automation and high throughput technologies, including PCR multiplexing, and genotyping without gel electrophoresis. Examples of automation and high throughput are the use of pipetting robots and capillary sequencers with faster run times.

The time and costs of producing large informative data sets can be greatly reduced by multiplex PCR. Multiplexing is the co-PCR amplification of more than one locus in a single PCR tube by using primer pairs from more than one locus. In addition to PCR multiplexing, laboratory time and cost can be reduced by the elimination of post-PCR handling of PCR products – i.e., the elimination of gel electrophoresis. Another approach to achieve high throughput (and thus reduce cost per locus) is to conduct numerous analyses in parallel – for example by using DNA chips or microarroays (see Section 4.6.4) to simultaneously assay hundreds or thousands of loci.

Reviewing the enormous and growing number of approaches for multiplexing and for rapid genotyping is beyond the scope of this book. We suggest readers search the recent literature and see review papers (e.g., Syvanen 2001), most of which relate to SNP geno-typing because SNPs are highly amenable to automated analysis (Morin et al. 2004). In the more distant future, perhaps in 5–10 years, DNA sequencing will likely become auto-mated and cheap enough to become widely used to produce large data sets with multiple loci and for numerous individuals.

4.6.4 DNA arrays

DNA arrays generally consist of thousands of DNA fragments (20–25 bases) bound in a grid pattern on a small glass slide or nylon filter paper the size of a credit card. Arrays are used most often to detect **gene expression** (i.e., the production of mRNA). Gene expres-sion studies allow us to identify genes associated with adaptations, e.g., to environmental change or disease pathogens. Recently, arrays also have been used for genotyping thou-sands of SNPs simultaneously (Jaccoud et al. 2001).

To construct a DNA array, DNA fragments are spotted single-stranded and fixed to a surface such as a glass slide. Expression analysis and genotyping are based on hybridization with a sample of denatured genomic DNA that is tagged with a radioactive or fluorescent label. At locations on the array where hybridization occurs, the labeled sample DNA will be detectable by a scanner. The output consists of a list of hybridization events, indicating the presence or the relative abundance of DNA sequences in the sample.

Robot technology is used to prepare most arrays by spotting the DNA fragments. For macroarrays, the DNA spot size is larger allowing only hundreds instead of thousands of spots of DNA per filter. Some arrays are prepared using a lithographic process and are called biochips or DNA chips. Even more DNA fragments can be put on chips than on arrays. Other types of arrays exist, e.g., for detecting gene expression. These arrays might contain DNA fragments of 500–5,000 bases in length.

4.7 Genetic variation in natural populations

The multiplicity of techniques presented in this chapter makes it possible to detect and study genetic variation in any species of choice. Some genetic variation has been discovered in virtually every species that has been studied. The Wollemi pine is a fascinating exception to this rule (Example 4.2).

As we mentioned at the beginning of this chapter, there is no single "best" technique to study variation in natural populations. The most appropriate technique to be used in a particular study depends on the question that is being asked. Generally, the relative amount of genetic variation detected by different techniques within a population or species is concordant. It is often informative to use more than one kind of marker. For example, using both mtDNA and nuclear markers allows assessment of female- versus male-mediated gene flow (see Guest Box 4; Section 9.5).

Substantial differences in the amount of genetic variation can occur even between different populations within the same species. From a conservation perspective, such intraspecific differences are more meaningful than differences between species because they may indicate recent reductions in genetic variation caused by human actions.

Example 4.2 The Wollemi pine: coming soon to a garden near you?

The discovery of this tree in 1994 has been described as the botanical find of the century. At the time of discovery, the Wollemi pine was thought to have been extinct for over 100 million years; there are no other extant species in this genus (Jones et al. 1995). There are currently less than 100 individuals known to exist in a "secret" and inaccessible canyon in Wollemi National Park, 150 km west of Sydney, Australia (Hogbin et al. 2000).

An initial study of 12 allozyme loci and 800 AFLP fragments failed to reveal any genetic variation (Hogbin et al. 2000). Recent study of 20 microsatellite loci also failed to detect any genetic variation in this species (Peakall et al. 2003). The exceptionally low genetic variation in this species combined with its known susceptibility to exotic fungal pathogens provides strong justification for current policies of strict control of access and the secrecy of their location.

The Wollemi pine reproduces both by sexual reproduction and asexual coppicing in which additional stems grow from the base of the tree. Some individual trees are more than 500 years old, and there are indications that coppicing can result in the longevity of a plant greatly exceeding the age of individual trunks (Peakall et al. 2003). It is possible that **genets** are thousands of years old.

There are currently plans to make Wollemi pine available as a horticultural plant in 2005 or 2006. The plant is distinct in appearance and somewhat resembles its close relative, the Norfolk Island pine, which is a popular ornamental tree throughout the world.

Table 4.1 Summary of genetic variation in four samples of brown bears from North America. The allozyme (34 loci) data are from K. L. Knudsen et al. (unpublished data); the microsatellite (eight loci) and mtDNA data are from Waits et al. (1998). The allozyme samples for Alaska/Canada are from the Western Brooks Range in Alaska, and the microsatellite and mtDNA samples for this sample are from Kluane National Park, Canada. H_e is the mean expected heterozygosity (see Section 3.3), $\bar{A}$ is the average number of alleles observed, and h is gene diversity. h is computationally equivalent to H_e, but is termed gene diversity because mtDNA is haploid so that individuals are not heterozygous (Nei 1987, p. 177).

Sample	Allozymes		Microsatellites		mtDNA	
	H_e	$\bar{A}$	H_e	$\bar{A}$	h	$\bar{A}$
Alaska/Canada	0.032	1.2	0.763	7.5	0.689	5
Kodiak Island	0.000	1.0	0.265	2.1	0.000	1
NCDE	0.014	1.1	0.702	6.8	0.611	5
YE	0.008	1.1	0.554	4.4	0.240	3

NCDE, Northern Continental Divide Ecosystem (including Glacier National Park); YE, Yellowstone Ecosystem (including Yellowstone National Park).

Table 4.1 shows differences in the amount of genetic variation found between different population samples of brown bears from North America as detected with allozymes, microsatellites, and mtDNA. The same relative pattern of variation is apparent at all three marker types. Not surprisingly, the isolated population of bears on Kodiak Island has relatively little genetic variation. There are approximately 3,000 bears on this island that have been isolated for approximately some 5,000–10,000 years. More surprising is the substantially lower genetic variation in bears from the Yellowstone ecosystem in comparison to their nearest neighboring population in the Northern Continental Divide Ecosystem. The Yellowstone population has been isolated for nearly 100 years. As we will see later, this reduction in genetic variation in YE bears may have important significance for the long-term viability of this population (see Section 14.3.1).

Guest Box 4 Multiple markers uncover marine turtle behavior
Nancy N. FitzSimmons

Applications of genetic markers to the study of marine turtle populations have allowed a phenomenal increase in our understanding of their migratory behavior and the geographic scope of populations. Genetic studies have provided strong evidence that, in most species, nesting females display strong natal homing. This results in regional breeding populations that have their own unique genetic structure (e.g., Meylan et al. 1990). Recently, genetic studies have been combined with satellite tracking to study the behavior of juvenile turtles in pelagic environments (Polovina et al. 2004). This has confirmed that in some species, juveniles traverse back and forth across entire ocean basins before selecting resident feeding grounds, often hundreds to thousands of kilometers from where they were born.

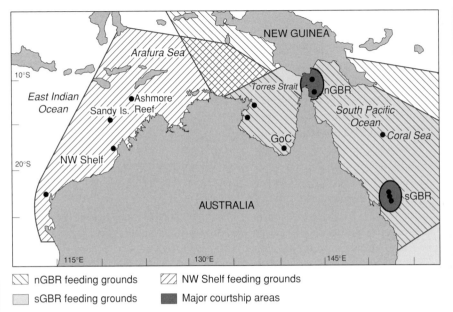

Figure 4.12 Breeding populations of green turtles in Australia. Shown are sample locations for genetic studies from the southern Great Barrier Reef (sGBR; three rookeries), Coral Sea, northern Great Barrier Reef (nGBR; two rookeries), Gulf of Carpentaria (GoC; three rookeries), and Ashmore Reef, Sandy Island, and Northwest Shelf (two rookeries). Overlapping feeding ground ranges are shown for the nGBR and sGBR breeding populations. Note that turtles from the sGBR that use the GoC or Torres Strait as feeding grounds must pass through a large congregation of nGBR breeding turtles when en route to their natal sGBR locations.

Most studies have relied upon data from mtDNA, originally using restriction digests of purified mtDNA, and, later, sequencing portions of the mtDNA, especially the control region (Bowen and Karl 1996). This reliance on mtDNA was justified because much of the focus for conservation action involves the nesting beach populations of females and it has been important to know the historical relationships and extent of gene flow among the nesting females.

What about the males? Because they rarely come ashore, studies of male marine turtles have been limited as it requires considerably more effort to capture them at feeding grounds or in the areas they congregate to mate. Genetic studies of males, or studies using nuclear markers, provide a means to test whether males, like females, display natal homing and the extent to which gene flow among populations is mediated by males.

For Australian nesting populations of green turtles, mtDNA data have identified seven breeding populations (Figure 4.12). These data confirm the operation of natal homing in females, most of which share feeding grounds with turtles from other populations. To assess the occurrence of male-mediated gene flow, the same breeding populations were analyzed for genetic variation at nine nuclear microsatellite loci (FitzSimmons et al. 1997b; N. N. FitzSimmons et al., unpublished data). In most cases, estimates of nuclear gene flow were much greater than expected based upon the mtDNA data; averaging 32 times greater gene flow attributed to males than females. However, these populations did exhibit significant genetic divergence, suggesting that natal homing also occurs in males.

It was also intriguing that estimates of gene flow from the nuclear markers did not correlate with that of mtDNA. Did this mean that natal homing behavior in males is influenced by geographic location? Alternatively, does geography influence opportunities for gene flow between populations irrespective of homing behavior? This was investigated by testing the degree of natal homing in males. Tissue samples were collected from mating males at breeding congregations in the southern and northern Great Barrier Reef and the Gulf of Carpentaria (FitzSimmons et al. 1997a). The mtDNA control region haplotypes were determined, and the frequencies for all males in each area were compared to those of the breeding females in the same area. Any differences would indicate that the breeding males, unlike the females, were not mating in their natal regions. In fact there were no differences, suggesting that male-mediated gene flow likely occurs through opportunistic matings between males and females from different population while they are en route to their natal regions to breed.

In some comparisons between marine turtle populations in Australia, greater structure has been observed using the microsatellite markers than with other nuclear markers, possibly indicating a more recent separation of populations and accumulation of novel mutations in these rapidly evolving markers. This is in sharp contrast to other nuclear markers including anonymous single copy nuclear loci (FitzSimmons et al. 1997b), and allozymes, which could not distinguish among populations (Norman et al. 1994). Current research includes the development of SNPs to help identify the origins of turtles at feeding grounds and those individuals adversely affected by human activities.

Problems 4.1 and 4.2

The illustrations opposite show hypothetical electrophoretic gel patterns for three microsatellite loci in an isolated population of snow leopards in Nepal. Each band represents a different allele. The alleles differ in the number of repeats that they contain. The number of repeats present in each allele is given to the right of each gel.

Individuals with two bands are heterozygotes, and those with a single band are homozygotes. A biologist is trying to determine who is the father of five progeny (P1 to P5 below) born to each of five females (F1 to F5 below); for example, P2 is the progeny of F2. There are only 10 adult males in this population (M1 to M10 below).

Problem 4.1

Designate alleles on the basis of the number of repeats they possess. Write down the genotype of each individual locus in the table provided below (e.g., *10/10, 5/8*, etc.) for each locus.

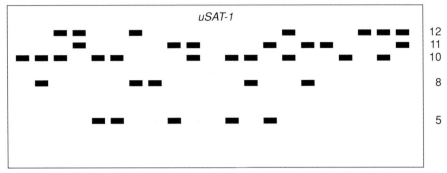

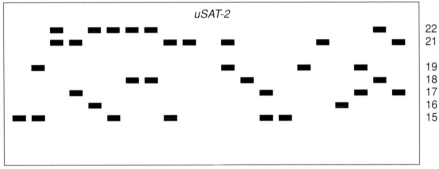

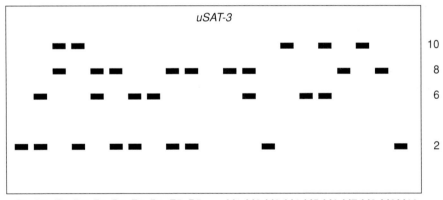

Problem 4.2

Do a paternity exclusion analysis for each of the five progeny based on these data. That is, try to identify the true father by eliminating potential fathers on the basis of genotypic incompatibilities.

No.	uSAT-1	uSAT-2	uSAT-3
F1			
P1			
F2			
P2			
F3			
P3			
F4			
P4			
F5			
P5			
M1			
M2			
M3			
M4			
M5			
M6			
M7			
M8			
M9			
M10			

For example, M1 cannot be the father of P1 based on genotypes at the *uSAT-1* locus. The mother of P1 (female F1) was homozygous *10/10* at *uSAT-1,* and therefore must have transmitted this allele to P1. P1 is *10/8,* and therefore the *8* allele had to be transmitted by the father of P1. M1 does not carry the *8* allele and therefore cannot be the father of P1.

If all males but one can be excluded, the remaining male must be the father. Who is the father of each of the five progeny? Are there any progeny for which more than one male is the potential father?

Problem 4.3

PCR null alleles are found at some microsatellite loci. These alleles have substitutions at one or both of the primer sites so that they do not amplify during PCR. Heterozygotes for a null allele and another allele appear to be homozygotes for that allele. For example, individual 9 in Figure 4.5 could actually be a *183/null* heterozygote rather than a *183/183* homozygote. Assume that you can perform experimental matings with hairy-nosed wombats and that all odd-numbered individuals are males and all even-numbered individuals are females. Design an experimental mating using the individuals shown in Figure 4.5 that would allow you to test if individual 9 is a *183/183* homozygote or a *183/null* heterozygote. Be sure to explain how you would use the progeny from the mating to determine if individual 9 is a *183/183* homozygote or a *183/null* heterozygote. How many progeny would you have to examine to be sure of the genotype of individual 9?

Problem 4.4

Identify the father of the three chicks shown in Figure 4.10. Assume that the same male fathered all three chicks. Note: some of the bands that are not identified as being diagnostic in this figure may also provide useful information about paternity.

Problems 4.5 and 4.6

Assume that the following 1,020 bp are the DNA sequence for part of a rainbow trout nuclear encoded growth hormone gene (*GH-2*) in two individuals that are homozygous for different alleles. The top sequence is real while the bottom sequence has been created for the purposes of these problems; these sequences are identical except for the bases in Individual 2 that are in bold and upper case. Do a hypothetical search for genetic variation. Assume that these 1,020 bp are the product of a PCR reaction. "Screen" these two individuals for RFLP by digesting this product with the restriction enzyme *Hum*fiI that cuts DNA in the middle of the sequence 5'-*AATT*-3'. By convention, the sequence listed is the coding strand beginning with the "upstream" or 5' end of the DNA. Note: an electronic version of these sequences that you can screen for "aatt" with your word processor is available on the web page.

INDIVIDUAL 1

```
  1 gtcaagttac agggttgtgt ctgtctgtgt gactgagtgt aactttgttc attcattatg
 61 tcctagacaa cagaggtttg tgtcgtctgt gttttgaccc tcatttgtca agtcatcgag
121 tacgtttttt gttttagga gtcacctctt cccgaactca tggaaagatt catgattgat
181 ttgacgcatt atactgattg ttccatagtc acatacaaaa acaggtccca tcggcgagag
241 gtggtacatg gagaaaatct catgtttcct cctgttgata cattaaaaca tgtgttctcc
301 atctataaaa acagtggccc caaacaagcg gcaacatact gaaccgacca ccacactttc
361 aagtgaagta atcatccttg gcaattaaga gaaaaaaatg ggacaaggta aaccagcttt
421 tattttattt ttttaagtgg gaagtcagtg taccatttaa taccatttaa ctttaacatt
481 aaatcactga ggcaggggcc aagaaggcag agaaagagtg aacaagtaat gtactgccat
541 gagggtataa tctacttaca cagaaccact tcctttaaca acctaaccat gtgatctatt
601 agatttacat ttgagttatt tagcagagac tcttatccag agcgacttac aggagcaatt
661 agggttaagt gccttgctca agggcacatc aacagatttc tcacctagtc agctcaggga
721 ttcaaaccag taacctttca gttactggcc caactctctt aatcgctagg ctaatgagaa
781 agatagcaaa ttgagaatat cttactattg agaatatctt actaacatgt cgcaacatca
841 tttgacttac tcgtttttat acatttctta ttttctgtca tctctctttt agtgtttctg
901 ctgatgccag tcttactggt cagttgtttc ctgggtcaag gggcggcgat ggaaaaccaa
961 cggctcttca acatcgcggt caaccgggtg caacacctcc acctattggc tcagaaaatg
```

INDIVIDUAL 2

```
  1 gtcaagAtac agggttgtgt ctgtctgtgt gactgagtgt aactttgttc attcattatg
 61 tcctagacaa cagaggtttg tgtcgtctgt gttttgaccc tcatttgtca agtcatcgag
121 tacgtttttt gttttagga gtcacGGctt cccgaactca tggaaaAatt catgattgat
181 ttgacgcatt atactgattg ttccatagtc acatacaaaa acaggtccca tcggcgagag
241 gtggtacatg gagaaaatct catgtttcct cctgttgata cattaaaaca tgtgttctcc
301 atctataaaa acagtggccc caaacaagcg gcaacatact gaaTcgacca ccacactttc
361 aagtgaagta atcatccttg gcaattaaga gaaaaaaatg ggacaaggta aaccagcttt
421 tattttattt ttttaagtgg gaagtcagtg taccatttaa taccatttaa ctttaacatt
481 taatcactga ggcaggggcc aagaaggcag agaaagagtg aacaagtaat gtacAgccat
541 gagggtataa tctacttaca cagaaccact tcctttaaca acctaaccat gtgatctatt
601 agatttacat ttgagttatt tagcagagac tcttatccag agcgacttac aggagcaatt
661 agggttaagt gccttgctca agggcacatc aacagatttc tcacctagtc agctcaggga
721 ttcaGaccag taacctttca gttactggcc caactctctt aatcgctagg ctaatgagaa
781 agatagcaaa ttgagaatat cttactattg agaatatctt actaacatgt cgcaacatca
841 tttgacttac tcgtttttat acatttctta ttttctgtca tctctctttt agtgtttctg
901 ctgaGgccag tcttactggt cagttgtttc ctgggtcaag gggcAgcgat ggaaaaccaa
961 cggctcttca acatcgcggt caaccgggtg caaTacctcc acctaGtggc tcagaaaatg
```

Problem 4.5

How many restriction fragments would you expect from this experiment with each individual? How large would you expect each fragment to be? Draw what you would expect a gel to look like after restricting these two PCR products with this restriction enzyme. Remember that the position on the gel will be determined solely by the size of the piece of DNA. The sample in the lane on the far right of the gel is a "size standard" that contains four fragments of 100, 200, 300, and 400 bp.

Problem 4.6

Draw the expected gel pattern for a progeny produced by crossing these two individuals.

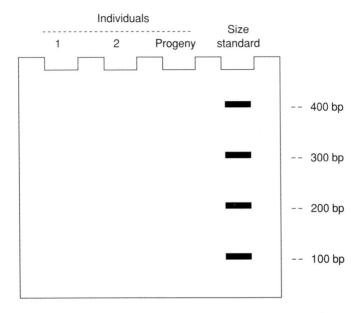

Problem 4.7

Sue Haig has been studying the genetic population structure of spotted owls for many years (e.g., Haig et al. 2004). Spotted owls, like many birds, have little genetic variation in comparison with other taxa. Over 400 RAPD fragments were examined but only some 20 of these varied between individuals. A comparison of males and females revealed that six of these fragments had different frequencies in males and females. All six fragments were more frequent in females than in males. On what chromosome do you think the region coding for these six fragments is located? Should these fragments be used to describe genetic population structure of spotted owls? Why or why not?

Problem 4.8

Birds sometimes produce twins that develop inside the same egg. The gel in the figure below shows minisatellite DNA from two emu twins (T$_1$ and T$_2$) produced on an emu farm (Bassett et al. 1999). The twins' parents ($♀$ and $♂$) and two other emus from the same farm are also shown. C is a chicken. Based on this gel, do you think these emu twins are identical or fraternal? Identical twins are produced by a single fertilized egg that develops into two genetically identical individuals. Fraternal twins are produced by two separate eggs that are fertilized by two sperm.

C ♂ T₁ T₂ ♀ C E E C

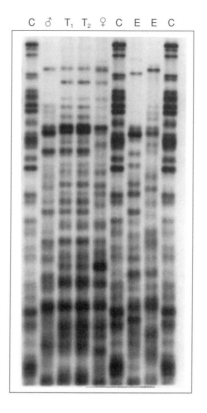

Problem 4.9

Pick a species of conservation concern that you are interested in and search the GenBank sequence database (http://www.ncbi.nih.gov/) to see how much sequence information is available for this species. Have sections of the mtDNA sequence of this species been published? How many sets of microsatellite primers have been described for this species? Have microsatellite primers been published for any congeneric species that may be useful for this species?

Problem 4.10

The entire 16,642 base pair sequence for the mtDNA molecule in rainbow trout can be found on this book's website in a WORD and a text file (*rainbow trout mtDNA*; Zardoya et al. 1995). Use your favorite word processing program to screen this molecule for *Eco*RI cut sites (5′-GAATTC-3′) by searching for the sequence GAATTC. Describe the expected number and size of the DNA fragments after this molecule is digested by *Eco*RI.

*Eco*RI is a six-based cutter. That is, it requires six consecutive bases to digest DNA. There are also four-based cutters. Try "digesting" this molecule with *Mbo*I (5′-GATC-3′). How many fragments would you expect after digestion by *Mbo*I? Don't describe the length of the fragments because there are too many!

Hint: there are no cut sites that overlap from one line to another.

Part II
Mechanisms of Evolutionary Change

5

Random Mating Populations: Hardy–Weinberg Principle

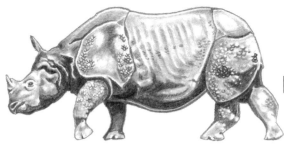

One-horned rhinoceros, Section 5.3

In a sexual population, each genotype is unique, never to recur. The life expectancy of a genotype is a single generation. In contrast, the population of genes endures.

James F. Crow (2001)

Today, the Hardy–Weinberg Law stands as a kind of Newton's First Law (bodies remain in their state of rest or uniform motion in a straight line, except insofar as acted upon by external forces) for evolution: Gene frequencies in a population do not alter from generation to generation in the absence of migration, selection, statistical fluctuation, mutation, etc.

Robert M. May (2004)

A description of genetic variation by itself, as in Chapters 3 and 4, will not help us understand the evolution and conservation of populations. We need to develop the theoretical expectations of the effects Mendelian inheritance in natural populations in order to understand the influence of natural selection, small population size, and other evolutionary factors that affect the persistence of populations and species. The strength of population genetics is the rich foundation of theoretical expectations that allows us to test the predictions of hypotheses to explain the patterns of genetic variation found in natural populations.

This chapter introduces the structure of the basic models used to understand the genetics of populations (Crow 2001). In later chapters, we will explore expected changes in allele and genotype frequencies in the presence of such evolutionary factors as natural selection or mutation. In this chapter, we will focus on the relationship between allele frequencies and genotype frequencies. In addition, we will examine techniques for estimating allele frequencies and for testing observed genotypic proportions with those expected.

Models

We will use a series of models to consider the pattern of genetic variation in natural populations and to understand the mechanisms that produce evolutionary change. Models allow us to simplify the complexity of the world around us. Models may be either conceptual or mathematical. Conceptual models allow us to simplify the world so that we can represent reality with words and in our thoughts. Mathematical models allow us to specify the relationship between empirical quantities that we can measure and parameters that we specify in our biological theory. These models are essential in understanding the factors that affect genetic change in natural populations, and in predicting the effects of human actions on natural populations.

In addition, models are very helpful in a variety of additional ways:

1   Models make us consider and define the parameters that need to be considered.
2   Models allow us to test hypotheses.
3   Models allow us to generalize results.
4   Models allow us to predict how a system will operate in the future.

The use of models in biology is sometimes criticized because genetic and ecological systems are complex, and simple models ignore many important properties of these systems. This criticism has some validity. Nevertheless, as a general rule of thumb, models that we develop to understand natural populations should be as simple as possible. That is, a hypothesis or model should not be any more complicated than necessary (**Ocham's razor**). There are several reasons for this. First, hypotheses and models are scientifically useful only if they can be tested and rejected. Simpler models are easier to reject, and, therefore, are more valuable. Second, simple models are likely to be more general and therefore more applicable to a wider number of situations.

5.1 The Hardy–Weinberg principle

We will begin with the simplest model of population genetics: a random mating population in which no factors are present to cause genetic change from generation to generation.

This model is based upon the fundamental framework of Mendelian segregation for diploid organisms that are reproducing sexually in combination with fundamental principles of probability (Box 5.1). These same principles apply to virtually all species, from elephants to pine trees to violets. We will make the following assumptions in constructing this model:

1 **Random mating.** "Random mating obviously does not mean promiscuity; it simply means . . . that in the choice of mates . . . there is neither preference for nor aversion to the union of persons similar or dissimilar with respect to a given trait or gene" (Wallace and Dobzhansky 1959). Thus, it is possible that a population may be randomly mating with regard to most loci, but be mating nonrandomly with regard to others that are influencing mate choice.

2 **No mutation.** We assume that genetic information is transmitted from parent to progeny (i.e., from generation to generation) without change. Mutations provide the genetic variability that is our primary concern in genetics. Nevertheless, mutation rates are generally quite small and are only important in population genetics from a long-term perspective, generally hundreds or thousands of generations. We will not consider the effects of mutations on changes in allele frequencies in detail since in conservation genetics we are more concerned with factors that can influence populations in a more immediate time frame.

3 **Large population size.** Many of the theoretical models that we will consider assume an infinite population size. This assumption may effectively be correct in some populations of insects or plants. However, it is obviously not true for many of the populations of concern in conservation genetics. Nevertheless, we will initially consider the ideal large population in order to develop the basic concepts of population genetics, and we will then consider the effects of small population size in later chapters.

4 **No natural selection.** We will assume that there is no differential survival or reproduction of individuals with different genotypes (that is, no natural selection). Again, this assumption will not be true at all loci in any real population, but it is necessary that we initially make this assumption in order to develop many of the basic concepts of population genetics. We will consider the effects of natural selection in later chapters.

5 **No immigration.** We will assume that we are dealing with a single isolated population. We will later consider multiple populations in which gene flow between populations is brought about through exchange of individuals.

There are two important consequences of these assumptions. **First**, the population will not evolve. Mendelian inheritance has no inherent tendency to favor one allele. Therefore, allele and genotype frequencies will remain constant from generation to generation. This is known as the **Hardy–Weinberg equilibrium**. In the next few chapters we will explore the consequences of relaxing these assumptions on changes in allele frequency from generation to generation. We will not be able to consider all possibilities. However, our goal is to develop an intuitive understanding of the effects of each of these evolutionary factors.

The **second** important outcome of the above assumptions is that genotype frequencies will be in binomial (Hardy–Weinberg) proportions. That is, **genotypic** frequencies after one generation of random mating will be a binomial function of **allele** frequencies. It is important to distinguish between the two primary ways in which we will describe the genetic characteristics of populations at individual loci: allele (gene) frequencies and genotypic frequencies.

Box 5.1 Probability

Genetics is a science of probabilities. Mendelian inheritance itself is based upon probability. We cannot know for certain which allele will be placed into a gamete produced by a heterozygote, but we know that there is a one-half probability that each of the two alleles will be transmitted. This is an example of a random, or **stochastic**, event. There are a few simple rules of probability that we will use to understand the extension of Mendelian genetics to populations.

The **probability** (P) of an event is the number of times the event will occur (a) divided by the total number of possible events (n):

$$P = a/n$$

For example, a die has six faces that are equally likely to land up if the die is tossed. Thus, the probability of throwing any particular number is one-sixth:

$$P = a/n = 1/6$$

We often are interested in combining the probabilities of different events. There are two different rules that we will use to combine probabilities.

The **product rule** states that the probability of the probability of two or more independent events occurring simultaneously is equal to the product of their individual probabilities. For example, what is the probability of throwing a total of 12 with a pair of dice? This can only occur by a six landing up on the first die and also on the second die. According to the product rule:

$$P = 1/6 \times 1/6 = 1/36$$

The **sum rule** states that the probability of two or more mutually exclusive events occurring is equal to the sum of their individual probabilities. For example, what is the probability of throwing either a five or six with a die? According to the sum rule:

$$P = 1/6 + 1/6 = 2/6 = 1/3$$

In many situations, we need to use both of these rules to compute a probability. For example, what is the probability of throwing a total of seven with a pair of dice?

Solution: There are six mutually exclusive ways that we can throw seven with two dice: $1 + 6$, $2 + 5$, $3 + 4$, $4 + 3$, $5 + 2$, and $6 + 1$. As we saw in the example for the product rule, each of these combinations has a probability of $1/6 \times 1/6 = 1/36$ of occurring. They are all mutually exclusive so we can use the sum rule. Therefore, the probability of throwing a seven is:

$$1/6 + 1/6 + 1/6 + 1/6 + 1/6 + 1/6 = 6/36 = 1/6$$

The Hardy–Weinberg principle greatly simplifies the task of describing the genetic characteristics of populations; it allows us to describe a population by the frequencies of the alleles at a locus rather than by the many different genotypes that can occur at a single diploid locus. This simplification becomes especially important when we consider multiple loci. For example, there are 59,049 different genotypes possible at just 10 loci that each has just two alleles. We can describe this tremendous genotypic variability by specifying only 10 allele frequencies if the populations is in Hardy–Weinberg proportions.

This principle was first described by a famous English mathematician G. H. Hardy (1908) and independently by a German physician Wilhelm Weinberg (1908). The principle was actually first used by an American geneticist W. E. Castle (1903) in a description of the effects of natural selection against recessive alleles. However, this aspect of the paper by Castle was not recognized until nearly 60 years later (Li 1967). A detailed and interesting history of the development of population genetics is provided by Provine (2001). There is great irony in our use of Hardy's name to describe a fundamental principle that has been of great practical value in medical genetics and now in our efforts to conserve biodiversity. Hardy (1967) saw himself as a "pure" mathematician whose work had no practical relevance: "I have never done anything 'useful'. No discovery of mine has made, or is likely to make, directly or indirectly, for good or ill, the least difference to the amenity of the world."

5.2 Hardy–Weinberg proportions

We will first consider a single locus with two alleles (A and a) in a population such that the population consists of the following numbers of each genotype:

AA	Aa	aa	Total
N_{11}	N_{12}	N_{22}	N

Each homozygote (AA or aa) contains two copies of the same allele while each heterozygote (Aa) contains one copy of each of its constituent. Therefore, the allele frequencies are:

$$p = \text{freq}(A) = \frac{(2N_{11} + N_{12})}{2N}$$

$$q = \text{freq}(a) = \frac{(N_{12} + 2N_{22})}{2N}$$

(5.1)

where $p + q = 1.0$.

Our assumption of random mating will result in random union of gametes to form zygotes. Thus, the frequency of any particular combination of gametes from the parents will be equal to the product of the frequencies of those gametes, which are the allele frequencies. This is shown graphically in Figure 5.1. Thus, the expected genotypic proportions are predicted by the binomial expansion:

$$(p + q)^2 = p^2 + 2pq + q^2$$
$$\quad\quad\quad AA \quad Aa \quad aa$$

(5.2)

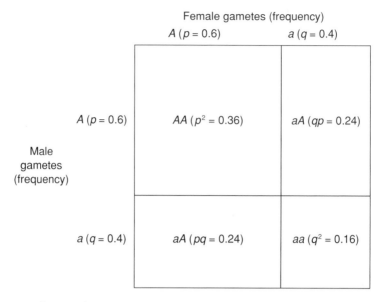

Figure 5.1 Hardy–Weinberg proportions at a locus with two alleles (A and a) generated by the random union of gametes produced by females and males. The area of each rectangle is proportional to the genotypic frequencies.

These proportions will be reached in one generation, providing all of the above assumptions are met and allele frequencies are equal in males and females. Additionally, these genotypic frequencies will be maintained forever, as long as these assumptions hold.

The Hardy–Weinberg principle can be readily extended to more than two alleles with two simple rules:

1   The expected frequency of homozygotes for any allele is the square of the frequency of that allele.
2   The expected frequency of any heterozygote is twice the product of the frequency of the two alleles present in the heterozygote.

In the case of three alleles the following genotypic frequencies are expected:

$$p = \text{freq}(A_1)$$
$$q = \text{freq}(A_2)$$
$$r = \text{freq}(A_3)$$

and

$$(p + q + r)^2 = \underset{A_1A_1}{p^2} + \underset{A_1A_2}{2pq} + \underset{A_2A_2}{q^2} + \underset{A_1A_3}{2pr} + \underset{A_2A_3}{2qr} + \underset{A_3A_3}{r^2}$$

(5.3)

5.3 Testing for Hardy–Weinberg proportions

Genotypic frequencies of samples from natural populations can be tested readily to see if they conform to Hardy–Weinberg expectations. However, there are a profusion of papers that discuss the sometimes hidden intricacies of testing for goodness-of-fit to Hardy–Weinberg proportions. Lessios (1992) has provided an interesting and valuable review of this literature. Guest Box 5 provides an example of how testing for departures from Hardy–Weinberg can provide important insights into the mating system, social behavior, and genetic structure of populations.

Dinerstein and McCracken (1990) described genetic variation using allozyme electrophoresis at 10 variable loci in a population of one-horned rhinoceros from the Chitwan valley of Nepal. The following numbers of each genotype were detected at a locus with two alleles (*100* and *125*) encoding the enzyme lactate dehydrogenase. Do these values differ from what we expect with Hardy–Weinberg proportions?

100/100	*100/125*	*125/125*	Total
$N_{11} = 5$	$N_{12} = 12$	$N_{22} = 6$	$N = 23$

We first need to estimate the allele frequencies in this sample. We do not know the true allele frequencies in this population, which consisted of some 400 animals at the time of sampling. However, we can estimate the allele frequency in this population based upon the sample of 23 individuals. The estimate of the allele frequency of the *100* allele obtained from this sample will be designated as $\hat{p}$ (called p hat) to designate that it is an estimate rather than the true value.

$$\hat{p} = \frac{2N_{11} + N_{12}}{2N} = \frac{10 + 12}{46} = 0.478$$

and

$$\hat{q} = \frac{N_{12} + 2N_{22}}{2N} = \frac{12 + 12}{46} = 0.522$$

We now can estimate the expected number of each genotype in our sample of 23 individuals genotype assuming Hardy–Weinberg proportions:

	100/100	*100/125*	*125/125*
Observed	5	12	6
Expected	($\hat{p}^2 N = 5.3$)	($2\hat{p}\hat{q}N = 11.5$)	($\hat{q}^2 N = 6.3$)

The agreement between observed and expected genotypic proportions in this case is very good. In fact, this is the closest fit possible in a sample of 23 individuals from a population with the estimated allele frequencies. Therefore, we would conclude that there is no indication that the genotype frequencies at this locus are not in Hardy–Weinberg proportions.

The chi-square method provides a statistical test to determine if the deviation between observed genotypic and expected Hardy–Weinberg proportions is greater than we would

Table 5.1 Critical values of the chi-square distribution for up to five degrees of freedom (v). The proportions in the table (corresponding to $\alpha = 0.05, 0.01$, etc.) represent the area to the right of the critical value of chi-square given in the table, as shown in the figure below. The null hypothesis is usually not rejected unless the probability associated with the calculated chi-square is less than 0.05.

Degrees of freedom	Probability (P)					
	0.90	0.50	0.10	0.05	0.01	0.001
1	0.02	0.46	2.71	3.84	6.64	10.83
2	0.21	1.39	4.60	5.99	9.21	13.82
3	0.58	2.37	6.25	7.82	11.34	16.27
4	1.06	3.86	7.78	9.49	13.28	18.47
5	1.61	14.35	9.24	11.07	15.09	20.52

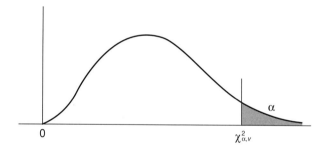

expect by chance alone. We first calculate the chi-square value for each of the genotypes and sum them into a single value:

$$X^2 = \sum \frac{(OBSERVED - EXPECTED)^2}{EXPECTED}$$

$$= \frac{(5 - 5.3)^2}{5.3} + \frac{(12 - 11.5)^2}{11.5} + \frac{(6 - 6.3)^2}{6.3}$$

$$= 0.02 + 0.02 + 0.01 = 0.05$$

The X^2 value becomes increasingly greater as the difference between the observed and expected values becomes greater.

The computed X^2 value is then tested by comparing it to a set of values (Table 5.1) calculated under the assumption that the null hypothesis we are testing is correct; in this case, our null hypothesis is that the population from which the samples was drawn is in Hardy–Weinberg proportions. We need one additional value to apply the chi-square test, the **degrees of freedom**. In using the chi-square test for Hardy–Weinberg proportions, the degrees of freedom is equal to the number of possible genotypes minus the number of alleles.

Number of alleles	Number of genotypes	Degrees of freedom
2	3	1
3	6	3
4	10	6
5	15	10

By convention, if the probability estimated by a statistical test is less than 0.05, then the difference between the observed and expected values is said to be significant. We can see in Table 5.1 that the chi-square value with one degree of freedom must be greater than 3.84 before we would conclude that the deviation between observed and expected proportions is greater than we would expect by chance with one degree of freedom. Our estimated X^2 value of 0.05 for the lactate dehyrogenase locus in the one-horned rhino is much smaller than this. Therefore, we would accept the null hypothesis that the population from which this sample was drawn was in Hardy–Weinberg proportions at this locus. Example 5.1 gives a situation where the null hypothesis of Hardy–Weinberg proportions can be rejected. An example of the chi-square test for Hardy–Weinberg proportions in the case of three alleles is given in Example 5.2.

Example 5.1 Test for Hardy–Weinberg proportions

Leary et al. (1993a) reported the following genotype frequencies at an allozyme locus (*mIDHP-1*) in a sample of bull trout from the Clark Fork River in Idaho:

Genotype	Observed	Expected	Chi-square
100/100	1	($\hat{p}^2N = 5.8$)	3.97
100/75	22	($2\hat{p}\hat{q}N = 12.5$)	7.22
75/75	2	($\hat{q}^2N = 6.8$)	3.38
Total	25	(25.1)	14.58

Estimated frequency of $100 = \hat{p} = [(2 \times 1) + 22]/50 = 0.480$
Estimated frequency of $75 = \hat{q} = [22 + (2 \times 2)]/50 = 0.520$
Degrees of freedom = 1

The calculated X^2 of 14.58 is greater than the critical value for $P < 0.001$ with 1 d.f. of 10.83 (see Table 5.1). Therefore, the probability of getting such a large deviation by chance alone is less than 0.001. Therefore we would reject the null hypothesis that the sampled population was in Hardy–Weinberg proportions at this locus.

There is a significant excess of heterozygotes in this sample of bull trout. We will return to this example in the next chapter to see the probable cause of this large deviation from Hardy–Weinberg proportions.

Example 5.2 Test for Hardy–Weinberg proportions at a locus (aminoacylase-1, *Acy-1*) with three alleles in the Polish brown hare (Hartl et al. 1992)

Genotype	Observed	Expected	Chi-square
100/100	4	$(\hat{p}^2N = 2.1)$	1.72
100/81	6	$(2\hat{p}\hat{q}N = 9.7)$	1.41
81/81	14	$(\hat{q}^2N = 11.0)$	0.82
100/66	4	$(2\hat{p}\hat{r}N = 4.0)$	0.00
81/66	7	$(2\hat{q}\hat{r}N = 9.2)$	0.53
66/66	3	$(\hat{r}^2N = 1.9)$	0.64
Total	38	(37.9)	5.12

Estimated frequency of $100 = \hat{p} = [(2 \times 4) + 6 + 4]/76 = 0.237$
Estimated frequency of $81 = \hat{q} = [6 + (2 \times 14) + 7]/76 = 0.539$
Estimated frequency of $66 = \hat{r} = [4 + 7 + (2 \times 3)]/76 = 0.224$
There are six genotypic classes and two independent allele frequencies at a locus with three alleles
Degrees of freedom $= 6 - 3 = 3$

The calculated X^2 of 5.12 is less than the critical value with 3 d.f. of 7.82 (see Table 5.1). Therefore, we accept the null hypothesis that the sampled population was in Hardy–Weinberg proportions at this locus.

5.3.1 Small sample sizes or many alleles

Sample sizes in conservation genetics are often smaller than our statistical advisors recommend because of the limitations imposed by working with rare species. The chi-square test is only an approximation of the actual probability distribution, and the approximation becomes poor when expected numbers are small. The usual rule-of-thumb is to not use the chi-square test when any expected number is less than five. However, some have argued that this rule is unnecessarily conservative and have suggested using smaller limits on expected values (three by Cochran (1954), and one by Lewontin and Felsenstein (1965)).

In addition, there is a systematic bias in small samples because of the discreteness of the possible numbers of genotypes. Levene (1949) has shown that in a finite sample of N individuals, the heterozygotes are increased by a fraction of $1/(2N - 1)$ and homozygotes are correspondingly decreased (see Crow and Kimura 1970, pp. 55–56). For example, if only one copy of a rare allele is detected in a sample then the only genotype containing the rare allele must be heterozygous. However, the simple binomial Hardy–Weinberg proportions will predict that some fraction of the sample is expected to be homozygous for the rare allele; however, this is impossible because there is only one copy of the allele in the sample. The above adjustment will correct for this bias.

Exact tests provide a method to overcome the limitation of small expected numbers with the chi-square test (Fisher 1935). Exact tests are performed by determining the probabilities of all possible samples assuming that the null hypothesis is true (Example 5.3). The probability of the observed distribution is then added to the sum of all less probable possible sample outcomes. Weir (1996, pp. 98–101) describes the use of the exact test, and Vithayasai (1973) presented tables for applying the exact test with two alleles.

Example 5.3 The exact test for Hardy–Weinberg proportions (Weir 1996)

In this case, we would reject the null hypothesis of Hardy–Weinberg proportions because our calculated chi-square value is greater than 3.84. However, an exact test indicates that we would expect to get a deviation as great or greater than the one we observed some 8% (0.082) of the time. Therefore, we would not reject the null hypothesis in this case using the exact test.

Genotypes

100/100	100/80	80/80		
21	19	0	$\hat{p} = 0.763$	$X^2 = 3.88$
(23.3)	(14.5)	(2.3)		

There are 10 possible samples of 40 individuals that would provide us with the same allele frequency estimates. We can calculate the exact probabilities for each of these possibilities if the sampled population was in Hardy–Weinberg proportions using the binomial distribution as shown below:

Possible samples

100/100	100/80	80/80	Probability	Cumulative probability	X^2
30	1	9	0.0000	0.0000	34.67
29	3	8	0.0000	0.0000	25.15
28	5	7	0.0001	0.0001	17.16
27	7	6	0.0023	0.0024	10.69
26	9	5	0.0205	0.0229	5.74
21	**19**	**0**	**0.0594**	**0.0823**	**3.88**
25	11	4	0.0970	0.1793	2.32
22	17	1	0.2308	0.4101	1.20
24	13	3	0.2488	0.6589	0.42
23	15	2	0.3411	1.0000	0.05

In practice, exact tests are performed using computer programs because calculating the exact binomial probabilities is extremely complicated and time consuming.

Testing for Hardy–Weinberg proportions at loci with many alleles, such as microsatellite loci, is also a problem because many genotypes will have extremely low expected numbers. There are $A(A - 1)/2$ heterozygotes and A homozygotes at a locus with A alleles. Therefore, there are the following possible numbers of genotypes at a locus in a population with A alleles:

$$\frac{A(A - 1)}{2} + A = \frac{A(A + 1)}{2} \tag{5.4}$$

For example, Olsen et al. (2000) found an average of 23 alleles at eight microsatellite loci in pink salmon in comparison to an average of 2.3 alleles at 24 polymorphic allozyme loci in the same population. There are a total of 279 genotypes possible with 23 alleles (expression 5.4). Exact tests for Hardy–Weinberg proportions are possible with more than two alleles (Louis and Dempster 1987; Guo and Thompson 1992). However, the number of possible genotypes increases very quickly with more than two alleles and computation time becomes prohibitive. Hernandez and Weir (1989) have described a method of approximating the exact probabilities.

Nearly exact tests are generally used to analyze data from natural populations using computer-based permutation or randomization testing. In the case of Example 5.3, a computer program would randomize genotypes by sampling, or creating, 40 diploid individuals from a pool of 61 copies of the *100* allele and 19 copies of the *80* allele. A chi-square value is then calculated for 1,000 or more of these randomized data sets and its value compared to the statistic obtained from the observed data set. The proportion of chi-square values from the randomized data sets that give a value as large or larger than the observed one provides an unbiased estimation of the probability that the null hypothesis is true.

5.3.2 Multiple simultaneous tests

In most studies of natural populations, multiple loci are examined from several populations resulting in multiple tests for Hardy–Weinberg proportions. For example, if we examine 10 loci in 10 population samples, 100 tests of Hardy–Weinberg proportions will be performed. If all of these loci are in Hardy–Weinberg proportions (that is, our null hypothesis is true at all loci in all populations), we expect to find five significant tests if we use the 5% significance level. Thus, simply applying the statistical procedure presented here would result in rejection of the null hypothesis of Hardy–Weinberg proportions approximately five times when our null hypothesis is true. This is called a type I error. (A type II error occurs when a false null hypothesis is accepted.)

There are a variety of approaches that can be used to treat this problem (see Rice 1989). One common approach is to use the so-called Bonferroni correction in which the significance level (say 5%) is adjusted by dividing it by the number of tests performed (Cooper 1968). Therefore in the case of 100 tests, we would use the adjusted nominal level of $0.05/100 = 0.0005$. The critical chi-square value for $P = 0.0005$ with one degree of freedom is 12.1. That is, we expect a chi-square value greater than 12.1 with one degree of freedom less than 0.0005 of the time if our null hypothesis is correct. Thus, we would reject the null hypothesis for a particular locus only if our calculated chi-square value was greater than 12.1. This procedure is known to be conservative and results in a loss of statistical power to detect multiple deviations from the null hypothesis. A procedure known as the sequential Bonferroni can be used to increase power to detect more than one deviation from the null hypothesis (Rice 1989).

It is also extremely important to examine the data to detect possible patterns for those loci that do not conform to Hardy–Weinberg proportions. For example, let us say that eight of our 100 tests have probability values less than 5%; this value is not much greater than our expectation of five. If the eight cases are spread fairly evenly among samples and loci and none of the individual probability values are less than 0.0005 obtained from the Bonferroni correction, then it is reasonable to not reject the null hypothesis that these samples are in Hardy–Weinberg proportions at these loci.

However, we may reach a different conclusion if all eight of the deviations from Hardy–Weinberg proportions occurred in the same sample, and all the deviations were in the same direction (e.g., a deficit of heterozygotes). This would suggest that this particular sample was taken from a population that was not in Hardy–Weinberg proportions. Perhaps this sample was collected from a group that consisted of two separate random mating populations (Wahlund effect; see Chapter 9).

Another possibility is that all eight deviations from Hardy–Weinberg proportions occurred at the same locus in eight different population samples, and all the deviations were in the same direction (e.g., a deficit of heterozygotes). This would suggest that there is something unusual about this particular locus. For example, the presence of a null allele, as we saw in Problem 4.3, would result in a tendency for a deficit of heterozygotes.

5.4 Estimation of allele frequencies

So far we have estimated allele frequencies when the number of copies of each allele in a sample can be counted directly from the genotypic frequencies. However, sometimes we cannot identify the alleles in every individual in a sample. The Hardy–Weinberg principle can be used to estimate allele frequencies at loci in which there is not a unique relationship between genotypes and phenotypes. We will consider two such situations that are often encountered in analyzing data from natural populations.

5.4.1 Recessive alleles

There are many cases in which heterozygotes cannot be distinguished from one of the homozygotes. For example, color polymorphisms and metabolic disorders in many organisms are caused by recessive alleles. The frequency of recessive alleles can be estimated if we assume Hardy–Weinberg proportions:

$$\hat{q} = \sqrt{\frac{N_{22}}{N}} \tag{5.5}$$

Dozier and Allen (1942) described differences in coat color in the muskrat of North America. A dark phase, the so-called blue muskrat, is generally rare relative to the ordinary brown form but occurs at high frequencies along the Atlantic coast between New Jersey and North Carolina. Breeding studies (Dozier 1948) have shown that the blue phase is caused by a recessive allele (b) which is recessive to the brown allele (B). A total of 9,895 adult muskrats were trapped on the Backwater National Wildlife Refuge, Maryland in

1941. The blue muskrat occurred at a frequency of 0.536 in this sample. If we assume Hardy–Weinberg proportions (see expression 5.2), then:

$$\hat{q}^2 = 0.536$$

and taking the square root of both sides of this relationship:

$$\hat{q} = \sqrt{0.536} = 0.732$$

The estimated frequency of the B allele is $(1 - 0.732) = 0.268$.

Example 5.4 demonstrates how the genotypic proportions in a population for a recessive allele can be tested for Hardy–Weinberg proportions.

Example 5.4 Color polymorphism in the eastern screech owl

We concluded in Chapter 2, based on the data below from Table 2.1, that the red morph of the eastern screech owl is caused by a dominant allele (R) at a single locus with two alleles; grey owls are homozygous for the recessive allele (rr).

Mating	No. of families	Progeny	
		Red	Gray
Red × red	8	23	5
Red × gray	46	68	63
Gray × gray	135	0	439
Total	189	91	507

We can estimate the allele frequency of the r allele at this locus by assuming that this population is in Hardy–Weinberg proportions and the total progeny observed.

$$\hat{q}^2 = (507/598) = 0.847$$
$$\hat{q}^2 = \sqrt{0.847} = 0.921$$
$$\hat{p} = 1 - \hat{q} = 0.079$$

We can also check to see if the progeny produced by matings with red birds is close to what we would expect if this population was in Hardy–Weinberg proportions. Remember red birds may be either homozygous (RR) or heterozygous (Rr). What proportion of grey progeny do we expect to be produced by matings between two red parents?

Three things must occur for a progeny to be grey: (1) the mother must be heterozygous (Rr); (2) the father must be heterozygous (Rr); and (3) the progeny must receive the recessive allele (r) from both parents:

Prob(progeny rr) = Prob(mother Rr) × Prob(father Rr) × 0.25

The proportion of normal birds in the population who are expected to be hetero-zygous is the proportion of heterozygotes divided by the total proportion of red birds:

Prob(parental bird (Rr) = $(2pq)/(p^2 + 2pq)$ = 0.959

Therefore, the expected proportion of grey progeny is:

0.959 × 0.959 × 0.25 = 0.230

This is fairly close to the observed proportion of 0.178 (5/28). Thus, we would con-clude that this population appears to be in Hardy–Weinberg proportions at this locus.

5.4.2 Null alleles

Null alleles at protein coding loci are alleles that do not produce a detectable protein prod-uct; null alleles at microsatellite loci are alleles that do not produce a detectable PCR (poly-merase chain reation) amplification product. Null alleles at allozyme loci result from alleles that produce either no protein product or a protein product that is enzymatically nonfunctional (Foltz 1986). Null alleles at microsatellite loci result from substitutions that prevent the primers from binding (Brookfield 1996). Heterozygotes for a null allele and another allele appear to be homozygotes on a gel. The presence of null alleles results in an apparent excess of homozygotes relative to Hardy–Weinberg proportions (see Problem 4.3). Brookfield (1996) discusses the estimation of null allele frequencies in the case of more than three alleles.

The familiar ABO blood group locus in humans presents a parallel situation to the case of a null allele in which all genotypes cannot be distinguished. In this case, the I^A and I^B alle-les are codominant, but the I^O allele is recessive (i.e., null). This results in the following relationship between genotypes and phenotypes (blood types):

Genotypes	Blood types	Expected frequency	Observed number
$I^A I^A, I^A I^O$	A	$p^2 + 2pr$	N_A
$I^B I^B, I^B I^O$	B	$q^2 + 2qr$	N_B
$I^A I^B$	AB	$2pq$	N_{AB}
$I^O I^O$	O	r^2	N_O

where p, q, and r are the frequencies of the I^A, I^B, and I^O alleles.

We can estimate allele frequencies at this locus by the expectation maximization (EM) algorithm that finds the allele frequencies that maximize the probability of obtaining the observed data from a sample of a population assumed to be in Hardy–Weinberg propor-tions (Dempster et al. 1977). This is an example of a **maximum likelihood estimate**, which has many desirable statistical properties (Fu and Li 1993).

We could estimate the frequency p directly, as in Example 5.2, if we knew how many individuals in our sample with blood type A were $I^A I^A$ and how many were $I^A I^O$:

$$\hat{p} = \frac{2N_{AA} + N_{AO} + N_{AB}}{2N} \qquad (5.6)$$

where N is the total number of individuals. However, we cannot distinguish the phenotypes of the $I^A I^A$ and $I^A I^O$ genotypes. The EM algorithm solves this ambiguity with a technique known as gene-counting. We start with guesses of the allele frequencies, and use them to calculate the expected frequencies of all genotypes (step E of the EM algorithm), assuming Hardy–Weinberg proportions. Then, we use these genotypic frequencies to obtain new estimates of the allele frequencies, using maximum likelihood (step M). We then use these new allele frequency estimates in a new E step, and so forth, in an iterative fashion, until the values converge.

We first guess the three allele frequencies (remember $p + q + r = 1.0$). The next step is to use these guesses to calculate the expected genotype frequencies assuming Hardy–Weinberg proportions. We next use gene-counting to estimate the allele frequencies from these genotypic frequencies. The count of the I^A alleles is twice the number of $I^A I^A$ genotypes plus the number of $I^A I^O$ genotypes. We expect p^2 of the total individuals with blood type A ($p^2 + 2pr$) to be homozygous $I^A I^A$, and $2pr$ of them to be heterozygous $I^A I^O$. These counts are then divided by the total number of genes in the sample ($2N$) to estimate the frequency of the I^A allele, as we did in expression 5.6. A similar calculation is performed for the I^B allele with the following result:

$$\hat{p} = \frac{\left[2\left(\dfrac{p^2}{p^2 + 2pr} \right)N_A + \left(\dfrac{2pr}{p^2 + 2pr} \right)N_A + N_{AB} \right]}{2N} = \frac{2\left(\dfrac{p + r}{p + 2r} \right)N_A + N_{AB}}{2N}$$

$$\hat{q} = \frac{2\left(\dfrac{q + r}{q + 2r} \right)N_B + N_{AB}}{2N} \qquad (5.7)$$

$$\hat{r} = 1 - \hat{p} - \hat{q}$$

These equations produce new estimates of p, q, and r that can be substituted into the right hand side of the equations in expression 5.7 to produce new estimates of p, q, and r. This iterative procedure is continued until the estimates converge. That is, until the estimated values on the left side are nearly equal to the values substituted into the right side.

5.5 Sex-linked loci

We so far have considered only autosomal loci in which there are no differences between males and females. However, the genotypes of genes on sex chromosomes will often differ between males and females. The most familiar situation is that of genes on the X

	Z-bearing eggs		W-bearing eggs
	Z^A (p)	Z^a (q)	W
Sperm Z^A (p)	$Z^A Z^A$ (p^2)	$Z^A Z^a$ (pq)	$Z^A W$ (p)
Sperm Z^a (q)	$Z^A Z^a$ (pq)	$Z^a Z^a$ (q^2)	$Z^a W$ (q)
	Males		Females

Figure 5.2 Expected genotypic proportions with random mating for a Z-linked locus with two alleles (*A* and *a*).

chromosome of mammals (and *Drosophila*) in which females are homogametic XX and males are heterogametic XY. In this case, genotype frequencies for females conform to the Hardy–Weinberg principle. However, the Y chromosome is largely void of genes so that males will have only one gene copy, and the genotype frequency in males will be equal to the allele frequencies. The situation is reversed in bird species: females are heterogametic ZW and males are homogametic ZZ (Ellegren 2000b). In this case, genotype frequencies for the ZZ males conform to the Hardy–Weinberg principle, and the genotype frequency in the ZW females will be equal to the allele frequencies (Figure 5.2).

Phenotypes resulting from rare recessive X-linked alleles will be much more common in males than in females because q^2 will always be less than q. The most familiar case of this is X-linked red–green color blindness in humans in which approximately 5% of males in some human groups lack a certain pigment in the retina of their eyes so they do not perceive colors as most people do. In this case, $q = 0.05$ and therefore we expect the frequency of color blindness in females to be $q^2 = (0.05)^2 = 0.0025$. Thus, we expect 20 times more red–green color blind males than females in this case.

A variety of other mechanisms for sex determination occur in other animals and plants (Bull 1983). Many plant species possess either XY or ZW systems. The use of XY or ZW indicates which sex is heterogametic (Charlesworth 2002). The sex chromosomes are identified as XY in species in which males are heterogametic and ZW in species in which females are heterogametic. Many reptiles have a ZW system (e.g., all snakes) (Graves and Shetty 2001). A wide variety of genetic sex determination systems are found in fish species (Devlin and Nagahama 2002) and in invertebrates. Some species have no detectable genetic mechanism for sex determination. For example, sex is determined by the temperature in which eggs are incubated in some reptile species (Graves and Shetty 2001).

The classic XY system of mammals and *Drosophila* with the Y chromosome being largely devoid of functional genes, as taught in introductory genetic classes, has been over-generalized. A broader taxonomic view suggests that mammals and *Drosophila* are exceptions and that both sex chromosomes contain many functional genes across a wide variety of animal taxa. Morizot et al. (1987) found that functional genes for the creatine kinase enzyme locus are present on both the Z and W chromosomes of Harris' hawk. Wright and Richards (1983) found that two of 12 allozyme loci that they mapped in the leopard frog were sex linked and that two functional gene copies of both loci are found in XY males. Functional copies of a peptidase locus are present on both the Z and W chromosomes in the salamander *Pleurodeles waltlii* (Dournon et al. 1988). Two allozyme loci in rainbow trout have functional alleles on both the X and Y chromosomes (Allendorf et al. 1994). Differences in allele frequencies between males and females for genes found on both sex chromosomes will result in an excess of heterozygotes in comparison to expected Hardy–Weinberg proportions in the heterogametic sex (Clark 1988; Allendorf et al. 1994).

5.6 Estimation of genetic variation

We often are interested in comparing the amount of genetic variation in different populations. For example, we saw in Table 4.1 that brown bears from Kodiak Island and Yellowstone National Park had less genetic variation than other populations for allozymes, microsatellites, and mtDNA. In addition, comparisons of the amount of genetic variation in a single population sampled at different times can provide evidence for loss of genetic variation because of population isolation and fragmentation due to habitat loss or other causes. In this section we will consider measures that have been used to compare the amount of genetic variation.

5.6.1 Heterozygosity

The average expected (Hardy–Weinberg) heterozygosity at n loci within a population is the best general measure of genetic variation:

$$H_e = 1 - \sum_{i=1}^{n} p_i^2 \qquad (5.8)$$

It is easier to calculate one minus the expected homozygosity, as in expression 5.8, than summing over all heterozygotes because there are fewer homozygous than heterozygous genotypes with three or more alleles. Nei (1987) has referred to this measure as gene diversity, and pointed out that it can be thought of as either the average proportion of heterozygotes per locus in a randomly mating population or the expected proportion of heterozygous loci in a randomly chosen individual. Gorman and Renzi (1979) have shown that estimates of H_e are generally insensitive to sample size and that even a few individuals are sufficient for estimating H_e if a large number of loci are examined. In general, comparisons of H_e among populations are not valid unless a large number of loci are examined.

There are a variety of characteristics of average heterozygosity that make it valuable for measuring genetic variation. It can be used for genes of different ploidy levels (e.g., haploid organelles) and in organisms with different reproductive systems. We will see in later chapters that there is considerable theory available to predict the effects of reduced population size on heterozygosity (Chapter 6), that average heterozygosity is also a good measure of the response of a population to natural selection (Chapter 11), and that it can also provide an estimation of the inbreeding coefficients of individuals (Chapter 14).

5.6.2 Allelic richness

The total number of alleles at a locus has also been used as a measure of genetic variation (see Table 4.1). This is a valuable complementary measure of genetic variation because it is more sensitive to the loss of genetic variation because of small population size than heterozygosity, and it is an important measure of the long-term evolutionary potential of populations (Allendorf 1986) (see Section 6.4).

The major drawback of the number of alleles is that, unlike heterozygosity, it is highly dependent on sample size. Therefore, comparisons between samples are not meaningful unless samples sizes are similar because of the presence of many low frequency alleles in natural populations. This problem can be avoided by using **allelic richness**, which is a measure of allelic diversity that takes into account sample size (El Mousadik and Petit 1996). This measure uses a rarefaction method to estimate allelic richness at a locus for a fixed sample size, usually the smallest sample size if a series of populations are sampled (see Petit et al. 1998). Allelic richness can be denoted by $R(g)$, where g is the number of genes sampled.

The **effective number of alleles** is sometimes used to describe genetic variation at a locus. However, this parameter provides no more information about the number of alleles present at a locus than does heterozygosity. The effective number of alleles is the number of alleles that if equally frequent would result in the observed heterozygosity or homozygosity. It is computed as $A_e = 1 / \Sigma\, p_i^2$ where p_i is the frequency of the ith allele. For example, consider two loci that both have an H_e of 0.50. The first locus has two equally frequent alleles ($p = q = 0.5$), and the second locus has five alleles at frequencies of 0.68, 0.17, 0.05, 0.05, and 0.05. Both of these loci will have the same value of $A_e = 2$.

5.6.3 Proportion of polymorphic loci

The proportion of loci that are polymorphic (P) in a population has been used to compare the amount of variation between populations and species at allozyme loci (see Table 3.5). Strictly speaking, a locus is polymorphic if it contains more than one allele. However, generally some standard definition is used to avoid problems associated with comparisons of samples that are different sizes. That is, the larger the sample, the more likely we are to detect a rare allele. A locus is usually considered to be polymorphic if the frequency of the most common allele is less than either 0.95 or 0.99 (Nei 1987). The 0.99 standard has been used most often, but it is not reasonable to use this definition unless all sample sizes are greater than 50 (which is often not the case).

This measure of variation is of limited value. In some circumstances it can provide a useful measure of another aspect of genetic variation that is not provided by heterozygosity or allelic richness. It has been most valuable in studies of allozyme loci with large sample sizes in which many loci are studied, and many of the loci are monomorphic. However, it is of much less value in studies of highly variable loci (e.g., microsatellites) in which most loci are polymorphic in most populations. In addition, microsatellite loci are often selected to be studied because they are highly polymorphic in the preliminary analysis.

Guest Box 5 Testing alternative explanations for deficiencies of heterozygotes in populations of brook trout in small lakes
Vincent Castric and Louis Bernatchez

Moderate departures from Hardy–Weinberg expected proportions have commonly been ignored in empirical studies of natural populations. Yet, when such departures are real, they may provide important insights into the species' mating system, social behavior, or population genetic structure. Recent advances in statistical population genetics now offer the potential to exploit individual multilocus genotypic information to test more rigorously for possible sources of heterozygote deficiencies.

Populations of brook trout from small lakes in Maine, USA, were found to exhibit stronger heterozygote deficits (higher F_{IS}) than populations from larger lakes at six microsatellite loci (Castric et al. 2001). Technical artifacts, such as null alleles (see Section 5.4.2) or selective amplification of shorter alleles, were unlikely to account for the relationship with lake size since they would be systematic for all samples (see Problem 4.3). Three biologically plausible explanations were subsequently tested by Castric et al. (2002).

First, it is known that brook trout may exhibit a morphologically subtle trophic polymorphism in north temperate lakes. If morphs are somewhat reproductively isolated and genetically divergent, sampling of both forms together would result in fewer heterozygotes than expected in a randomly mating population (Wahlund effect). Using an individual-based maximum likelihood method, we found that the presence of two genetically divergent subpopulations in our sample could not account for the observed heterozygote deficiencies (Figure 5.3a), and so rejected this explanation.

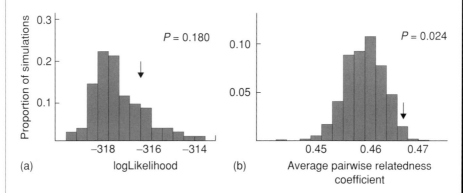

(a) logLikelihood

(b) Average pairwise relatedness coefficient

Figure 5.3 Statistical analysis of genotypes sampled from Clish Lake to test for the causes of a deficit of heterozygotes at six microsatellite loci in brook trout from small lakes. (a) Evidence against the Wahlund effect being the sole explanation for the observed heterozygote deficiency. The maximum likelihood partition in the observed population (arrow) is not significantly more likely than the distribution of maximum likelihood partitions based on randomized populations. (b) Evidence that the observed relatedness among sampled individuals from the same sample from Clish Lake is higher than expected with random mating. The F_{IS} of this sample was 0.153 ($P = 0.007$). From Castric et al. (2002).

Second, mating among relatives could perhaps be more frequent in small lakes and lead to heterozygote deficiencies. In two of the lakes, significantly more fish had low individual multilocus heterozygosity than expected with random mating. Thus, more inbred fish apparently were sampled than expected in a randomly mating population. This suggested that small lakes may bias the reproductive system of this fish towards more frequent matings among relatives.

Third, sampling of genetically related fish would also lead to departures from Hardy—Weinberg proportions. In the same two lakes, the distribution of pairwise individual relatedness coefficients departed from its random expectation (Figure 5.3b), and suggested that perhaps kin groups were sampled.

These results showed that lake size not only affects the number of individuals in a population, but also can affect the mating system.

Problem 5.1

The gel below shows genetic variation at a microsatellite locus (*COCL4*) in 16 majestic mountain whitefish sampled from the Bear River, Utah, USA (A. Whiteley, unpublished data). There are two alleles in this population (*154* and *150*). The two individuals marked with arrows are controls with known genotypes from other populations. Calculate the allele frequencies in this sample of 16 fish and test for conformance to Hardy—Weinberg proportions.

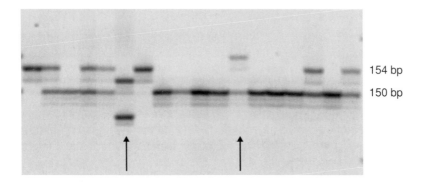

154 bp

150 bp

Problem 5.2

Test the chromosomal inversion types reported in orangutans (see Table 3.3) for conformance to Hardy—Weinberg proportions. You should do a total of four tests, one for each of the two polymorphisms (chromosomes 2 and 9) in both wild- and zoo-born animals. Note: there is an Excel program (*HW program*) on the book website that you can use to do these calculations.

Problem 5.3

The following genotypic data were collected at 34 allozyme loci from brown bears on the mainland of Alaska and Canada (see Table 4.1). Complete the table below. Is there any indication that these genotypes differ from expected Hardy–Weinberg proportions? The most common allele at each locus is designated as 1.

Locus	Genotypes						Allele frequencies			X^2	df	Probability
	11	12	22	13	23	33	1	2	3			
GPI-2	81	2	0	–	–	–	0.988	0.012	0.000			
LDH-A	77	6	0	–	–	–						
PGD	39	30	7	1	3	3						
PGM-1	55	26	2	–	–	–						
TPI-1	67	16	0	–	–	–						
29-loci	83	–	–	–	–	–						

Calculate the mean expected heterozygosity (H_e), and the proportion of loci that are polymorphic (P) for this population sample at these 34 loci. Note: your answer should be similar to the value presented for this sample in Table 4.1.

Problem 5.4

O'Donald and his colleagues have studied the genetics and evolution of color phases of the Arctic skua (*Stercorarius parasiticus*) for many years on Fair Isle, Scotland (O'Donald and Davis 1959, 1975). Some birds have pale plumage with a white neck and body while other birds have a dark brown head and body. The breeding adults and their chicks (normally two per brood) are caught and classified just before flying.

The following results were obtained through 1951–1958:

	Chicks	
Parental types	Pale	Dark
Pale × pale	29	0
Pale × dark	52	86
Dark × dark	25	240
Total	106	326

Describe a single locus model that fits the above inheritance results. Assume that the chicks represent a random sample of the population and use your genetic model to estimate allele frequencies. What proportion of chicks resulting from matings between dark parents are expected to be pale if your genetic model is correct and the population is in Hardy–Weinberg proportions? Does this agree with the observations? (See Example 5.4.)

Problem 5.5

Groombridge et al. (2000) have described the loss of genetic variation in kestrels (*Falco punctatus*) on the Indian Ocean Island of Mauritius by comparing geno-types at 10 microsatellite loci in 26 individuals from museum specimens up to 150 years old and 75 individuals from the extant Mauritius population. Only three of these 10 loci are polymorphic in the extant population (*5*, *82-2*, and *Fu-2*). We will examine these data in more detail in the next chapter. There is an Excel file (*Mauritius kestrel data*) on the course website that presents the original data from these authors.

(a) Test if the genotype frequencies from the extant population are in Hardy–Weinberg proportions using the genotypic data presented in Sheet 1. (Note: do not test the historical sample for Hardy–Weinberg proportions because the sample size is so small and there are so many alleles at some loci.)

(b) Use the allele frequencies presented in Sheet 2 to calculate the average expected heterozygosity (H_e) of the historical and current population of Mauritius kestrels at these 10 loci.

Problem 5.6

Seven of 50 black bears sampled on Princess Royal Island off the coast of British Columbia were white (Marshall and Ritland 2002). As discussed in Chapter 2, white bears are homozygous for a recessive allele at a single autosomal locus (*aa*).

(a) Estimate the frequency of the *a* allele in black bears on this island.

(b) What is the probability that the first cub produced by a mating between a white (*aa*) and a black bear (*AA* or *Aa*) in this population will be white?

Problem 5.7

The Gouldian finch of northern Australia is polymorphic for facial color. Most birds are black-faced, but some birds are red-faced. Inheritance results have indicated that black-faced coloring is caused by a recessive allele (*r*) that is Z-linked (Buckley 1987). The following phenotypes were observed in birds from the Yinberrie Hills (Franklin and Dostine 2000):

	Black	Red	% red
Males	97	45	31.7
Females	93	23	19.8
Total	190	68	26.4

Are the observed frequencies compatible with the proposed model of black-faced coloring being caused by a Z-linked recessive allele?

Problem 5.8

Spruell et al. (1999a) described a null allele at a microsatellite locus (*SSA197*) in pink salmon on the basis of inheritance analysis. A total of 16 alleles were found at this locus, including the null allele. The following genotypes resulted from combining (or binning) all alleles smaller than 150 base pairs into one allelic class (*A*) and binning all alleles greater than 150 base pairs into another allelic class (*B*). Let us call the null allele homozygotes (*O*) so we can estimate allele frequencies using the approach presented in Section 5.4 for the ABO blood group locus.

$$AA \quad AB \quad BB \quad OO$$
$$7 \quad 11 \quad 11 \quad 2$$

Use the Excel program *ABO estimation* on the course website to estimate the frequency of the null allele in this sample.

Problem 5.9

We saw in Section 5.3 that an examination of the pattern across samples and loci can be important for identifying possible causes for observed deviations from Hardy–Weinberg proportions. Let us again consider a situation where eight deviations ($P < 0.05$) from Hardy–Weinberg proportions are detected in a study of 10 loci examined in 10 population samples. Assume that all eight of these deviations are found in the sample population and there is an excess of heterozygotes in all cases. Suggest a possible explanation for this observation.

Land snail, Example 6.2

6

Small Populations and Genetic Drift

The race is not always to the swift, nor the battle to the strong, for time and chance happens to us all.

Ecclesiastes 9:11

Whether our concern is the wild relatives of cultivated plants or wild animals, the conservationist is faced with the ultimate sampling problem – how to preserve genetic variability and evolutionary flexibility in the face of diminishing space and with very limited economic resources. Inevitably we are concerned with the genetics and evolution of small populations, and with establishing practical guidelines for the practicing conservation biologist.

Sir Otto H. Frankel and Michael E. Soulé (1981, p. 31)

Genetic change will not occur in populations if all the assumptions of the Hardy–Weinberg equilibrium are met (Section 5.1). In this and the next several chapters we will see what happens when the assumptions of Hardy–Weinberg equilibrium are violated. In this chapter, we will examine what happens when we violate the assumption of infinite population size. That is, what will be the effect on allele and genotype frequencies when population size (N) is finite?

All natural populations are finite so genetic drift will occur in all natural populations, even very large ones. For example, consider a new mutation that occurs in an extremely large population of insects that numbers in the millions. Whether or not the single copy of a new mutation is lost from this population will be determined primarily by the sampling process that determines what alleles are transmitted to the next generation. Thus, the fate of this rare allele in even an extremely large population will be determined by genetic drift.

Understanding genetic drift and its effects is extremely important for conservation. Fragmentation and isolation due to habitat loss and modification has reduced the population size of many species of plants and animals throughout the world. We will see in future chapters how genetic drift is expected to affect genetic variation is these populations. More importantly, we will consider how genetic drift may reduce the fitness of individuals in these populations and limit the evolutionary potential of these populations to evolve by natural selection.

6.1 Genetic drift

Genetic drift is random change in allele frequencies from generation to generation because of sampling error. That is, the finite number of genes transmitted to progeny will be an imperfect sample of the allele frequencies in the parents (Figure 6.1). The mathematical treatment of genetic drift began with R. A. Fisher (1930) and Sewall Wright (1931) who considered the effects of binomial sampling in small populations. This model is therefore often referred to as the **Wright–Fisher** or Fisher–Wright model. However, Fisher and Wright strongly disagreed on the importance of drift in bringing about evolutionary change. Genetic drift is often called the "Sewall Wright effect" in recognition that the importance of drift in evolution was largely introduced by Wright's arguments.

It is often helpful to consider extreme situations in order to understand the expected effects of relaxing assumptions on models. Consider the example of a plant species capable of self-fertilization with a constant population size of $N = 1$, consisting of a single individual of genotype Aa; the allele frequency in this generation is 0.5. We cannot predict what the allele frequency will be in the next generation because the genotype of the single individual in the next generation will depend upon which alleles are transmitted via the chance elements of Mendelian inheritance. However, we do know that the allele frequency in the next generation has to be 0.0, 0.50, or 1.0 because the only three possible genotypes are AA, Aa, or aa. In fact, based upon Mendelian expectations there is a 50% probability that allele frequency will become either zero or one in the next generation.

Genetic drift is an example of a **stochastic** process in which the actual outcome cannot be predicted because it is affected by random elements (chance). Tossing a coin is one example of a stochastic process. One-half of the time, we expect a head to result, and one-half of the time we expect a tail. However, we do not know what the outcome of any specific coin toss will be. We can mimic or simulate the effects of genetic drift by using a

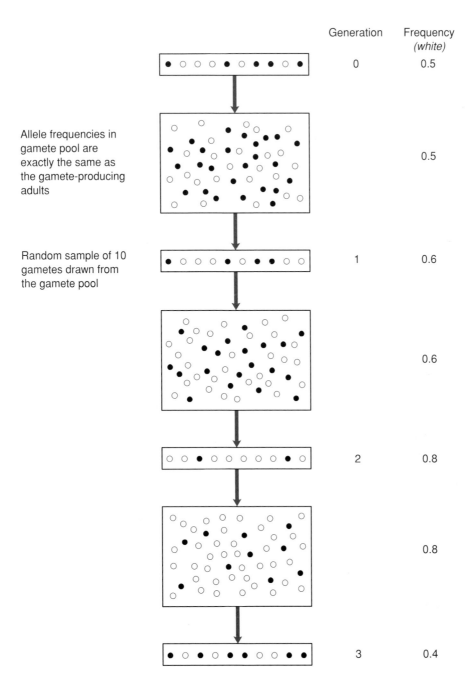

Figure 6.1 Random sampling of gametes resulting in genetic drift in a population. Allele frequencies in the gamete pools (large boxes) are assumed to reflect exactly the allele frequencies in the adults of the parental generation (small boxes). The allele frequencies fluctuate from generation to generation because the population size is finite ($N = 5$). From Grauer and Li (2000).

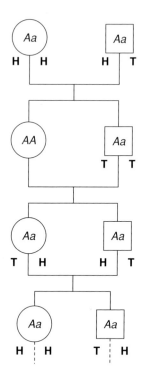

Figure 6.2 Simulation of genetic drift in a population consisting of a single female (circle) and male (square) each generation. A coin is tossed twice to simulate the two gametes produced by each heterozygote. A head (**H**) indicates that the *A* allele is transmitted and a tail (**T**) indicates the *a* allele. Homozygotes always transmit the allele for which they are homozygous.

series of coin tosses. Consider a population initially consisting of two heterozygous (*Aa*) individuals, one male and one female. Heterozygotes are expected to transmit the *A* and *a* alleles with equal probability to each gamete. A coin is tossed to specify which allele is transmitted by heterozygotes; an outcome of head (H) represents an *A* allele; and a tail (T) represents an *a*. No coin toss is needed for homozygous individuals since they will always transmit the same allele.

The results of one such simulation using these rules are shown in Table 6.1 and Figure 6.2. In the first generation, the female transmitted the *A* allele to both progeny because both coin tosses resulted in heads. The male transmitted an *A* allele to his daughter and an *a* allele to his son because the coin tosses resulted in a head and then a tail. Thus, the allele frequency (*p*) changed from 0.5 in the initial generation to 0.75 in the first generation, and the expected heterozygosity in the population changed as well. This process is continued until the seventh generation when both individuals become homozygous for the *a* allele, and, thus, no further gene frequency changes can occur.

Table 6.1 shows one of many possible outcomes of genetic drift in a population with two individuals. However, we are nearly certain to get a different result if we start over again. In addition, it would be helpful to simulate the effects of genetic drift in larger

Table 6.1 Simulation of genetic drift by coin tossing in a population of one female and one male over seven generations. A coin is tossed twice to specify which alleles are transmitted by heterozygotes; an outcome of head (H) represents an A allele; and a tail (T) represents an a. The first toss represents the allele transmitted to the female in the next generation and the second toss the male (as shown in Figure 6.2). p is the frequency of the A allele. The observed and expected heterozygosities (assuming Hardy–Weinberg proportions) are also shown.

Generation	Mother	Father	p	H_0	H_e
0	Aa (HH)	Aa (HT)	0.50	1.000	0.500
1	AA	Aa (TT)	0.75	0.500	0.375
2	Aa (TH)	Aa (HT)	0.50	1.000	0.500
3	Aa (HH)	Aa (TH)	0.50	1.000	0.500
4	Aa (TT)	AA	0.75	0.500	0.375
5	aA (HT)	aA (TT)	0.50	1.000	0.500
6	Aa (TT)	aa	0.25	0.500	0.375
7	aa	aa	0.00	0.000	0.000

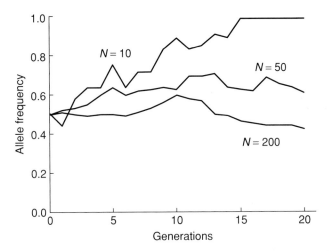

Figure 6.3 Results of computer simulations of changes in allele frequency by genetic drift for each of three population sizes (N) with an initial allele frequency of 0.5.

populations. In principle, this can be done by tossing a coin; however, it quickly becomes extremely time consuming.

A better way to simulate genetic drift is through computer simulations. Computational methods are available to produce a random number that is uniformly distributed between zero and one. This random number can be used to determine which allele is transmitted by a heterozygote. For example, if the random number is in the range of 0.0 to 0.5, we can specify that the A allele is transmitted; similarly, a random number in the range of 0.5 to 1.0 would specify an a allele. Models such as this are often referred to as Monte Carlo simulations in reference to the gambling tables in Monte Carlo. Figure 6.3 shows changes in allele frequencies in three populations of different sizes as simulated with a computer. The smaller the population size, the greater are the changes in allele frequency due to drift (compare N of 10 and 200).

The sampling process that we have examined here has two primary effects on the genetic composition of small populations:

1   Allele frequencies will change.
2   Genetic variation will be lost.

We can model genetic changes in small populations either by changes in allele frequencies or increase in homozygosity caused by inbreeding. As allele frequencies change because of genetic drift, heterozygosity is expected to decrease (and homozygosity increase). For example, heterozygosity became zero in generation 16 with N of 10 because only one allele remained in the population. Once such a "fixation" of one allele or another occurs, it is permanent; only mutation (Chapter 12) or gene flow (Chapter 9) from another population can introduce new alleles. We will consider the effects of genetic drift on both allele frequencies and genetic variation in the next two sections.

6.2 Changes in allele frequency

We cannot predict the direction of change in allele frequencies from generation to generation because genetic drift is a random process. The frequency of any allele is equally likely to increase or decrease from one generation to the next because of genetic drift. Although we cannot predict the direction of change in allele frequency, we can describe the expected magnitude of the change in allele frequency. In general, the smaller the population, the greater the change in allele frequency that is expected (Figure 6.3).

The change in allele frequencies from one generation to the next because of genetic drift is a problem in sampling. A finite sample of gametes is drawn from the parental generation to produce the next generation. Both the sampling of gametes and the coin toss can be described by the binomial sampling distribution (see Appendix Section A2.1). The variance of change in allele frequency from one generation to the next is thus the binomial sampling variance:

$$V_q = \frac{pq}{2N}$$

Given that the current allele frequency is p with a population size of N, there is approximately a 95% probability that the allele frequency in the next generation will be in the interval:

$$p' = p \pm 2 \sqrt{\frac{(pq)}{(2N)}} \tag{6.1}$$

For example, with an allele frequency of 0.50 and an N of 10, the allele frequency in the next generation will be in the interval 0.28 to 0.72 with 95% probability (expression 6.1). In contrast, with a p of 0.5 and an N of 200, this interval is only 0.45–0.55.

6.3 Loss of genetic variation: the inbreeding effect of small populations

Genetic drift is expected to cause a loss of genetic variation from generation to generation. **Inbreeding** occurs when related individuals mate with one another. Inbreeding is one consequence of small population size; see Chapter 13 for a detailed consideration of inbreeding. For example, in an animal species with $N = 2$, the parents in each generation will be full sibs (that is, brother and sister). Matings between relatives will cause a loss of genetic variation. The **inbreeding coefficient** (f) is the probability that the two alleles at a locus within an individual are identical by descent (that is, identical because they are derived from a common ancestor in a previous generation). We will consider several different inbreeding coefficients that have specialized meaning (e.g., F_{IS}, F_{ST}, etc. in Chapter 9 and F in Chapter 13). We will use the general inbreeding coefficient f in this chapter as defined above, along with its counterpart heterozygosity (h), which is equal to $1 - f$.

In general, the increase in homozygosity due to genetic drift will occur at the following rate per generation:

$$\Delta f = \frac{1}{2N} \tag{6.2}$$

This effect was first discussed by Gregor Mendel who pointed out that only half of the progeny of a heterozygous self-fertilizing plant will be heterozygous; one-quarter will be homozygous for one allele and the remaining one-quarter will be homozygous for the other allele (Table 6.2). This is as predicted by expression 6.2 ($N = 1$, $\Delta f = 0.50$).

We have seen that the expected rate of loss of heterozygosity per generation is $\Delta f = 1/2N$; therefore, after t generations:

$$f_t = 1 - (1 - \frac{1}{2N})^t \tag{6.3}$$

Table 6.2 This table appeared in Mendel's original classic paper in 1865. He was considering the expected genotypic ratios in subsequent generations from a single hybrid (i.e., heterozygous) individual that reproduced by self-fertilization. Mendel assumed that each plant in each generation had four offspring. The homozygosity (Homo) and heterozygosity (Het) columns did not appear in the original paper.

Generation	AA	Aa	aa	Ratio AA : Aa : aa	Homo	Het
1	1	2	1	1 : 2 : 1	0.500	0.500
2	6	4	6	3 : 2 : 3	0.750	0.250
3	28	8	28	7 : 2 : 7	0.875	0.125
4	120	16	120	15 : 2 : 15	0.938	0.062
5	496	32	496	31 : 2 : 31	0.939	0.031
n				$2^n - 1 : 2 : 2^n - 1$	$1 - (1/2)^n$	$(1/2)^n$

f_t is the expected increase in homozygosity at generation t and is known by a variety of names (e.g., **autozygosity**, **fixation index**, or the inbreeding coefficient) depending upon the context in which it is used.

It is often more convenient to keep track of the amount of variation remaining in a population using h (heterozygosity), where:

$$f = 1 - h \tag{6.4}$$

Therefore, the expected decline in h per generation is:

$$\Delta h = -\frac{1}{2N} \tag{6.5}$$

so that after one generation:

$$h_{t+1} = (1 - \frac{1}{2N})h_t \tag{6.6}$$

The heterozygosity after t generations can be found by:

$$h_t = (1 - \frac{1}{2N})^t h_0 \tag{6.7}$$

where h_0 is the initial heterozygosity (Example 6.1).

Example 6.1 Mauritius kestrels bottleneck

Kestrels on the Indian Ocean Island of Mauritius went through a bottleneck of one female and one male in 1974 (Nichols et al. 2001). The population had fewer than 10 birds throughout the 1970s, and there were less than 50 birds in this population for many years because of the widespread use of pesticides from 1940 to1960. However, this population grew to nearly 500 birds by the mid-1990s. Nichols et al. (2001) examined the loss in genetic variation in this population at 10 microsatellite loci by comparing living birds to 26 ancestral birds from museum skins that were up to 170 years old. The heterozygosity of the restored population was 0.099 compared to heterozygosity in the ancestral birds of 0.231. The amount of heterozygosity expected to remain in Mauritius kestrels after one generation of a bottleneck of $N = 2$ can be estimated using expression 6.6:

$$(1 - \frac{1}{2N})h_t = (1 - \frac{1}{4})(0.231) = 0.173$$

We can use expression 6.7 to see that the amount of heterozygosity in the restored population of Mauritius kestrels is approximately the same as would we would expect after a bottleneck of two individuals for three generations:

$$(1 - \frac{1}{2N})^t h_0 = (0.75)^3 (0.231) = 0.097$$

The actual bottleneck in Mauritius kestrels was almost certainly longer than three generations with more birds than two birds each generation. However, the expressions in this chapter all assume discrete generations and cannot be applied directly to species such as the Mauritius kestrels that have overlapping generations.

Figure 6.4 shows this effect at a locus with two alleles and an initial frequency of 0.5 in a series of computer simulations of eight populations that consist of 20 individuals each. These 20 individuals possess 40 gene copies at any given locus. Forty gametes must be drawn from these 40 parental gene copies to form the next generation. The genotype of any one selected gamete does not affect the probability of the next gamete that is drawn; this is similar to a coin toss where one outcome does not affect the probability of the next toss.

Two of the eight populations simulated became fixed for the A allele and one became fixed for the a allele. Both of the alleles were retained by five of the populations after 20 generations. The heterozygosity in each of the populations is shown in Figure 6.5. There are large differences among populations in the decline in heterozygosity over time. Nevertheless, the mean decline in heterozygosity for all eight populations is very close to that predicted with expression 6.6.

The heterozygosity at any single locus with two alleles is equally likely to increase or decrease from one generation to the next (except in the case when the allele frequencies are at 0.5). This may seem counterintuitive in view of expression 6.6, which describes a monotonic decline in heterozygosity. Heterozygosity at a locus with two alleles is at a

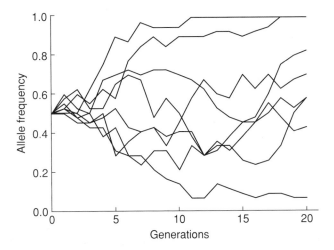

Figure 6.4 Computer simulations of genetic drift at a locus having two alleles with initial frequencies of 0.5 in eight populations of 20 individuals each.

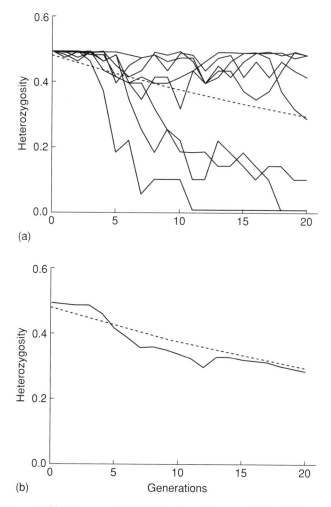

Figure 6.5 (a) Expected heterozygosities $(2pq)$ in the eight populations $(N=20)$ undergoing genetic drift as shown in Figure 6.3. The dashed line shows the expected change in heterozygosity using expression 6.6. (b) Mean heterozygosity for all loci (solid line) and the expected heterozygosity using expression 6.6 (dashed line).

maximum when the two alleles are equally frequent $(p=q=0.5)$ (Figure 6.6). The frequency of any particular allele is equally likely to increase or decrease due to genetic drift. Thus, heterozygosity will increase if the allele frequency drifts towards 0.5, and it will decrease if the allele frequency drifts toward 0 or 1. However, the expected net loss is greater than the net gain in heterozygosity in each generation by $1/2N$.

6.4 Loss of allelic diversity

We have so far measured the loss of genetic variation caused by small population size by the expected reduction in heterozygosity (h). There are other ways to measure genetic

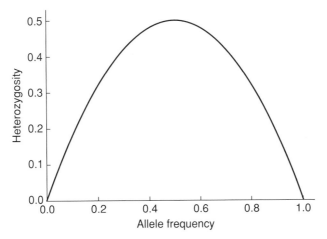

Figure 6.6 Expected heterozygosity (2pq) at a locus with two alleles as a function of allele frequency.

variation and its loss. A second important measure of genetic variation is the number of alleles present at a locus (A). There are advantages and disadvantages to both of these measures.

Heterozygosity has been widely used because it is proportional to the amount of genetic variance at a locus, and it lends itself readily to theoretical considerations of the effect of limited population size on genetic variation. In addition, the expected reduction in heterozygosity because of genetic drift is independent of the number of alleles present. Finally, estimates of heterozygosity from empirical data are relatively insensitive to sample size, whereas estimates of the number of alleles in a population are strongly dependent upon sample size. Therefore, comparisons of heterozygosities in different species or populations are generally more meaningful than comparisons of the number of alleles detected.

Nevertheless, heterozygosity has the disadvantage of being relatively insensitive to the effects of bottlenecks (Allendorf 1986). The difference between heterozygosity and n is greatest with extremely small populations (Figure 6.7). For example, a population with two individuals is expected to lose only 25% (1/2N = 25%) of its heterozygosity. Thus, 75% of the heterozygosity in a population will be retained even through such an extreme **bottleneck**. However, two individuals can possess a maximum of four different alleles. Thus, considerably more of the allelic variation may be lost during a bottleneck if there are many alleles present at a locus.

The effect of a bottleneck on the number of alleles present is more complicated than the effect on heterozygosity because it is dependent on the number and frequencies of alleles present (Allendorf 1986) (Question 6.1). The probability of an allele being lost during a bottleneck of size N is:

$$(1 - p)^{2N} \tag{6.8}$$

where p is the frequency of the allele. This is the probability of sampling all of the gametes to create the next generation (2N) without selecting at least one copy of the allele in question. Rare alleles (say $p < 0.10$) are especially susceptible to loss during a bottleneck.

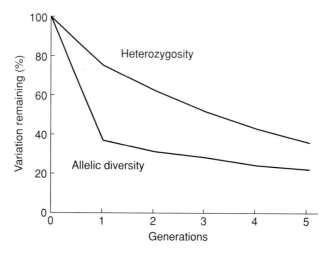

Figure 6.7 Simulated loss of heterozygosity and allelic diversity at eight microsatellite loci during a bottleneck of two individuals for five generations. The original allele frequencies are from a population of brown bears from the Western Brooks Range of Alaska. Redrawn from Luikart and Cornuet (1998).

However, the loss of rare, potentially important, alleles will have little effect on heterozygosity. For example, an allele at a frequency of 0.01 has a 60% chance of being lost following a bottleneck of 25 individuals (expression 6.8). Figure 6.8 shows the probability of the loss of rare alleles during a bottleneck of N individuals.

> **Question 6.1** We have seen that the proportion of heterozygosity remaining after a bottleneck of a single generation is $(1 - 1/2N)$, regardless of the number of alleles present and their frequencies. However, the amount of allelic diversity expected to remain after a bottleneck does depend on allele frequencies. Show that the number of alleles expected to remain after a bottle of $N = 2$ differs at a locus with two alleles if one of the alleles is rare (say $q = 0.1$ and $p = 0.9$) in comparison to when the two alleles are equally frequent ($p = q = 0.5$).

In general, if a population is reduced to N individuals for one generation, then the expected total number of alleles (A') remaining is:

$$E(A') = A - \sum_{j=1}^{A} (1 - p_j)^{2N} \qquad (6.9)$$

where A is the initial number of alleles and p_j is the frequency of the jth allele. For example, consider a locus with two alleles at frequencies of 0.9 and 0.1 and a bottleneck of just two individuals. In this case:

$$E(A') = 2 - (1 - 0.9)^4 - (1 - 0.1)^4 = 1.34$$

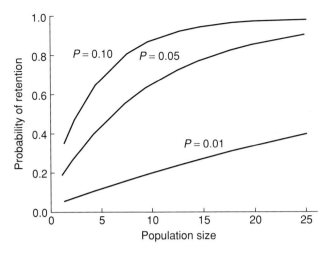

Figure 6.8 Probability of retaining a rare allele ($P = 0.01, 0.05$, or 0.10) after a bottleneck of size N for a single generation (expression 6.8).

Thus, on the average, we expect to lose one of these two alleles nearly two-thirds of the time. In contrast, there is a much greater expected probability of retaining both alleles at a locus with two alleles if the two alleles are equally frequent:

$$E(A') = 2 - (1 - 0.5)^4 - (1 - 0.5)^4 = 1.88$$

Thus, the expected loss of alleles during a bottleneck depends upon the number and frequencies of the alleles present. This is in contrast to heterozygosity, which is lost at a rate of $1/2N$ regardless of the current heterozygosity.

The loss of alleles during a bottleneck will have a drastic effect on the overall genotypic diversity of a population. As we saw in Section 4.2, the number of genotypes grows very quickly as the number of alleles increases. For example, the number of possible genotypes at a locus with 2, 5, and 10 alleles is 3, 15, and 55, respectively. Thus, the loss of alleles during a bottleneck will greatly reduce the genotypic diversity in a population that is subject to natural selection.

6.5 Founder effect

The founding of a new population by a small number of individuals will cause abrupt changes in allele frequency and loss of genetic variation (Example 6.2). Such severe "bottlenecks" in population size are a special case of genetic drift. Perhaps surprisingly, however, even extremely small bottlenecks have relatively little effect on heterozygosity. For example, with sexual species the smallest possible bottleneck is $N = 2$. Even in this extreme case, the population will only lose 25% of its heterozygosity (see expression 6.5). Stated in another way, just two individuals randomly selected from any population, regardless of size, will contain 75% of the total heterozygosity in the original population. We can also use expression 6.5 to estimate the size of the founding population if we know how much heterozygosity has been lost through the founding bottleneck (Question 6.2).

Example 6.2 Effects of founding events on allelic diversity in a snail

The land snail *Theba pisana* was introduced from Europe into western Australia in the 1890s. A colony was founded in 1925 on Rottnest Island with animals taken from the mainland population near Perth. Johnson (1988) reported the allele frequencies at 25 allozyme loci. Figure 6.9 shows the loss of rare alleles caused by the bottleneck associated with the founding of a population in Perth on the

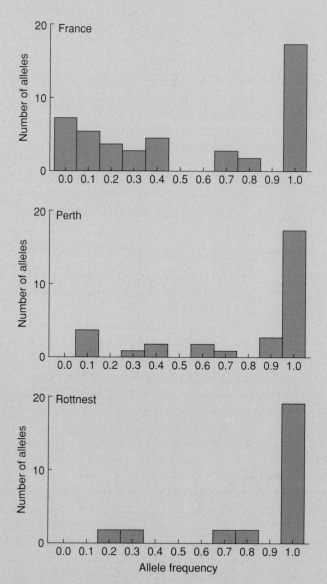

Figure 6.9 Effects of bottlenecks on the number of rare alleles at 25 allozyme loci in the land snail *Theba pisana* that was introduced from Europe into western Australia in the 1890s. Data from Johnson (1988).

mainland and in the second bottleneck associated with the founding of a population on nearby Rottnest Island. The height of each bar represents the number of alleles in that sample that had the frequency specified on the *x*-axis. For example, there were eight alleles that had a frequency of less than 0.05 in the founding French population. However, there were no alleles in either of the two Australian populations at a frequency of less than 0.05.

The distribution of allele frequencies, such as plotted in Figure 6.9, can be used to detect bottlenecks even when data are not available from the pre-bottlenecked population (Luikart et al. 1998). Rare alleles (frequency less than 0.05) are expected to be common in samples from populations that have not been bottlenecked in their recent history, such as observed in the French sample (see Chapter 12). The complete absence of such rare alleles in Australia would have suggested that these samples came from recently bottlenecked populations even if the French sample was not available for comparison.

Question 6.2 We will see in Example 6.3 that the founding of a new population of Laysan finches on Southeast Island resulted in a reduction in average heterozygosity of 8% in comparison to the founding population on Laysan Island. Assume that all of the loss in heterozygosity was caused by the initial founding event; that is, assume that there has been no additional loss of heterozygosity on Southeast Island because of small population size. What is the number of founders that would be expected to produce the observed result of an 8% reduction in heterozygosity?

A laboratory experiment with guppies demonstrates this effect clearly (Nakajima et al. 1991). Sixteen separate **subpopulations** were derived from a large random mating laboratory colony of guppies by mating a female with a single male. After four generations, each of these subpopulations contained more than 500 individuals. Approximately 45 fish were then sampled from each subpopulation and genotyped at two protein loci that were polymorphic in the original colony (Table 6.3).

The mean heterozygosity at both loci in these 16 subpopulations was 0.358, in comparison to heterozygosity in the original colony of 0.482. Thus, the mean heterozygosity in the subpopulations, following a bottleneck of two individuals, was 26% lower than in the population from which the subpopulations were founded. This agrees very closely with the prediction from expression 6.5 that predicts a 25% reduction following a bottleneck of two individuals.

The total amount of heterozygosity lost during a bottleneck depends upon how long it takes the population to return to a "large" size (Nei et al. 1975). That is, species such as guppies, in which individual females may produce 50 or so progeny, may quickly attain large enough population sizes following a bottleneck so that little further variation is lost following the initial bottleneck. However, species with lower population growth rates may persist at small population sizes for many generations during which heterozygosity is further eroded.

Table 6.3 Allele frequencies (p) and heterozygosities (h) at two loci in 16 subpopulations of guppies founded by a single female and male. H_e is the mean heterozygosity at the two loci.

Subpopulation	AAT-1		PGM-1		
	p	h	p	h	H_e
1	0.521	0.499	0.677	0.437	0.468
2	0.738	0.387	0.600	0.480	0.433
3	0.377	0.470	0.131	0.227	0.349
4	0.915	0.156	0.939	0.114	0.135
5	0.645	0.458	0.638	0.461	0.460
6	0.571	0.490	0.548	0.495	0.492
7	0.946	0.102	0.833	0.278	0.190
8	0.174	0.287	0.341	0.449	0.368
9	0.617	0.473	0.500	0.500	0.486
10	0.820	0.295	0.640	0.461	0.378
11	0.667	0.444	0.917	0.152	0.298
12	0.219	0.342	0.531	0.498	0.420
13	1.000	0	0.838	0.272	0.136
14	0.250	0.375	0.853	0.251	0.313
15	0.375	0.469	0.740	0.385	0.427
16	0.152	0.258	0.582	0.486	0.372
Average	0.562	0.344	0.644	0.372	0.358
Original colony	0.581	0.487	0.605	0.478	0.482

The growth rate of a population following a bottleneck can be modeled using the so-called logistic growth equation that describes the size of a population after t generations based upon the initial population size (N_0), the intrinsic growth rate (r), and the equilibrium size of the population (K):

$$N(t) = \frac{K}{1 + be^{-rt}} \qquad (6.10)$$

The constant e is the base of the natural logarithms (approximately 2.72), and b is a constant equal to $(K - N_0)/N_0$.

We can estimate the total expected loss in heterozygosity in the guppy example depending upon the rate of population growth of the subpopulations. The initial size of the subpopulations (N_0) was 2, and we assume the equilibrium size (K) was 500. We can then examine three different intrinsic growth rates (r): 1.0, 0.5, and 0.2. An r of 1.0 indicates that population size is increasing by a factor of 2.72 (e) each generation when population size is far below K. Similarly, rs of 0.5 and 0.2 indicate growth rates of 1.65 and 1.22 at small population sizes, respectively. A detailed discussion of use of the logistic equation to describe population growth can be found in chapters 15 and 16 of Ricklefs and Miller (2000).

Expression 6.10 can be used to predict the expected population size in each generation following the bottleneck. We expect heterozygosity to be eroded at a rate of $1/2N$ in each

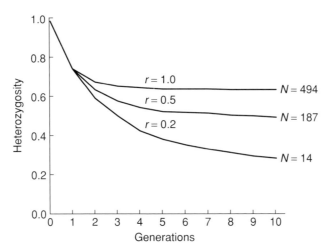

Figure 6.10 Expected heterozygosities in subpopulations of guppies going through a bottleneck of two individuals and growing at different rates (r) according to the logistic growth equation (expression 6.10). N is the expected population size for each growth rate at the 10th generation.

of these generations. Figure 6.10 shows the expected loss in heterozygosity in our guppy example for 10 generations following the bottleneck. As expected, populations having a relatively high growth rate ($r = 1.0$) will lose little heterozygosity following the initial bottleneck. However, heterozygosity is expected to continue to erode even 10 generations following the bottleneck in populations with the slowest growth rate. In general, bottlenecks will have a greater and more long lasting effect on the loss of genetic variation in species with smaller intrinsic growth rates (e.g., large mammals) than species with high intrinsic growth rates (e.g., insects).

Founder events and population bottlenecks will have a greater effect on the number of alleles in a population than on heterozygosity (see Figure 6.6). Some classes of loci in vertebrates have been found to have many nearly equally frequent alleles. For example, Gibbs et al. (1991) have described 37 alleles at the hypervariable major histocompatibility (*MHC*) locus in a sample of 77 adult blackbirds from Ontario (Figure 6.11). As we have seen, two birds chosen at random from this population are expected to contain 75% of the heterozygosity. However, two birds can at best possess four of the 37 different *MHC* alleles. Thus, at least 33 of the 37 detected alleles (89%) will be lost in a bottleneck of two individuals (Question 6.3). Thus, bottlenecks of short duration may have little effect on heterozygosity but will reduce severely the number of alleles present at some loci (Example 6.3).

Question 6.3 Gibbs et al. (1991) have described 37 nearly equally frequent alleles at the hypervariable major histocompatibility (*MHC*) locus in a sample of 77 adult blackbirds from Ontario (Figure 6.11). How much genetic variation would you expect to be retained at this locus if this population went through a very small population bottleneck (say 10 breeding individuals) and then quickly recovered to over 100 individuals? Consider both heterozygosity and allelic diversity.

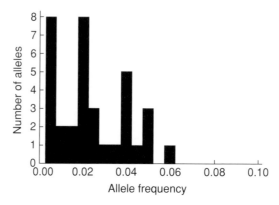

Figure 6.11 Distribution of frequencies of alleles at the highly variable *MHC* locus detected in red-wing blackbirds. The height of each bar represents the number of the alleles that had the frequency specified on the *x*-axis. Thus, just over 20% of the 37 alleles had frequencies of less than 1% in the sample. Compare this to Figure 6.9 which is the equivalent plot of allele frequencies at much less variable allozyme loci in a land snail; note the difference in the scale of the allele frequency axis. Data from Gibbs et al. (1991).

Example 6.3 Founding events in the Laysan finch

The Laysan finch is an endangered Hawaiian honeycreeper found on several islands in the Pacific Ocean (Tarr et al. 1998). The species underwent a bottle-neck of approximately 100 birds on Laysan Island after the introduction of rabbits in the early 1900s (Figure 6.12). The population recovered rapidly after eradica-tion of the rabbits and has fluctuated around a mean of 10,000 birds since 1968. In 1967, the US Fish and Wildlife Service translocated 108 finches to Southeast Island, one of several small islets approximately 300 km northwest of Laysan that comprise Pearl and Hermes Reef. The translocated population declined to 30–50 birds and then rapidly increased to some 500 birds on Southeast Island. Several smaller populations have since become established in other islets within the reef. Two birds colonized Grass Island in 1968 and six more finches were moved to this islet in 1970. The population of birds on Grass Island has fluctuated between 20 and 50 birds. In 1973, a pair of finches founded a population on North Island. The population of birds on North Island has fluctuated between 30 and 550 birds.

Tarr et al. (1998) assayed variation at nine microsatellite loci to examine the effects of the founder events and small population sizes in these four populations (Table 6.4). Their empirical results are in close agreement with theoretical expecta-tions. All three newly founded populations have fewer alleles than the founding population on Laysan. The average heterozygosity on Southeast Island is approx-imately 8% less than on Laysan Island; the heterozygosities on the two other islands are approximately 30% less than the original founding population on Laysan.

However, heterozygosities at four of the nine loci are actually greater in the post-bottleneck population on Southeast Island than on Laysan; see the discus-sion in Section 6.3. Thus, it is important to examine many loci to detect and quant-ify the effects of bottlenecks in populations on heterozygosity.

Table 6.4 Numbers of alleles and observed heterozygosities at nine microsatellite loci in Laysan finches in four island populations. The number in parentheses after the name of the island is the sample size. The populations on Southeast, North, and Grass Islands were all founded from birds from Laysan.

Locus	Laysan (44)		Southeast (43)		North (43)		Grass (36)	
	Alleles	Heterozygosity	Alleles	Heterozygosity	Alleles	Heterozygosity	Alleles	Heterozygosity
Tc.3A2C	2	0.558	2	0.535	2	0.535	2	0.528
Tc.4A4E	2	0.386	2	0.605	2	0.209	2	0.556
Tc.5A1B	3	0.372	3	0.233	1	0	2	0.583
Tc.5A5A	3	0.409	2	0.071	2	0.372	2	0.278
Tc.1A4D	3	0.659	3	0.744	3	0.698	2	0.528
Tc.11B1C	3	0.636	3	0.674	3	0.628	3	0.194
Tc.11B2E	3	0.614	3	0.488	1	0	2	0.500
Tc.11B4E	4	0.614	4	0.628	2	0.256	3	0.444
Tc.12B5E	5	0.568	4	0.442	3	0.372	1	0
All loci	3.11	0.535	2.89	0.491	2.11	0.341	2.11	0.401
		100%		92%		64%		75%

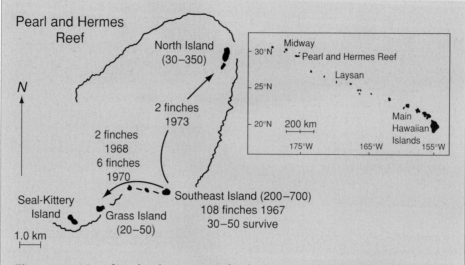

Figure 6.12 Map of Pearl and Hermes Reef and colonization history of the three finch populations from Example 6.3. Ranges of population size on the islands are shown in parentheses. The inset is the Hawaiian Archipelago. Redrawn from Tarr et al. (1998).

6.6 Genotypic proportions in small populations

We saw in the guppy example (see Table 6.3) that the separation of a large random mating population into a number of subpopulations can cause a reduction in heterozygosity, and a corresponding increase in homozygosity. However, genotypes within each subpopulation will be in Hardy–Weinberg proportions as long as random mating occurs within the subpopulations. It may seem paradoxical that heterozygosity is decreased in small populations, but the subpopulations themselves remain in Hardy–Weinberg proportions. The explanation is that the reduction in heterozygosity is caused by changes in allele frequency from one generation to the next, while Hardy–Weinberg genotypic proportions will occur in any one generation as long as mating is random.

In fact, there actually is a tendency for an **excess** of heterozygotes in small populations of animals and plants with separate sexes (Example 6.4). Different allele frequencies in the two sexes will cause an excess in heterozygotes relative to Hardy–Weinberg proportions (Robertson 1965; Kirby 1975; Brown 1979). An extreme example of this is that of hybrids produced by males from one strain (or species) and females from another so that all progeny are heterozygous at any loci where the two strains differ. In this case, however, genotypic proportions will return to Hardy–Weinberg proportions in the next generation.

Example 6.4 Small populations of bull trout

The expected excess of heterozygotes in small populations can sometimes be used to detect populations with a small number of individuals. For example, Leary et al. (1993a) examined bull trout from a hatchery population at four polymorphic allozyme loci and found a strong tendency for an excess of heterozygotes (Table 6.5; see Example 5.1). On the average, there was a 38% excess of heterozygotes at four polymorphic loci; a 25% excess of heterozygotes would be expected if all of the progeny came from just two individuals (see expression 6.8).

Further examination of genotype frequencies suggests that these fish were produced by a very small number of parents. The exceptionally high proportion of heterozygotes (0.88) at *mIDHP-1* suggests that most fish came from a single-pair mating between individuals homozygous for the two different alleles at this locus; all progeny from such a mating will be heterozygous at the locus. Allele frequencies at the four polymorphic loci also support the inference that most of these fish resulted from a single-pair mating. The only allele frequencies possible in a full-sib family are 0.00, 0.25, 0.50, 0.75, and 1.00 because the two parents possess four copies of each gene (see Table 6.6); the allele frequencies at all four loci are near these values.

When we asked about the source of these fish after this genetic evaluation, we were told that the fish sampled were produced from at most three wild females and two wild males that were taken from the Clark Fork River in Idaho.

Table 6.5 Observed (and expected) genotypic proportions in bull trout sampled from a hatchery population. $\hat{p}(1)$ is the estimated frequency of the *1* allele. *, $P < 0.05$; ***, $P < 0.001$.

	Genotype				
Locus	*11*	*12*	*22*	*$\hat{p}(1)$*	*F*
GPI-A	10	15	0	0.700	−0.43*
	(12.2)	(10.5)	(2.2)		
IDDH	24	1	0	0.980	0.00
	(24.1)	(1.0)	(0.0)		
mIDHP-1	1	22	2	0.480	−0.76***
	(5.8)	(12.5)	(6.8)		
IDHP-1	12	13	0	0.740	−0.35
	(13.7)	(9.6)	(1.7)		
Mean					−0.38

In small populations, allele frequencies are likely to differ between the sexes just due to chance. On average, the frequency of heterozygotes in the progeny population will exceed Hardy–Weinberg expectations by a proportion of:

$$\frac{1}{8N_m} + \frac{1}{8N_f} \tag{6.11}$$

Table 6.6 Expected Mendelian genotypic proportions at a locus with two alleles in a population with a single female and a single male. F_{IS} equals $[1 - (H_0/H_e)]$ and is a measure of the deficit of heterozygotes observed relative to the expected Hardy–Weinberg proportions (see Chapter 9). A negative F_{IS} indicates an excess of heterozygotes.

Mating	AA	Aa	aa	Freq(A)	F_{IS}
AA × AA	1.00	0	0	1.00	–
AA × Aa	0.50	0.50	0	0.75	−0.33
AA × aa	0	1.00	0	0.50	−1.00
Aa × Aa	0.25	0.50	0.25	0.50	0.00
Aa × aa	0	0.50	0.50	0.25	−0.33
aa × aa	0	0	1.00	0.00	–

where N_m and N_f are the numbers of male and female parents (Robertson 1965). This result holds regardless of the number of alleles at the locus concerned.

Let us consider the extreme case of a population with one female and one male ($N = 2$) and two alleles (Table 6.6). There are six possible types of matings. Mating between identical homozygotes (either *AA* or *aa*) will produce monomorphic progeny. Progeny produced by matings between two heterozygotes will result in expected Hardy–Weinberg proportions. However, the other three matings will result in an excess of heterozygotes. The extreme case is a mating between opposite homozygotes, which will produce all heterozygous progeny. On the average, there will be a 25% excess of heterozygotes in populations produced by a single male and a single female (expression 6.11).

With more than two alleles, there will be a deficit of each homozygote and an overall excess of heterozygotes. However, some heterozygous genotypes may be less frequent than expected by Hardy–Weinberg proportions, despite the overall excess of heterozygotes.

6.7 Fitness effects of genetic drift

We have considered in some detail how genetic drift is expected to affect allele frequencies and reduce the amount of genetic variation in small populations. We will now take an initial look at the effects that this loss of genetic variation is expected to have on the population itself (see Guest Box 6). That is, how will the loss of genetic variation expected in small populations affect the capability of a population to persist and evolve? We will take a more in depth look at these effects in later chapters.

6.7.1 Changes in allele frequency

Large changes in allele frequency from one generation to the next are likely in small populations due to chance. This effect may cause an increase in frequency of alleles that have harmful effects. Such deleterious alleles are continually introduced by mutation but are kept at low frequencies by natural selection. Moreover, most of these harmful alleles are recessive so that their harmful effects on the phenotype are only expressed in homozygotes. It is estimated that every individual in a population harbors several of these harmful recessive alleles in a heterozygous condition without any phenotypic effects (see Chapter 13).

Let us consider the possible effect of a population bottleneck of two individuals. As we have seen, most rare alleles will be lost in such a small bottleneck. However, any allele for which one of the two founders is heterozygous will be found in the new population at a frequency of 25%. Thus, rare deleterious alleles present in the founders will jump in frequency to 25%. Of course, at most loci the two founders will not carry a harmful allele. However, every individual carries harmful alleles at **some** loci. Therefore, we cannot predict which particular harmful alleles will increase in frequency following a bottleneck, but we can predict that several harmful alleles that were rare in the original population will be found at much higher frequencies. And if the bottleneck persists for several generations, these harmful alleles may become more frequent in the new population.

This effect is commonly seen in domestic animals such as dogs in which breeds often originated from a small number of founders. Different dog breeds usually have some characteristic genetic abnormality that is much more common within the breed than in the species as a whole (Hutt 1979). For example, different kinds of hemolytic anemia are common in several dog breeds (e.g., basenjis, beagles, and Alaskan malamutes).

Dalmatians were originally developed from a few founders that were selected for their running ability and distinctive spotting pattern. Dalmatians are susceptible to kidney stones because they excrete exceptionally high amounts of uric acid in their urine. This difference is due to a recessive allele at a single locus (Trimble and Keeler 1938). Apparently, one of the principal founders of this breed carried this recessive allele, and it subsequently drifted to high frequency in this breed.

6.7.2 Loss of allelic diversity

We have seen in Section 6.4 that genetic drift will have a much greater effect on the allelic diversity of a population than on heterozygosity if there are many alleles present at a locus. Evidence from many species indicates that loci associated with disease resistance often have many alleles (Clarke 1979). The best example of this is the major histocompatibility complex (MHC) in vertebrates (Edwards and Hedrick 1998). The MHC in humans consists of four or five tightly linked loci on chromosome 6 (Vogel and Motulsky 1986). Many alleles occur at all of these loci; for example, there are 10 or more nearly equally frequent alleles at the *A* locus and 15 or more at the *B* locus.

MHC molecules assist in the triggering of the immune response to disease organisms. Individuals heterozygous at MHC loci are relatively more resistant to a wider array of pathogens than are homozygotes (see Hughes 1991). Most vertebrate species that have been studied have been found to harbor an amazing number of MHC alleles (Hughes 1991) (see Figure 6.11). Thus, the loss of allelic diversity at MHC loci is likely to render small populations of vertebrates much more susceptible to disease epidemics (Paterson et al. 1998; Gutierrez-Espeleta et al. 2001).

6.7.3 Inbreeding depression

The harmful effects of inbreeding have been known for a long time. Experiments with plants by Darwin and others demonstrated that loss of vigor generally accompanied continued selfing and that crossing different lines maintained by selfing restored the lost vigor. Livestock breeders also generally accepted that continued inbreeding within a herd or flock could lead to a general deterioration that could be restored by outcrossing. The first published experimental report of the effects of inbreeding in animals were with rats (Crampe 1883; Ritzema-Bos 1894).

The implication of these results for wild populations did not go unnoticed by Darwin. It occurred to him that deer kept in British parks might be affected by isolation and "long-continued close interbreeding". He was especially concerned because he was aware that the effects of inbreeding may go unnoticed because they accumulate slowly. Darwin inquired about this effect and received the following response from an experienced gamekeeper:

> . . . the constant breeding in-and-in is sure to tell to the disadvantage of the whole herd, though it may take a long time to prove it; moreover, when we find, as is very constantly the case, that the introduction of fresh blood has been of the greatest use to deer, both by improving their size and appearance, and particularly by being of service in removing the taint of "rickback" if not other diseases, to which deer are sometime subject when the blood has not been changed, there can, I think, be no doubt but that a judicious cross with a good stock is of the greatest consequence, and is indeed essential, sooner or later, to the prosperity of every well-ordered park. (Darwin 1896, p. 99)

Despite Darwin's concern and warning, these early lessons from agriculture were largely ignored by those responsible for the management of wild populations of game and by captive breeding programs of zoos for nearly 100 years (see Voipio 1950 for an exception).

A seminal paper in 1979 by Kathy Ralls and her colleagues had a dramatic effect on the application of genetics to the management of wild and captive populations of animals. They used zoo pedigrees of 12 species of mammals to show that individuals from matings between related individuals tended to show reduced survival relative to progeny produced by matings between unrelated parents. The pedigree inbreeding coefficient (F) is the expected increase in homozygosity for inbred individuals; it is also the expected decrease in heterozygosity throughout the genome of inbred individuals (see Chapter 13). One of us (FWA) can clearly remember being excitedly questioned in the hallway by our departmental mammalogist who had just received his weekly issue of *Science* and could not believe the data of Ralls and her colleagues. Subsequent studies (Ralls and Ballou 1983; Ballou 1997; Lacy 1997) have supported their original conclusions (Figure 6.13).

Inbreeding depression may result from either increased homozygosity or reduced heterozygosity (Crow 1948). That is, a greater number of deleterious recessive alleles will be expressed in inbred individuals because of their increased homozygosity. In addition, fitness of inbred individuals will be reduced at loci at which the heterozygotes have a selective advantage over all homozygous types (heterozygous advantage or overdominance). Both of these mechanisms are likely to contribute to inbreeding depression, but it is thought that increased expression of deleterious recessive alleles is the more important mechanism (Charlesworth and Charlesworth 1987; Hedrick and Kalinowski 2000). Inbreeding depression is considered in detail in Chapter 13.

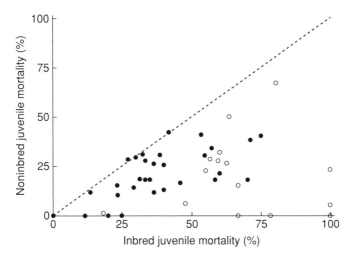

Figure 6.13 Effects of inbreeding on juvenile mortality in 44 captive populations of mammals (16 ungulates, 16 primates, and 12 small mammals). The line shows equal mortality in inbred and noninbred progeny. The preponderance of points below the line (42 of 44) indicates that inbreeding generally increased juvenile mortality. The open circles indicate populations in which juvenile mortality was significantly different ($P < 0.05$; exact test). Data from Ralls and Ballou (1983).

Guest Box 6 The inbreeding effect of small population size reduces population growth rate in mosquitofish
Paul L. Leberg and David L. Rogowski

Although domestic breeding programs, experiments, and observations from zoos indicate that inbreeding can have a detrimental effect on the well being of individuals, it is not clear to what extent the depression of individual survival, growth, and fecundity, translates to reduced viability of populations. Conservation biologists are most interested in the fates of populations; if the effects of inbreeding on population viability are trivial, scarce financial resources would be best expended on issues other than inbreeding avoidance and genetic management (e.g., restoration of gene flow or immigration). Numerous investigations have demonstrated that inbreeding has affected population growth (which is closely associated with viability when population sizes are small) in the laboratory; however, most of these have been conducted with *Drosophila* or houseflies that are not very similar to many threatened taxa. Studies of the effects of inbreeding on population growth are very difficult in vertebrates, and there have been few attempts to examine this question under conditions reflecting the complexity of natural systems.

Using large pools, set up as small ponds with predators, prey populations, and nutrient cycles, it is possible to study the role of founder numbers and relatedness on population growth in fish under conditions that are more complex than those typical of laboratory studies (Leberg 1991, 1993, unpublished data). Mosquitofish are similar in size and life history to the guppy; their short life span and rapid reproduction makes studies of population processes easier than in longer lived

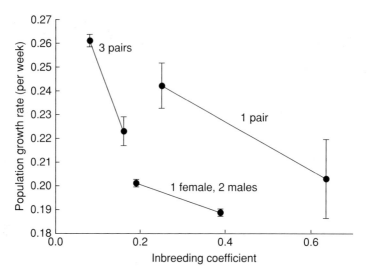

Figure 6.14 Relationship of founder relatedness to growth (±1 SE) of mosquitofish populations, for three numbers of founders. Pairs of points connected by lines indicate populations founded with the same number of founders; however founders differed in their degree of relatedness as indicated by the fixation index (higher index values indicate that founders are more related). The number of individuals used to found the experimental populations is located above each pair of points.

vertebrates. Furthermore, several species of mosquitofish are endangered making them of conservation interest, although our experiments are based on very abundant taxa. Growth rates were estimated as the log of the ratio of two population censuses divided by the number of weeks between samples. While related to the intrinsic rate of increase (r) of the population size (expression 6.10), the growth rates presented here are only surrogates for r.

Population growth rates for populations were much lower in populations founded with related individuals (higher inbreeding coefficients, expression 6.3) than unrelated individuals (Figure 6.14). Direct comparisons should be made only between points connected by lines as these populations had the same number and sex ratio of founders and differed only in the relatedness of the founders. It is unclear whether the negative effects of inbreeding on population growth were primarily due to decreased survival or to reduced fecundity rates. However, decreases in fecundity must have played some role as the proportions of juveniles (in a population), average brood sizes, and the size of males were reduced in the most inbred populations. As inbreeding can decrease population growth rates (Mills and Smouse 1994; Newman and Pilson 1997), it has the potential to reduce the recovery of populations from reductions in size, thus increasing the potential of more inbred matings. Inbreeding and reduced population growth rates can have important implications for the viability of populations, especially in smaller populations that have a greater risk of being extirpated as a result of stochastic processes.

Problem 6.1

Define genetic drift. What are the two main effects of genetic drift on the genetic composition of populations?

Problem 6.2

Why is allelic diversity lost faster than heterozygosity in small populations? What is the primary characteristic of an allele that will determine its probability of being lost in a bottleneck?

Problem 6.3

Simulate genetic drift in a population ($N = 2$) with one female (Aa) and one male (Aa) by flipping a coin as shown in Figure 6.2 until the frequency of the A allele becomes one or zero. Present your results as shown in Table 6.1. How many generations did it take before one allele or the other became fixed? Which allele became fixed? If you did this simulation 100 times, how many times would you expect the a allele to become fixed?

Problem 6.4

Simulate genetic drift in a population with one female and one male by flipping a coin, except this time begin with the female being AA and the male being Aa. Present your results as shown in Table 6.1. How many generations did it take before one allele or the other became fixed? Which allele became fixed? If you did this simulation 100 times, how many times would you expect the a allele to become fixed?

Problem 6.5

There are a variety of computer programs available for simulating the effects of genetic drift. *Populus* (Alstad 2001) is a particularly useful educational software package for simulating genetic drift and other population genetic processes that we will use. Open the Mendelian Genetics/Genetic Drift module of *Populus* and determine if your answers to the two previous questions are correct. That is, what is the relationship between the initial allele frequency and the probability of an allele eventually becoming fixed in a population? Set the number of loci to the maximum of 10 and vary the population size and initial allele frequency.

Problem 6.6

We saw in Example 6.3 that heterozygosities at individual loci may increase even though overall heterozygosity will be reduced as predicted by expression 6.5. Set the initial allele frequency with *Populus* at 0.25 with 10 loci and explore the observed pattern of genetic drift with a population size of 25 for 100 generations. You should see that heterozygosity will increase after one generation at approximately half of the loci. Remember, in this case heterozygosity will increase if the allele frequency increases and will decrease if the allele frequency decreases. How many generations does it take before the heterozygosity at all 10 loci is less than in the initial population? How many loci do you think would be necessary to monitor in a population to get an accurate measure of the size of a population?

Problem 6.7

Kimura and Ohta (1969) have shown that the expected time until fixation occurs in a population is approximately $4N$ generations. Set the initial allele frequency with *Populus* at 0.5 with 10 loci and explore if this expectation seems to hold. It may be helpful to begin with very small population sizes and then explore results with larger populations. Is this result sensitive to the initial allele frequency?

Problem 6.8

The following allele frequencies were estimated at two microsatellite loci in brown bears from Scandinavia (Waits et al. 2000):

Locus	Alleles							
Mu51	*102*	*110*	*112*	*114*	*116*	*118*	*120*	*122*
	0.080	0.241	0.098	0.011	0.006	0.098	0.408	0.057
Mu61	*205*	*207*	*211*					
	0.472	0.301	0.227					

Assume that a new population of brown bears is founded on an island in the Baltic Sea from two bears randomly selected from this population. Use expression 6.7 to predict the number of alleles expected at these two loci in this newly founded population. The expected heterozygosity at 19 microsatellite loci in this population was 0.66. What do you expect the heterozygosity to be at these loci in the newly founded population after a generation of random mating? Compare the proportion of alleles expected to be lost at these two loci to the proportion of heterozygosity expected to be lost in a bottleneck of $N = 2$.

Problem 6.9

Kurt Vonnegut (1985) wrote a science fiction novel (*Galápagos*) about a time in the future when all humans are descended from two men and two women who happened to be on a cruise to the Galápagos Islands in 1986. What proportion of the total heterozygosity in a population do we expect to be retained after a bottle-neck of four individuals (see expression 6.5)? One could argue on this basis that even extreme bottlenecks, such as four individuals, would not have a major effect on populations or species. Do you agree with this argument? Why not?

Problem 6.10

Johnson (1988) reported the following allele frequencies at 25 allozyme loci in the land snail *Theba pisana*, which was introduced from Europe into western Australia in the 1890s (see Example 6.2). Describe the amounts of genetic variation in these three samples (heterozygosity and allelic diversity). Assume that all of the loss in heterozygosity was caused by the initial founding events; that is, assume that there has been no additional loss of heterozygosity in the mainland Australia or Rottnest Island populations because of small population size. What is the number of founders in the two founding events that would explain the differences in heterozygosities in these three populations?

Locus	Allele	France	Australia	
			Perth	**Rottnest**
G6PDH	105	0.03	–	–
	100	0.97	1.00	1.00
IDH-1	110	0.14	–	–
	100	0.79	0.92	0.77
	90	0.07	0.08	0.23
LDH	100	0.97	1.00	1.00
	77	0.03	–	–
LAP-1	106	0.35	–	–
	103	0.26	–	–
	100	0.39	1.00	1.00
LGP-2	127	0.11	–	–
	118	0.09	–	–
	109	0.09	–	–
	100	0.69	1.00	1.00
	82	0.02	–	–
LTP-1	100	0.78	0.88	1.00
	96	0.22	0.12	–
LTP-2	105	–	0.06	–
	100	0.40	0.29	0.80
	95	0.44	0.65	0.20
	90	0.16	–	–
LTP-3	111	0.02	–	–
	100	0.70	0.60	1.00
	91	0.26	0.40	–
	84	0.02	–	–

			Australia	
Locus	Allele	France	Perth	Rottnest
MDH-3	100	0.97	1.00	1.00
	70	0.03	–	–
PGM-2	130	0.07	0.10	0.34
	117	0.21	–	–
	100	0.72	0.90	0.66
PGM-3	225	0.03	–	–
	175	0.21	0.60	0.71
	155	0.34	–	–
	140	0.02	–	–
	100	0.40	0.40	0.29
14 loci	100	1.00	1.00	1.00

7

Effective Population Size

Medium ground finch, Example 7.2

Effective population size is whatever must be substituted in the formula (1/2N) to describe the actual loss in heterozygosity.

Sewall Wright (1969)

Effective population size (N_e) is one of the most fundamental evolutionary parameters of biological systems, and it affects many processes that are relevant to biological conservation.

Robin Waples (2002)

We saw in the previous chapter that we expect heterozygosity to be lost at a rate of $1/2N$ in finite populations (see expression 6.2). However, this expectation holds only under certain conditions that will rarely apply to real populations. For example, such factors as the number of individuals of reproductive age rather than the total of all ages, the sex ratio, and differences in reproductive success among individuals must be considered. Thus, the actual number of individuals in a natural population (so-called census size, N_c) is not sufficient for predicting the rate of genetic drift. We will use the concept of effective population size to deal with the discrepancy between the demographic size and population size relevant to the rate of genetic drift in natural populations.

Perhaps the most important assumption of our model of genetic drift has been the absence of natural selection. That is, we have assumed that the genotypes under investigation do not affect the fitness (survival and reproductive success) of individuals. We would not be concerned with the retention of genetic variation in small populations if the assumption of genetic neutrality were true for all loci in the genome. However, the assumption of neutrality, and the use of neutral loci, allows us to predict the effects of finite population size with great generality. In Chapter 8 we will consider the effects of incorporating natural selection into our basic models of genetic drift.

7.1 Concept of effective population size

Our consideration in the previous chapter of genetic drift dealt only with "ideal" populations. **Effective population size** (N_e) is defined as the size of the ideal population (N) that will result in the same amount of genetic drift as in the actual population being considered. The basic ideal population consists of "N diploid individuals reconstituted each generation from a random sample of $2N$ gametes" (Wright 1939, p. 298). In an ideal population individuals produce both female and male gametes (**monoecy**) and self-fertilization is possible. Under these conditions, heterozygosity will decrease exactly by $1/2N$ per generation.

We can see this by considering an ideal population of N individuals (say 10) in which each individual is heterozygous for two unique alleles (Figure 7.1). All of these 10 individuals will contribute equally to the gamete pool to create each individual in the next generation. Thus, each allele will be at a frequency of $1/2N = 0.05$ in the gamete pool. A new individual will only be homozygous if the same allele is present in both gametes. For the purposes of our calculations, it does not matter which allele is sampled first because all alleles are equally frequent. Let us say the first gamete chosen is $\alpha15$. This individual will be homozygous only if the next gamete sampled is also $\alpha15$. What is the probability that the next gamete sampled is $\alpha15$? This probability is simply the frequency of the $\alpha15$ allele in the gamete pool, which is $1/2N = 0.05$ because all 20 alleles (2×10) are at equal frequency in the gamete pool (Figure 7.1). Therefore, the expected homozygosity is $1/2N$, and the expected heterozygosity of each individual in the next generation is $1 - (1/2N) = 0.95$.

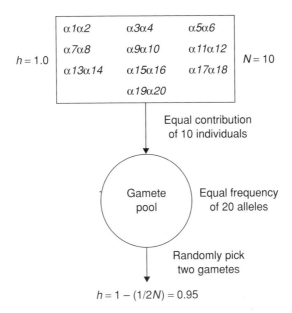

Figure 7.1 Diagram of reduction in heterozygosity (h) in an ideal population consisting of 10 individuals that are each heterozygous for two unique alleles ($h = 1$). Two gametes are picked from the gamete pool to create each individual in the next generation. Let us say the first gamete chosen is $\alpha 15$. This individual will be heterozygous unless the next gamete sampled is also $\alpha 15$. What is the probability that the next gamete sampled is $\alpha 15$? This is the frequency of the $\alpha 15$ allele in the gamete pool, which is $1/2N = 0.05$ because of the equal contribution of individuals to the gamete pool. Therefore, the expected heterozygosity of each individual in the next generation is $1 - (1/2N) = 0.95$.

This conceptual model becomes more complicated if self-fertilization is prevented, or if the population is **dioecious** and is equally divided between males and females. In these two cases, the decrease in heterozygosity due to sampling individuals from the gamete pool will skip a generation because both gametes in an individual cannot come from the same parent. Nevertheless, the mean rate of loss per generation over many generations is similar in this case; heterozygosity is lost at a rate more closely approximated by $1/(2N + 1)$ (Wright 1931; Crow and Denniston 1988). The difference between these two expectations, $1/2N$ and $1/(2N + 1)$, is usually ignored because the difference is insignificant except when N is very small.

For our general purposes, the ideal population consists of a constant number of N diploid individuals ($N/2$ females and $N/2$ males) in which all parents have an equal probability of being the parent of any individual progeny. We will consider the following effects of violating the following assumptions of such idealized populations on the rate of genetic drift:

1 Equal numbers of males and females.
2 All individuals have an equal probability of contributing an offspring to the next generation.
3 Constant population size.
4 Nonoverlapping (discrete) generations.

We have already examined two expected effects of genetic drift: changes in allele frequency (Section 6.2) and a decrease in heterozygosity (Section 6.3). Thus, there are at least two possible measures of the effective population size (i.e., the rate of genetic change due to drift). First, the "variance effective number" (N_{eV}) is whatever must be substituted in expression 6.1 to predict the expected changes in allele frequency; second, the "inbreeding effective number" (N_{eI}) is whatever must be substituted in expression 6.2 to predict the expected reduction in heterozygosity (Example 7.1). Crow (1954) and Ewens (1982) have described effective population numbers that predict the expected rate of decay of the proportion of polymorphic loci (P); we will only consider the first two kinds of effective population size (N_{eV} and N_{eI}) because they have more relevance for understanding the loss of genetic variation in populations.

Example 7.1 Effective population size of grizzly bears

Harris and Allendorf (1989) estimated the effective population size of grizzly bear populations using computer simulations based upon life history characteristics (survival, age at first reproduction, litter size, etc.). They estimated N_e by comparing the actual loss of heterozygosity in the simulated populations to that expected in ideal populations of $N = 100$ (Figure 7.2). Over a wide range of conditions, the effective population size was approximately 25% of the actual population size.

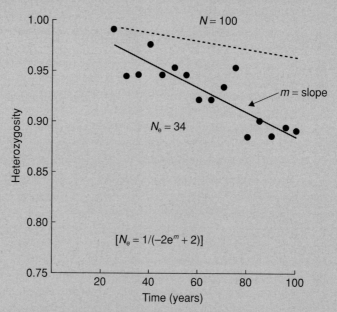

Figure 7.2 Estimation of effective size of grizzly bear populations ($N_c = 100$) by computer simulation. The dashed line shows the expected decline in heterozygosity over 10 generations (100 years) in an ideal population using expression 6.7. The solid line shows the decline in heterozygosity in a simulated population. The decline in heterozygosity in the simulated population is equal to that expected in an ideal population of 34 bears; thus, $N_e = 34$. (m is the slope of the regression of the log of heterozygosity on time.) From Harris and Allendorf (1989).

Crow and Denniston (1988) have clarified the distinction between these two measures of effective population size. In many cases, a population has nearly the same effective population number for either effect. Specifically, their values are identical in constant size populations in which the age and sex distributions are unchanging. We will first consider N_e to be the inbreeding effective population size under different circumstances because this number is most widely used, and we will then consider when these two numbers will differ.

7.2 Unequal sex ratio

Populations often have unequal numbers of males and females contributing to the next generation. The two sexes, however, contribute an equal number of genes to the next generation regardless of the total of males and females in the population. Therefore, the amount of genetic drift attributable to the two sexes must be considered separately. Consider the extreme case of one male mating with 100 females. In this case, all progeny will be half-sibs because they share the same father. In general, the rarer sex is going to have a much greater effect on genetic drift, so that the effective population size will seldom be much greater than twice the size of the rarer sex.

What is the size of the ideal population that will lose heterozygosity at the same rate as the population we are considering which has different numbers of females and males? We saw in Section 6.3 that the increase in homozygosity due to genetic drift is caused by an individual being homozygous because its two gene copies were derived from a common ancestor in a previous generation. The inbreeding effective population size in a monoecious population in which selfing is permitted may be defined as the reciprocal of the probability that two uniting gametes come from the same parent. With separate sexes, or if selfing is not permitted, uniting gametes must come from different parents; thus, the effective population size is the probability that two uniting gametes come from the same grandparent.

The probability that the two uniting gametes in an individual came from a male grandparent is $1/4$. (One-half of the time uniting gametes will come from a grandmother and a grandfather, and $1/4$ of the time both gametes will come from a grandmother.) Given that both gametes come from a grandfather, the probability that both come from the *same* male is $1/N_m$, where N_m is the number of males in the grandparental generation. Thus, the combined probability that both uniting gametes come from the same grandfather is $(1/4 \times 1/N_m) = 1/4N_m$. The same probabilities hold for grandmothers. Thus, the combined probabilities of uniting gametes coming from the same grandparent is then:

$$\frac{1}{N_e} = \frac{1}{4N_f} + \frac{1}{4N_m} \tag{7.1}$$

This is more commonly represented by solving for N_e with the following result:

$$N_e = \frac{4N_f N_m}{N_f + N_m} \tag{7.2}$$

As we would expect, if there are equal numbers of males and females ($N_f = N_m = 0.5N$), then this expression reduces to $N_e = N$.

Table 7.1 Sex ratios of the tropical tree *Triplaris americana* at four study sites in Costa Rica (Melampy and Howe 1977). N_c is the census population size, which in this case is the number of trees present at a site.

Sites	Females	Males	N_c	N_e	N_e/N_c
1	61	41	102	98.1	0.96
2	58	42	100	97.4	0.97
3	56	44	100	98.6	0.99
4	47	12	59	38.2	0.65

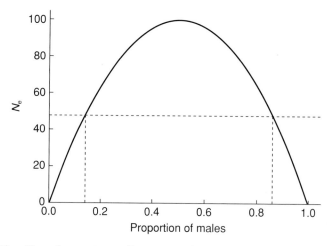

Figure 7.3 The effect of sex ratio on effective population size for a population with a total of 100 males and females using expression 7.2. The dashed lines indicate the sex ratios at which N_e will be reduced by half because of a skewed sex ratio.

Melampy and Howe (1977) described skewed sex ratios in the tropical tree *Triplaris americana* from four sites in Costa Rica. We can use expression 7.2 to predict what effect the observed excess of females would have on the effective population size (Table 7.1). There is a substantial reduction in N_e of this tree only at site 4 where males comprise approximately 20% of the population.

In general, a skewed sex ratio will not have a large effect on the $N_e : N$ ratio unless there is a great excess of one sex or the other. Figure 7.3 shows this for a hypothetical population with a total of 100 individuals. N_e is maximum (100) when there are equal numbers of males and females, but declines as the sex ratio departs from 50 : 50. However, small departures from 50 : 50 have little effect on N_e. The dashed lines in this figure show that the $N_e : N$ ratio will only be reduced by half if the least common sex is less than 15% of the total population. In the most extreme situations, the N_e will be approximately four times the rarer sex:

N_f	N_m	N_e
1,000	1	4.0
1,000	2	8.0
1,000	3	12.0
1,000	4	15.9
1,000	5	19.9

The estimated N_e will not be a very good indicator of the loss of allelic diversity when sex ratios are extremely skewed (Question 7.1).

Question 7.1 Consider the hypothetical situation where a single tree is left standing after the clearcut of an entire population of ponderosa pine. This lone tree holds over 5,000 seeds that were fertilized by pollen from a random sample of the entire pre-clearcut population of 500 trees. What proportion of the heterozygosity and allelic diversity present in the pre-clearcut population would you expect to be present in a new population founded from these 5,000 seeds?

7.3 Nonrandom number of progeny

Our idealized model assumes that all individuals have an equal probability of contributing progeny to the next generation. That is, a random sample of $2N$ gametes is drawn from a population of N diploid individuals. In real populations, parents seldom have an equal chance of contributing progeny because they differ in fertility and in the survival of their progeny. The variation among parents results in a greater proportion of the next generation coming from a smaller number of parents. Thus, the effective population size is reduced.

It is somewhat surprising just how much variation in reproductive success there actually is even when all individuals have equal probability of reproducing. Figure 7.4 shows the expected frequency of progeny number in a very large, stable population in which the mean number of progeny is two and all individuals have equal probability of reproducing. Take, for example, a stable population of 20 individuals (10 males and 10 females). On average, each individual will have two progeny. However, approximately 12% of all individuals will not contribute any progeny! Consider that the probability of any male **not** fathering a particular child in this population is $(0.90 = 9/10)$. The probability of a male not contributing any of the 20 progeny therefore is $(0.90)^{20}$, or approximately 12%. The same statistical reasoning holds for females as well. Thus, on average, two or three of the 20 individuals in this population will not contribute any genes to the next generation while an expected one out of the 20 individuals (5% from Figure 7.4) will produce five or more progeny.

We can adjust for nonrandom progeny contribution following Wright (1939). Consider N individuals that contribute varying numbers of gametes (k) to the next generation of the

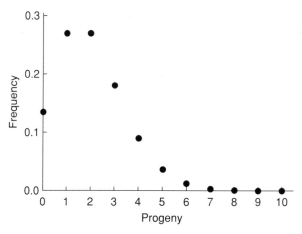

Figure 7.4 Expected frequency of number of progeny per individual in a large stable population in which the mean number of progeny per individual is two and all individuals have equal probability of reproducing.

same size (N) so that the mean number of gametes contributed per individual is $\bar{k} = 2$. The variance of the number of gametes contributed to the next generation is:

$$V_k = \frac{\sum_{i=1}^{N} (k_i - 2)^2}{N} \tag{7.3}$$

The proportion of cases in which two random gametes will come from the same parent is:

$$\frac{\sum_{i=1}^{N} k_i(k_i - 1)}{2N(2N - 1)} = \frac{2 + V_k}{4N - 2} \tag{7.4}$$

As we saw in the previous section, the effective population size may be defined as the reciprocal of the probability that two gametes come from the same parent. Thus, we may write the effective population size as:

$$N_e = \frac{4N - 2}{2 + V_k} \tag{7.5}$$

Random variation of k will produce a distribution that approximates a Poisson distribution. A Poisson distribution has a mean equal to the variance; thus, $V_k = \bar{k} = 2$ and $N_e = N$ for the idealized population. However, as the variability among parents (V_k) increases the effective population size decreases. An interesting result is that the effective population size will be larger than the actual population size if $V_k < 2$. In the extreme where each

Table 7.2 Estimation of effective population size in three hypothetical populations of constant size $N = 10$ with extreme differences in individual reproductive success. Each population consists of five pairs of mates. In Population A, only one pair of mates reproduces successfully. In Population B, each of the five pairs produces two offspring so that there is no variance in reproductive success. There is an intermediate amount of variability in reproductive success in Population C.

	A			B			C		
i	k_i	$k_i - \bar{k}$	$(k_i - \bar{k})^2$	k_i	$k_i - \bar{k}$	$(k_i - \bar{k})^2$	k_i	$k_i - \bar{k}$	$(k_i - \bar{k})^2$
1	10	8	64	2	0	0	0	−2	4
2	10	8	64	2	0	0	0	−2	4
3	0	−2	4	2	0	0	3	1	1
4	0	−2	4	2	0	0	3	1	1
5	0	−2	4	2	0	0	2	0	0
6	0	−2	4	2	0	0	2	0	0
7	0	−2	4	2	0	0	1	−1	1
8	0	−2	4	2	0	0	1	−1	1
9	0	−2	4	2	0	0	4	2	4
10	0	−2	4	2	0	0	4	2	4
			160			0			20

parent produces exactly two progeny, $N_e = 2N - 1$. Thus, in captive breeding where we can control reproduction, we may nearly double the effective population size by making sure that all individuals contribute equal numbers of progeny.

This potential near doubling of effective population size occurs because there are two sources of genetic drift: reproductive differences among individuals and Mendelian segregation in heterozygotes. These two sources contribute equally to genetic drift. Thus, eliminating the first source of drift will double the effective population size. Unfortunately, there is no way to eliminate the second source of genetic drift (Mendelian segregation), except by nonsexual reproduction (cloning, etc.).

The following example considers three hypothetical populations of constant size $N = 10$ with extreme differences in individual reproductive success (Table 7.2). Each population consists of five pairs of mates. In Population A, only one pair of mates reproduces successfully. In Population B, each of the five pairs produces two offspring so that there is no variance in reproductive success. There is an intermediate amount of variability in reproductive success in Population C.

We can estimate N_e for each of these populations using expressions 7.3 and 7.5 as shown in Table 7.3. Thus, Population B is expected to lose only approximately 3% ($1/2N_e = 0.026$) of its heterozygosity per generation while Populations A and C are expected to lose 24 and 5%, respectively. There are very few examples in natural populations where the reproductive success of individuals is known so that N_e can be estimated using this approach (Example 7.2).

Table 7.3 Estimation of effective population size for three hypothetical populations with high, low, and intermediate variability in family size using expression 7.5.

	$\Sigma (k_i - \bar{k})^2$	V_k	N_e	$1/2N_e$
Population A	160	16	2.11	0.237
Population B	0	0	19.00	0.026
Population C	20	2	9.50	0.053

Example 7.2 Effective population size of Darwin's finches

Grant and Grant (1992a) reported the lifetime reproductive success of two species of Darwin's ground finches on Daphne Major, Galápagos: the cactus finch and the medium ground finch. They followed survival and lifetime reproductive success of four cohorts born in the years 1975–1978. Figure 7.5 shows the lifetime reproductive success of the 1975 cohort for both species. The variance in reproductive success for both species was much greater than expected in ideal population (see Figure 7.2). Over one-half of the birds in both species did not produce any recruits to the next generation, and several birds produced eight or more recruits. Eighteen cactus finches produced 33 recruits ($\bar{k} = 1.83$) distributed with a

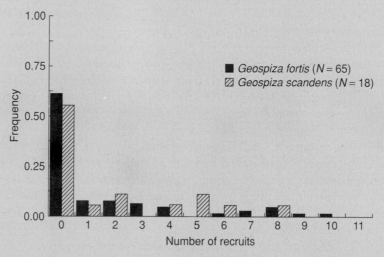

Figure 7.5 Lifetime reproductive success of the 1975 cohort of the cactus finch and medium ground finches on Isla Daphne Major, Galápagos. The x-axis shows the number of recruits (progeny that breed) produced. Thus, over 50% of the breeding birds for both species did not produce any progeny that lived to breed. From Grant and Grant (1992a).

variance (V_k) of 6.74; 65 medium ground finches produced 102 recruits ($\bar{k} = 1.57$) distributed with a variance (V_k) of 7.12. The average number of breeding birds (census population sizes) for these years was approximately 94 cactus finches and 197 medium ground finches. The estimated N_e based on these data is 38 cactus finches and 60 medium ground finches. Thus, the $N_e : N_c$ ratio for these two species is 38/94 = 0.40 and 60/197 = 0.30.

Expression 7.5 assumes that the variance in progeny number is the same in males and females. However, the variation in progeny number among parents is likely to be different for males and females. For many animal species, the variance of progeny number for males is expected to be larger than that for females. For example, according to the 1990 *Guinness Book of World Records*, the greatest number of children produced by a human mother is 69; in great contrast, the last Sharifian Emperor of Morocco is estimated to have fathered some 1,400 children! We can take such differences between the sexes into account as shown

$$N_e = \frac{8N - 4}{V_{km} + V_{kf} + 4} \tag{7.6}$$

The estimation of effective population size with nonrandom progeny number becomes much more complex if we relax our assumption of constant population size. In the case of separate sexes, the following expression may be used:

$$N_e = \frac{(N_{t-2}\bar{k}) - 2}{\bar{k} - 1 + \dfrac{V_k}{\bar{k}}} \tag{7.7}$$

where N_{t-2} is N in the grandparental generation (Crow and Denniston 1988).

7.4 Fluctuating population size

Natural populations may fluctuate greatly in size. The rate of loss of heterozygosity ($1/2N$) is proportional to the reciprocal of population size ($1/N$). Generations with small population sizes will dominate the effect on loss of heterozygosity. This is analogous to the sex with the smallest population size dominating the effect on loss of heterozygosity (see Section 7.2). Therefore, the average population size over many generations is a poor metric for the loss of heterozygosity over many generations.

For example, consider three generations for a population that goes though a severe bottleneck, say $N_1 = 100$, $N_2 = 2$, and $N_3 = 100$. A very small proportion of the heterozygosity will be lost in generations 1 and 3 ($1/200 = 0.5\%$); however, 25% of the heterozygosity will be lost in the second generation. The exact heterozygosity remaining after these three generations can be found as shown:

$$h = \left(1 - \frac{1}{200}\right)\left(1 - \frac{1}{4}\right)\left(1 - \frac{1}{200}\right) = 0.743$$

The average population size over these three generations is $(100 + 2 + 100)/3 = 67.3$. Using expression 6.7 we would expect to lose only approximately 2% of the heterozygosity over three generations with a population size of 67.3, rather than the 25.7% heterozygosity that is actually lost.

We can estimate the effective population size over these three generations by using the mean of the reciprocal of population size $(1/N)$ in successive generations, rather than the mean of N itself. This is known as the harmonic mean. Thus,

$$\frac{1}{N_e} = \frac{1}{t}\left(\frac{1}{N_1} + \frac{1}{N_2} + \frac{1}{N_3} + \ldots + \frac{1}{N_t}\right) \tag{7.8}$$

After a little algebra, this becomes:

$$N_e = \frac{t}{\sum\left(\frac{1}{N_i}\right)} \tag{7.9}$$

Generations with the smallest N have the greatest effect. A single generation of small population size may cause a large reduction in genetic variation. A rapid expansion in numbers does not affect the previous loss of genetic variation; it merely reduces the current rate of loss. This is known as the "bottleneck" effect as discussed in Section 6.5.

We can use expression 7.9 to predict the expected loss of heterozygosity in the example that we began this section with:

$$N_e = \frac{3}{\left(\frac{1}{100} + \frac{1}{2} + \frac{1}{100}\right)} = 5.77$$

We expect to lose 23.8% of the heterozygosity in a population where $N_e = 5.77$ over three generations (see expression 6.7). This is very close to the exact value of 25.7% that we calculated previously.

7.5 Overlapping generations

We so far have considered only populations with discrete generations. However, most species have overlapping generations. Hill (1979) has shown that the effective number in the case of overlapping generation is the same as that for a discrete-generation population having the same variance in lifetime progeny numbers and the same number of individuals entering the population each generation. Thus, the occurrence of overlapping generations itself does not have a major effect on effective population size. However, this result assumes a constant population size and a stable age distribution. Crow and Denniston (1988) concluded that Hill's results would be approximately correct for populations that are growing or contracting, as long as the age distribution is fairly stable.

7.6 Variance effective population size

The two measures of population size (N_{eI} and N_{eV}) will differ when the population size is changing. In general, the inbreeding effective population size (N_{eI}) is more related to the number of parents since it is based on the probability of two gametes coming from the same parent. The variance effective population size (N_{eV}) is more related to the number of progeny since it is based on the number of gametes contributed rather than the number of parents (Crow and Kimura 1970, p. 361).

Consider the extreme of two parents that have a very large number of progeny. In this case, the allele frequencies in the progeny will be an accurate reflection of the allele frequencies in the parents; therefore, N_{eV} will be nearly infinite. However, all the progeny will be full-sibs and thus their progeny will show the reduction in homozygosity expected in matings between full-sibs; thus, N_{eI} is very small. In the other extreme, if each parent has exactly one offspring then there will be no tendency for inbreeding in the populations, and, therefore, N_{eI} will be infinite. However, N_{eV} will be small.

Therefore, if a population is growing, the inbreeding effective number is usually less than the variance effective number (Waples 2002). If the population size is decreasing, the reverse is true. In the long run, these two effects will tend to cancel each other and the two effective numbers will be roughly the same (Crow and Kimura 1970; Crow and Denniston 1988).

7.7 Cytoplasmic genes

Genetic variation in cytoplasmic gene systems (e.g., mitochondria and chloroplasts) has come under active investigation in the last few years because of advances in techniques to analyze differences in DNA sequences. We will consider mitochondrial DNA (mtDNA) because so much is known about genetic variation of this molecule. The principles we will consider also apply to genetic variation in chloroplast DNA (cpDNA). However, cpDNA is paternally inherited in some plants (Harris and Ingram 1991).

There are three major differences between mitochondrial and nuclear genes that are relevant for this comparison:

1   Individuals usually possess many mitochondria that share a single predominant mtDNA sequence. That is, individuals are effectively haploid for a single mtDNA type.
2   Individuals inherit their mtDNA genotype from their mother.
3   There is no recombination between mtDNA molecules.

The effective population size for mtDNA is smaller than that for nuclear genes because each individual has only one haplotype and uniparental inheritance (Birky et al. 1983).

For purposes of comparison, we will use h to compare genetic drift at mtDNA with nuclear genes even though mtDNA is haploid so that individuals are not heterozygous. It may seem inappropriate to use h as a measure of variation for mtDNA since it is haploid and individuals therefore cannot be heterozygous for mtDNA. Nevertheless, h is called gene diversity in this context and is a valuable measure of the variation present within a population (Nei 1987, p. 177). It can be thought of as the probability that two randomly sampled individuals from a population will have the same mtDNA genotype, and has also been called gene diversity (Nei 1987, p. 177).

The probability of sampling the same mtDNA haplotype in two consecutive gametes is $1/N_f$, where N_f is the number of females in the population. And since $N_f = 0.5N$,

$$\Delta h = \frac{1}{N_f} = \frac{1}{0.5N} \tag{7.10}$$

In the case of a 1 : 1 sex ratio, there are four times as many nuclear genes as mitochondrial genes (N_f):

$$\frac{N_e(\text{nuc})}{N_e(\text{mt})} = \frac{2(N_f + N_m)}{N_f} = 4 \tag{7.11}$$

Things become more interesting with an unequal sex ratio. If we use expression 7.2 for N_e for nuclear genes, then the ratio between the effective number of nuclear genes ($2N_e$) to the effective number of mitochondrial genes is:

$$\frac{N_e(\text{nuc})}{N_e(\text{mt})} = \frac{\dfrac{2(4N_fN_m)}{(N_f + N_m)}}{N_f}$$

which, after a little bit of algebra, becomes

$$\frac{8N_m}{(N_f + N_m)} \tag{7.12}$$

In general, drift is of more importance and bottlenecks have greater effects for genes in mtDNA than for nuclear genes (Example 7.3). Figure 7.7 shows the relative loss of variation during a bottleneck of a single generation for a nuclear and mitochondrial gene.

Example 7.3 Effects of a bottleneck in the Australian spotted mountain trout

Ovenden and White (1990) demonstrated that genetic variation in mtDNA is much more sensitive to bottlenecks than nuclear variation in the southern Australian spotted mountain trout from Tasmania (Figure 7.6). These fish spawn in fresh water, and the larvae are immediately washed to sea where they grow and develop. The juvenile fish re-enter fresh water the following spring where they remain until they spawn. Landlocked populations of spotted mountain trout also occur in isolated lakes that were formed by the retreat of glaciers some 3,000–7,000 years ago.

Ovenden and White found 58 mtDNA genotypes identified by the presence or absence of restriction sites in 150 fish collected from 14 coastal streams. There is evidence of a substantial exchange of individuals among the 14 coastal stream populations. In contrast, they found only two mtDNA genotypes in 66 fish collected

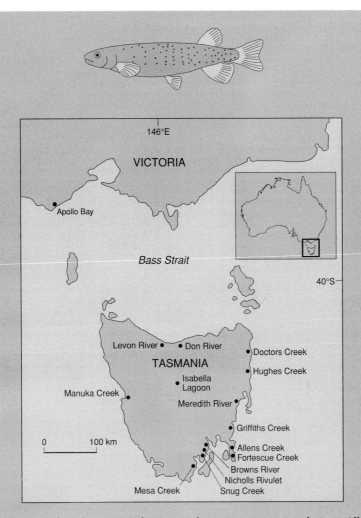

Figure 7.6 Map of southern Australian spotted mountain trout populations. Allens Creek and Fortescue Creek are coastal populations that exhange migrants with each other and other coastal populations. Isabella Lagoon is an isolated landlocked population. Map from Ovenden and White (1990) and trout from McDowall (1990).

from landlocked populations in isolated lakes. However, the lake populations and coastal populations had nearly identical heterozygosities at 22 allozyme loci (Table 7.4). As expected, the allelic diversity of the lake populations was smaller than the coastal populations.

The reduced genetic variation in mtDNA in the landlocked populations is apparently due to a bottleneck associated with their founding and continued isolation. Oveden and White suggested that the founding bottleneck may have been exacerbated by natural selection for the landlocked life history in these populations. Regardless of the mechanism, the reduced N_e of the landlocked populations has had a dramatic effect on genetic variation in mtDNA but virtually no effect on nuclear heterozygosity.

Table 7.4 Expected heterozygosity (H_e), diversity (h) at mtDNA, and average number of alleles ($\bar{A}$) per locus in three populations of the southern Australian spotted mountain trout from Tasmania at 22 allozyme loci and mtDNA. The Allens Creek and Fortesque Creek populations are coastal populations that are connected by substantial exchange of individuals. The Isabella Lagoon population is an isolated landlocked population.

	Nuclear loci		mtDNA	
Sample	$\bar{A}$	H_e	$\bar{A}$	h
Allens Creek	1.9	0.123	28	0.946
Fortescue Creek	1.9	0.111	25	0.922
Isabella Lagoon	1.3	0.104	2	0.038

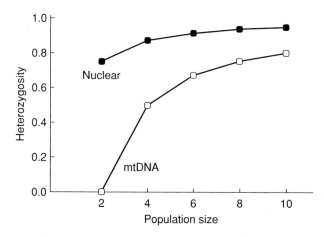

Figure 7.7 Amount of heterozygosity or diversity remaining after a bottleneck of a single generation for a nuclear and mitochondrial gene with equal numbers of males and females. For example, there is no mitochondrial variation left after a bottleneck of two individuals because only one female is present. In contrast, 75% of the nuclear heterozygosity will remain after a bottleneck of two individuals (see expression 6.5).

7.8 Gene genealogies and lineage sorting

So far we have described genetic changes in populations due to genetic drift by changes in allele frequencies from generation to generation. There is an alternative approach to study the loss of genetic variation in populations that can be seen most easily in the case of mtDNA in which each individual receives the mtDNA haplotype of its mother. We can trace the transmission of mtDNA haplotypes over many generations. That is, we can trace the genealogy of the mtDNA genotype of each individual in a population (Figure 7.8). We can see in the example shown in Figure 7.8 that only two of the original 18 haplotypes

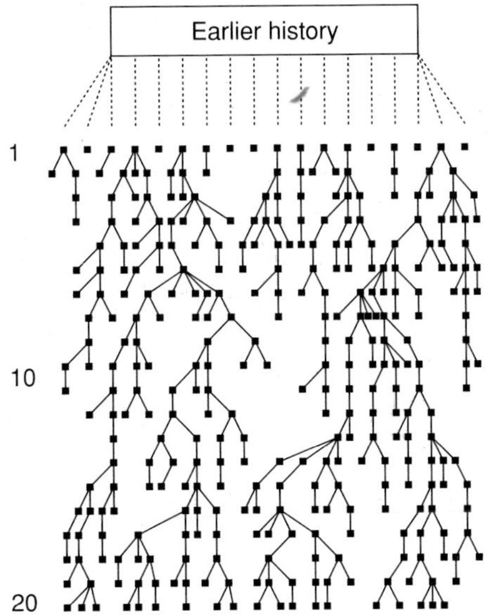

Figure 7.8 The allelic lineage sorting process of mtDNA haplotypes in a population. Each node represents an individual female and branches lead to daughters. The tree was generated by assuming a random distribution of female progeny with a mean of one daughter per female. From Avise (1994).

remain in a population after just 20 generations due to a process called stochastic lineage sorting.

The **gene genealogy** approach also can be applied to nuclear genes, although it is somewhat more complex because of diploidy and recombination. The recent development of the application of genealogical data to the study of population-level genetic processes is the major advance in population genetics theory in the last 50 years (Fu and Li 1999; Schaal and Olsen 2000). This development has been based upon two major advances, one technical and one conceptual. The technological advance is the collection of DNA sequence data that allow tracing and reconstructing gene genealogies. The conceptual advance that has contributed to the theory to interpret these results is called "**coalescent** theory".

Lineage sorting, as in Figure 7.8, will eventually lead to the condition where all alleles in a population are derived from (i.e., coalesce to) a single common ancestral allele. The time to coalescence is expected to be shorter for smaller populations. In fact, the mean time to coalescence is equal N_e. Coalescent theory provides a powerful framework to study the effects of genetic drift, natural selection, mutation, and gene flow in natural populations (Rosenberg and Nordborg 2002).

7.9 Limitations of effective population size

The effective population size can be used to predict the expected rate of loss of heterozygosity or change in allele frequencies resulting from genetic drift. In practice, however, we generally need to know the rate of genetic drift in order to estimate effective population size. Thus, effective population size is perhaps best thought of as a standard, or unit of measure, rather than as a predictor of the loss of heterozygosity. That is, if we know the rate of change in allele frequency or loss of heterozygosity in a given population, we can use the expected rates expected in an ideal population to represent that rate (see Examples 7.1 and 7.4, and Guest Box 7). We will consider the estimation of effective population size in more detail in Chapter 14.

Example 7.4 Effective population size in a marine fish.

A comparison of genetic variation at seven microsatellite loci in New Zealand snapper from the Tasman Bay has shown that the N_e may be four orders of

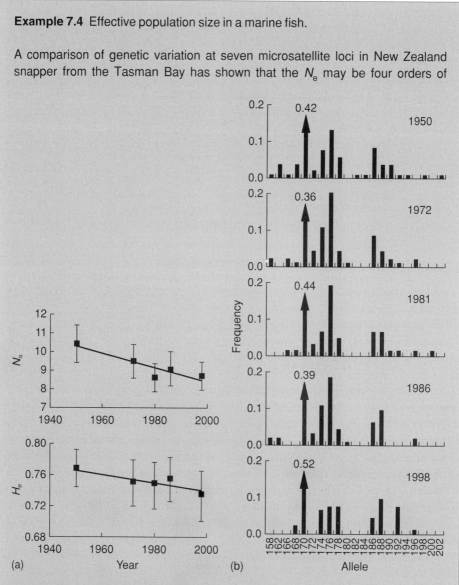

Figure 7.9 Loss of genetic variation of New Zealand snapper from Tasman Bay. (a) The decline in the number of alleles (N_a) and expected heterozygosity (H_e) at seven microsatellite loci. (b) The loss of alleles at the *GA2B* locus between 1950 and 1998. The frequency of the most common allele is shown above the arrow. From Hauser et al. (2002).

magnitude smaller than N_c in this population (Hauser et al. 2002). Collection of a time series of scale samples began in 1950 just after the commencement of a commercial fishery on this population. Genetic variation, as measured by both the number of alleles and heterozygosity, declined between 1950 and 1998 in this population (Figure 7.9).

The estimated N_e over this time period in this population based on the reduction in heterozygosity and temporal changes in allele frequency are 46 and 176, respectively. The minimum estimated population size during this period was 3.3 million fish in 1985; thus, N_e/N_c is of the order of 0.0001! These results support the conclusion of Hedgecock (1994) that the $N_e : N_c$ ratio may be very small in a variety of marine species. This suggests that even very large exploited marine fish populations may be in danger of losing substantial genetic variation.

Perhaps the greatest value of effective population size is heuristic. That is, we can better our understanding of genetic drift by comparing the effects of different violations of the assumptions of ideal populations on N_e (e.g., Figure 7.2). Similarly, in applying the concept of effective population size to managing populations, certain specific effective population sizes are often used as benchmarks. For example, it has been suggested that an N_e of at least 50 is necessary to avoid serious loss of genetic variation in the short term (Soulé 1980; Allendorf and Ryman 2002).

7.9.1 Allelic diversity and N_e

Effective population size is not a very good indicator of the loss of allelic diversity within populations. That is, two populations that go through a bottleneck of the same N_e may lose very different amounts of allelic diversity. This difference is greatest when the bottleneck is caused by an extremely skewed sex ratio. Bottlenecks generally have a greater effect on heterozygosity than allelic diversity. However, as we saw in Question 7.1, a population with an extremely skewed sex ration may experience a substantial reduction in heterozygosity with very little loss of allelic diversity.

The duration of a bottleneck (intense versus diffuse) will also affect heterozygosity and allelic diversity differently (England et al. 2003). Consider two populations that fluctuate in size over several generations with the same N_e, and therefore the same loss of heterozygosity. A brief but very small bottleneck (intense) will cause substantial loss of allelic diversity. However, a diffuse bottleneck spread over several generations can result in the same loss of heterozygosity, but will cause a much smaller reduction in allelic diversity.

In summary, populations that experience the same rate of decline of heterozygosity may experience very different rates of loss of allelic diversity. Therefore, we must consider more than just N_e when considering the rate of loss of genetic variation in populations.

7.9.2 Generation interval

The rate of loss of genetic variation through time depends upon both N_e and mean generation interval because $1/(2N_e)$ is the expected rate of loss **per** generation. The generation interval (g) is the average age of reproduction, not the age of first reproduction (Harris and

Table 7.5 Expected loss in heterozygosity after 1,000 years for five species of reptiles on North Brother Island in Cook Strait, New Zealand. N_e estimates for each species are 20% of the estimated census size (Keall et al. 2001). The estimated generation interval (g) was then used to estimate the number of generations (t) and predict the proportion of heterozygosity (h) remaining after 1,000 years using expression 6.3.

Species	N_c	N_e	g (years)	t	h
Tuatara	350	70	50	20	0.866
Duvuacel's gecko	1,440	288	15	67	0.890
Common gecko	3,738	747	5	200	0.875
Spotted skink	3,400	680	5	200	0.863
Common skink	4,930	986	5	200	0.904

Allendorf 1989). In conservation biology, we are usually concerned with the loss of genetic variation over some specified period of time because that is the measure of time used in developing policies. For example, "endangered" has been defined as having greater than a 5% chance of becoming extinct within 100 years. Therefore, it is equally important to consider generation interval and N_e when predicting the expected rate of decline of heterozygosity in natural populations.

Conditions that reduce N_e often lengthen the generation interval. Therefore, populations with smaller effective population sizes may actually retain more genetic variation over time than populations with larger effective population sizes (Ryman et al. 1981). This relationship also is often true for differences between species. For example, Keall et al. (2001) estimated the census population size of five species of reptiles on North Brother Island in Cook Strait, New Zealand (Table 7.5). The generation interval for these five species was estimated based upon their life history (age of first reproduction, longevity, etc.; C. H. Daugherty, personal communication). As expected, the species with larger body size (e.g., tuatara) have smaller population sizes and longer generation intervals. The loss of heterozygosity over time is strikingly similar in these five species although they have very different population sizes. That is, the longer generation interval balances the smaller population sizes of the species with larger body size.

7.10 Effective population size in natural populations

The ratio of effective to census population size ($N_e : N_c$) in natural populations is of general importance for the conservation of populations. Census size is generally much easier to estimate than N_e. Therefore, establishing a general relationship between N_c and N_e would allow us to predict the rate of loss of genetic variation in a wide variety of species (Waples 2002).

Frankham (1995) provided a comprehensive review of estimates of effective population size in over 100 species of animals and plants. He concluded that estimates of N_e/N_c averaged approximately 10% in natural populations for studies in which the effects of unequal

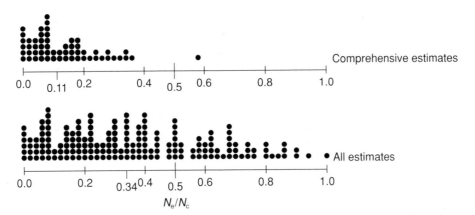

Figure 7.10 Distribution of estimates of N_e/N_c in natural populations. Comprehensive estimates that include unequal sex ratio, variance in reproductive success, and fluctuations in population size are above, and estimates that included only one or two of these effects are below. The means of the estimates (0.11 and 0.34) are indicated below each line. From Frankham (1995).

sex ratio, variance in reproductive success, and fluctuations in population size were included (Figure 7.10). However, Waples (2002) concluded that Frankham (1995) overestimated the contribution of temporal changes by computing the $N_e : N_c$ ratio as a harmonic mean divided by an arithmetic mean. The empirical estimates of N_e that do not include the effect of temporal changes (Frankham 1995) suggest that 20% of the adult population size is perhaps a better general value to use for N_e.

The actual value of N_e/N_c in a particular population or species will differ greatly depending upon demography and life history. For example, Hedgecock (1994) has argued that the high fecundities and high mortalities in early life history stages of many marine organisms can lead to exceptionally high variability in mortality in different families. Thus, N_c may be many orders of magnitude greater than N_e in some populations (see Example 7.4).

Guest Box 7 Estimation of effective population size in Yellowstone grizzly bears
Craig R. Miller and Lisette P. Waits

Grizzly bears have been extirpated from over 99% of their historical range south of the Canadian border (Allendorf and Servheen 1986). During the last century, bears of the Yellowstone ecosystem became isolated from bears in Canada and northern Montana (Figure 7.11). Further, at least 220 bear mortalities occurred between 1967 and 1972 resulting from garbage dump closures and the removal of bears habituated to garbage (Craighead et al. 1995). Assessments of genetic variation with allozymes, mtDNA, and nuclear microsatellite DNA all indicated that the Yellowstone population has significantly lower variability than all other North American mainland populations (see Table 4.1).

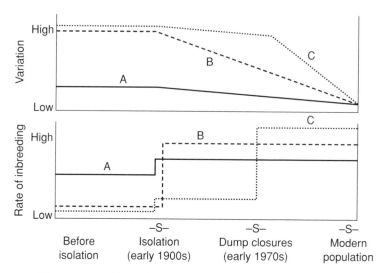

Figure 7.11 Three potential hypotheses (A–C) explaining the low level of genetic diversity observed in the modern Yellowstone grizzly bear population. Genetic samples taken at times indicated by "-S-" were used to resolve among hypotheses. From Miller and Waits (2003).

In the modern Yellowstone population, estimated heterozygosity at eight nuclear microsatellite loci is approximately 20% lower than in the nearby Glacier population (Paetkau et al. 1998). If we assume that Yellowstone bears historically had the same heterozygosity as Glacier bears, we can estimate N_e in Yellowstone since isolation using the expectation that heterozygosity declines at a rate of $1/2N_e$ per generation (recall $H_t = H_0(1 - 1/2N_e)^t$, and expression 6.7). With approximately eight generations since the time of isolation (i.e., $t = 8$), this implies an N_e of only 22. This is depicted in Figure 7.11 as hypothesis B. More troubling is the possibility that most of the postulated decline in heterozygosity occurred following dump closure, implying a very small N_e and a rapid increase in the rate of inbreeding (hypothesis C).

If N_e in Yellowstone has been this small, then genetic drift, inbreeding, and loss of quantitative genetic variation may reduce the population's viability (see Chapter 14). We estimated the effective population size during the 20th century to distinguish among hypotheses B and C, and a third possibility that variation in Yellowstone was historically low and that N_e has remained moderate across the last century (hypothesis A) (Miller and Waits 2003).

DNA was extracted from museum specimens (bones, teeth, and skins) taken from the periods 1910–1920 and 1960–1970, and individuals were genotyped at the same eight loci as above. We used the changes in allele frequency over time to estimate the harmonic mean N_e using maximum likelihood. For both the periods 1915–1965 and 1965–1995, the estimated N_e is approximately 80 (95% CI approximately 50–150). Allelic diversity has declined significantly ($P < 0.05$), but only slightly (Miller and Waits 2003). Estimates of population size in Yellowstone between the 1960s and 1990s are surrounded by large uncertainty, but a summary suggests a harmonic mean N of around 280 individuals. Combining these values yields an estimate of $N_e/N_c = 27\%$; this is similar to an estimate of 25% obtained from a simulation approach (see Example 7.1).

Hence it appears that hypothesis A is supported most by our data. The lower genetic variation in Yellowstone bears appears to predate the decline of population size in the 20th century. With N_e apparently near 80, we argue that the need for gene flow into Yellowstone is not pressing yet. Management should focus mostly on habitat protection, restoring natural connectivity, and limiting human-caused mortalities. If natural connectivity cannot be achieved within a few additional generations (20–30 years), we recommend the translocation of a small number of individuals into Yellowstone (or perhaps artificial insemination using nonresident males, if such technology becomes available for bears). Translocations might be warranted sooner if population vital rates (e.g., survival and reproduction) decline substantially. This example illustrates the usefulness of temporally spaced samples and historical museum specimens to estimate N_e and provide information of great relevance to conservation.

Problem 7.1

Only 10% of the bulls do all of the breeding in some colonies of the southern elephant seal. What effect do you expect this mating system to have on the rate of loss of heterozygosity in these colonies? As an example, compare the expected loss of heterozygosity in an ideal population of 100 individuals (50 males and 50 females) in which only five of the males contribute offspring to the next generation.

Problem 7.2

Bison at the National Bison Range in Montana have been found to be polymorphic at a blood group locus with two common alleles and one allele that is rare ($r = 0.01$). A geneticist has suggested that the heterozygosity, and hence genetic variation, in the herd could and should be increased by preferentially using bulls that have the rare allele. Do you think this suggestion is a good way to increase genetic variation in this herd? Why or why not?

Problem 7.3

A population consisting of 1,000 individuals is reduced to a single female and a single male. The population then doubles in size each generation until it reaches its former size. What proportion of the heterozygosity in the original population would you expect to be lost because of this bottleneck?

Problem 7.4

We saw in Section 7.3 that there are two sources of genetic drift: variability in reproductive success and Mendelian segregation in heterozygotes. If we eliminate variability in reproductive success we can nearly double the effective population size. Why is it impossible in a random mating population to eliminate Mendelian segregation so that we could further increase effective population size? That is, couldn't we just screen all the progeny at a particular locus to insure that the numbers of alleles in the progeny are the same as in the parental generation? There is at least one way to eliminate Mendelian segregation (although it has other serious drawbacks); can you describe such a way?

Problem 7.5

The tuatara has temperature-dependent sex determination (TSD) so that sexual differentiation of the gonads is sensitive to the incubation temperature of eggs during a critical period of embryonic development. In the case of tuatara, higher temperatures during development result in more males being produced (Cree et al. 1995). What effect do you think TSD may have on effective population sizes over long periods of time?

Problem 7.6

Assume that the lone pine tree from Question 7.1 is heterozygous A_1A_2 at a particular locus. What is the probability that a tree in the post-clearcut population will be homozygous for the A_1 allele? What is the probability that a tree in the post-clearcut population will be homozygous for the A_2 allele? What is the probability that a tree in the post-clearcut population will be homozygous for either of the two alleles? Does your answer agree with the solution provided in Question 7.1? Hint: remember that each of the 5,000 seeds will inherit either the A_1 or A_2 allele from the maternal tree.

Problem 7.7

The authors of Guest Box 7 have concluded that the need for gene flow into the grizzly bear population of Yellowstone National Park is not pressing yet. The authors of this book do not agree with this conclusion. What do you think?

Problem 7.8

How many of the original 18 mtDNA haplotypes remain in the population shown in Figure 7.8 after 20 generations of lineage sorting?

8

Natural Selection

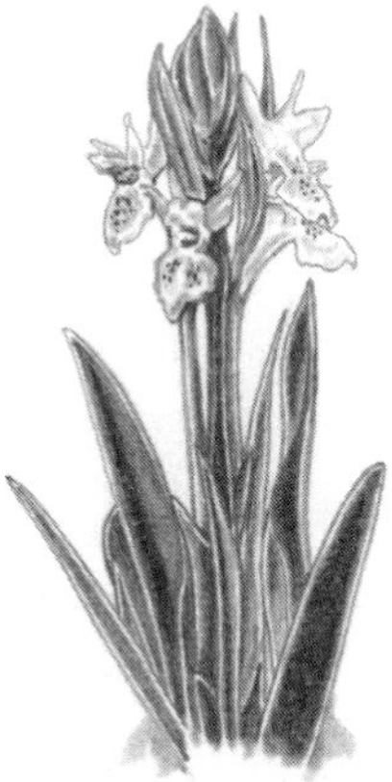

Orchid Dactylorhiza sambucina, *Example 8.2*

I have called this principle, by which each slight variation, if useful, is preserved, by the term Natural Selection.

Charles Darwin (1859)

Then comes the question, Why do some live rather than others? If all the individuals of each species were exactly alike in every respect, we could only say it is a matter of chance. But they are not alike. We find that they vary in many different ways. Some are stronger, some swifter, some hardier in constitution, some more cunning.

Alfred Russel Wallace (1923)

We have so far assumed that different genotypes have an equal probability of surviving and passing on their alleles to future generations. That is, we have assumed that natural selection is not operating. If this assumption were actually true in real populations, we would not be concerned with genetic variation in conservation because genetic changes would not affect a population's longevity or its evolutionary future. However, as we saw in Chapter 6, there is ample evidence that the genetic changes that occur when a population goes through a bottleneck often result in increased frequencies of alleles that reduce an individual's probability of surviving to reproduce.

In addition, some alleles and genotypes affect greater survival and reproductive success under different environmental conditions. Therefore, genetic differences between local populations may be important for continued persistence of those populations. In addition, individuals that are moved by human actions between populations or environments may not be genetically suited to survive and reproduce in their new surroundings. Perhaps worse from a conservation perspective, gene flow caused by such translocations may reduce the adaptation of local populations.

For example, many native species of legumes (*Gastrolobium* and *Oxylobium*) in western Australia naturally synthesize large concentrations of the fluoroacetate, which is the active ingredient in 1080 (a poison used to remove mammalian pests) (King et al. 1978). Native marsupials in western Australia have been found to be resistant to 1080 because they have been eating plants that contain fluoroacetate for thousands of years. Therefore, 1080 is not effective against native mammals in western Australia, which means it can be used as a specific poison for introduced foxes and feral cats that are a serious problem. However, members of the same 1080-resistant mammal species (e.g., brush-tailed possums) that occur to the east, beyond the range of the fluoroacetate-producing legumes, are susceptible to 1080 poisoning. Therefore, translocating brush-tailed possums into western Australia from eastern populations may not be successful because the introduced individuals would not be "adapted" to consume the local vegetation.

Many of the best examples of local adaptation are from plant species because it is possible to do reciprocal transplantations and measure components of fitness (Joshi et al. 2001). Nagy and Rice (1997) performed reciprocal transplant experiments with coastal and inland California populations of the native annual *Gilia capitata*. They compared performance for four traits: seedling emergence, early vegetative size (leaf length), probability of surviving to flowering, and number of inflorescences. Native plants significantly outperformed non-natives for all characters except leaf length. Figure 8.1 shows the results for the proportion of plants that survived to flowering. On average, the native inland plants had over twice the rate of survival compared to non-native plants grown on the inland site; the native coastal plants had 5–10 times great survival rates compared to non-native plants grown on the coastal site.

The adaptive significance of the vast genetic variation that we can now detect using the techniques of biochemical and molecular genetics remains controversial (Gillespie 1992; Mitton 1997). Most of the models that we use to interpret data and predict effects in natural populations assume selective neutrality. This is done not because we believe that all genetic variation is neutral; rather, neutrality is assumed because we sometimes have no choice if we want to use the rich theory of population genetics to interpret data and make predictions because most models assume the absence of natural selection.

Sewall Wright developed powerful theoretical models that allow us to predict the effects of small populations on genetic variability. These models assume selective neutrality. For example, heterozygosity will be lost at a rate of $1/2N$ per generation in the ideal

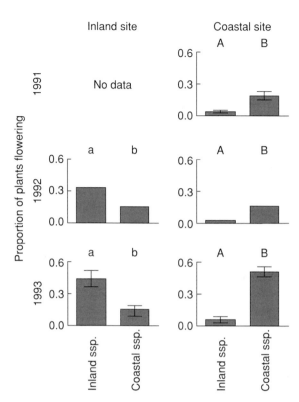

Figure 8.1 Reciprocal transplant experiment with *Gilia capitata* showing local adaptation for the proportion of plants that survived to flowering. The native subspecies had significantly greater survival than the non-native subspecies in each of the five experiments. For example, approximately 45% of the seeds from the inland subspecies survived to flowering in 1993 at the inland site; however, only some 15% of the seeds from the non-native subspecies survived to flowering in the same experiment. a and b are the proportion of the inland and costal plants, respectively, that survived at the inland site; A and B are the proportion of the inland and costal plants, respectively, that survived at the coastal site. From Nagy and Rice (1997).

population. What is the expected rate of loss of heterozygosity if the genetic variability is affected by natural selection? The answer depends upon the pattern and intensity of natural selection in operation. And since no one has ever been successful in measuring these values at any locus in any natural populations, we cannot predict the expected rate of loss of heterozygosity unless we ignore the effects of natural selection. And worse yet, since natural selection acts differently on each locus, there is not one answer, but rather there is a different answer for each of the thousands of variable loci that are likely to exist in any population. In this chapter, we will begin to consider the effects of natural selection on allele and genotype frequencies.

8.1 Fitness

Natural selection is the differential success of genotypes in contributing to the next generation. In the simplest conceptual model, there are two major life history components that may bring about selective differences between genotypes: **viability** and **fertility**. Viability is the probability of survival to reproductive age, and fertility is the average number of off-spring per individual that survive to reproductive maturity for a particular genotype.

The effect of natural selection on genotypes is measured by fitness. Fitness is the average number of offspring produced by individuals of a particular genotype. Fitness can be calculated as the product of viability and fertility, as defined above, and we can define fitness for a diallelic locus as:

Genotype	Viability	Fertility	Fitness
AA	v_{11}	f_{11}	$(v_{11})(f_{11}) = W_{11}$
Aa	v_{12}	f_{12}	$(v_{12})(f_{12}) = W_{12}$
aa	v_{22}	f_{22}	$(v_{22})(f_{22}) = W_{22}$

These are absolute fitnesses; that is, they are based on the total number of expected progeny from each genotype. It is often convenient to use relative fitnesses to predict genetic changes caused by natural selection. Relative fitnesses are estimated by the ratios of absolute fitnesses. For example, in the data below, fitnesses have been standardized by dividing by the fitness of the genotype with the highest fitness (AA). Thus, the relative fitness of heterozygotes is 0.67 because, on average, heterozygotes have 0.67 times as many progeny as AA individuals (1.80/2.70 = 0.67).

Genotype	Viability	Fertility	Absolute fitness	Relative fitness
AA	0.90	3.00	2.70	1.00
Aa	0.90	2.00	1.80	0.67
aa	0.45	2.00	0.90	0.33

8.2 Single locus with two alleles

We will begin by modeling changes caused by differential survival (viability selection) in the simple case of a single locus with two alleles. Consider a single diallelic locus with differential reproductive success in a large random mating population in which all of the other assumptions of the Hardy–Weinberg model are valid. We would expect the following result after one generation of selection:

Genotype	Zygote frequency	Relative fitness	Frequency after selection
AA	p^2	w_{11}	$(p^2 w_{11})/w$
Aa	$2pq$	w_{12}	$(2pq w_{12})/w$
aa	q^2	w_{22}	$(q^2 w_{22})/w$

where w is used to normalize the frequencies following selection so that they sum to one. This is the average fitness of the population, and it is the fitness of each genotype weighted by its frequency.

$$\bar{w} = p^2 w_{11} + 2pq w_{12} + q^2 w_{22} \tag{8.1}$$

After one generation of selection the frequency of the A allele is:

$$p' = \frac{p^2 w_{11}}{\bar{w}} + \left(\frac{1}{2}\right)\left(\frac{2pq w_{12}}{\bar{w}}\right) = \frac{p^2 w_{11} + pq w_{12}}{\bar{w}} \tag{8.2}$$

and, similarly, the frequency of the a allele is:

$$q' = \frac{pq w_{12} + q^2 w_{22}}{\bar{w}} \tag{8.3}$$

It is often convenient to predict the change in allele frequency from generation to generation, Δp, caused by selection. We get the following result if we solve for Δp in the current case:

$$\Delta p = \frac{pq}{\bar{w}}[p(w_{11} - w_{12}) + q(w_{12} - w_{22})] \tag{8.4}$$

We can see that the magnitude and direction of change in allele frequency is dependent on the fitnesses of the genotypes and the allele frequency.

Expression 8.4 can be used to predict the expected change in allele frequency after one generation of selection for any array of fitnesses. The allele frequency in the following generation will be:

$$p' = p + \Delta p \tag{8.5}$$

We will use this model to study the dynamics of selection for three basic modes of natural selection with constant fitnesses:

1 Directional selection.
2 Heterozygous advantage (overdominance).
3 Heterozygous disadvantage (underdominance).

8.2.1 Directional selection

Directional selection occurs when one allele is always at a selective advantage. The advantageous allele under directional selection may either be dominant, intermediate, or recessive to the alternative allele as shown below:

Dominant	$w_{11} = w_{12} > w_{22}$
Intermediate	$w_{11} > w_{12} > w_{22}$
Recessive	$w_{11} > w_{12} = w_{22}$

The advantageous allele will increase in frequency and will be ultimately fixed by natural selection under all three modes of directional selection (Figure 8.2). Thus, the eventual or equilibrium outcome is independent of the dominance of the advantageous allele. However, the rate of change of allele frequency does depend on dominance relationships

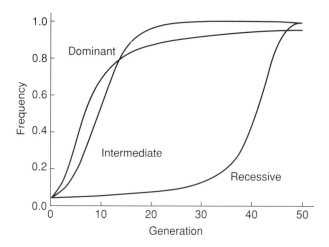

Figure 8.2 Change in allele frequency under directional selection when the homozygote for the favored allele has twice the fitness of the homozygote for the unfavored allele (1.00 vs. 0.50). The heterozygote either has the same fitness as the favored allele (1.0; dominant), the same fitness as the unfavored allele (recessive; 0.50), or has intermediate fitness (0.75). The initial frequency of the favored allele is 0.03.

as well as the intensity of selection. For example, selection on a recessive allele is ineffective when the recessive allele is rare because most of the copies of that allele occur in heterozygotes and are therefore "hidden" from selection.

8.2.2 Heterozygous advantage (overdominance)

Heterozygous advantage occurs when the heterozygote has the greatest fitness:

$$w_{11} < w_{12} > w_{22}$$

This mode of selection is expected to maintain both alleles in the population as a **stable polymorphism**. This pattern of selection is often called overdominance. In the case of dominance, the phenotype of the heterozygote is equal to the phenotype of one of the homozygotes. In overdominance, the phenotype (i.e., fitness) of the heterozygote is greater than the phenotype of either homozygotes (Example 8.1).

Let us examine the simple case of heterozygous advantage in which the two homozygotes have equal fitness:

	AA	Aa	aa
Fitness	$1 - s$	1.0	$1 - s$

where s (the selection coefficient) is greater than 0 and less than or equal to 1. We can examine the dynamics of this case of selection by plotting the values of Δp as a function of allele frequency (Figure 8.3). When p is less than 0.5, selection will increase p, and when p

Example 8.1 Natural selection at an allozyme locus

Patarnello and Battaglia (1992) have described an example of heterozygous advantage at a locus encoding the enzyme glucose phosphate isomerase (GPI) in a copepod (*Gammarus insensibilis*) that lives in the Lagoon of Venice. Individuals were collected in the wild, acclimated in the laboratory at room temperature, and then held at a high temperature (27°C) for 36 hours. Individuals with different genotypes differed significantly in their survival at this temperature (Table 8.1; $P < 0.005$). Heterozygotes survived better than either of the homozygotes.

A persistent problem in measuring fitnesses of individual genotypes is whether any observed differences are due to the locus under investigation or to other loci that are linked to that locus (Eanes 1987). *In vitro* measurements show that heterozygotes at the *GPI* locus in *Gammarus insensibilis* have greater enzyme activity than either homozygote over a wide range of temperatures. In addition, the *80/80* homozygote has the greatest mortality and the lowest enzyme activity. Patarnello and Battaglia (1992) have argued that the observed differences are caused by the *GPI* genotype on the basis of these enzyme kinetic properties and other considerations.

Table 8.1 Differential survival of GPI genotypes in the copepod *Gammarus insensibilis* held in the laboratory for 36 hours at high temperature (27°C). From Patarnello and Battaglia (1992).

	Genotype		
	100/100	*100/80*	*80/80*
Alive	48	90	12
Dead	47	53	27
Total	95	143	39
Relative survival	0.803	1.000	0.490

is greater than 0.5, selection will decrease p. Thus, 0.5 is a stable equilibrium; that is, when p is perturbed from 0.5, it will return to that value.

Any overdominant fitness set will produce a stable intermediate equilibrium allele frequency ($p^\star$). However, the value of $p^\star$ depends upon the relative fitnesses of the homozygotes. If we solve expression 8.4 for $\Delta p = 0$ we get the following result:

$$p^\star = \frac{w_{12} - w_{22}}{2w_{12} - w_{11} - w_{22}} \tag{8.6}$$

Thus, the equilibrium allele frequency will be near 0.5 if the two homozygotes have nearly equal fitnesses. However, if one homozygote has a great advantage over the over, that allele will be much more frequent at equilibrium.

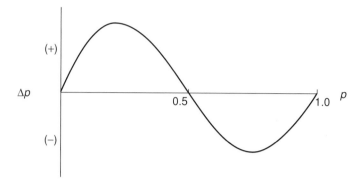

Figure 8.3 Expected change in allele frequency (Δp) as a function of allele frequency (p) in the case of heterozygous advantage when the homozygotes have equal fitness.

8.2.3 Heterozygous disadvantage (underdominance)

Underdominance occurs when the heterozygote is least fit:

$$w_{11} > w_{12} < w_{22}$$

An examination of Δp as a function of p reveals that underdominance will produce what is called an unstable equilibrium (Figure 8.4). The $p^\star$ value is found using the same formula as for overdominance (expression 8.6). However, this equilibrium is unstable because allele frequencies will tend to move away from the equilibrium value once they are perturbed. Underdominance, therefore, is not a mode of selection that will maintain genetic variation in natural populations.

We saw in Chapter 3 that heterozygotes for chromosomal rearrangements often have reduced fertility because they produce unbalanced or aneuploid gametes. Foster et al. (1972) examined the behavior of translocations in population cages of *Drosophila melanogaster*. They set up cages in which homozygotes for the chromosomal rearrange-

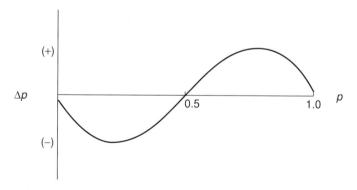

Figure 8.4 Expected change in allele frequency (Δp) as a function of allele frequency (p) in the case of heterozygous disadvantage.

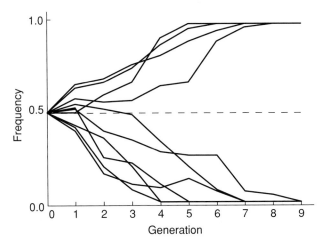

Figure 8.5 Population cage results with *Drosophila melanogaster* showing change in frequency of a chromosomal translocation in 10 populations when the two homozygotes have equal fitness that is approximately twice that of heterozygotes (Foster et al. 1972). Populations were founded by 20 individuals and population sizes fluctuated between 100 and 400 flies.

ments had equal fitness; in this case, the unstable equilibrium frequency is expected to be 0.5 (see expression 8.6). As predicted by this analysis, the populations quickly went to fixation for whichever chromosomal type became most frequent in the early generations because of genetic drift (Figure 8.5).

8.2.4 Selection and Hardy–Weinberg proportions

The absence of departures from Hardy–Weinberg proportions is sometimes taken as evidence that a particular locus is not affected by natural selection. However, this interpretation is incorrect for several reasons. First, differences in fecundity will not affect Hardy–Weinberg proportions. Thus, only differential survival can be detected by testing for Hardy–Weinberg proportions. Second, even strong differences in survival may not cause departures from Hardy–Weinberg proportions. For example, Lewontin and Cockerham (1959) have shown that at a locus with two alleles, differential survival will not cause a departure from Hardy–Weinberg proportions if the product of the fitnesses of the two homozygotes is equal to the square of the fitness of the homozygotes. Finally, the goodness-of-fit test for Hardy–Weinberg proportions has little power to detect departures from Hardy–Weinberg proportions caused by differential survival (see Problem 8.9).

8.3 Multiple alleles

Analysis of the effects of natural selection becomes more complex when there are more than two alleles at a locus because the number of genotypes increases dramatically with a modest increase in the number of alleles; remember, there are 55 possible genotypes with just 10 alleles at a single locus. Nevertheless, our model of selection can be readily extended to three alleles (A_1, A_2, and A_3):

Genotype	A_1A_1	A_1A_2	A_2A_2	A_1A_3	A_2A_3	A_3A_3
Fitness	w_{11}	w_{12}	w_{22}	w_{13}	w_{23}	w_{33}
Frequency	p^2	$2pq$	q^2	$2pr$	$2qr$	r^2

The average fitness of the population is:

$$\bar{w} = p^2 w_{11} + 2pq w_{12} + q^2 w_{22} + 2pr w_{13} + 2qr w_{23} + r^2 w_{33} \tag{8.7}$$

And the expected allele frequencies in the next generation are:

$$p' = \frac{(p^2 w_{11} + pq w_{12} + pr w_{13})}{\bar{w}}$$

$$q' = \frac{(pq w_{12} + q^2 w_{22} + qr w_{23})}{\bar{w}} \tag{8.8}$$

$$r' = \frac{(pr w_{13} + qr w_{23} + r^2 w_{33})}{\bar{w}}$$

We can find any equilibria that exist for a particular set of fitnesses by setting $p' = p = p^\star$ and solving these equations. The following conditions emerge after a bit of maths:

$$p^\star = \frac{z_1}{z} \quad \text{where} \quad z_1 = (w_{12} - w_{22})(w_{13} - w_{33}) - (w_{12} - w_{23})(w_{13} - w_{23})$$

$$q^\star = \frac{z_2}{z} \quad \text{where} \quad z_2 = (w_{23} - w_{33})(w_{12} - w_{11}) - (w_{23} - w_{13})(w_{12} - w_{13})$$

$$r^\star = \frac{z_3}{z} \quad \text{where} \quad z_3 = (w_{13} - w_{11})(w_{23} - w_{22}) - (w_{13} - w_{12})(w_{23} - w_{12})$$

where:

$$z = z_1 + z_2 + z_3 \tag{8.9}$$

If these equations give negative values for the allele frequencies that means there is no three-allele equilibrium (i.e., at least one allele will be lost due to selection). The equilibrium will be stable if the equilibrium is a maximum for average fitness (see expression 8.6) and will be unstable if it is a minimum for average fitness. In general, a three-allele equilibrium will be stable if z_1, z_2, and z_3 are greater than zero and:

$$(w_{11} + w_{22}) < (2w_{13}) \tag{8.10}$$

There are no simple rules for a locus with three alleles as there are for a dialleic locus. However, the following statements may be helpful:

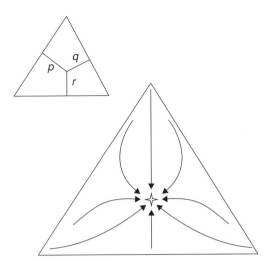

Figure 8.6 Expected trajectory of allele frequency change in the case of heterozygous advantage with three alleles plotted on triangular coordinate paper. All homozygotes have a fitness of 0.9, and all heterozygotes have a fitness of 1.0. As shown in the upper left, allele frequencies are represented by the relative lengths of the three perpendicular lines from any point to the three sides of the triangle.

1   There is at most one stable equilibrium for two or more alleles.
2   A stable equilibrium will be globally stable; that is, it will be reached from any starting point containing all three alleles.
3   If a stable polymorphism exists, the mean fitness of the population exceeds that of any homozygote. If such a homozygote existed, it would become fixed in the population.
4   Heterozygous advantage (i.e., all heterozygotes have greater fitness than all homozygotes) is neither necessary nor sufficient for a stable polymorphism.

The dynamics of selection acting on three alleles can be shown by plotting the trajectories of allele frequencies on triangular coordinate paper. Figure 8.6 shows allele frequency change when all of the homozygotes have a fitness of 0.9 and all heterozygotes have a fitness of 1.0. In this case, we would expect a stable equilibrium to occur when all three alleles are equally frequent at a frequency of 0.33.

Templeton (1982) has described a very interesting set of fitnesses for three alleles at the human β-chain hemoglobin locus (Table 8.2). Figure 8.7 shows the expected trajectories of gene frequencies at this locus. The stable two-allele polymorphism with the A and S alleles is a familiar example of heterozygous advantage (using expression 8.6, $p^\star = 0.89$). However, the fitness of the homozygotes for the C allele is greater than the AS heterozygotes. Nevertheless, the C allele will be selected against when it is rare because the AC and SC genotypes both have relatively low fitnesses. Thus, the C allele will be removed by selection from a population if it occurs as a new mutation in a population with the A and C alleles present. The only way the C allele can successfully invade a population is if it increases in frequency through genetic drift so that the CC genotype becomes frequent enough to outweigh the disadvantage of the C allele when heterozygous. However, recent data have

Table 8.2 Estimated relative fitness at the β-hemoglobin locus in West African human populations (Templeton 1982).

Genotype	Fitness	Phenotype
AA	0.9	Malarial susceptibility
AS	1.0	Malarial resistance
SS	0.2	Sickle-cell anemia
AC	0.9	Malarial susceptibility
SC	0.7	Malarial susceptibility
CC	1.3	Superior malarial resistance

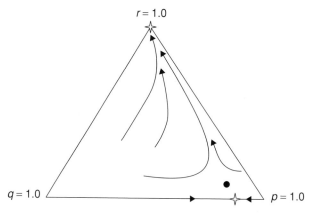

Figure 8.7 Expected allele frequency trajectories for the fitnesses of the hemoglobin locus shown in Table 8.2. There are two stable equilibria indicated by stars at the top of the triangle ($r = 1.0$) and towards the bottom right ($r = 0.0$) of the triangle. The equilibrium indicated by the black circle at the bottom right is unstable. $p = \text{freq}(A)$; $q = \text{freq}(S)$; $r = \text{freq}(C)$.

suggested that *C* allele heterozygotes do not have decreased fitness (Modiano et al. 2001), and, therefore, would be expected to replace the *A* and *S* alleles in malarial regions.

8.3.1 Heterozygous advantage and multiple alleles

Overdominance was once thought to be the major mechanism maintaining genetic variation in natural populations (see discussion in Lewontin 1974, pp. 23–31). However, modern molecular techniques have revealed that many alleles exist at some loci in natural populations. For example, Singh et al. (1976) discovered 37 different alleles at a locus coding for xanthine dehydrogenase in a sample of 73 individuals collected from 12 natural populations of *Drosophila pseudoobscura*.

Can overdominance maintain many alleles at a single locus? This question was approached in a classic paper by Lewontin et al. (1978). The following discussion is based on their paper. They estimated the proportion of randomly chosen fitness sets that would maintain all alleles through overdominance. For a locus with two alleles, heterozygous advantage is both necessary and sufficient to maintain both alleles. If fitnesses are selected

Table 8.3 Proportion of randomly chosen fitness that maintains all *A* alleles in a stable equilibrium (Lewontin et al. 1978). The third column shows the proportion of fitness sets expected to maintain all *A* alleles considering only those fitness sets in which all heterozygotes have greater fitness than all homozygotes.

n	All fitness sets	Heterozygous advantage
2	0.33	1.00
3	0.04	0.71
4	0.0024	0.34
5	0.00006	0.10
6	0	0.01

at random, the heterozygotes will have the greatest fitness one-third of the time because there are three genotypes (Table 8.3). However, it becomes increasingly unlikely that all heterozygotes will have greater fitness than all homozygotes as the number of alleles increases. In addition, heterozygous advantage (i.e., all heterozygotes have greater fitness than all homozygotes) is neither necessary nor sufficient to maintain an *A* allele polymorphism when *A* is greater than two. In fact, fitness sets capable of maintaining an *A* allele polymorphism quickly become extremely unlikely as *n* increases (Table 8.3). For example, heterozygous advantage is sufficient to always produce a stable polymorphism with two alleles. However, only 34% of all fitness sets with four alleles in which all heterozygotes have greater fitness than all homozygotes will maintain all four alleles. Thus, overdominance with constant fitness is not an effective mechanism for maintaining many alleles at individual loci in natural populations (Kimura 1983).

Spencer and Marks (1993) have revisited this issue with different results. Rather than randomly assigning fitness as done by Lewontin et al. (1978), they simulated evolution by allowing new mutations with randomly assigned fitnesses to occur within a large population and then determined how many alleles could be maintained in the population by viability selection. They found that up to 38 alleles were sometimes maintained by selection in their simulated populations. In general, they found many more alleles could be maintained by this type of selection than predicted by Lewontin et al. (1978).

Spencer and Marks (1993) argued that their approach, which examines how a polymorphism may be constructed by evolution, is a complementary approach to understanding evolutionary dynamics when used along with traditional models that focus only on conditions that maintain equilibrium. Nevertheless, the conclusions of Lewontin et al. (1978) are still likely to be valid, even if the approach of Spencer and Marks (1993) is more realistic. One major drawback of the results of Spencer and Marks (1993) is that their models do not include genetic drift, and, as we will see in Section 8.5, heterozygous advantage is only effective in maintaining alleles that are relatively common in a population at equilibrium.

Hedrick (2002) has considered the maintenance of many alleles at a single locus by "balancing selection" at the *MHC* locus. He then assumed resistance to pathogens is conferred by specific alleles and the action of each allele is dominant. He concluded that this model of selection could maintain stable multiple allele polymorphisms, even in the absence of any intrinsic heterozygous advantage, because heterozygotes will have higher fitness in the presence of multiple pathogens.

8.4 Frequency-dependent selection

We have so far assumed that fitnesses are constant. However, fitnesses are not likely to be constant in natural populations (Kojima 1971). Fitnesses are likely to change under different environmental conditions. Fitnesses may also change when allele frequencies change; this is called frequency-dependent selection. This type of selection is a potentially powerful mechanism for maintaining genetic variation in natural populations (Clarke and Partridge 1988).

8.4.1 Two alleles

Let us begin with the simple case where the fitness of a genotype is a direct function of its frequency. For example,

$$AA \qquad Aa \qquad aa$$
$$1 - p^2 \qquad 1 - 2pq \qquad 1 - q^2 \qquad\qquad\qquad (8.11)$$

With this model of selection, a genotype becomes less fit as it becomes more common in a population. The change in allele frequency at any value of p can be calculated with expression 8.4. We can predict the expected effects of this pattern of selection by examination of the plot of Δp versus allele frequency; we will get the same plot as Figure 8.3. In this case, there is an equilibrium at $p^\star = 0.5$ where Δp is zero. Is this equilibrium stable or unstable? When p is less than 0.5, $w_{11} > w_{22}$ and therefore p will increase; when p is greater than 0.5, $w_{11} < w_{22}$ and p will decrease. This is a stable equilibrium.

Note that the homozygotes have a fitness of 0.75, and the heterozygote has a fitness of 0.5 at equilibrium. Therefore, this is a stable polymorphism in which the heterozygote has a disadvantage at equilibrium. We can see that our rules for understanding the effects of selection with constant fitnesses are not likely to be helpful in understanding the effects of frequency-dependent selection. In general, frequency-dependent selection will produce a stable polymorphism whenever the rare phenotype has a selective advantage. However, there is no general rule about the relative fitnesses at the equilibrium.

8.4.2 Multiple alleles: self-incompatibility locus in plants

In contrast to heterozygous advantage, frequency-dependent selection can be extremely powerful for maintaining multiple alleles. The self-incompatibility locus (S) of many flowering plants is an extreme example of this (Wright 1965a; Vieira and Charlesworth 2002; Castric and Vekemans 2004). In the simplest system, pollen grains can only fertilize plants that do not have the same S allele as carried by the pollen. Homozygotes cannot be produced at this locus, and at least three alleles must be present at this locus.

The expected equilibrium with three alleles will be a frequency of 0.33 for each allele because fitnesses are equivalent for all three alleles. At equilibrium, any pollen grain will be able to fertilize one-third of the plants in the population (Table 8.4). However, a fourth allele produced by mutation (S_4) would have a great selective advantage because it will be able to fertilize every plant in the population. Thus, we would expect the fourth allele to increase in frequency until it reaches a frequency equal to the other three alleles.

Table 8.4 Genotypes possible at the self-incompatibility locus (S) in a species of flowering plants with three alleles.

Parental genotypes		Progeny frequencies		
Ovule	Pollen	S_1S_2	S_1S_3	S_2S_3
S_1S_2	S_3	0.00	0.50	0.50
S_1S_3	S_2	0.50	0.00	0.50
S_2S_3	S_1	0.50	0.50	0.00

Any new mutation at the S locus is expected to have an initial selective advantage because of its rarity regardless of the existing number of alleles. However, we also would expect rare alleles to be susceptible to loss because of genetic drift. Therefore, the total number of S alleles will be an equilibrium between mutation and genetic drift.

Emerson (1939, 1940) described 45 nearly equal-frequency S alleles in a narrow endemic plant (*Oenothera organensis*) that occurs in an area of approximately 50 km^2 in the Organ Mountains, New Mexico. Emerson originally thought that the total population size of this species was approximately 500 individuals. More recent surveys indicate that the total population size may be as great as 5,000 individuals (Levin et al. 1979). Regardless of the actual population size, this is an enormous amount of variability at a single locus. As expected, because of its small population size, this species has very little genetic variation at other loci as measured by protein electrophoresis (Levin et al. 1979).

8.4.3 Frequency-dependent selection in nature

There is mounting evidence that frequency-dependent selection is an important mechanism for maintaining genetic variation in natural populations. We have already examined the ability of frequency-dependent selection in maintaining a large number of alleles at self-sterility loci. You are encouraged to read the review by Clarke (1979); additional references on frequency-dependent selection can be found in a collection of papers edited by Clarke and Partridge (1988). Frequency-dependent selection often results from mechanisms of sexual selection, predation and disease, and ecological competition (Example 8.2).

Example 8.2 Frequency-dependent selection in an orchid

Gigord et al. (2001) have presented an elegant example of frequency-dependent selection in the orchid *Dactylorhiza sambucina*. This species has a dramatic flower color polymorphism; both yellow- and purple-flowered individuals occur throughout the range of the species in Europe. Laboratory experiments suggested that behavioral responses by pollinators to the lack of reward might result in a reproductive advantage for rare color morphs. This was confirmed in an experiment that demonstrated that rare color morphs had a selective advantage in natural populations (Figure 8.8).

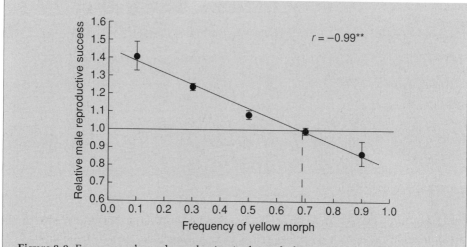

Figure 8.8 Frequency-dependent selection in the orchid *Dactylorhiza sambucina*. Relative male reproductive success of the yellow morph increases as the frequency of the yellow morph decreases. Male reproductive success was estimated by the average proportion of pollinia (mass of fused pollen produced by many orchids) removed from plants by insect pollinators. The horizontal line corresponds to equal reproductive success between the two morphs. The intersection between the regression line and the horizontal line is the value of predicted morph frequencies at equilibrium (represented by vertical dashed line). From Gigord et al. (2001)).

8.5 Natural selection in small populations

We will now consider what happens when we combine the effects of genetic drift and natural selection. In other words, we will see what effect finite population size has on the models of natural selection that we have studied previously. There are two general effects of adding genetic drift to these models. First, natural selection becomes less effective because the random changes caused by drift can swamp the effects of increased survival or fertility. In general, changes in allele frequency are determined primarily by genetic drift rather than by natural selection when the product of the effective population size and the selection coefficient ($N_e s$) is less than one (Li 1978). Thus, a deleterious allele that reduces fitness by 5% will act as if it were selectively neutral in a population with an N_e of $20 (20 \times 0.05 = 1.00)$.

Second, the effects of natural selection become less predictable. The results of our models are deterministic so that we always get the same result if we begin with the same fitnesses and the same initial allele frequency. However, the stochasticity due to genetic drift makes it more difficult to predict what the effects of natural selection will be.

8.5.1 Directional selection

Genetic drift will make directional selection less effective. This may be harmful in small populations in two ways. First, the effects of random genetic drift can outweigh the effects of natural selection so that alleles that have a selective advantage may be lost in small populations. Second, alleles that are at a selective disadvantage may go to fixation in small populations through genetic drift.

Wright (1931, p. 157) first suggested that small populations would continue to decline in vigor slowly over time because of the accumulation of deleterious mutations, which natural selection would not be effective in removing because of the overpowering effects of genetic drift. A number of theoretical papers have considered the expected rate and importance of this effect for population persistence (Lynch and Gabriel 1990; Gabriel and Bürger 1994; Lande 1995). As deleterious mutations accumulate, population size may decrease further and thereby accelerate the rate of accumulation of deleterious mutations. This feedback process has been termed **mutational meltdown** (Lynch et al. 1993).

8.5.2 Underdominance and drift

Most chromosomal rearrangements (translocations, inversions, etc.) cause reduced fertility in heterozygotes because of the production of aneuploid gametes. Homozygotes for such chromosomal mutations, however, may have increased fitness. Thus, chromosomal mutations generally fit a pattern of underdominance and will always be initially selected against, regardless of their selective advantage when homozygous. However, we know that chromosomal rearrangements are sometimes incorporated into populations and species. In fact, rearrangements are thought to be an important factor in reproductive isolation and speciation.

How can we reconcile our theory with our knowledge from natural populations? That is, how can chromosomal rearrangements be incorporated into a population when they will always be initially selected against? The answer is, of course, genetic drift. If random changes in allele frequency perturb the population across the threshold of the unstable p^*, then natural selection will act to "fix" the chromosomal rearrangement. Thus, we would expect faster rates of chromosomal evolution in species with small local deme sizes.

In fact, it has been proposed that the rapid rate of chromosomal evolution and speciation in mammals is due to their social structuring and reduced local deme sizes (Wilson et al. 1975). A paper by Russell Lande (1979) examined the theoretical relationship between local deme sizes and rates of chromosomal evolution. As discussed in Chapter 3, chromosomal variability is of special importance for conservation because the demographic characteristics that make a species a likely candidate for being threatened are the same characteristics that favor the evolution of chromosomal differences between groups. Therefore, reintroduction or translocation programs may reduce the average fitness of a population if individuals are exchanged among chromosomally distinct groups.

8.5.3 Heterozygous advantage and drift

We have seen that heterozygous advantage in a two-allele system will always produce a stable polymorphism with infinite population size. However, overdominance may actually accelerate the loss of genetic variation in finite populations if the equilibrium allele frequency is near 0 or 1 (Robertson 1962).

Consider the following fitness set:

AA	Aa	aa
$1 - s_1$	1	$1 - s_2$

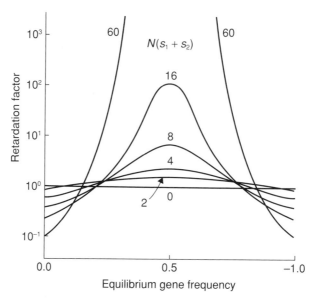

Figure 8.9 Relative effectiveness of heterozygous advantage to maintain polymorphism. The retardation factor is the reciprocal of the rate of decay of genetic variation relative to the neutral case, and N is actually N_e. Values of less than one indicate a more rapid rate of loss of genetic variation than expected with selective neutrality. Thus, even strong natural selection [e.g., $N(s_1 + s_2) = 60$] is not effective at maintaining a polymorphism if the equilibrium allele frequency is less than 0.20 or greater than 0.80.

The following equilibrium allele frequency results if we substitute these fitness values into expression 8.6:

$$p^\star = \frac{s_2}{s_1 + s_2} \tag{8.12}$$

When $p^\star$ is less than 0.2 or greater than 0.8, this mode of selection will actually lose genetic variation more quickly than the neutral case where $s_1 = s_2 = 0$ unless selection is very strong or the population size is very large (Figure 8.9). $N_e(s_1 + s_2)$ is used as a measure of the effectiveness of selection in this analysis; the effectiveness of selection increases as effective population size (N_e) increases and the intensity of selection ($s_1 + s_2$) increases. For example, $N_e(s_1 + s_2)$ will equal 60 when $(s_1 + s_2) = 0.2$ and $N_e = 300$ or $(s_1 + s_2) = 0.4$ and $N_e = 150$. Thus, heterozygous advantage is only effective at maintaining fairly common alleles (frequency > 0.2) except in large populations.

8.6 Natural selection and conservation

An understanding of natural selection is important for the management and conservation of populations. Adaptation to captive conditions is a major concern for captive breeding programs of plants and animals (Frankham et al. 1986; Ford 2002) (see Chapter 18). We

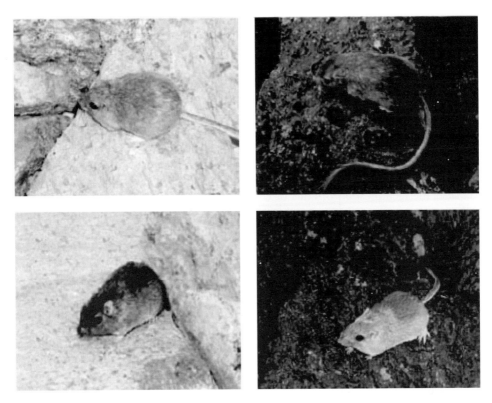

Figure 8.10 Light and dark phenotypes of rock pocket mice on light-colored rocks and dark lava. From Nachman et al. (2003).

saw in Chapter 1 that the size of pink salmon on the west coast of North America has declined dramatically in just 25 years, apparently in response to selective fishing pressure. There are many examples of rapid responses to natural selection in wild populations. Stockwell et al. (2003) have recently reviewed the potential importance of short-term responses to natural selection in conservation biology.

Evidence for natural selection on morphological traits is widespread in natural populations (see Guest Box 8). However, detecting the effects of natural selection at individual loci has proven to be a very difficult problem ever since the discovery of widespread molecular polymorphisms in natural populations (Lewontin 1974; Watt 1995; but see Section 9.6.3).

Nachman et al. (2003) have presented an elegant example of the action of natural selection on an individual locus resulting in local adaptation. Rock pocket mice are generally light-colored and match the color of the rocks on which they live. However, mice that live on dark lava are dark colored (melanic), and this concealing coloration provides protection from predation (Figure 8.10). These authors examined several candidate loci that were known to result in changes in pigmentation in other species. They found mutations in the melanocortin-1 receptor gene that were responsible for the dark coloration in one population of lava-dwelling mice that were melanic. However, they found no evidence of mutations at this locus in another melanic population. Thus, the similar adaptation of dark coloration apparently has evolved by different genetic mechanisms in different populations (see Guest Box 12).

There is substantial evidence for natural selection acting on *MHC* loci in many species (Edwards and Hedrick 1998). Nevertheless, how this information should be applied in a conservation perspective has been controversial. Hughes (1991) recommended that "all captive breeding programs for endangered vertebrate species should be designed with the preservation of MHC allelic diversity as their main goal". There are a variety of potential problems associated with following this recommendation (Gilpin and Wills 1991; Miller and Hedrick 1991; Vrijenhoek and Leberg 1991). The primary problem is that "selecting" individuals on the basis of their MHC genotype could reduce genetic variation throughout the rest of the genome (Lacy 2000a). We will revisit these issues in later chapters when we consider the identification of units of conservation (Chapter 16) and captive breeding (Chapter 18).

Frequency-dependent selection has special importance for conservation because of the many functionally distinct alleles that are maintained by frequency-dependent selection at some loci. We have seen that allelic diversity is much more affected by bottlenecks than is heterozygosity (see Section 6.4). Reinartz and Les (1994) concluded that some one-third of the remaining 14 natural populations of *Aster furactus* in Wisconsin, USA had reduced seed sets because of a diminished number of *S* alleles. Young et al. (2000a) have considered the effect of loss of allelic variation at the *S* locus on the viability of small populations (see Guest Box 3). In addition, frequency-dependent selection probably contributes to the large number of alleles present at some loci associated with disease resistance (e.g., MHC; see Section 6.7). Thus, the loss of allelic diversity caused by bottlenecks is likely to make small populations more susceptible to epidemics.

Many local adaptations of native populations will be difficult to detect because they will only be manifest during periodic episodes of extreme environmental conditions, such as winter storms (Example 8.3), drought, or fire (Gutschick and BassiriRad 2003). Weins (1977) has argued that short-term studies of fitness and other population characteristics are of limited value because of the importance of "ecological crunches" in variable environments. For example, Rieman and Clayton (1997) suggest that the complex life histories (e.g., mixed migratory behaviors) of bull trout are adaptations to periodic disturbances such as fire that may affect populations only every 25–100 years.

Example 8.3 Intense natural selection on cliff swallows during winter storms

Brown and Brown (1998) reported dramatic selective effects of body size on the survival of cliff swallows in a population from the Great Plains of North America (Figure 8.11). Cliff swallows in these areas are sometimes exposed to periods of cold weather in late spring that reduce the availability of food. Substantial mortality generally results if the cold spell lasts 4 or more days. A once in a hundred year 6-day cold spell occurred in 1996 that killed approximately 50% of the cliff swallows in southwestern Nebraska.

Comparison of survivors and dead birds revealed that larger birds were much more likely to survive (Figure 8.11a). Mortality patterns did not differ in males and

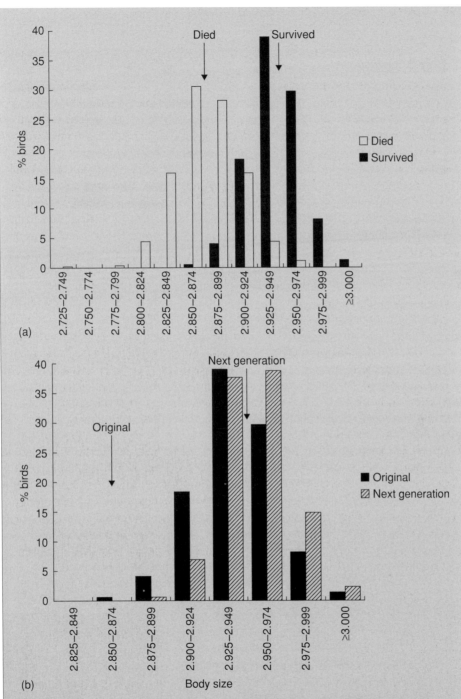

Figure 8.11 Intense natural selection on cliff swallows during a harsh winter storm. (a) Larger birds were much more likely to survive the storm than smaller birds. Body size is a multivariate measure that includes wing length, tail length, tarsus length, and culmen length and width. (b) Adult progeny in the next generation were much larger than the mean of the population before the storm event. Thus, natural selection increased the size of this population. Arrows indicate means. From Brown and Brown (1998).

females, but older birds were less likely to survive. Morphology did not differ with age. Nonsurvivors were not in poorer condition before the storm suggesting that selection acted on size and not condition. Larger birds apparently were favored in extreme cold weather due to the thermal advantage of larger size and the ability to store more body fat.

Examination of the adult progeny of the survivors indicated that mean body size of the population responded to the selective event caused by the storm. The body size of progeny was significantly greater than the body size of the population before the storm (Figure 8.11b). Thus body size had a substantially high heritability (see Section 2.2; Chapter 11).

Guest Box 8 Rapid adaptation and conservation
Craig A. Stockwell and Michael L. Collyer

Recent work has demonstrated that adaptive evolution often occurs on contemporary time scales (years to decades), making it of particular relevance to conservation planning (Ashley et al. 2003; Stockwell et al. 2003). Reports of rapid evolution span a variety of species, traits, and situations, suggesting that rapid adaptation is the norm rather than the exception (Stockwell et al. 2003). Furthermore, rapid adaptation is often associated with the same anthropogenic factors responsible for the current extinction crisis, including overharvest, habitat degradation, habitat fragmentation, and exotic species (Stockwell et al. 2003). Rapid evolution has crucial importance to conservation biology.

Here, we briefly discuss the implications of rapid adaptation for conservation biology in the context of exotic species and actively managed species. First, many case studies of rapid adaptation involve non-native species. For instance, introduced fish populations have been shown to undergo rapid adaptation in response to novel predator regimes and breeding environments (Stockwell et al. 2003). Further, exotic species may create novel selection pressures for native biota (see Chapter 19). Such is the case with the soapberry bug that apparently evolved shorter beak length in response to the smaller seed pods of the exotic flat-potted golden rain tree (Carroll et al. 2001). Finally, invasion dynamics may be influenced by the evolution of exotics as they encounter novel selection pressure(s) during invasion (García-Ramos and Rodríguez 2002).

Rapid adaptation can also result in the evolution of less preferred phenotypes. For instance, selective harvest has been associated with evolution of smaller body size in harvested populations of fish (Olsen et al. 2004a) and bighorn sheep (see Guest Box 11; Coltman et al. 2003). Likewise, rapid adaptation of smaller egg size has been observed for a captive population of chinook salmon (Heath et al. 2003). In this case, wild populations supplemented with this stock have also shown a decrease in egg size (Heath et al. 2003).

Rapid adaptation may also occur for so-called **refuge populations** that are established as a hedge against extinction. For instance, a recently established population

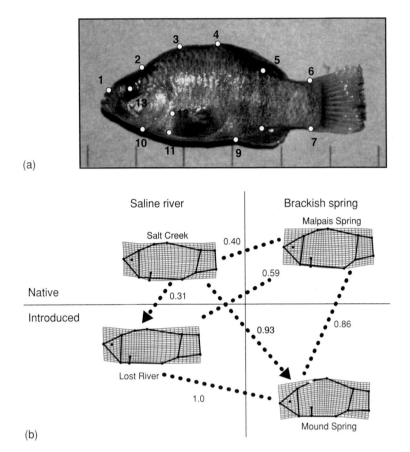

(a)

(b)

Figure 8.12 Body shape variation among native and recently established populations of White Sands pupfish. (a) A male pupfish is shown with the 13 anatomical landmarks used to calculate morphological distances from a generalized Procrustes analysis. These distances (on a relative scale) are shown in (b) along with deformation grids that depict shape change from an overall mean form. The directions of the arrows indicate that introduced populations were derived from Salt Creek. Experimental research (Collyer 2003) demonstrated that body shape variation in this system is strongly heritable, revealing that the evolutionary shape divergence of the Mound Spring population exceeded native shape divergence in as little as three decades. Modified from Collyer et al. (2005).

of the White Sands pupfish has undergone rapid adaptation in body shape (Figure 8.12). The native (Salt Creek) population is characterized by slender body shapes, indicative of adaptive streamlining because of high water flow during periodic flash floods at Salt Creek. The introduced population at Mound Spring has evolved a deep-bodied shape, presumably in response to the absence of high flows (Collyer et al. 2005). This evolution took place in less than three decades. The functional importance of body shape may preclude the refuge population's usefulness as a genetic replicate of the native population.

In all of these cases, evolutionary responses have occurred on limited time scales (often less than 50 years). Thus, for managed species, evolution may occur within the timeframe of a typical management plan. This in turn can influence the dynamics of population persistence. If selection is particularly harsh, a population may go extinct even as it evolves due to demographic stochasticity (Gomulkiewicz and Holt 1995). Further, rapid adaptation is likely to occur in the context of restoration efforts due to the fact that populations often encounter novel environmental conditions at restoration sites (Stockwell et al. 2005). These observations collectively suggest that an evolutionary approach to conservation is especially timely.

Problem 8.1

The rate of evolution of a population adapting to a new environment is positively correlated with the genetic variability of a population. How does one measure "rate of evolution", "adaptation", and "genetic variability"?

Problem 8.2

A population is in Hardy–Weinberg proportions at a particular locus. Do you think that it is it valid to conclude that little or no natural selection is occurring at that locus? Remember to consider the effects of both differential survival and differential fertility.

Problem 8.3

What is the expected equilibrium allele frequency (p^*) in each of the following cases after an infinite number of generations in a population with two alleles? The initial allele frequency is p_0. Assume that only natural selection is affecting allele frequencies.

(a) $w_{11} = w_{12} = w_{22}$; $p_0 = 0.40$.
(b) $w_{11} > w_{12} = w_{22}$; $p_0 = 0.01$.
(c) $w_{11} = w_{22} = 0.9$, $w_{12} = 1.0$; $p_0 = 0.41$.
(d) $w_{11} = 0.9$, $w_{12} = 1.0$, $w_{22} = 0.5$; $p_0 = 0.41$.
(e) $w_{11} = w_{22} = 1.0$, $w_{12} = 0.5$; $p_0 = 0.61$.

Problem 8.4

How would your answers for Problem 8.3 change if genetic drift was also acting because of small population size (say $N_e = 20$)? If more than one equilibrium is possible, identify which equilibrium frequency is more likely to be reached.

Problem 8.5

A single individual in a population is heterozygous for a chromosomal translocation. Homozygotes for both the common and translocated chromosome have
fitnesses of 1.0, but the fitness of the heterozygotes is reduced by 50%. How will
the fate of this translocation in the population differ if $N = 10$, as compared to, say,
$N = 1,000$?

Problem 8.6

Is heterozygous advantage (overdominance) more, less, or equally effective in
maintaining genetic variation in a large (say $N_e > 1,000$) as compared to a small
(say $N_e < 100$) population?

Problem 8.7

We saw in Chapter 7 that greater variability in reproductive success will decrease
effective population size and correspondingly increase the effects of genetic drift.
Intense natural selection at a particular locus will cause an increase in the variability in reproductive success among individuals. This leads to an apparent paradox:
increased natural selection causes increased drift. We often think of selection and
drift as opposing forces. This paradox can be resolved by considering the locus of
action. That is, intense natural selection at a single locus will cause a reduction in
effective population size that will increase the effects of genetic drift over the
entire genome. Consider the possible implications of this effect for conservation.

Problem 8.8

There is evidence from a variety of species that adaptation to captive conditions
can occur in just a few generations. It is likely that such rapid changes are caused
by the increase in frequency of alleles that exist in wild populations at low frequency (say less than 0.05). What kind of genotype–phenotype relationship do
you believe these alleles, which are responsible for adaptation to captivity, are
likely to demonstrate: recessive or dominant?

Problem 8.9

The program *HW Power* by Pedro J. N. Silva (2002) is available on the course
website. This program allows the evaluation of the power of standard tests for
Hardy–Weinberg proportions with two alleles. Familiarize yourself with this program by running different conditions. Now use this program to find the power of
testing for Hardy–Weinberg proportions to detect differential survival with heterozygous advantage. For example, how large a sample is necessary from a

population to have a 50% chance of detecting an excess of heterozygotes when the fitness of both homozygotes is by reduced 15%? Is your answer dependent on the allele frequency?

Problem 8.10

Supplementation of populations is sometimes part of recovery plans for declining populations or species. Assume that a population of melanic pocket mice living on dark lava is declining, and that supplementation from an outside population is recommended. Do you think it would be better to use an adjacent population of light-colored mice or a distant population of dark-colored mice as the source population of individuals to be used for supplementation? Why?

9

Population Subdivision

Grevillea barklyana, *Example 9.1*

There is abundant geographical variation in both morphology and gene frequency in most species. The extent of geographic variation results from a balance of forces tending to produce local genetic differentiation and forces tending to produce genetic homogeneity.

Montgomery Slatkin (1987)

The term "species" includes any subspecies of fish or wildlife or plants, and any distinct population segment of any species of vertebrate fish or wildlife which breeds when mature.

US Endangered Species Act of 1973

So far we have considered only random mating (i.e., **panmictic**) populations. Natural populations of most species are subdivided or "structured" into separate local random mating units that are called **demes**. The subdivision of a species into separate subpopulations means that genetic variation within species exists at two primary levels:

1   Genetic variation within local populations.
2   Genetic diversity between local populations.

We saw in Chapter 3 that there are large differences between species in the proportion of total genetic variation that is due to differences among populations (F_{ST}). For example, Schwartz et al. (2002) found very little genetic divergence ($F_{ST} = 0.033$) at nine microsatellite loci among 17 Canada lynx population samples collected from northern Alaska to central Montana (over 3,100 km). However other species of vertebrates, including carnivores, can be highly structured over a relatively short geographic distance (Figure 9.1). For example, Spruell et al. (2003) found 20 times this amount of genetic divergence among bull trout populations within the Pacific Northwest of the United States ($F_{ST} = 0.659$). Even separate spawning populations of bull trout just a few kilometers apart within a small tributary of Lake Pend Oreille in Idaho had twice the amount of genetic divergence ($F_{ST} = 0.063$) than the widespread population samples of lynx (Spruell et al. 1999b).

Understanding the patterns and extent of genetic divergence among populations is crucial for protecting species and developing effective conservation plans (see Guest Box 9). For example, translocation of animals or plants to supplement suppressed populations may have harmful effects if the translocated individuals are genetically different from the recipient population (Storfer 1999; Edmands 2002). In addition, developing priorities for

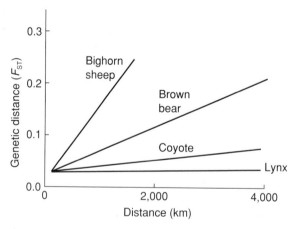

Figure 9.1 General relationship between geographic distance and genetic distance at microsatellite loci for four species of mammals. Lynx and coyotes show little genetic differentiation over thousands of kilometers; wolves (not shown) are similar to coyotes in this respect. However, less mobile species have significant differences in allele frequencies between populations over only a few hundreds of kilometers. Bighorn sheep, for example, live on mountain tops and tend not to disperse across deep valleys and forests that often separate mountain ranges. Modified from Forbes and Hogg (1999); additional unpublished data from M. Schwartz.

the conservation of a species requires an understanding of adaptive genetic differentiation among populations. Perhaps most importantly, an understanding of genetic population structure is essential for identifying units to be conserved. For example, as stated in the quote at the beginning of the chapter, distinct populations can be listed under the US Endangered Species Act and receive the same protection as biological species.

The use of the terms **migration** and **dispersal** is somewhat confusing. The classic population genetics literature uses migration to refer to the movement of individuals from one genetic population to another (i.e., genetic exchange among breeding groups). This exchange is generally referred to as dispersal in the ecology literature. Migration in the ecology literature refers to movement of individuals during their lifetime from one geographic region to another. For example, anadromous salmon undertake long migrations from their natal stream to the ocean where they feed for several years before migrating to their natal streams for reproduction. In the genetic sense, migration of salmon refers to an individual returning to a spawning population other than its natal population.

In this chapter, we will consider populations that are subdivided into a series of partially isolated **subpopulations** that are connected by some amount of genetic exchange (migration). We will first consider how genetic variation is distributed at neutral loci within subdivided populations because of the effects of two opposing processes: gene flow and genetic drift. We will next consider the effects of natural selection on the distribution of genetic variation within species. Finally, we will consider the application of this analysis to the observed distribution of genetic variation in natural populations.

9.1 *F*-statistics

The oldest and most widely used metrics of genetic differentiation are *F*-statistics. Sewall Wright (1931, 1951) developed a conceptual and mathematical framework to describe the distribution of genetic variation within a species that used a series of inbreeding coefficients: F_{IS}, F_{ST}, and F_{IT}. F_{IS} is a measure of departure from Hardy–Weinberg proportions within local demes or subpopulations; as we have seen, F_{ST} is a measure of allele frequency divergence among demes or subpopulations; and F_{IT} is a measure of the overall departure from Hardy–Weinberg proportions in the entire base population (or species) due to both nonrandom mating within local subpopulations (F_{IS}) and allele frequency divergence among subpopulations (F_{ST}).

In general, inbreeding is the tendency for mates to be more closely related than two individuals drawn at random from the population. It is crucial to define inbreeding relative to some clearly specified base population. For example, using the entire species as the base population, a mating between two individuals within a local population will produce "inbred" progeny because individuals from the same local populations are likely to have shared a more recent common ancestor than two individuals chosen at random from throughout the range of a species. As we will see, F_{ST} is a measure of this type of inbreeding.

These parameters were initially defined by Wright for loci with just two alleles. They were extended to three or more alleles by Nei in 1977, who used the parameters G_{IS}, G_{ST}, and G_{IT} in what he termed the analysis of gene diversity. F and G are now often used interchangeably in the literature.

F-statistics are a measure of the deficit of heterozygotes relative to expected Hardy–Weinberg proportions in the specified base population. That is, *F* is the proportion by which heterozygosity is reduced relative to heterozygosity in a random mating population with the same allele frequencies:

$$F = 1 - (H_0/H_e) \tag{9.1}$$

where H_0 is the observed proportion of heterozygotes and H_e is the expected Hardy–Weinberg proportion of heterozygotes.

F_{IS} is a measure of departure from Hardy–Weinberg proportions within local subpopulations:

$$F_{IS} = 1 - (H_0/H_S) \tag{9.2}$$

where H_0 is the observed heterozygosity averaged over all subpopulations, and H_S is the expected heterozygosity averaged over all subpopulations. F_{IS} will be positive if there is a deficit of heterozygotes and negative if there is an excess of heterozygotes. Inbreeding within local populations, such as selfing, will cause a deficit of heterozygotes (Example 9.1). As we saw in Chapter 6, a small effective population size can cause an excess of heterozygotes and result in negative F_{IS} values.

Example 9.1 Selfing in a Australian shrub

Ayre et al. (1994) studied genetic variation in the rare Australian shrub *Grevillea barklyana*, which reproduces by both selfing and outcrossing. They found a significant ($P < 0.001$) deficit of heterozygotes at the *Gpi* locus in a sample of progeny from one of their four populations:

Genotypes				
A/A	*A/B*	*B/B*		
112	43	31	$\hat{p} = 0.718$	$\hat{F}_{IS} = 0.429$
(95.9)	(75.3)	(14.8)		

We can estimate the proportion of selfing that can explain these results by solving for *S* in expression 9.7:

$$S = \frac{2F_{IS}}{(1 + F_{IS})}$$

This results in an estimated 60% of the progeny in this population being produced by selfing and the remaining 40% by random mating, if we assume that the deficit of heterozygotes is caused entirely by selfing.

Example 9.2 The Wahlund effect in a lake of brown trout

The approach of Nei (1977) can be used to compute F-statistics with genotypic data from natural populations. For example, two nearly equal size demes of brown trout occurred in Lake Bunnersjöarna in northern Sweden (Ryman et al. 1979). One deme spawned in the inlet, and the other deme spawned in the outlet. The fish spent almost all of their life in the lake itself rather than in the inlet and outlet streams. These two demes were nearly fixed for two alleles (*100* and *null*) at the *LDH-A2* locus. Genotype frequencies for a hypothetical sample taken from the lake itself of 100 individuals, made up of exactly 50 individuals from each deme, are shown below:

	100/100	*100/null*	*null/null*	Total	p	*2pq*
Inlet deme	50	0	0	50	1.000	0.000
Outlet deme	1	13	36	50	0.150	0.255
Lake sample	51	13	36	100	0.575	0.489
(expected)	(33.1)	(48.9)	(18.1)			

The mean expected heterozygosity within these two demes (H_S) is 0.128 (the mean of 0.000 and 0.255). Thus, the value of F_{ST} for this population at this locus is 0.738:

$$F_{ST} = 1 - (H_S/H_T) = 1 - (0.128/0.489) = 0.728$$

That is, the heterozygosity of the sample of fish from the lake is approximately 74% lower than we would expect if this population was panmictic.

F_{ST} is a measure of genetic divergence among subpopulations:

$$F_{ST} = 1 - (H_S/H_T) \tag{9.3}$$

where H_T is the expected Hardy–Weinberg heterozygosity if the entire base population were panmictic (Example 9.2). H_T is the expected Hardy–Weinberg proportion of heterozygotes using the allele frequencies averaged over all subpopulations. F_{ST} ranges from zero, when all subpopulations have equal allele frequencies, to one, when all the subpopulations are fixed for different alleles. F_{ST} is sometimes called the **fixation index**.

F_{IT} is a measure of the total departure from Hardy–Weinberg proportions that includes departures from Hardy–Weinberg proportions within local populations and divergence among populations:

$$F_{IT} = 1 - (H_0/H_T)$$

These three F-statistics are related by the expression:

$$F_{IT} = F_{IS} + F_{ST} - (F_{IS})(F_{ST}) \qquad (9.4)$$

This approach will be used in this chapter to describe the effects of population subdivision on the genetic structure of populations.

9.1.1 Subdivision and the Wahlund effect

This deficit of heterozygotes relative to Hardy–Weinberg proportions, caused by the subdivision of a population into separate demes, is often referred to as the **Wahlund principle**. For example, a large deficit of heterozygotes was found at many loci when brown trout captured in Lake Bunnersjöarna (Example 9.2) were initially analyzed without knowledge of the two separate subpopulations (Ryman et al. 1979). Wahlund was a Swedish geneticist who first described this effect in 1928. He analyzed the excess of homozygotes and deficit of heterozygotes in terms of the variance of allele frequencies among S subpopulations:

$$Var(q) = \frac{1}{S} \sum (q_i - \bar{q})^2 \qquad (9.5)$$

When $Var(q) = 0$, all subpopulations have the same allele frequencies and the population is in Hardy-Weinberg (HW) proportions. As $Var(q)$ increases, the allele frequency differences among subpopulations increases and the deficit of heterozygotes increases. In fact,

$$F_{ST} = \frac{Var(q)}{pq} \qquad (9.6)$$

so that we can express the genotypic array of the population in terms of either F_{ST} or $Var(q)$:

Genotype	HW	Wright	Wahlund
AA	p^2	$p^2 + pqF_{ST}$	$p^2 + Var(q)$
Aa	$2pq$	$2pq - 2pqF_{ST}$	$2pq - 2Var(q)$
aa	q^2	$q^2 + pqF_{ST}$	$q^2 + Var(q)$

These two approaches for describing the genotypic effects of population subdivision (Wright and Wahlund) are analogous to the two ways we modeled genetic drift in Chapter 6: either an increase in homozygosity or a change in allele frequency.

The Wahlund effect can readily be extended to more than two alleles (Nei 1965). However, the variance in frequencies will generally differ for different alleles. The frequency of particular heterozygotes may be greater or less than expected with Hardy–Weinberg proportions. Nevertheless, there will always be an overall deficit of heterozygotes due to the Wahlund effect.

9.1.2 When is F_{IS} not zero?

Generally the first step in analyzing genotypic data from a natural population is to test for Hardy–Weinberg proportions. As we have seen, F_{IS} is a measure of departure from expected Hardy–Weinberg proportions. A positive value indicates an excess of homozygotes, and a negative value indicates a deficit of homozygotes. Interpreting the causes of an observed excess or deficit of homozygotes can be difficult.

The most general cause of an excess of homozygotes is nonrandom mating or population subdivision. In the case of the Wahlund effect, the presence of multiple demes within a single population sample will produce an excess of homozygotes at all loci for which the demes differ in allele frequency.

Inbreeding within a single deme will produce a similar genotypic effect. That is, the tendency for related individuals to mate will also produce an excess of homozygotes. Perhaps the simplest example of this is a plant that reproduces by both self-pollination and outcrossing. Assume that a proportion S (i.e., the selfing rate) of the matings in a population are the result of selfing and the remainder $(1 - S)$ result from random mating. The equilibrium value of F_{IS} in this case will be:

$$F_{IS}^{\star} = \frac{S}{(2 - S)} \tag{9.7}$$

For example, consider a population in which half of the progeny are produced by selfing and half by outcrossing ($S = 0.5$). In this case, F_{IS} will be 0.33. (See Example 9.1.)

Null alleles at allozyme and microsatellite loci are another possible source of an excess of homozygotes (see Section 5.4 for a description of null alleles).

Perhaps the best way to discriminate between nonrandom mating (either inbreeding or including multiple populations in a single sample) or a null allele to explain an excess of homozygotes, is to examine if the effect appears to be locus specific or population specific. All loci that differ in allele frequency between demes will have a tendency to show an excess of homozygotes. Assume you examine 10 loci in 10 different population samples ($10 \times 10 = 100$ total tests), and that you detect a significant ($P < 0.05$) excess of homozygotes for 12 tests. If eight of the 12 deviations are in a single population, this would suggest that this population sample consisted of more than one deme. In contrast, a homozygote excess due to a null allele should be locus specific. In the same example as above, if eight of the deviations were at just one of the 10 loci, this would suggest that a null allele at appreciable frequency was present at that locus.

It may also be possible to discriminate between inbreeding or including multiple populations in a single sample (the Wahlund effect) to explain an observed excess of homozygotes caused by nonrandom mating. Inbreeding will reduce the frequency of all heterozygotes. However, as discussed in the previous section, some heterozygotes will be in excess and some will be in deficit in the case of more than two alleles.

A deficit of homozygotes (excess of heterozygotes) may also occur under some circumstances. We saw in Section 6.6 that we expect a slight excess of heterozygotes in small randomly mating populations. Natural selection may also cause an excess of heterozygotes if heterozygotes have a greater probability of surviving than homozygotes (see Section 8.2 and Table 8.1). However, the differential advantage of heterozygotes has to be very great to have a detectable effect on genotypic proportions (see Problem 9.6).

Differences in allele frequency between the sex chromosomes will result in an excess of heterozygotes in comparison to expected Hardy–Weinberg proportions in the heterogametic sex for sex-linked loci (Clark 1988; Allendorf et al. 1994). For example, Berlocher

(1984) observed the following genotypic frequencies at a sex-linked allozyme locus (*Pgm*) with two alleles (*100* and *82*) in the walnut husk fly for which males are XY and females are XX:

	100/100	100/82	82/82
Females	25	0	0
Males	4	21	0

Based on these genotypic data, only the *100* allele is present on the X chromosome, but the *82* allele is at an estimated frequency of 0.84 (21/25) on the Y chromosome.

9.2 Complete isolation

Let us consider a large random mating population that is subdivided into many completely isolated demes. Let us consider the effect of this subdivision on a single locus with two alleles. Assume all Hardy–Weinberg conditions are valid except for small population size within each individual isolated subpopulation. Genetic drift will occur in each of the isolated demes; eventually, each deme will become fixed for one allele or other.

What is the effect of this subdivision on our two measures of the genetic characteristics of populations: allele frequencies and genotype frequencies? If the initial allele frequency of the A allele in the large, random mating population was p, the allele frequency in our large, subdivided population will still be p because we expect p of the isolates to become fixed for the A allele, and $(1 - p)$ of the isolates to become fixed for the a allele. Thus, subdivision (nonrandom mating) itself has no effect on overall allele frequencies.

We can see this effect in the guppy example from Table 6.3 where 16 subpopulations were founded by a single male and female from a large population. Genetic drift within each subpopulation acted to change allele frequencies at the two loci: *AAT-1* and *PGM-1* for four generations. However, the average allele frequencies over the 16 subpopulations at both loci are very close to the frequencies in the large founding population. Therefore, allele frequencies in the populations as a whole were not affected by subdivision.

However, the subdivision into 16 separate subpopulations did affect the genetic structure of this population. We can use the *F*-statistics approach developed in the previous section to describe this effect at the *AAT-1* locus. In this case, H_S is the mean expected heterozygosity averaged over the 16 subpopulations (0.344), and H_T is the Hardy–Weinberg heterozygosity (0.492) using the average allele frequency averaged over all subpopulations (0.562). Therefore:

$$F_{ST} = 1 - (H_S/H_T) = 1 - (0.344/0.492) = 0.301$$

In words, the average heterozygosity of individual guppies in this population has been reduced by 31% because of the subdivision and subsequent genetic drift within the subpopulations.

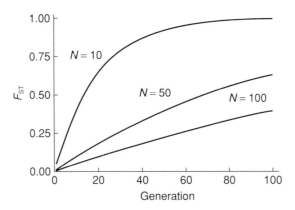

Figure 9.2 Expected increase in F_{ST} over time (generations) among completely isolated populations of different population sizes using expression 9.8.

We know from Section 6.3 that heterozygosity will be lost by genetic drift at a rate of $1/2N_e$ per generation. Therefore, we expect F_{ST} among completely isolated populations to increase as follows (modified from expression 6.3):

$$F_{ST} = 1 - (1 - \frac{1}{2N_e})^t \tag{9.8}$$

where N_e is the effective population size of each subpopulation and t is the number of generations (Figure 9.2). The application of this expression to real populations is limited because it assumes a large number of equal size subpopulations and constant population size.

9.3 Gene flow

In most cases, there will be some genetic exchange (**gene flow**) among demes within a species. We must therefore consider the effects of such partial isolation on the genetic structure of species. Let us first consider the simple case of two demes (A and B) of equal size that are exchanging individuals in both directions at a rate m. Therefore, m is the proportion of individuals reproducing in one deme that were born in the other deme. In this case:

$$q'_A = (1 - m)q_A + mq_B$$
$$q'_B = mq_A + (1 - m)q_B \tag{9.9}$$

For example, consider two previously isolated populations that begin to exchange migrants at a rate of $m = 0.10$ (10% exchange). Assume that the allele frequency in population A is 1.0, and in population B it is 0.0. The above model can be used to predict the effects of gene flow between these two populations as shown in Figure 9.3.

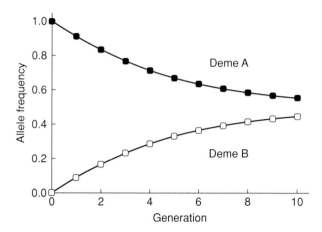

Figure 9.3 Expected changes in allele frequencies in two demes that are exchanging 10% of their individuals each generation ($m = 0.10$) using expression 9.9.

Equilibrium will be reached when $q_A = q_B$, and q^* will be the average of the initial allele frequencies in the two demes; in this case, $q^* = 0.5$. In general, there are two primary effects of gene flow:

1 Gene flow reduces genetic differences between populations.
2 Gene flow increases genetic variation within populations.

Gene flow among populations is the cohesive force that holds together geographically separated populations into a single evolutionary unit – the species. In the rest of this chapter we will consider the interaction between the homogenizing effects of gene flow and the action of genetic drift and natural selection that cause populations to diverge.

9.4 Gene flow and genetic drift

In the absence of other evolutionary forces, any gene flow between populations will bring about genetic homogeneity. With lower amounts of gene flow it will take longer, but eventually all populations will become genetically identical. However, we saw in Section 6.1 that genetic drift causes isolated subpopulations to become genetically distinct. Thus, the actual amount of divergence between subpopulations will be a balance between the homogenizing effects of gene flow making subpopulations more similar and the disruptive effects of drift causing divergence among subpopulations. We examine this using a series of models for different patterns of gene flow. All of these models will necessarily be much simpler than the actual patterns of gene flow in natural populations.

9.4.1 Island model

We will begin with the simplest model that combines the effects of gene flow and genetic drift. Assume that a population is subdivided into a series of demes, each of size N, that exchange individuals at a rate of m. That is, each generation an individual has probability m

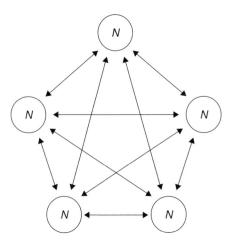

Figure 9.4 Pattern of exchange among five subpopulations under the island model of migration. Each subpopulation of size N exchanges migrants with the other subpopulations with equal probability. The total proportion of migrants into a subpopulation is m so that an average of $(1/4\,mN)$ migrants are exchanged between each pair of subpopulations each generation.

of breeding in a deme other than that of his or her birth. Let us further assume that a migrant is equally likely to immigrate into any of the other demes. This model is called the **island model of migration** (Figure 9.4).

As before, we will measure divergence among subpopulations (demes) using F_{ST}. Genetic drift within each deme will act to increase divergence among demes, i.e., increase F_{ST}. However, migration between demes will act to decrease F_{ST}. As long as $m > 0$, there will be some steady-state (equilibrium) value of F_{ST} at which the effects of drift and gene flow will be balanced.

Sewall Wright (1969) has shown that at equilibrium under the island model of migration with an infinite number of demes:

$$F_{ST} = \frac{(1 - m)^2}{[2N - (2N - 1)(1 - m)^2]} \tag{9.10}$$

Fortunately, if m is small this approaches the much simpler:

$$F_{ST} \approx \frac{1}{(4mN + 1)} \tag{9.11}$$

This approximation provides an accurate estimation of the amount of divergence under the island model. For example, the expected equilibrium value of F_{ST} with one migrant per generation ($mN = 1$) using expression 9.11 is 0.200; the value resulting from the simulation shown in Figure 9.5 with 20 subpopulations ($F_{ST} = 0.215$) is very close to this expected value. One important result of this analysis is that very little gene flow is necessary for populations to be genetically connected (Question 9.1).

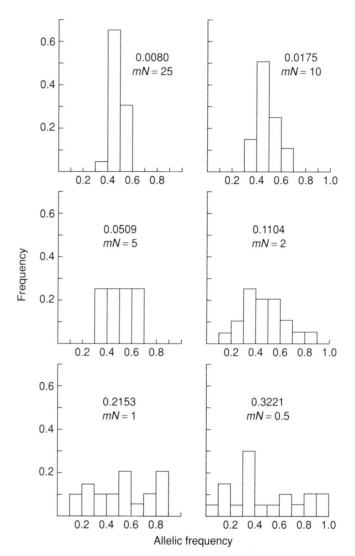

Figure 9.5 Relationship between migration, F_{ST}, and allelic divergence. Computer simulations were carried out with 20 subpopulations ($N=200$) and different expected amounts of migration. The first value on each graph (e.g., 0.0080) is the F_{ST} for that particular distribution of allele frequencies: $mN=25$; $mN=10$; $mN=5$; $mN=2$; $mN=1$; and $mN=0.5$. From Allendorf and Phelps (1981).

Question 9.1 Is one migrant per generation sufficient to insure that two or more populations are genetically identical?

Expression 9.11 also provides a surprisingly simple result: the amount of divergence among demes depends only on the number of migrant individuals (mN), and not the proportion of exchange among demes (m). Thus, we expect to find approximately the same

Figure 9.6 Pattern of exchange among subpopulations under the single-dimension stepping-stone model of migration. Each subpopulation of size N exchanges $m/2$ individuals with each adjacent subpopulation.

amount of divergence among demes of size 200 with $m = 0.025$, as we do with demes of size 50 with $m = 0.1$ ($0.025 \times 200 = 0.1 \times 50 = 5$ migrants per generation).

The dependence of divergence on only the number of migrants irrespective of population size may seem counterintuitive. Remember, however, that the amount of divergence results from the opposing forces of drift and migration. The larger the demes are, the slower they are diverging through drift; thus, proportionally fewer migrants are needed to counteract the effects of drift. Small demes diverge rapidly through drift, and thus proportionally more migrants are needed to counteract drift.

9.4.2 Stepping-stone model

In natural populations, migration is often greater between subpopulations that are near each other (Slatkin 1987). This violates the assumption of equal probability of exchange among all pairs of subpopulations with the island model of migration. The stepping-stone model of migration was introduced (Kimura and Weiss 1964) to take into account both short-range migration (which occurs only between adjacent subpopulations) and long-range migration (which occurs at random between subpopulations). Linear stepping-stone models (Figure 9.6) are useful for modeling populations with a one-dimensional linear structure, as occurs along a river, river valley, valley, or a mountain ridge, for example. Two-dimensional stepping-stone models are useful for modeling populations with a grid structure (or 2D checker board pattern) across the landscape.

The mathematical treatment of the stepping-stone model is much more complex than the island model. In general, migration in the stepping-stone model is less effective at reducing differentiation caused by drift because subpopulations exchanging genes tend to be genetically similar to each other. Therefore, there will be greater differentiation (i.e., greater F_{ST}) among subpopulations with the stepping-stone than the island model for the same amount of genetic exchange (m). In addition, in the stepping-stone model, adjacent subpopulations should be more similar to each other than geographically distant populations (**isolation by distance**) (see Figure 9.1). With the island model of migration, genetic divergence will be independent of geographic distance (Figure 9.7).

9.4.3 Continuous distribution model

In some species, individuals are distributed continuously across large landscapes (e.g. coniferous tree species across boreal forests) and are not subdivided by sharp barriers to gene flow (Figure 9.8). Nonetheless, gene flow can be limited to relatively short distances leading to genetic differentiation because of isolation by distance (Wright 1943). It is impossible to identify and sample discrete population units because no sharp boundaries exist. In this case, the **neighborhood** is defined as the area from which individuals can be

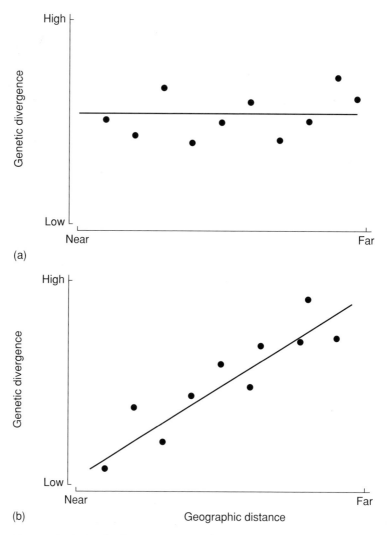

Figure 9.7 Expected relationship between genetic divergence and geographic distance with the island (a) and stepping-stone (b) model of migration. The stepping-stone model results in isolation by distance because there is greater gene flow between adjacent subpopulations.

considered to be drawn at random from a panmictic population. We can estimate the geographic distance at which individuals will become genetically differentiated due to limited gene flow. For example, if the mean gene flow distance is 1 km then we would expect substantial genetic differentiation between individuals separated by, say, 5–10 km (Manel et al. 2003).

9.5 Cytoplasmic genes and sex-linked markers

Maternally inherited cytoplasmic genes and sex-linked markers generally show different amounts of differentiation among populations than autosomal loci for several reasons.

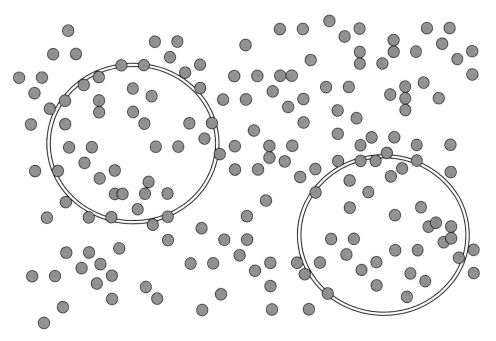

Figure 9.8 Continuous distribution of individuals where no sharp boundaries separate individuals (gray dots) into discrete groups. Nonetheless, genetic isolation arises over geographic distance because nearby individuals are more likely to mate with each other than with individuals that are farther away (isolation by distance). We can place a circle of the appropriate neighborhood size anywhere (see two circles) and individuals inside will represent a panmictic group in Hardy–Weinberg proportions.

First, they usually have a smaller effective population size than autosomal loci and therefore show greater divergence due to genetic drift. In addition, differences in migration rates between males and females can cause large differences between cytoplasmic genes and sex-linked markers compared to autosomal loci.

9.5.1 Cytoplasmic genes

The expected amount of allele frequency differentiation with a given amount of gene flow is different for mitochondrial and nuclear genes because of haploidy and uniparental inheritance. We can calculate F_{ST} to compare the amount of allelic differentiation for nuclear and mitochondrial genes. However, since mtDNA is haploid, individuals are hemizygous rather than homozygous or heterozygous. We expect more differentiation at mtDNA than for nuclear genes because of their smaller effective size. That is, the greater genetic drift with smaller effective population size will bring about greater differentiation populations that are connected by the same amount of gene flow. If migration rates are equal in males and females, then we expect the following differentiation for mtDNA with the island model of migration (Birky et al. 1983):

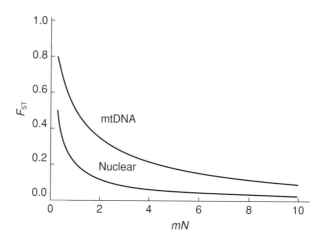

Figure 9.9 Expected values of F_{ST} with the island model of migration for a nuclear locus (expression 9.11) and mtDNA (expression 9.12).

$$F_{ST} \approx \frac{1}{(mN + 1)} \tag{9.12}$$

This expression is sometimes written to consider only females:

$$F_{ST} \approx \frac{1}{(2mN_f + 1)} \tag{9.13}$$

where N_f is the number of females in the population. Expressions 9.12 and 9.13 are identical if there are an equal number of males and females in the population ($N_f = N_m$) because $(2N_f) = N$.

Thus, with equal migration rates in males and females, we expect approximately two to four times as much allele frequency differentiation at mitochondrial genes than at nuclear genes (Figure 9.9). We can see this effect in the study of sockeye salmon shown in Section 9.6.3; the F_{ST} at mtDNA was greater than the F_{ST} at all but one of the 20 nuclear loci examined.

This difference in F_{ST} for a nuclear locus and for mtDNA is expected to be greater for species in which the emigration rates of males are greater than those of females (Example 9.3). Female green turtles deposit their eggs in rookeries on oceanic beaches. Analysis of mtDNA variation shows large genetic differences even between nearby beaches. This suggests that the females are extremely **philopatric** (Bowen et al. 1992). That is, females have a strong tendency to deposit their eggs in the same rookeries in which they began their own life. However, analysis of nuclear genes indicates substantial gene flow among rookeries that apparently results from females mating with males at sea before they return to their natal beaches (Karl et al. 1992). FitzSimmons et al. (1997b) found similar results with green turtles on the Australian coast (see Guest Box 4). Rookeries south and north of the Great Barrier Reef were nearly fixed for different haplotypes at mtDNA ($F_{ST} = 0.83$; $P < 0.001$), but lacked any divergence at eight nuclear loci ($F_{ST} = 0.014$; NS).

Example 9.3 Divergence at nuclear loci and cpDNA in the white campion

McCauley (1994) compared the distribution of genetic variation at seven allozyme loci and chloroplast DNA in white campion (Table 9.1). As expected, a much greater proportion of the variation was distributed among populations for the cpDNA marker ($F_{ST} = 0.674$) compared to the nuclear loci ($F_{ST} = 0.134$). However, this difference is even greater than expected with the island model of migration at equilibrium (see Figure 9.9). $F_{ST} = 0.134$ is expected to result from 1.6 migrants per generation with the island model (Figure 9.9 and expression 9.11). This amount of gene flow should result in an F_{ST} for cpDNA of 0.385 (expression 9.12). This value is outside the 95% confidence interval for the estimated F_{ST} for cpDNA.

The simplest explanation for this discordance between nuclear loci and cpDNA is that most of the gene flow is from pollen rather than seeds. Therefore, migration rates will be greater for nuclear genes than for maternally inherited genes such as cpDNA. This same effect has been seen in many plants, especially ones that are wind pollinated (Ouborg et al. 1999).

Table 9.1 Estimates of F_{ST} from 10 populations of white campion at seven allozyme loci and cpDNA. The 95% confidence limits are provided in the bottom two rows (McCauley 1994).

Locus	F_{ST}
GPI	0.125
IDH	0.083
LAP	0.172
MDH	0.230
PGM	0.145
6-PGD	0.042
SKDH	0.083
Allozymes	0.134 (0.073–0.195)
cpDNA	0.674 (0.407–0.941)

9.5.2 Sex-linked markers

Genes on the Y chromosome of mammals present a parallel situation to mitochondrial genes. The Y chromosome is haploid and is only transmitted through the father. Thus, the expectations that we just developed for cytoplasmic genes also apply to Y-linked genes except that we must substitute the number of males for females. Comparison of the patterns of differentiation at autosomal, mitochondrial, and Y-linked genes can provide valuable insight into the evolutionary history of species and current patterns of gene flow (Example 9.4).

Example 9.4 Y chromosome isolation in a shrew hybrid zone

A Y chromosome microsatellite locus and 10 autosomal microsatellite loci were typed across a hybrid zone between the Cordon and Valais races of the common shrew in western France (Balloux et al. 2000). There is a contact where the two races occur on either side of a stream. Gene flow is somewhat limited, but the two races show relatively little divergence at the autosomal microsatellite loci ($F_{ST} = 0.02$) (Brünner and Hausser 1996; Balloux et al. 2000).

Almost all gene flow in these shrews appears female mediated, and male hybrids are generally unviable. No alleles were shared across the hybrid zone at the Y-linked locus (Figure 9.10). However, the F_{ST} value between races at the Y-linked microsatellite loci is just 0.19; this low value does not reflect the absence of alleles shared between races because of the high within-race heterozygosity at this locus (this effect is discussed in Section 9.7). R_{ST} is an analogue of F_{ST} that takes the relative size of microsatellite alleles (i.e., allelic state) into consideration. The strong divergence is reflected in a value R_{ST} (0.98). It is important to incorporate allele length (mutational) information when H_S is high, because mutations likely contribute to population differentiation when populations are long isolated as in this example (see Section 9.7).

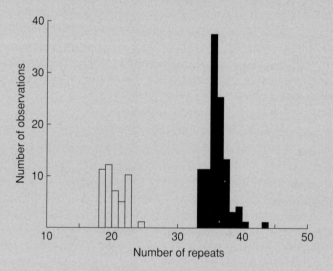

Figure 9.10 Allele frequency distribution of a Y-linked microsatellite locus (*L8Y*) in the European common shrew. Males of the Cordon race are represented by black bars, and Valais males by white bars. From Balloux et al. (2000).

9.6 Gene flow and natural selection

We will now examine the effects of natural selection on the amount of genetic divergence expected among subpopulations in an island model of migration (Allendorf 1983). We previously concluded that the amount of divergence, as measured by F_{ST}, is dependent only upon the product of migration rate and deme size (mN). Does this simple principle hold when we combine the effects of natural selection with the island model of migration? As we will see shortly, the answer is no.

9.6.1 Heterozygous advantage

Assume the following fitnesses hold within each deme:

	AA	Aa	aa
Fitness	$1 - s$	1.0	$1 - s$

Because of its complexity, one of the best ways to analyze this system is using computer simulations. We will examine the results of simulations combining natural selection with the island model of migration in which there are 20 subpopulations. Natural selection will act to maintain a stable equilibrium of $p^* = 0.5$ in each deme. Thus, this model of selection will reduce the amount of divergence among demes (Table 9.2). The greater the value of s, the greater will be the reduction in F_{ST} (Table 9.2). Even relatively weak selection can have a marked effect on F_{ST}; for example, see $s = 0.01$ and $mN = 0.5$ in Table 9.2. It is also apparent that genetic divergence among demes is no longer only a function of mN. For a given value of mN, natural selection becomes more effective, and thus F_{ST} is reduced, as population size increases.

Table 9.2 Simulation results (except top row) of steady-state F_{ST} values for 20 demes and selective neutrality ($s = 0$) or heterozygous advantage in which both homozygous phenotypes have a reduction in fitness of s. Each value is the mean of 20 repeats. Expected values were calculated by $F_{ST} = 1/(4mN + 1)$.

	mN						
	0.5	**1**	**2**	**5**	**10**	**25**	**N**
Expected	0.3333	0.2000	0.1111	0.0476	0.0244	0.0099	
$s = 0.00$	0.3070	0.2043	0.1245	0.0418	0.0198	–	25
	0.3350	0.1826	0.1077	0.0484	0.0264	0.0120	50
	0.3216	0.1884	0.1061	0.0437	0.0251	0.0095	100
$s = 0.01$	0.2826	0.1640	0.0666	0.0499	0.0220	–	25
	0.2431	0.1534	0.0824	0.0406	0.0232	0.0117	50
	0.1782	0.1236	0.0930	0.0383	0.0355	0.0109	100
$s = 0.05$	0.1930	0.1259	0.0714	0.0441	0.0237	–	25
	0.1327	0.1072	0.0620	0.0341	0.0238	0.0092	50
	0.0827	0.1072	0.0432	0.0242	0.0185	0.0110	100
$s = 0.10$	0.1217	0.1039	0.0533	0.0429	0.0216	–	25
	0.0938	0.0763	0.0503	0.0307	0.0207	0.0087	50
	0.0410	0.0290	0.0317	0.0217	0.0103	0.0070	100

Table 9.3 Simulation results of steady-state F_{ST} values for 20 demes with differential directional selection. Each value is the mean of 20 repeats. One homozygous genotype has a reduction in fitness of t in 10 demes; the other homozygous genotype has the same reduction in fitness in the other 10 demes. Heterozygotes have a reduction in fitness of one-half t in all demes.

	mN						**N**
	0.5	1	2	5	10	25	
Expected	0.3333	0.2000	0.1111	0.0476	0.0244	0.0099	
$t = 0.00$	0.3070	0.2043	0.1245	0.0418	0.0198	–	25
	0.3350	0.1826	0.1077	0.0484	0.0264	0.0120	50
	0.3216	0.1884	0.1061	0.0437	0.0251	0.0095	100
$t = 0.01$	0.3343	0.1703	0.1070	0.0556	0.0220	–	25
	0.2979	0.1192	0.1000	0.0381	0.0256	0.0099	50
	0.2997	0.1850	0.1146	0.0354	0.0229	0.0105	100
$t = 0.05$	0.3560	0.1857	0.1204	0.0497	0.0217	–	25
	0.4618	0.2679	0.1489	0.0550	0.0265	0.0113	50
	0.5950	0.4230	0.1982	0.0632	0.0207	0.0118	100
$t = 0.10$	0.4700	0.2446	0.1632	0.0473	0.0289	–	25
	0.6242	0.3653	0.2611	0.0771	0.0356	0.0128	50
	0.8054	0.6575	0.4432	0.1589	0.0632	0.0193	100

9.6.2 Divergent directional selection

Assume the following relative fitnesses in a population consisting of 20 demes:

	AA	Aa	aa
Demes 1–10	1	$1 - t/2$	$1 - t$
Demes 11–20	$1 - t$	$1 - t/2$	1

This pattern of divergent directional selection will act to maintain allele frequency differences among demes so that large differences can be maintained even with extensive genetic exchange. Again, selection is more effective with larger demes (Table 9.3).

9.6.3 Comparisons among loci

Gene flow and genetic drift are expected to affect all loci uniformly throughout the genome. However, the effects of natural selection will affect loci differently depending upon the intensity and pattern of selection. As we have noted above, even fairly weak natural selection can have a substantial effect on divergence. Therefore, surveys of genetic differentiation at many loci throughout the genome can be used to detect outlier loci that are candidates for the effects of natural selection.

Detecting locus-specific effects is critical because only genome-wide effects inform us reliably about population demography and phylogenetic history, whereas locus-specific effects can help identify genes important for fitness and adaptation. An example of a locus-specific effect is differential directional selection whereby one allele is selected for in one environment but the allele is disadvantageous in a different environment. This selection would generate a large allele frequency difference (high F_{ST}) only at this locus relative to neutral loci throughout the genome. For example, just a 10% selection coefficient favoring different alleles in two environments can generate large differences with this pattern of selection between the selected locus ($F_{ST} = 0.66$) and neutral loci ($F_{ST} = 0.20$), as shown in Table 9.3 with local population sizes of $N = 100$.

It is crucial to identify outlier loci, not only because such loci might be under selection and help us to understand adaptive differentiation, but also because outlier loci can severely bias estimates of population parameters (e.g., F_{ST} or the number of migrants). Most estimates of population parameters assume that loci are neutral. For example, Allendorf and Seeb (2000) found with sockeye salmon that a single outlier locus with extremely high F_{ST} could bias high estimates of the mean F_{ST} from 0.09 to 0.20 (Figure 9.11). This bias more than doubles the F_{ST} estimate!

In another example, Wilding et al. (2001) genotyped 306 AFLP loci in an intertidal snail (the rough periwinkle) collected along rocky ocean shorelines. Fifteen of the 306 loci had an F_{ST} substantially higher than expected for neutral loci in a comparison of two morphological forms (H and M) that were collected along the same shoreline (Figure 9.12). Interestingly, these same 15 loci also were found to be outliers at other shoreline locations,

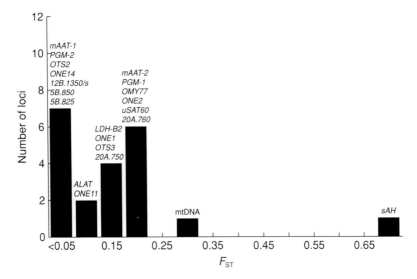

Figure 9.11 Genome-wide versus locus-specific effects, and the identification of outlier loci that are candidates for being under selection. Gene flow and genetic drift lead to similar genome-wide allele frequency differentiation (F_{ST}) among populations of sockeye salmon for 19 nuclear loci with an F_{ST} less than 0.20. One nuclear locus (sAH) has a much greater F_{ST} and is a candidate for being under natural selection. From Allendorf and Seeb (2000).

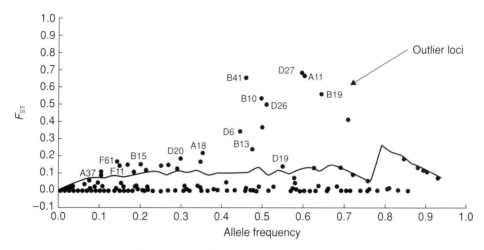

Figure 9.12 Locus-specific F_{ST} estimates between two morphological forms (H and M) of the rough periwinkle collected along the same shoreline at Thornwick Bay, UK, show outlier loci from the neutral expectation. Fifteen AFLP loci (dots above solid line) had exceptionally high F_{ST} values (> 0.20) compared to the mean observed F_{ST} (< 0.04) and to the null distribution of "neutral" F_{ST} values (~0.0–0.2). The solid line is the upper 99% percentile of the null distribution of F_{ST} for the simulated neutral loci. Very few outliers were found when comparing same-morphology populations (i.e., H vs. H, or M vs. M) from different geographic areas. From Wilding et al. (2001).

supporting the hypothesis that these 15 loci are under selection. Furthermore, when these 15 loci are used in phylogeny reconstruction, the phylogeny is concordant with morphology and habitat type rather than geographic distance between the locations of the populations. This illustrates the importance of removing outlier loci when inferring historical relationships between populations. Similarly, estimates of the time since populations diverged should be based only on neutral markers, as should estimates of gene flow and migration rates (Luikart et al. 2003).

9.7 Limitations of F_{ST} and other measures of subdivision

The measure of subdivision, F_{ST}, has limitations when using loci with high levels of variation, such as microsatellites. F_{ST} often is biased downwards when variation **within** subpopulations (H_S) is high. For example, if $H_S = 0.90$, F_{ST} cannot be higher than approximately 0.10 (1 − 0.90 = 0.10; see expression 9.3). The source of this bias is obvious: when variation within populations is high, the proportion of the total variation distributed between populations can never be very high (Hedrick 1999).

Another limitation of classic F_{ST} measures (and related measures like G_{ST}) is that they do not consider the identity of alleles (i.e., genealogical degree of relatedness). For example, in the common shrew example in Example 9.4, the F_{ST} for a Y-linked microsatellite is only 0.19 across a hybrid zone between races even though the two races share no alleles at this locus. An examination of Figure 9.10 clearly shows that all of the alleles on either side of the hybrid zone are more similar to each other than to any of the alleles on the other side of the hybrid zone. A measure related to F_{ST}, called R_{ST}, uses information on the length of alleles at microsatellite loci, and is much higher in this case ($R_{ST} = 0.98$). This example illustrates the importance of computing different summary statistics, in this case, F_{ST} and R_{ST}.

R_{ST} is analogous to F_{ST}, and is defined as the proportion of variation in allele length that is due to differences among populations. R_{ST} assumes that each mutation changes an allele's length by only one repeat unit, e.g., a mutation adds or removes one dinucleotide "CA" unit (see Example 4.1; see also stepwise mutation model in Section 12.1.2). This is important because if mutations cause only a one-step change, then any populations with alleles differing by few steps will have experienced substantial recent gene flow, whereas populations with alleles differing by many steps will have had little or no gene flow (such that isolation has allowed accumulation of many mutational steps between populations). This is the pattern that we see in Figure 9.10.

Another measure of differentiation that uses information on allele genealogical relationships is phi-st (Φ_{ST}), which is computed by Excoffier's AMOVA framework (Excoffier et al. 1992). Measures using genealogical information (like Φ and R_{ST}) use the degree of differentiation between alleles as a weighting factor that increases the metric (e.g., F_{ST}) proportionally to the number of mutational differences between alleles.

Another widely used measure of population genetic differentiation is Nei's genetic distance (D; Nei 1972). This measure will increase linearly with time for completely isolated populations and the infinite allele model of mutation with selective neutrality. Nonetheless, D is often used and appears to perform relatively well for nonisolated populations (Paetkau et al. 1997). Nei's (1978) unbiased D provides a correction for sample size. This correction is not so important for comparison between species, but can be for conservation in cases where intraspecific populations are being compared. Without this correction, poorly sampled populations will on average appear to be the most divergent. Another reliable and widely used measure of genetic distance is Cavalli-Sforza and Edwards' chord distance (Cavalli-Sforza and Edwards 1967). There are numerous other genetic distance measures (e.g., see Paetkau et al. 1997) that are less widely used and beyond the scope of this book.

9.7.1 Hierarchical structure

Populations are often substructured at multiple hierarchical levels, e.g., locally and regionally. For example, several subpopulations (demes) might exist on each side of a barrier such as a river or mountain ridge. Here, two hierarchical levels are: (1) the local deme level; and (2) the regional group of demes on either side of the river (Figure 9.13). It is useful to identify such hierarchical structures and to quantify the magnitude of differentiation at each level to help guide conservation management (e.g., identification of management units and evolutionary significant units; see Chapter 16). For example, if regional populations are highly differentiated but local demes within regions are not, managers should often prioritize translocations between local demes and not between regional populations.

Hierarchical structure is often quantified using hierarchical F-statistics that partition the variation into local and regional components, i.e., the proportion of the total differentiation due to differences between subpopulations within regions (F_{SR}), and the proportion of differentiation due to differences between regions (F_{RT}). Hierarchical structure is also

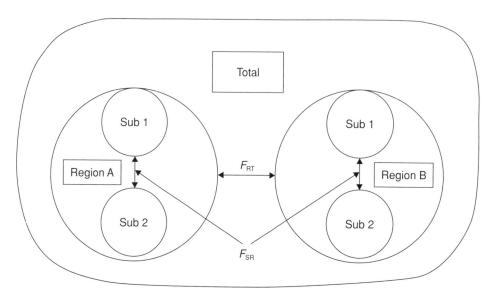

Figure 9.13 Organization of hierarchical population structure with two levels of subdivision: subpopulations within regions (F_{SR}) and regions within the total species (F_{RT}). Each region has two subpopulations. F_{SR} is the proportion of the total diversity due to differences between subpopulations within regions. F_{RT} is the proportion of the total diversity due to differences between regions.

often quantified using AMOVA (analysis of molecular variance) (Excoffier et al. 1992), which is analogous to the standard statistical approach ANOVA (analysis of variance).

9.8 Estimation of gene flow

Gene flow is important to measure in conservation biology because low or reduced gene flow can lead to local inbreeding and inbreeding depression, whereas high or increased gene flow can limit local adaptation and cause outbreeding depression. Measuring and monitoring gene flow can help to maintain viable populations (and metapopulations) in the face of changing environments and habitat fragmentation. Recent research shows that renewed gene flow (following isolation) can result in "**genetic rescue**", through heterosis in "hybrid" offspring (Tallmon et al. 2004). Finally, rates of gene flow in animals are correlated with rates of dispersal; thus knowing rates of gene flow can help predict the likelihood of recolonization of vacant habitats following extirpation or overharvest (i.e., "demographic rescue"). However, over 90% of the gene flow in plant species is due to pollen movement and not seed dispersal (Petit et al. 2005).

Rates of gene flow can be estimated in two general ways using molecular markers. First, indirect estimates of average migration rates (Nm) can be obtained from: (1) allele frequency differences (F_{ST}) among populations; (2) the proportion of private alleles in populations; or (3) a likelihood-based approach using information both on allele frequencies and private alleles (see below). The migration rate estimate is an average over the past tens to hundreds of generations (see below).

Second, direct estimates of current dispersal rates can be obtained using genetic tagging and a mark–recapture approach that directly identifies individual immigrants by identifying their "foreign" genotypes (i.e., genotypes unlikely to originate from the local gene pool). This approach can give estimates of migration rates in the current generation. We now discuss the indirect and direct (assignment test) approaches, in turn, below.

9.8.1 F_{ST} and indirect *Nm*

We can estimate the average number of migrants per generation (*Nm*) by using the island model of migration (see Figure 9.4; Guest Box 9). For example, an F_{ST} of 0.20 yields an estimate of one migrant per generation (*Nm* = 1) under the island model of migration (Figure 9.9, nuclear markers). Less differentiation (F_{ST} = 0.10) leads to a higher estimate of gene flow (*Nm* = 2). Expression 9.12 can be rearranged to allow estimation of the average number of migrants (*Nm*) from F_{ST}, under the island model, as follows:

$$Nm \approx \frac{(1 - F_{ST})}{4F_{ST}} \tag{9.14}$$

The following assumptions are required for interpreting estimates of *Nm* from the simple island model:

1 An infinite number of populations of equal size (see also Section 9.4.1).
2 That *N* and *m* are the same and constant for all populations (thus migration is symmetric).
3 Selective neutrality and no mutation.
4 That populations are at migration–drift equilibrium (a dynamic balance between migration and drift).
5 Demographic equality of migrants and residents (e.g., all have the same probability of reproduction).

The assumptions of this simple model are unlikely to hold in natural populations. This has lead to criticism of the usefulness of *Nm* estimates from the island model approach (Whitlock and McCauley 1999). Nonetheless, performance evaluations using both simulations and analytical theory, suggest that the approach gives reasonable estimates of *Nm* even when certain assumptions are violated (Slatkin and Barton 1989; Mills and Allendorf 1996).

A major limitation to estimating *Nm* from F_{ST} (and from other methods below) is that F_{ST} must be moderate to large ($F_{ST} > 0.05 - 0.10$). This is because the variance in estimates of F_{ST} (and thus confidence intervals on *Nm* estimates) is high at low F_{ST}. Confidence intervals on one *Nm* estimate could range, for example, from less than 10 up to 1,000 (depending on the number and variability of the loci used). This high variance is unfortunate because managers often need to know if, for example, *Nm* is 5 versus 50, because 50 would be high enough to allow recolonization and demographic rescue on an ecological time scale, whereas 5 might not. The high variance of *Nm* estimates at low F_{ST}, along with the model assumptions, means that we often cannot interpret *Nm* estimates literally; instead often we use *Nm* to roughly assess the approximate magnitude of migration rates (e.g., "high" versus "low").

Another limitation of indirect approaches is that few natural populations are at equilibrium, primarily because many generations are required to reach equilibrium. For example, if a population becomes fragmented, but N remains large, drift will be weak. In this case, many generations are required for F_{ST} to increase to the equilibrium level (even with zero migration)! The approximate time required to approach equilibrium is given by the following expression: $1/[2m + 1/(2N)]$. We see that if N is large (and m is small), the time to equilibrium is large. Thus, in large population fragments (with no migration), F_{ST} will increase slowly and the reduced gene flow will not be detectable by indirect methods until after many generations of isolation. In such a case, direct estimates of gene flow (below) are preferred, to complement the indirect estimates. Nonetheless, for conservation genetic purposes, fragments with large N are relatively less crucial to detect because they are relatively less susceptible to rapid genetic change.

If N is small, then drift will be rapid and we might detect increased F_{ST} after only a few generations. Such a scenario of severe fragmentation is obviously the most important to detect for conservation biologists. It will also be the most likely to be detectable using an indirect (e.g., F_{ST}-based) genetic monitoring approach.

In summary, although Nm estimates from F_{ST} must be interpreted with caution, they can provide useful information about gene flow and population differentiation. Nonetheless, the use of different and complementary methods (several indirect plus direct methods) is recommended (Neigel 2002).

9.8.2 Private alleles and *Nm*

Another indirect estimator of Nm is the private alleles method (Slatkin 1985). A private allele is one found in only one population. Slatkin showed that a linear relationship exists between Nm and the average frequency of private alleles. This method works because if gene flow (Nm) is low, populations will have numerous private alleles that arise through mutation, for example. The time during which a new allele remains private depends only on migration rates, such that the proportion of alleles that are private decreases as migration rate increases. If gene flow is high, private alleles will be uncommon.

This method could be less biased than the F_{ST} island model method (above), when using highly polymorphic markers, because it apparently is less sensitive to problems of homoplasy created by back mutations, than is the F_{ST} method (Allen et al. 1995). Homoplasy is most likely when using loci with high mutation rates and back mutation, like some microsatellites (e.g., evolving under the stepwise mutation model).

For example, Allen et al. (1995) studied grey seals and obtained estimates of Nm of 41 using the F_{ST} method, 14 using the R_{ST} method, and 5.6 from the private allele method. The lowest Nm estimate might arise from the private alleles method because this method could be less sensitive to homoplasy, which causes underestimation of F_{ST} or R_{ST}, and thus overestimation of Nm (Allen et al. 1995). The values of Nm from this study must be interpreted with caution as the assumptions of the island model are probably not met and Nm values are fairly high and thus have a high variance. Furthermore, the reliability of the private alleles method has not been thoroughly investigated for loci with potential homoplasy (e.g., microsatellites).

In another study, Nm estimates from allozyme markers were highly correlated with dispersal capability among 10 species of ocean shore fish (Waples 1987). Three estimators of Nm were compared: Nei and Chesser's F_{ST}-based method (F_{STn}), Weir and Cockerham's F_{ST}-based method (F_{STw}), and the private alleles method. The two F_{ST}-based estimators gave highly correlated estimates of Nm, whereas the private alleles method gave less correlated estimates. This lower correlation could result from a low incidence of private alleles in some species. These species were studied with up to 19 polymorphic allozymes with heterozygosities ranging from 0.009 to 0.087 (mean 0.031). Low polymorphism markers might be of little use with the private alleles method because very few private alleles might exist. More studies are needed comparing the performance of different Nm estimators (e.g., likelihood-based methods below) and different marker types (microsatellites versus allozymes or single nucleotide polymorphisms).

9.8.3 Maximum likelihood and the coalescent

A maximum likelihood estimator of Nm was published by Beerli and Felsenstein (2001). This method is promising because, unlike classic methods (above), it does not assume symmetric migration rates or identical population sizes. Furthermore, likelihood-based methods use all the data in their raw form (Appendix Section A4), rather than a single summary statistic, such as F_{ST}. The statistic F_{ST} does not use information such as the proportion of alleles that are rare. Thus, the likelihood method should give less biased and more precise estimates of Nm than classic moments-based methods (Beerli and Felesenstein 2001). Indeed, a recent empirical study in garter snakes (Bittner and King 2003) suggests that coalescent methods are likely to give more reliable estimates of Nm than F_{ST}-based methods, because the F_{ST}-based methods are more biased by lack of migration–drift equilibrium and changing population size.

Beerli and Felsenstein (2001) state that "Maximum likelihood methods for estimating population parameters, as implemented in MIGRATE and GENETREE will make the classical F_{ST}-based estimators obsolete . . .". While this is likely true for some scenarios, new methods and software should be used cautiously (and in conjunction with the classic methods), at least until performance evaluations have thoroughly validated the new methods (e.g., see Appendix Section A3.2). A problem with evaluating the performance of the many likelihood-based methods is they are computationally slow. For example, it can take days or weeks of computing time to obtain a **single** Nm estimate (e.g., using 10–20 loci per population). This makes the validation of methods difficult because validation requires hundreds of estimates for each of numerous simulated scenarios (i.e., different migration rates and patterns, population sizes, mutation dynamics, and sample sizes). The software program MIGRATE (Beerli and Felesenstein 2001) for likelihood-based estimates of Nm is freely available (see this book's website) (see also GENETREE from Bahlo and Griffiths 2000).

The **coalescent** modeling approach (a "backward looking" strategy of simulating genealogies) is usually used in likelihood-based analysis in population genetics (see Appendix Section A5). The coalescent is useful because it provides a convenient and computationally efficient way to generate random genealogies for different gene flow patterns and rates. The efficiency of constructing coalescent trees is important, because likelihood (see Appendix Section A4) involves comparisons of enormous numbers of different genealogies in order to find those genealogies (and population models) that maximize the likelihood of the observed data. The coalescent also facilitates the extraction of genealogical information from data (e.g., divergence patterns between microsatellite alleles or DNA sequences), by easily incorporating both random drift and mutation into population models. Traditional estimators of gene flow sometimes do not use genealogical information, and are based on "forward looking" models for which simulations are slow and probability computations are difficult.

9.8.4 Assignment tests and direct estimates of *Nm*

Direct estimates of migration (*Nm*) can be obtained by directly observing migrants moving between populations. Direct estimates of *Nm* have been obtained traditionally by marking many individuals after birth and following them until they reproduce or by tracking pollen dispersal by looking for the spread of rare alleles or morphological mutants in seeds or seedlings. The number of dispersers that breed in a new (non-natal) population then becomes the estimated *Nm*.

An advantage of direct estimates is that they detect migration patterns of the current generation without the assumption of population equilibrium (migration–mutation–drift equilibrium). This allows up-to-date monitoring of movement and more reliable detection of population fragmentation (reduced dispersal) without waiting for populations to approach equilibrium (see above).

An important limitation of direct estimates is that they might not detect pulses of migrants that can occur only every 5–10 years, as in species where dispersal is driven by cyclical population demography or periodic weather conditions. Unlike direct estimates, indirect estimates of *Nm* estimate the average gene flow over many generations and thus will incorporate effects of pulse migration. For example, 10 migrants every 10 generations will have the same impact on indirect *Nm* estimates as will one migrant per generation for each of 10 generations.

Another limitation of direct estimates is that they often cannot estimate rates of "evolutionarily effective" gene flow. Direct estimates of *Nm* only assume that an observed migrant will reproduce and pass on genes (with the same probability as a local resident individual). However, migrants might have a reduced mating success if they cannot obtain a local territory, for example. Alternatively, migrants might have exceptionally high mating success if there is a "rare male" or "foreign individual" advantage. Furthermore, immigrants could produce offspring more fit than local individuals if heterosis occurs following crossbreeding between immigrants and residents. Heterosis can lead to more gene flow than expected from neutral theory, for any given number of migrants (see "genetic rescue", Section 15.5). Because direct observation of migrants generally does not detect local mating success (effective gene flow), direct observations generally only estimate dispersal and not gene flow (i.e., migration), unless we assume observed migrants reproduce.

Unfortunately, direct estimates of *Nm* are difficult to obtain using traditional field methods of capture–mark–recapture. Following individuals from their birth place until reproduction is extremely difficult or impossible for many species.

Assignment tests offer an attractive alternative to the traditional capture–mark–recapture approach to estimate direct estimates of *Nm*. For example, we can genotype many individuals in a single population sample, and then determine the proportion of "immigrant" individuals, i.e., individuals with a foreign genotype that is unlikely to have originated locally. For example, a study of a galaxiid fish (the inanga) revealed that one individual sampled in New Zealand had an extremely divergent ("foreign") mtDNA haplotype, which was very similar to the haplotypes found in Tasmania (Figure 9.14). It is likely the individual originated in Tasmania and migrated to New Zealand. The inanga spawns in fresh water, but spends part of its life history in the ocean.

One problem with using only mtDNA is we cannot estimate male-mediated migration rates (because mtDNA in maternally inherited). Further, the actual migrant could have been the mother or grandmother of the individual sampled. We could test if the migrant

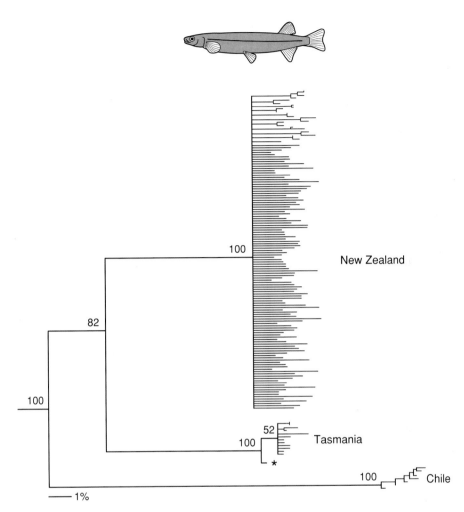

Figure 9.14 Detection of a migrant between populations of inanga using a phylogram derived from mtDNA control region sequences. One mtDNA type (marked with a star) sampled in New Zealand was very similar to the mtDNA types found in Tasmania. This suggests that a small amount of gene flow occurs between the New Zealand populations and Tasmania. Drawing of inanga from McDowall (1990); phylogram from Waters et al. (2000).

or its mother was the actual immigrant by genotyping many autosomal markers (e.g., microsatellites). For example, if a parent was the migrant then only half of the individual's genome (alleles) would have originated from another population (and not the Y chromosome). We can estimate the proportion of an individual's genome arising from each or two parental population via admixture analysis (see Chapter 20).

Assignment tests based on multiple autosomal makers are useful for identifying immigrants. For example, for a candidate immigrant, we first remove the individual from the data set and then compute the expected frequency of its genotype (p^2) in each candidate population of origin by using the observed allele frequencies (p) from each population (Figure 9.15). If the likelihood for one population is far higher than the other,

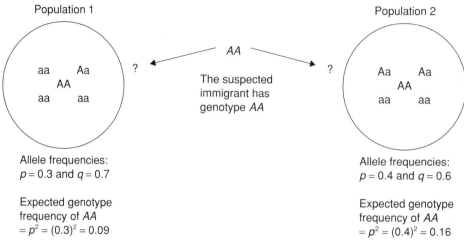

Figure 9.15 Simplified example of using an assignment test to identify an immigrant (*AA*). We first remove the individual in question from the data set and then compute its expected genotype frequency (p^2) in each population using the observed allele frequencies for each population (p^1 and p^2, respectively), and assuming Hardy–Weinberg proportions. If the individual with the genotype *AA* was captured in Population 1 but its expected genotype frequency is far higher in Population 2, then we could conclude the individual is an immigrant. The beauty of assignment tests is they are relatively simple but potentially powerful if many loci (each with many alleles) are used. Note that obtaining the multilocus likelihoods generally requires multiplication of single-locus probabilities (multiplication rule), and thus requires independent loci in gametic equilibrium.

we "assign" the individual to the most likely population. The likelihood can be computed as the frequency of the genotype in the population (expected under Hardy–Weinberg proportions).

The power of assignment tests increases with the number of loci (and the polymorphism level of the loci). Computing the multilocus assignment likelihood requires the multiplication together of single-locus probabilities (multiplication rule), and thus requires the assumption of independence among loci (e.g., no linkage disequilibrium).

9.9 Population subdivision and conservation

Understanding the genetic population structure of species is essential for conservation and management (see Guest Box 9). The techniques to study genetic variation and the genetic models that we have presented in this chapter allow us to rather quickly understand the genetic population structure of any species of interest. The application of this information, however, is often not straightforward and is sometimes controversial. For example, how "distinct" does a population have to be to be considered a **distinct population segment** (**DPS**) in order to be listed under the Endangered Species Act (USA)? The application of genetic information to identify appropriate units for conservation and management is considered in detail in Chapter 16.

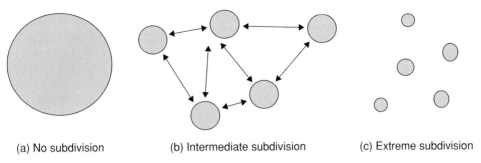

(a) No subdivision   (b) Intermediate subdivision   (c) Extreme subdivision

Figure 9.16 Range of possible degrees of population subdivision. Intermediate degrees of subdivision generally yield the highest adaptive potential with possibilities for local adoption to local environments, yet with occasional gene flow and large enough local effective size to prevent rapid inbreeding and loss of variation.

Population subdivision influences the evolutionary potential of a species, that is, the ability of a species to evolve and adapt to environmental change. To understand this, it is helpful to consider extremes of subdivision (Figure 9.16). For example, a species with no subdivision would have such high gene flow that local adaptation would not be possible. Thus, the total range of types of multilocus genotypes would be limited. On the other hand, if subdivision is extreme then new beneficial mutations that arise will not readily spread across the species. Furthermore, subpopulations may be so small that genetic drift overwhelms natural selection. Thus local adaptation is limited and random change in allele frequencies dominates so that harmful alleles may drift to high frequency or go to fixation. An intermediate amount of population subdivision will result in substantial genetic variation both within and between local populations; this population structure has the greatest evolutionary potential.

Guest Box 9 Hector's dolphin population structure and conservation
C. Scott Baker and Franz B. Pichler

Population subdivisions are important to help define the "unit to conserve" for rare or threatened species or the "unit to manage" for exploited species. The problem of delimiting population units is especially difficult for marine species where there is often an absence of obvious geographic barriers.

Hector's dolphins are endemic to the coastal waters of New Zealand, with a total estimated abundance of about 7,000. They appear phylopatric, and are mostly concentrated along the central regions of the east and west coasts of South Island (Figure 9.17). In North Island, Hector's dolphins are rare, perhaps less than 100 individuals, and are found currently in only a small part of their former range (Dawson et al. 2001). As a coastal species, Hector's dolphins are prone to entanglement and drowning in gillnets. The extent of this incidental mortality or "bycatch" and the potential for local population decline has become a critical and contentious issue in the management of local fisheries, particularly in North Island.

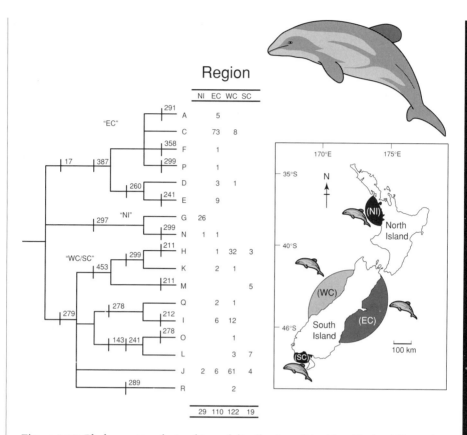

Figure 9.17 Phylogenetic relationship and distribution of the New Zealand Hector's dolphin mtDNA types (haplotypes) based on 440 base pairs of control region sequences from 280 individuals. A parsimony cladogram (e.g., see Section 16.3) shows the number of inferred mutational substitutions (vertical bars) distinguishing each haplotype and the sample frequencies of each haplotype for each of the four regional populations, the North Island (NI) and East Coast (EC), West Coast (WC) and South Coast (SC) of the South Island. Modified from Pichler (2002) and Pichler and Baker (2000).

We initiated a range-wide survey of mtDNA diversity among Hector's dolphins to help designate the appropriate population units for setting limits to bycatch and to better understand long-term gene flow among local populations. Our initial sample collection started with dolphins found beachcast or taken as fisheries' bycatch but was expanded to include historical specimens (teeth and bone dating from 1870) and samples from free-ranging dolphins collected using a "skin scrubby" (a nylon scrub pad on the end of a pole, used to collect sloughed skin from bow-riding dolphins; Harlin et al. 1999) or a small biopsy dart.

Our results have shown a striking degree of differentiation in mtDNA lineages among four regional populations: North Island and the east, west, and south coasts of South Island (Figure 9.17). The frequencies of the 17 identified mtDNA haplotypes differed significantly among regions ($P < 0.001$), and many were found in only one region. Based on Wright's fixation index ($F_{ST} = 0.55$), average long-term migration (mN_f) among the four regions was estimated to be less than one female per generation. Maternal gene flow was highest between the larger east and west coast populations of South Island ($mN_f \sim 3$; $F_{ST} = 0.337$). The North Island population was the most isolated, and diversity was far reduced relative to the South Island populations. Historical samples from North Island included three haplotypes but two of these were not found in the contemporary population sample (Pichler and Baker 2000) and might have been misattributed or misplaced, as beach-cast specimens, by wind and current (Baker et al. 2002a). Analysis of six microsatellite loci also showed low diversity and strong differentiation of the North Island population ($F_{ST} = 0.4545$) but considerably weaker differentiation among the three South Island populations, where only the west and east coast ones differed significantly ($F_{ST} = 0.038$) (Pichler 2002).

The scale of mtDNA differentiation across the species-wide range of Hector's dolphins has important implications for the conservation and evolutionary potential of population units. The strong frequency-based differences for mtDNA (and moderate differences for microsatellites) among the three South Island populations confirm that these should be managed as independent demographic units. Although connected on a long-term or evolutionary time scale, a loss to local abundance or genetic diversity would not be replaced by immigration or gene flow from other populations on a management time scale. The North Island population, however, is completely isolated from the South Island. Based on its fixed mtDNA difference and quantitative morphological differences, A. N. Baker et al. (2002a) proposed to recognize the North Island population as a subspecies, *Cephalorhynchus hectori maui*, or "Maui's" dolphin after the Polynesian demigod who fished up the North Island.

Maui's dolphin is now recognized as "critically endangered" by the World Conservation Union (IUCN) and the New Zealand government has closed part of the North Island coast to all gillnet fishing in an effort to protect this rare dolphin. This study of Hector's dolphin illustrates the usefulness of molecular analyses (along with morphological data) for identifying population units for conservation (see also Chapter 16).

Problem 9.1

Chromosomal polymorphisms generally result in reduced fitness of heterozygotes (heterozygous disadvantage; see Section 8.2). What type of population structure (see Figure 9.16) would you expect to result in relatively large chromosomal differences among subpopulations?

Problem 9.2

Statements are often made in the literature that one migrant per generation among populations ($mN = 1$) is sufficient to cause geographically separate populations to be "effectively panmictic". Do you agree with these statements? If not, how much migration do you think is necessary to bring about effective panmixia? (Hint: see Figure 9.5.)

Problem 9.3

Tallmon et al. (2000) examined genetic variation at six polymorphic allozyme loci in the long-toed salamander from 34 lakes in the Selway-Bitterroot Wilderness on the Montana–Idaho border. The following diagram shows the frequency of the *100* allele at the *PMI* locus in five of these populations; this locus had only two alleles (*100* and *120*) in these five populations so the frequency of the *120* allele is one minus the frequency of the *100* allele shown below. That is, the frequency of the *120* allele in the sample from Terilyn Lake was $1.000 - 0.608 = 0.392$.

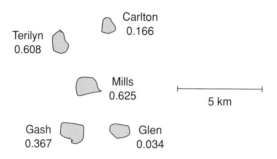

Calculate the F_{ST} for this locus in these samples using expression 9.3. Use your F_{ST} value and expression 9.14 to estimate the approximate numbers of migrants per generation into each of the subpopulations. Does the geographic pattern of allele frequencies at *PMI* seem to fit the expectations of the island model of migration (as is assumed by expression 9.3) or is there evidence of isolation by distance in these data? (See Figure 9.7.)

Problem 9.4

In Section 9.2, we calculated F_{ST} at the *AAT-1* locus in the guppies from Table 6.3 to be 0.301. Calculate F_{ST} at the *PGM-1* locus in this same population.

Problem 9.5

What is the expected F_{ST} in the guppy population from Table 6.3 after thousands of generations of complete isolation?

Problem 9.6

How many individuals must be sampled to detect a deviation from Hardy–Weinberg proportions that results from heterozygote advantage (overdominance selection)? To address this question we can conduct a power analysis (e.g., Appendix Section A3.2), and compute the power of statistical tests to detect a heterozygote excess. Go to the class website and download the program *HW Power*. Click on the icon to launch the program. Select the "Natural selection" model, and type in fitness values that are 10% higher for the heterozygote ("W12"). Run the program until you find the sample size that provides a 50% chance (0.50 probability, i.e., power = 0.50) of detecting a significant deviation at the $P < 0.05$ level of significance (see Appendix Section A3.2).

(a) What is the sample size?
(b) What is the effect on power of changing the allele frequency ("p") from 0.40 to 0.10?
(c) If you increase the amount of heterozygote advantage to 30%, how does power change?

Problem 9.7

Compute the statistical power of the chi-square test to detect inbreeding of 10%. As in Problem 9.6, launch the program *HW Power* and select the "Inbreeding model" and "f: 0.10").

(a) What is the sample size necessary to achieve a power of 0.50 to detect 10% inbreeding?
(b) What sample size is necessary to achieve a power of 0.80? (Note: 0.80 is considered by many statisticians as a "reasonably high power" for a statistical test or experiment, such that the experiment is generally worth conducting; see Appendix Section A3.2.)

Problem 9.8

Why does an intermediate amount of population substructure offer the greatest adaptive potential? Can you think of a species with no or little genetic substructure, even though it is widespread geographically?

Problem 9.9

The following genotypes were found at the *PER* locus in the annual ryegrass *Lolium multiforum* sampled at two different life history stages (Mitton 1989). He observed a significant deficit of heterozygotes in seedlings while the adult plants were in Hardy–Weinberg proportions.

	11	*12*	*22*
Seedlings	278	626	758
Adults	22	87	76

Estimate the F_{IS} and test these two samples for Hardy–Weinberg proportions. Estimate the proportion of progeny produced by selfing in this population by assuming that the observed deficit of homozygotes in the seedlings sample is caused by selfing (see Example 9.1). What is the most likely explanation for the difference between these two life history stages? That is, why are the adults in Hardy–Weinberg proportions even though there is a substantial excess of homozygotes in the seedlings?

10

Multiple Loci

Field cricket, Section 10.5, Example 10.3

It is now generally understood that, as a consequence of selection, random genetic drift, co-ancestry, or gene flow, alleles at different loci may not be randomly associated with each other in a population. While this effect is generally regarded as a consequence of linkage, even genes on different chromosomes may be held temporarily or permanently out of random association by forces of selection, drift and nonrandom mating.

Richard C. Lewontin (1988)

Population geneticists recently have devoted much attention to the topic of gametic disequilibrium. The analysis of multiple-locus genotypic distributions can provide a sensitive measure of selection, genetic drift, and other factors that influence the genetic structure of populations.

David W. Foltz et al. (1982)

We have so far considered only individual loci. Population genetic models become much more complex when two or more loci are considered simultaneously. Fortunately, many of our genetic concerns in conservation can be dealt with from the perspective of individual loci. Nevertheless, there are a variety of situations in which we must concern ourselves with the interactions between multiple loci. For example, genetic drift in small populations can cause nonrandom associations between loci to develop. Therefore, the consideration of multilocus genotypes can provide another method of detecting the effects of genetic drift in natural populations.

In addition, the genotype of individuals over many loci (say 10 or more) may be used to identify individuals genetically because the genotype of each individual (with the exception of identical twins or clones) is genetically unique if enough loci are considered. This genetic "fingerprinting" capability has many potential applications in understanding populations, estimating population size (Chapter 14), and in applying genetics to problems in **forensics** (Chapter 20).

The nomenclature of multilocus genotypes is particularly messy and often inconsistent. It is difficult to find any two papers (even by the same author!) that use the same gene symbols and nomenclature for multilocus genotypes. Therefore, we have made a special effort to use the simplest possible nomenclature and symbols that are consistent as possible with previous usage in the literature.

The term **linkage disequilibrium** is commonly used to describe the nonrandom association between alleles at two loci (Lewontin and Kojima 1960). However, this term is misleading because, as we will see, unlinked loci may be in so-called "linkage disequilibrium". Things are complicated enough without using misnomers that lead to additional confusion when considering multilocus models. The term **gametic disequilibrium** is a much more descriptive and appropriate term to use in this situation. We have chosen to use gametic disequilibrium in order to reduce confusion.

We will first examine general models describing associations between loci and their evolutionary dynamics from generation to generation. We will then explore the various evolutionary forces that cause nonrandom associations between loci to come about in natural populations (genetic drift, natural selection, population subdivision, and hybridization). Finally, we will compare various methods for estimating associations between loci in natural populations.

10.1 Gametic disequilibrium

We now focus our interest on the behavior of two autosomal loci considered simultaneously under all of our Hardy–Weinberg equilibrium assumptions. We know that each locus individually will reach a neutral equilibrium in one generation under Hardy–Weinberg conditions. Is this true for two loci considered jointly? We will see shortly that the answer is no.

Allele frequencies are insufficient to describe genetic variation at multiple loci. Fortunately, however, we do not have to keep track of all possible genotypes. Rather we can use the gamete frequencies to describe nonrandom associations between alleles at different loci. For example, in the case of two loci that each have two alleles, there are just two allele frequencies, but there are nine different genotype frequencies. However, we can describe this system with just four gamete frequencies.

Table 10.1 Genotypic array for two loci showing the expected genotypic frequencies in a random mating population.

	AA	**Aa**	**aa**
BB	$(G_1)^2$	$2G_1G_3$	$(G_3)^2$
Bb	$2G_1G_2$	$2G_1G_4 + 2G_2G_3$	$2G_3G_4$
bb	$(G_2)^2$	$2G_2G_4$	$(G_4)^2$

Let G_1, G_2, G_3, and G_4 be the frequencies of the four gametes AB, Ab, aB, and ab respectively as shown below. If the alleles at these loci are associated randomly then the frequency of any gamete type will be the product of the frequencies of its two alleles:

$$
\begin{array}{ll}
\text{Gamete} & \text{Frequency} \\
AB & G_1 = (p_1)(p_2) \\
Ab & G_2 = (p_1)(q_2) \\
aB & G_3 = (q_1)(p_2) \\
ab & G_4 = (q_1)(q_2)
\end{array}
\tag{10.1}
$$

where $(p_1; q_1)$ and $(p_2; q_2)$ are the frequencies of the alleles $(A; a)$ and $(B; b)$, at locus 1 and 2 respectively. The expected frequencies of two-locus genotypes in a random mating population can then be found as shown in Table 10.1.

D is used a measure of the deviation from random association between alleles at the two loci (Lewontin and Kojima 1960). D is known as the coefficient of gametic disequilibrium and is defined as:

$$D = (G_1G_4) - (G_2G_3) \tag{10.2}$$

or:

$$D = G_1 - p_1p_2 \tag{10.3}$$

If alleles are associated at random in the gametes (as in expression 10.1), then the population is in gametic equilibrium and $D = 0$. If D is not equal to zero, the alleles at the two loci are not associated at random with respect to each other and the population is said to be in gametic disequilibrium (Example 10.1). For example, if a population consists only of a 50 : 50 mixture of the gametes A_1B_1 and A_2B_2, then:

$G_1 = 0.5$

$G_2 = 0.0$

$G_3 = 0.0$

$G_4 = 0.5$

and,

$$D = (0.5)(0.5) - (0.0)(0.0) = +0.25$$

Example 10.1 Genotypic frequencies with and without gametic disequilibrium

Let us consider two loci at which allele frequencies are $p_1 = 0.4(q_1 = 1 - p_1 = 0.6)$ and $p_2 = 0.7(q_2 = 1 - p_2 = 0.3)$ in two populations. The two loci are randomly associated in one population, but show maximum nonrandom association in the other. The gametic frequency values below show the case of random association of alleles at the two loci (gametic equilibrium, $D = 0$) and the case of maximum positive disequilibrium ($D = +0.12$; see Section 10.1.1 for an explanation of the maximum value of D).

Gamete	$D = 0$	D (max)
AB	$(p_1)(p_2) = 0.28$	0.40
Ab	$(p_1)(q_2) = 0.12$	0.00
aB	$(q_1)(p_2) = 0.42$	0.30
ab	$(q_1)(q_2) = 0.18$	0.30

In a random mating population, the following genotypic frequencies will result in each case as shown below. The expected genotypic frequencies with $D = 0$ are shown without brackets, and the expected genotypic frequencies with maximum positive gametic disequilibrium are shown in square brackets:

	AA	Aa	aa	Total
BB	0.08	0.24	0.18	0.49
	[0.16]	[0.24]	[0.09]	[0.49]
Bb	0.07	0.20	0.15	0.42
	[0.0]	[0.24]	[0.18]	[0.42]
bb	0.01	0.04	0.03	0.09
	[0.0]	[0.0]	[0.09]	[0.09]
Total	0.16	0.48	0.36	
	[0.16]	[0.48]	[0.36]	

Notice that each locus is in Hardy–Weinberg proportions in the populations with and without gametic disequilibrium.

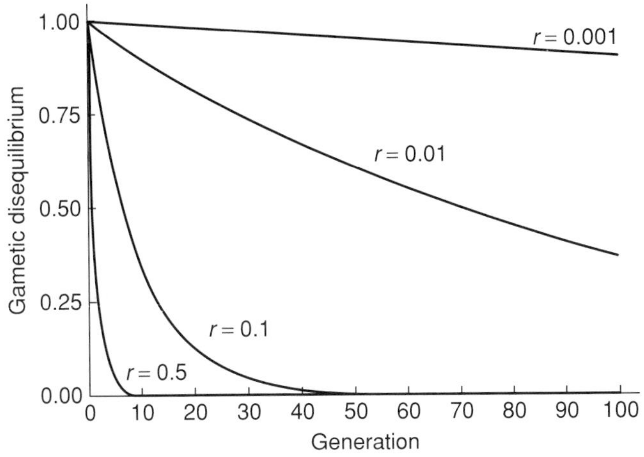

Figure 10.1 Decay of gametic disequilibrium (D_t/D_0) with time for various amounts of recombination (r) between the loci from expression 10.3.

The amount of gametic disequilibrium (i.e., the value of D) will decay from generation to generation as a function of the rate of recombination (r) between the two loci:

$$D' = D(1 - r) \tag{10.4}$$

So that after t generations:

$$D_{t'} = D_0(1 - r)^t \tag{10.5}$$

If the two loci are not linked (i.e., $r = 0.5$), the value of D_t will be halved each generation until equilibrium at $D = 0$. Linkage ($r < 0.5$) will delay the rate of decay of gametic disequilibrium. Nevertheless, D eventually will be equal to zero, as long as there is some recombination ($r > 0.0$) between the loci. However, if the two loci are tightly linked, it will take many generations for them to reach gametic equilibrium (Figure 10.1).

We therefore expect that nonrandom associations of genotypes between loci (i.e., gametic disequilibrium) would be much more frequent between tightly linked loci. For example, Zapata and Alvarez (1992) summarized observed estimates of gametic disequilibrium between five allozyme loci in several natural populations of *Drosophila melanogaster* on the second chromosome (Figure 10.2). The effective frequency of recombination is the mean of recombination rates in females and males assuming no recombination in males. Only pairs of loci with less than 15% recombination showed consistent evidence of gametic disequilibrium.

10.1.1 Other measures of gametic disequilibrium

D is a poor measure of the relative amount of disequilibrium at different pairs of loci because the possible values of D are constrained by allele frequencies at both loci. The largest possible positive value of D is either $p_1 q_2$ or $p_2 q_1$, whichever value is smaller; and the

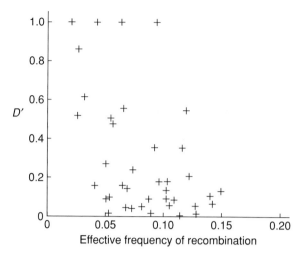

Figure 10.2 Observed estimates of gametic disequilibrium between five allozyme loci in several natural populations of *D. melanogaster* on the second chromosome. The effective frequency of recombination is the mean of recombination rates in females and males assuming no recombination in males. From Zapata and Alvarez (1992).

largest negative value of D is the lesser value of $p_1 p_2$ or $q_1 q_2$. We can see that the largest positive value of D occurs when G_1 is maximum. p_1 is equal to G_1 plus G_2, and p_2 is equal to G_1 plus G_3. Therefore, the largest possible value of G_1 is the smaller of p_1 and p_2. We can see this in Example 10.1 in which the largest positive value of D occurs when G_1 is equal to p_1 which is less than p_2. Once the values of G_1, p_1, and p_2 are set, all of the other gamete frequencies must follow.

This allele frequency constraint of D reduces its value for comparing the amount of gametic equilibrium for the same loci in different populations or for different pairs of loci in the same population. For example, consider two pairs of loci in complete gametic disequilibrium. In case 1 both loci are at allele frequencies of 0.5, while in case 2 both loci are at frequencies of 0.8. The following gamete frequencies result:

	Frequencies	
Gamete	Case 1	Case 2
AB	0.5	0.9
Ab	0.0	0.0
aB	0.0	0.0
ab	0.5	0.1

The value of D in case 1 will be +0.25, while it will be +0.09 in case 2.

Several other measures of gametic disequilibrium have been proposed that are useful for various purposes (Hedrick 1987). A useful measure of gametic disequilibrium should have the same range regardless of allele frequencies. This will allow comparing the amount of disequilibrium among pairs of loci with different allele frequencies.

Lewontin (1964) suggested using the parameter D' to circumvent the problem of the range of values being dependent upon the allele frequencies:

$$D' = \frac{D}{D_{max}} \qquad (10.6)$$

Thus, D' ranges from 0 to 1 for all allele frequencies. However, even D' is not independent of allele frequencies, and, therefore, is not an ideal measure of gametic disequilibrium (Lewontin 1988). Nevertheless, recent analysis has concluded that the D' coefficient is a useful tool for the estimation and comparison of the extent of overall disequilibrium among many pairs of multiallelic loci (Zapata 2000).

The correlation coefficient (R) between alleles at the two loci has also been used to measure gametic disequilibrium:

$$R = \frac{D}{(p_1 q_1 p_2 q_2)^{1/2}} \qquad (10.7)$$

R has a range of values between -1.0 and $+1.0$. However, this range is reduced somewhat if the two loci have different allele frequencies. Both D' and R will decay from generation to generation by a rate of $(1-r)$, as does D, because they are both functions of D.

10.1.2 Associations between cytoplasmic and nuclear genes

Just as with multiple nuclear genes, non-random associations between nuclear loci and mtDNA genotypes may occur in populations.

$$
\begin{array}{cc}
\text{Gamete} & \text{Frequency} \\
AM & G_1 \\
Am & G_2 \\
aM & G_3 \\
am & G_4 \\
\end{array}
\qquad (10.8)
$$

Again, D is a measure of the amount of gametic equilibrium and is defined as in expression 10.2. D between nuclear and cytoplasmic genes will decay at a rate of one-half per generation, just as for two unlinked nuclear genes. That is,

$$D' = D(0.5) \qquad (10.9)$$

and, therefore,

$$D_t = D(0.5)^t \qquad (10.10)$$

10.2 Small population size

Nonrandom associations between loci will be generated by genetic drift in small populations. We can see this readily in the extreme case of a bottleneck of a single individual capable of reproducing by selfing because a maximum of two gamete types can occur within a single individual. In general, we can imagine the four gamete frequencies to be analogous to four alleles at a single locus. Changes in frequency from generation to generation caused by drift will often result in nonrandom associations between alleles at different loci. The expected value of D due to drift is zero. Nevertheless, drift-generated gametic disequilibria may be very great and are equally likely to be positive or negative in sign. For example, genome-wide investigations in humans have found that large blocks of gametic disequilibrium occur throughout the genome in human populations. These blocks of disequilibrium are thought to have arisen during an extreme population bottleneck that occurred some 25,000–50,000 years ago (Reich et al. 2001).

Gametic disequilibrium produced by a single generation of drift may take many generations to decay. Therefore, we would expect substantially more drift-generated gametic disequilibrium between closely linked loci. In fact, the expected amount of disequilibrium for closely linked loci is:

$$E(R^2) \approx \frac{1}{1 + 4Nr} \tag{10.11}$$

where R^2 is the square of the correlation coefficient (R) between alleles at the two loci (see expression 10.5) (Hill and Robertson 1968; Ohta and Kimura 1969). For unlinked loci, the following value of R^2 is expected (Weir and Hill 1980):

$$E(R^2) \approx \frac{1}{3N} \tag{10.12}$$

10.3 Natural selection

Let us examine the effects of natural selection with constant fitnesses at two loci each with two alleles. We will designate the fitness of a genotype to be w_{ij}, where i and j are the two gametes that join to form a particular genotype. There are two genotypes that are heterozygous at both loci (AB/ab and Ab/aB); we will assume that both double heterozygotes have the same fitness (i.e., $w_{23} = w_{14}$).

	AA	Aa	aa
BB	w_{11}	w_{13}	w_{33}
Bb	w_{12}	$w_{23} = w_{14}$	w_{34}
bb	w_{22}	w_{24}	w_{44}

The frequency of the AB gamete after one generation of selection will be:

$$G_{1'} = \frac{G_1(G_1 w_{11} + G_2 w_{12} + G_3 w_{13} + G_4 w_{14}) - r w_{14} D}{\bar{w}} \tag{10.13}$$

where $\bar{w}$ is the average fitness of the population. We can simplify this expression by $\bar{w}_i$ defining to be the average fitness of the ith gamete.

$$\bar{w}_i = \sum_{j=1}^{4} G_j w_{ij} \tag{10.14}$$

and then,

$$\bar{w} = \sum_{i=1}^{4} G_i \bar{w}_i \tag{10.15}$$

and,

$$G_{1'} = \frac{G_1 \bar{w}_1 - r w_{14} D}{\bar{w}} \tag{10.16}$$

We can derive similar recursion equations for the other gamete frequencies:

$$G_{2'} = \frac{G_2 \bar{w}_2 + r w_{14} D}{\bar{w}}$$

$$G_{3'} = \frac{G_3 \bar{w}_3 + r w_{14} D}{\bar{w}} \tag{10.17}$$

$$G_{4'} = \frac{G_4 \bar{w}_4 - r w_{14} D}{\bar{w}}$$

There are no general solutions for selection at two loci. That is, there is no simple formula for the equilibria and their stability. However, a number of specific models of selection have been analyzed. The simplest of these is the additive model where the fitness effects of the two loci are summed to yield the two-locus fitnesses. Another simple case is the multiplicative model where the two-locus fitnesses are determined by the product of the individual locus fitnesses. In both of these cases, heterozygous advantage at each locus is necessary and sufficient to insure stable polymorphisms at both loci.

A detailed examination of the effects of natural selection at two loci is beyond the scope of our consideration. Interested readers are directed to appropriate population genetics sources (e.g., Hartl and Clark 1997; Hedrick 2005). We will consider two situations of selection at multiple loci that are particularly relevant for conservation.

10.3.1 Genetic hitchhiking

Natural selection at one locus can affect closely linked loci in many ways. Let us first consider the case where directional selection occurs at one locus (B) and the second locus is selectively neutral (A). The following fitness set results:

	A_1A_1	A_1A_2	A_2A_2
B_1B_1	w_{11}	w_{11}	w_{11}
B_1B_2	w_{12}	w_{12}	w_{12}
B_2B_2	w_{22}	w_{22}	w_{22}

where $w_{11} < w_{12} < w_{22}$.

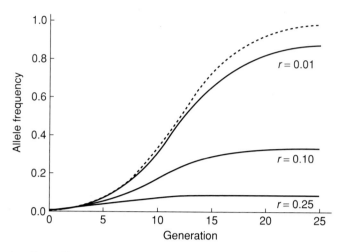

Figure 10.3 The effect of hitchhiking on a neutral locus that is initially in complete gametic disequilibrium with a linked locus that is undergoing directional selection ($w_{11} = 1.0$; $w_{12} = 0.75$; $w_{22} = 0.5$). r is the amount of recombination between the two loci. The dashed line shows the expected change at the selected locus.

Imagine that the favored B_2 allele is a new mutation at the B locus. In this case, the selective advantage of the B_2 allele frequencies may carry along either the A_1 or A_2 allele, depending upon which allele is initially associated with the B_2 mutation. This is known as genetic **hitchhiking** and will result in a so-called **selective sweep**. The magnitude of this effect depends on the selection differential, the amount of recombination (r), and the initial gametic array (Figure 10.3). A selective sweep will reduce the amount of variation at loci that are tightly linked to the locus under selection.

For example, low activity alleles at the glucose-6-phosphate dehydrogenase locus in humans are thought to reduce risk from the parasite responsible for causing malaria (Tishkoff et al. 2001). The pattern of gametic disequilibrium between these alleles and closely linked microsatellite loci suggests that these alleles have increased rapidly in frequency by natural selection since the onset of agriculture in the past 10,000 years.

10.3.2 Associative overdominance

Selection at one locus also can affect closely linked neutral loci when the genotypes at the selected locus are at an equilibrium allele frequency. Consider the case of heterozygous advantage where, using the previous fitness array, $w_{11} = w_{22} = 1.0$ and $w_{12} = (1 + s)$. The effective fitnesses at the A locus are affected by selection at the B locus (s) and D; the marginal fitnesses are the average fitness at the A locus considering the two-locus genotypes. These would be the estimated fitnesses at the A locus if only that locus were observed. If D is zero then all the genotypes at the A locus will have the same fitness. However, if there is gametic disequilibrium (i.e., D is not equal to zero) then heterozygotes at the A locus will experience a selective advantage because of selection at the B locus.

This effect has been called **associative overdominance** (Ohta 1971) or **pseudo-overdominance** (Carr and Dudash 2003). This pattern of selection has also been called marginal overdominance (Hastings 1981). However, **marginal overdominance** has more generally been used for the situation where genotypes experience multiple environments and different alleles are favored in different environments (Wallace 1968). This can lead to an overall greater fitness of heterozygotes even though they do not have a greater fitness in any single environment.

Heterozygous advantage is not necessary for linked loci to experience associative over-dominance. Heterozygous individuals at a selectively neutral locus will have higher average fitnesses than homozygotes if the locus is in gametic disequilibrium with a locus having deleterious recessive alleles (Ohta 1971).

We can see this with the genotypic arrays in Example 10.1. Let us assume that the *b* allele is a recessive lethal (i.e., fitness of the *bb* genotype is zero). In the case of gametic equilibrium ($D = 0$), exactly q_2^2 ($0.3 \times 0.3 = 0.09$) of genotypes at the *A* locus have a fitness of zero. Thus, the mean or marginal fitness at the *A* locus is $1 - 0.09 = 0.91$. However, in the case of maximum positive disequilibrium, only the *aa* genotypes have reduced fitness because the *AA* and *Aa* genotypes do not occur in association with the *bb* genotype. Thus, the fitness of *AA*, *Aa*, and *aa* are 1, 1, and 0.75. There are many more *aa* than *AA* homozygotes in the population; therefore, *Aa* heterozygotes have greater fitness than the mean of the homozygotes.

Associative overdominance is one possible explanation for the pattern seen in many species in which individuals that are more heterozygous at many loci have greater fitness. Example 10.2 presents an example where associative overdominance is most likely responsible for **heterozygosity–fitness correlations** (HFCs) in great reed warblers.

10.3.3 Genetic draft

We saw in Section 10.3.1 that directional selection at one locus can reduce the amount of genetic variation at closely linked loci following a selective sweep. This is a special case of a more general effect in which selection at one locus will reduce the effective population size of linked loci. This has been termed the **Hill–Robertson effect** (Hey 2000) because it was first discussed in a paper that considered the effect of linkage between two loci under selection (Hill and Robertson 1966). Observations with *Drosophila* have found that regions of the genome with less recombination tend to be less genetically variable as would be expected with the Hill–Robertson effect (Begun and Aquadro 1992; Charlesworth 1996).

This effect has potential importance for conservation genetics. For example, we would expect a strong Hill–Robertson effect for mtDNA where there is no recombination. A selective sweep of a mutant with some fitness advantage could quickly fix a single haplotype and therefore greatly reduce genetic variation. Therefore, low variation at mtDNA may not be a good indicator of the effective population size experienced by the nuclear genome.

Gillespie (2001) has presented an interesting consideration of the effects of hitchhiking on regions near a selected locus. He has termed this effect **genetic draft** and has suggested that the stochastic effects of genetic draft may be more important than genetic drift in large populations. In general, it would reduce the central role thought to be played by effective population size in determining the amount of genetic variation in large populations. The potential effects of genetic draft seem to not be important for the effective population sizes usually of concern in conservation genetics.

Example 10.2 Associative overdominance and HFCs in great reed warblers

Individuals that are more heterozygous at many loci have been found to have greater fitness in many species (Britten 1996; David 1998; Hansson and Westerberg 2002). Such heterozygosity–fitness correlations (HFCs) have three possible primary explanations. First, the association may be a consequence of differences in inbreeding among individuals within a population. Inbred individuals will tend to be less heterozygous and experience inbreeding depression. Second, the loci being scored may be in gametic disequilibrium with loci that affect the traits being studied, resulting in associative overdominance. Lastly, the associations may be due to heterozygous advantage at the loci being studied. This latter explanation seems unlikely for loci such as microsatellites that are generally assumed to be selectively neutral. There is some evidence that HFCs at allozyme loci may be due to the loci themselves (Thelen and Allendorf 2001).

Hansson et al. (2004) distinguished between inbreeding versus associative overdominance in great reed warblers by testing for HFC within pairs of siblings with the same pedigree. This comparison eliminated the reduced genome-wide heterozygosity of inbred individuals as an explanation because full siblings have the same pedigree inbreeding coefficient (F). Fifty pairs of sibling were compared in which only one individual survived to adult age. Paired siblings were confirmed to have the same genetic parents (by molecular methods) and were matched for sex, size (length of the innermost primary feather), and body mass (when 9 days old).

The surviving sib tended to have greater multilocus heterozygosity at 19 microsatellite loci (Figure 10.4; $P < 0.05$). In addition, the surviving sibs also had significantly greater d^2 values ($P < 0.01$). This measure is the squared difference in number of repeat units between the two alleles, $d^2 = $ (number of repeats at allele A – number of repeats at allele B)^2. The difference in repeat score between alleles carries information about the amount of time that has passed since they shared a common ancestral allele (see the coalescent in Appendix Section A5). This assumes a single-step model of mutation. Heterozygotes with smaller values possess two alleles that are likely to have shared a common ancestral allele more recently than heterozygotes with larger d^2 values (see Figure 12.1). Therefore, heterozygotes with lower d^2 possess two alleles marking chromosomal segments that are more likely to carry the same deleterious recessive allele responsible for associative overdominance. The strong relationship between d^2 and recruitment suggests that associative overdominance is responsible for the observed HFC.

The studied population of great reed warblers was small and recently founded. Thirty-five of 162 pairwise tests for gametic disequilibrium were significant (uncorrected for multiple tests, $P < 0.05$), suggesting widespread gametic disequilibrium in this population because of its recent founding and small size (see Section 10.2). These authors conclude that associative overdominance is likely to be responsible for the HFC that they have observed. They also argue that gametic disequilibrium is likely to be responsible for many observations of HFC in other species, especially in cases of recently founded or small populations.

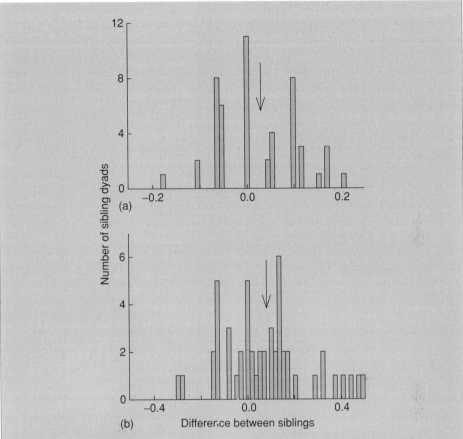

Figure 10.4 Difference between surviving and nonsurviving great reed warbler siblings (50 matched pairs) in: (a) multilocus heterozygosity (MLH; $P < 0.05$) and (b) mean d^2 ($P < 0.01$). Arrows indicate mean difference. The greater MLH and d^2 of surviving birds apparently results from associative overdominance. From Hansson et al. (2004).

10.4 Population subdivision

Population subdivision will generate nonrandom associations (gametic disequilibrium) between alleles at multiple loci if the allele frequencies differ among subpopulations at both loci. This is an extension to two loci of the Wahlund principle, the excess of homozygotes caused by population subdivision at a single locus, to two loci (Sinnock 1975) (see Section 9.2). In general, for k equal-sized subpopulations:

$$D = \bar{D} + cov(p_1, p_2) \tag{10.18}$$

where $\bar{D}$ is the average D value within the k subpopulations (Nei and Li 1973; Prout 1973).

This effect is important when two or more distinct subpopulations are collected in a single sample. For example, many populations of fish living in lakes consist of several genetically distinct subpopulations that reproduce in different tributary streams. Thus, a single random sample taken of the fish living in the lake will comprise several separate demes. Makela and Richardson (1977) have described the detection of multiple genetic subpopulations by an examination of gametic disequilibrium among many pairs of loci.

Cockerham and Weir (1977) have introduced a so-called composite measure of gametic disequilibrium that partitions gametic disequilibrium into two components: the usual measure of gametic disequilibrium, D, plus an added component that is due to the nonrandom union of gametes caused by population subdivision (D_B).

$$D_C = D + D_B \tag{10.19}$$

In a random mating population, D and D_C will have the same value. We will see in the next section that the composite measure is of special value when estimating gametic disequilibrium from population samples. Campton (1987) has provided a helpful and exceptionally lucid discussion of the derivation and use of the composite gametic disequilibrium measure.

10.5 Hybridization

Hybridization between populations, subspecies, or species will result in gametic disequilibrium. Figure 10.5 outlines the resulting genotypes and gametes in the first two generations in this case. The F_1 hybrid will be heterozygous for all loci at which the two taxa differ. The gametes produced by the F_1 hybrid will depend upon the linkage relationship of the two loci. If the two loci are unlinked, then all four gametes will be produced in equal frequencies because of recombination.

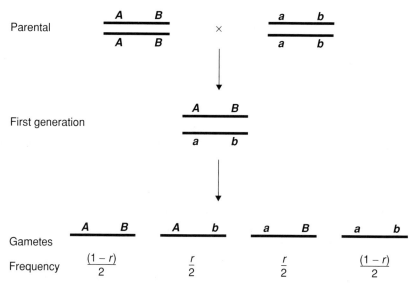

Figure 10.5 Outline of gamete formation in a hybrid between two parental taxa fixed for different alleles at two loci. Taxon one is *AABB*, and taxon two is *aabb*. The gametes produced by F_1 hybrids are influenced by the rate of recombination (r). These four gametes will be equally frequent (25% each) for unlinked loci ($r = 0.5$). There will be an excess of parental gametes (*AB* and *ab* in this case) if the loci are linked ($r < 0.5$).

Table 10.2 Expected genotype frequencies and coefficient of gametic disequilibrium (D) in a random mating hybrid swarm.

		Genotype frequencies			
Genotypes	Parental	First generation	Second generation	Third generation	Equilibrium
AABB	0.500	0.250	0.141	0.098	0.063
AABb			0.094	0.118	0.125
AAbb			0.016	0.035	0.063
AaBB			0.094	0.118	0.125
AaBb		0.500	0.312	0.267	0.250
Aabb			0.094	0.118	0.125
aaBB			0.016	0.035	0.063
aaBb			0.094	0.118	0.125
aabb	0.500	0.250	0.141	0.098	0.063
D	–	+0.250	+0.125	+0.063	0.000

Table 10.2 shows the genotypes produced by hybridization between two taxa that are fixed for different alleles at two unlinked loci. This assumes that the two taxa are equally frequent and mate at random. We can see here that gametic disequilibrium (D) will be reduced by exactly one-half each generation. For unlinked loci, recombination will eliminate the association between loci in heterozygotes. However, only one-half of the population in a random mating population will be heterozygotes in the first generation. Recombination in the two homozygous genotypes will not have any effect. Therefore, gametic disequilibrium (D) will be reduced by exactly one-half each generation. A similar effect will occur in later generations even though more genotypes will be present. That is, recombination will only affect the frequency of gametes produced in individuals that are heterozygous at both loci ($AaBb$).

Gametic disequilibrium will decay at a rate slower than one-half per generation if the loci are linked. Tight linkage will greatly delay the rate of decay of D. For example, it will take 69 generations for D to be reduced by one-half if there is 1% recombination between loci (see expression 10.5).

Gametic disequilibrium also will decay at a slower rate if the population does not mate at random because there is positive assortative mating of the parent types. This will reduce the frequency of double heterozygotes in which recombination can act to reduce gametic disequilibrium. We can see this using expression 10.8. In this case, the D component of the composite measure (D_C) will decline at the expected rate, but D_B will persist depending upon the amount of assortative mating. Random mating in a hybrid population can be detected by testing for Hardy–Weinberg proportions at individual loci.

These two alternative explanations of persisting gametic disequilibrium in a hybrid can be distinguished. Assortative mating will affect all pairs of loci (including cytoplasmic and nuclear associations) while the effect of linkage will differ between pairs depending upon their rate of recombination. Example 10.3 describes the multilocus genotypes in a natural hybrid zone between two species of crickets. In this case, most genotypes are similar to the parental taxa and gametic disequilibrium persists over all loci because of assortative mating. Guest Box 10 presents a true hybrid swarm in which mating is at random (all loci are in Hardy-Weinberg proportions), but gametic disequilibrium persists at linked loci (Example 10.4).

Example 10.3 Cytonuclear disequilibrium in a hybrid zone of field crickets

Hybrid zones occur where two genetically distinct taxa are sympatric and hybridize to form at least partially fertile progeny. Observations of the distribution

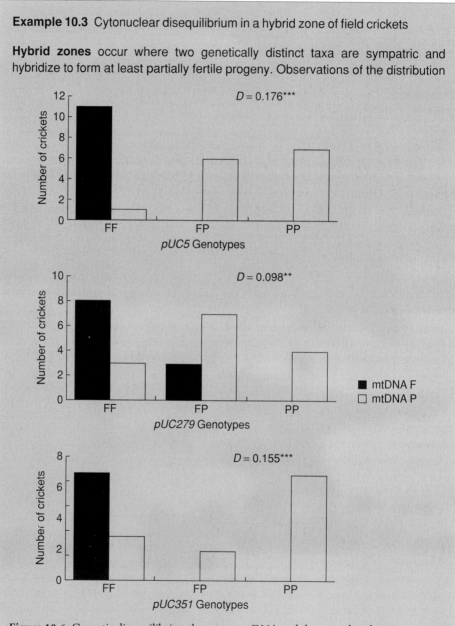

Figure 10.6 Gametic disequilibrium between mtDNA and three nuclear loci in a hybrid zone between two species of field crickets, *Gryllus pennsylvanicus* (P) and *G. firmus* (F). The mtDNA from *G. firmus* is significantly more frequent for homozygotes (FF) for the *G. firmus* nuclear allele at all three loci. ***, $P < 0.001$; **, $P < 0.01$. From Harrison and Bogdanowicz (1997).

of multilocus genotypes within hybrid zones and the patterns of introgression across hybrid zones can provide insight into the patterns of mating and the fitnesses of hybrids that may contribute to barriers to gene exchange between taxa.

Harrison and Bogdanowicz (1997) describe gametic disequilibrium in a hybrid zone between two species of field crickets, *Gryllus pennsylvanicus* and *G. firmus*. These two species hybridize in a zone that extends from New England to Virginia in the United States. Analyses of four anonymous nuclear loci, allozymes, mtDNA, and morphology at three sites in Connecticut indicate that nonrandom associations between nuclear markers, between nuclear and mtDNA (Figure 10.6), and between genotypes and morphology persist primarily because of more frequent matings between parental types. That is, the crickets at these three sites in this hybrid appear to be primarily parental with a few F_1 individuals and even fewer later generation hybrids.

These two species of field crickets are genetically similar. There are no fixed diagnostic differences at allozyme loci, and more than 50 anonymous nuclear loci had to be screened to find four that were diagnostic. These two taxa meet the criteria for species according to some **species concepts** but not others. Regardless, as we will see in Chapter 16, the long-term persistence of parental types throughout an extensive hybrid zone indicates that these species are clearly distinct biological units.

Example 10.4 Gametic disequilibrium in a hybrid swarm

Forbes and Allendorf (1991) studied gametic disequilibrium in a hybrid swarm of cutthroat trout (see Guest Box 10). They observed the following genotypic distribution between two closely linked diagnostic allozyme loci. At both loci, the upper-case allele (*A* and *B*) designates the allele fixed in the Yellowstone cutthroat trout and the lower-case allele (*a* and *b*) is fixed in westslope cutthroat trout. There is a large excess of parental gamete types. The allele frequencies at the two loci are $p_1 = 0.589$ and $p_2 = 0.518$. The expected genotypes if $D = 0$ are presented in parentheses:

		LDH-A2		
ME-4	*AA*	*Aa*	*aa*	Total
BB	7	0	0	7
	(2.6)	(3.6)	(1.3)	
Bb	3	12	0	15
	(4.8)	(6.8)	(4.9)	
bb	0	1	5	6
	(1.2)	(2.2)	(1.0)	
Total	10	13	5	

The estimated value of D in this case is 0.213 and $D' = 1.000$ using the EM method described in Section 10.6.1. The estimated gamete frequencies are presented below:

Gamete	$D = 0$	$D = 0.213$
AB	$(p_1)(p_2) = 0.305$	0.518
Ab	$(p_1)(q_2) = 0.284$	0.071
aB	$(q_1)(p_2) = 0.213$	0.000
ab	$(q_1)(q_2) = 0.198$	0.411

We will examine hybridization and its genotypic effects again in Chapter 17 when we consider the effects of hybridization on conservation.

10.6 Estimation of gametic disequilibrium

There is no simple way to estimate gametic equilibrium values from population data (Kalinowski and Hedrick 2001). As described in the next section, even the simplest case of two alleles at a pair of loci is complicated. Estimation becomes more difficult when we consider that most loci have more than two alleles, and we often have genotypes from many loci. There are a total of $n(n-1)/2$ pairwise combinations of loci if we examine n loci. So with 10 loci, there are a total of 45 combinations of two-locus gametic equilibrium values to estimate.

10.6.1 Two loci with two alleles each

Let us consider the simplest case of two alleles at a pair of loci (see genotypic array in Table 10.1). The gamete types (e.g., AB or Ab) cannot be observed directly but must be inferred from the diploid genotypes. For example, AABB individuals can only result from the union of two AB gametes, and AABb individuals can only result from the union of an AB gamete and an Ab gamete. Similar inferences of gametic types can be made for all individuals that are homozygous at one or both loci. In contrast, gamete frequencies cannot be inferred from double heterozygotes (AaBb) because they may result from either union of AA and bb gametes or Ab and aB gametes. Consequently, gametic disequilibrium cannot be calculated directly from diploids.

Several methods are available to estimate gametic disequilibrium values in natural populations when the two gametic types of double heterozygotes cannot be distinguished. The simplest way is to ignore them, and simply estimate D from the remaining eight genotypic classes. The problem with this method is that double heterozygous individuals may represent a large proportion of the sample (Example 10.4), and their exclusion from the estimate will result in a substantial loss of information.

The best alternative is the expectation maximization (EM) algorithm that provides a maximum likelihood estimate of gamete frequencies assuming random mating (Hill

1974). We previously used the EM approach in the case of a null allele where not all genotypes could be distinguished at a single locus (see Section 5.4.2). This approach uses an iteration procedure along with the maximum likelihood estimate of the gamete frequencies:

$$\hat{G}_1 = \left(\frac{1}{2N}\right)\left[2N_{11} + N_{12} + N_{21} + \frac{N_{22}\hat{G}_1(1 - \hat{p}_1 - \hat{p}_2 + \hat{G}_1)}{\hat{G}_1(1 - \hat{p}_1 - \hat{p}_2 - \hat{G}_1) + (\hat{p}_1 - \hat{G}_1)(\hat{p}_2 - \hat{G}_1)}\right] \tag{10.20}$$

where N is the sample size, N_{11} is the number of $AABB$ genotypes observed, N_{12} is the number of $AaBB$ genotypes observed, and N_{21} is the number of $AABb$ genotypes observed. This expression is not as opaque as it first appears. The first three sums in the right-hand parentheses are the observed numbers of the G_1 gametes in genotypes that are homozygous at least one locus. The fourth value is the expected number of copies of the G_1 gamete in the double heterozygotes.

We need to make an initial estimate of gamete frequencies and then iterate using this expression. Our initial estimate can either be the estimate of gamete frequencies with $D = 0$, or we can use the procedure described in the previous paragraph to initially estimate D from the remaining eight genotypic classes. The other three gamete frequencies can be solved directly once we estimate G_1 and the single locus allele frequencies. Iteration can sometimes converge on different gamete values depending upon the initial gamete frequencies (Excoffier and Slatkin 1995). Kalinowski and Hedrick (2001) present a detailed consideration of the implications of this problem when analyzing data sets with multiple loci.

It is crucial to remember that the EM algorithm assumes random mating and Hardy–Weinberg proportions. The greater the deviation from expected Hardy–Weinberg proportions, the greater the probability that this iteration will not converge on the maximum likelihood estimate. Stephens et al. (2001) have provided an algorithm to estimate gamete frequencies that assumes that the gametes in the double heterozygotes are likely to be similar to the other gametes in the samples. This method is likely to be less sensitive to nonrandom mating in the population being sampled.

10.6.2 More than two alleles per locus

The numbers of possible multilocus genotypes expands rapidly when we consider more than two alleles per locus. For example, there are six genotypes and nine gametes at a single locus with three alleles. Therefore, there are $6 \times 6 = 36$ diploid genotypes and $9 \times 9 = 81$ possible combinations of gametes at two loci each with three alleles. D values for each pair of alleles at two loci can be estimated and tested statistically (Kalinowksi and Hedrick 2001). The iteration procedure is more likely to converge to a value other than the maximum likelihood solution as the number of alleles per locus increases. Therefore, it is important to start the iteration from many different starting points with highly polymorphic samples.

So far we have considered genotypes at a pair of loci. There is much more information available if we consider the distribution of genotypes over many loci simultaneously. This is not a simple problem (Waits et al. 2001)! And, there are much larger data sets that are being used. For example, studies with humans sometimes use screens of over 1,000 microsatellite loci to detect regions of the genome that may contain genes associated with particular diseases.

Guest Box 10 Dating hybrid populations using gametic disequilibrium
Stephen H. Forbes

Newly formed hybrids have nonrandom association of alleles contributed by the two parental taxa, and this can be used to assess the age of a hybrid population (see Table 10.2). Persistence of these associations (gametic disequilibrium, or GD) between unlinked loci indicates recent admixture, while decay of GD between closely linked loci indicates a much earlier hybridization event. Given random mating, large N_e, large sample size, and known linkage relationships, the GD for a single pair of loci would reliably indicate the age of a hybrid population. In practice, however, hybridization may be geographically localized, with limited numbers of founders, limited N_e in following generations, and limited sample sizes. When a limited number of gametes contribute to the next generation, stochastic variation in the frequencies of two-locus genotypes contributes to variance in GD estimates. This variance cannot be reduced by sampling more individuals, but can only be reduced by using many pairs of loci.

We demonstrated these issues using GD analysis in hybrid populations of cutthroat trout (Forbes and Allendorf 1991). The westslope and Yellowstone cutthroat trout subspecies differ genetically as much as do most trout species. Diagnostic allozyme loci with fixed allelic differences between the subspecies were genotyped in hybrid populations formed by the introduction of Yellowstone cutthroat trout into native westslope populations occupying small lakes and creeks. We began with the assumption that most pairs of arbitrarily chosen loci are unlinked. As will be seen, if some are linked, this will be evident in the data.

All trout in both populations were hybrids beyond the F_1 generation. In Cataract Creek (Figure 10.7), the mean GD was near zero, and the distributions of all pairwise tests had equivalent numbers of positive and negative values, suggesting complete decay of GD among unlinked loci. The variance among locus pairs, however, was substantial; in fact, there were significant positive and negative individual tests (chi-square values > 3.84). A key point, however, is that negative GD values (associations between nonparental alleles) can be generated only by the genetic drift and sampling variances, and only after the initial GD due to admixture has largely decayed. In contrast to Cataract Creek, the GD distribution in Forest Lake (Figure 10.7) had a more positive skew. This indicates a population not yet at gametic equilibrium for unlinked markers. Note the effect of sample size, however, in both populations. The skew towards positive D values is much greater in sample sizes greater than 100 fish.

Hybrid populations at equilibrium for unlinked markers may still retain high GD for linked loci, and three locus pairs here showed evidence of linkage. Pair A showed significant disequilibrium in all Forest Lake samples, but none in Cataract Creek. This is consistent with the unlinked markers indicating that Cataract Creek

Cataract Creek

D	Q	1983 (N = 28)	1987 (N = 58)	1988 (N = 127)
	−6		♦	♦
	−5			
−D	−4	♦	♦	
	−3	♦♦	♦	
	−2	♦	♦♦	
	−1	♦♦♦	♦♦♦	♦♦♦♦♦♦A
	0	— ♦♦♦♦♦♦♦♦♦♦♦♦♦	— ♦♦♦♦♦♦♦♦♦♦♦♦♦A	— ♦♦♦♦♦♦♦♦♦♦♦♦♦
	1	♦♦♦♦♦A	♦♦♦	♦♦♦
	2	♦♦	♦♦	♦
	3		♦	♦♦♦
+D	4			♦
	5	♦♦	♦	♦
	6			
	23	C		
	54		C	
	113			C

Forest Lake

D	Q	1983 (N = 33)	1987 (N = 62)	1988 (N = 140)
	−6			
	−5			
−D	−4			
	−3		♦	
	−2	♦	♦	♦
	−1	♦♦♦♦♦	♦♦♦♦	♦♦
	0	— ♦♦♦♦♦♦♦♦♦♦♦♦♦♦♦♦♦♦♦	— ♦♦♦♦♦♦♦♦♦♦♦	— ♦♦♦♦♦♦♦♦♦♦♦♦♦♦
	1	♦♦♦♦♦♦	♦♦♦♦♦♦♦♦♦♦	♦♦♦♦
	2	♦	♦♦♦	♦♦♦♦♦♦♦♦
	3	♦		♦
+D	4	B	♦♦	♦♦♦
	5	♦	♦	♦♦A
	6		B	
	7			
	8	A		
	9		A	
	22			B

Figure 10.7 Gametic disequilibrium at nine diagnostic loci in two hybrid swarms of native westslope cutthroat trout and introduced Yellowstone cutthroat trout in two locations, Cataract Creek and Forest Lake. Each diamond indicates the amount of gametic disequilibrium for each pair of loci. Q is distributed as a chi-square with one degree of freedom and plotted with the sign of D. Positive associations (+D) are due to persistence in the hybrids of allele combinations from the parental types. Negative associations (−D) can only be generated be genetic drift or sampling variance where association due to admixture has largely decayed. The pairs of ioci labeled A, B, and C are linked. From Forbes and Allendorf (1991).

is the older hybrid population, and it suggests that pair A is not closely linked. Pairs B and C both showed high GD across temporally distinct samples, indicating closer linkage. Pair B was previously shown to be linked in experimental matings ($r = 0.32$ in females, $r = 0.05$ in males). Neither pair B nor pair C is informative in both populations, apparently because one locus in each pair is not strictly diagnostic in all parental stocks. Nonetheless, decay of GD for pair A contrasts the age of Cataract Creek from that of Forest Lake, and still older hybrid populations than Cataract Creek could be distinguished by GD decay for closely linked pairs such as B and C.

Detecting recent hybridization using unlinked loci does not call for detailed genetic maps, but dating older hybridization does call for more linkage data. Here, one linkage was known from prior mapping in experimental matings, and two more were inferred (but not accurately measured) from the population data. Increasingly, however, improving genetic maps for many taxa will enable the choice of linked loci with known map distances.

Problem 10.1

The effects of genetic hitchhiking are likely to be especially important in populations that have gone though a bottleneck. Why?

Problem 10.2

Can the amount of recombination between two loci be estimated by examining the genotypic proportions in a population?

Problem 10.3

If a population is completely A_1B_2/A_2B_1 what is the value of D?

Problem 10.4

The number of generations needed for D to go halfway to zero is found by noting when $D_t = 0.5\,D_0$. For a given amount of recombination, we need to solve the equation:

$$(1 - r)^t = 0.5$$
$$t\ln(1 - r) = \ln(0.5)$$
$$t = \ln(0.5)/\ln(1 - r)$$

With unlinked loci, $t = \ln(0.5)/\ln(1 - 0.5) = 1$ and D is therefore halved in a single generation. Vary the fraction of recombination and evaluate the half-life as a function of r and plot your results on the graph below:

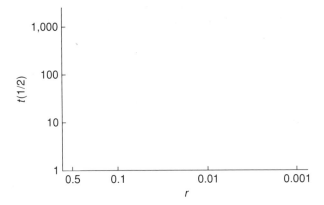

Problem 10.5

What would you consider a small fraction of recombination? (Hint: examine your answer to the previous problem and Figure 10.2.)

Problem 10.6

Given heterozygous advantage at one locus (*A*) and selective neutrality at the other (*B*) as shown below. What is the marginal fitness at locus *B* when $D = +0.25$ and the allele frequencies at locus *A* are at equilibrium? How about if $D = -0.25$?

	AA	Aa	aa
BB	$w_{11} = 0.8$	$w_{13} = 1.0$	$w_{33} = 0.8$
Bb	$w_{11} = 0.8$	$w_{13} = 1.0$	$w_{33} = 0.8$
Bb	$w_{11} = 0.8$	$w_{13} = 1.0$	$w_{33} = 0.8$

Problem 10.7

Given that strong gametic disequilibrium was observed between alleles at two loci in different populations, how would you design a research program to determine the cause of this association?

Problem 10.8

Several studies have found a positive correlation between multilocus heterozygosity and measures of fitness in a variety of plant and animal species (Britten 1996). How can such correlations be explained? How could you distinguish among alternative possible explanations for such correlations?

Problem 10.9

We saw in Example 10.3 that parental types persisted in a hybrid zone between two species of field crickets because of assortative mating between parental types. We should be able to see this by testing for Hardy–Weinberg proportions for the genotype frequencies presented at three loci in Figure 10.6. For example, at locus *pUC5*, the genotype numbers are 11 FF, 6 FP, and 7 PP. Test all three loci in this figure for Hardy–Weinberg proportions. Do your results support the authors' conclusion of assortative mating between parental types?

Problem 10.10

Example 10.4 and Guest Box 10 consider hybrid swarms of trout that appear to mating at random. Test the two loci shown in Example 10.4 for Hardy–Weinberg proportions. Does the population from which this sample was taken appear to be mating at random? Estimate *D* from this genotypic table using the eight genotypic classes for which gametes can be unambiguously defined. How does your estimate compare to the EM estimate provided in the example?

11

Quantitative Genetics

Bighorn sheep, Guest Box 11

Most of the major genetic concerns in conservation biology, including inbreeding depression, loss of evolutionary potential, genetic adaptation to captivity, and outbreeding depression, involve quantitative genetics.

Richard Frankham (1999)

An overview of theoretical and empirical results in quantitative genetics provides some insight into the critical population sizes below which species begin to experience genetic problems that exacerbate the risk of extinction.

Michel Lynch (1996)

Most phenotypic differences among individuals within natural populations are quantitative rather than qualitative. Some individuals are larger, stronger, or can run faster than others. Such phenotypic differences cannot be classified by the presence or absence of certain characteristics, such as plumage color or bands on a gel. The inheritance of quantitative traits is usually complex, and many genes are involved (i.e., polygenic). In addition to genetics, the environment to which individuals are exposed will also affect their phenotype (see Figure 2.2). The single locus genetic models that we have been using until this point are inadequate for understanding this variation. Instead of considering the effects of one gene at a time, we will examine multifactorial inheritance and partition the genetic basis of such phenotypic variation into various sources of variation by statistical procedures.

Quantitative genetics began shortly following the rediscovery of Mendel's principles to resolve the controversy of whether discrete Mendelian factors (genes) could explain the genetic basis of continuously varying characters (Lynch and Walsh 1998). The theoretical basis of quantitative genetics was developed primarily by R. A. Fisher (1918) and Sewall Wright (1921b). The empirical aspects of quantitative genetics were developed primarily from applications to improve domesticated animals and agricultural crops (Lush 1937; Falconer and Mackay 1996).

The models of quantitative genetics have been applied to understanding genetic variation in natural populations only in the last 20 years or so (Roff 1997). The abundance of molecular markers now available makes it possible to identify quantitative trait loci (QTLs) – the specific regions that are responsible for variation in continuous traits (Barton and Keightley 2002). Understanding the evolutionary effects of QTLs will allow us to improve our understanding of how genes influence phenotypic variation and improve our understanding of the evolutionary importance of genetic variation.

The principles of quantitative genetics also can be applied to a variety of problems in conservation (reviews by Barker 1994; Lande 1996; Lynch 1996; Storfer 1996; Frankham 1999). We saw in Chapter 2 that pink salmon on the west coast of North America have become smaller at sexual maturity over a period of 25 years (see Figure 2.3). This apparently resulted from the effects of a size-selective fishery in which larger individuals had a higher probability of being caught (Ricker 1981). In addition, understanding the quantitative genetic basis of traits is essential for predicting genetic changes that are likely to occur in captive propagation programs as populations become "adapted" to captivity (see Chapter 18).

This chapter provides a conceptual overview of the application of the approaches of quantitative genetics to problems in conservation. Our emphasis is on the interpretation of results of quantitative genetic experiments with model species in the laboratory and on more recent studies of quantitative genetic variation in natural populations. Detailed consideration of quantitative genetic principles can be found in Falconer and Mackay (1996) and Lynch and Walsh (1998).

11.1 Heritability

There are three major types of quantitative characters that are affected by a combination of polygenic inheritance and the environment:

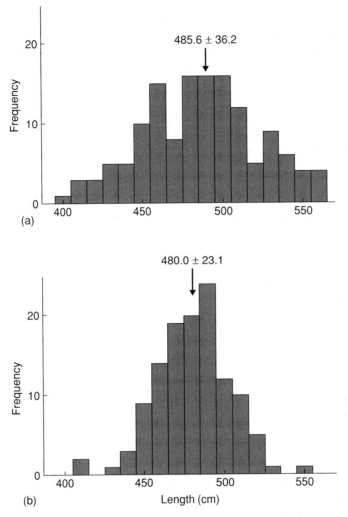

Figure 11.1 Body length of (a) male and (b) female pink salmon at sexual maturity in a population from Alaska. There is no difference in the mean length of males and females in this population. However, males have a significantly greater variance and standard deviation of body length. From Funk et al. (2005).

1 **Continuous characters**. These characters are continuously distributed in populations (e.g., weight, height, and body temperature). For example, Figure 11.1 shows the body length of pink salmon at sexual maturity in an experimental population from Alaska (Funk et al. 2005).

2 **Meristic characters**. The values of these characters are restricted to integers; that is they are countable (e.g., number of vertebrae, number of fingerprint ridges, and clutch size). For example, Figure 11.2 shows the distribution of the total number (left plus right) of gill rakers on the upper gill arch in the same population of pink salmon as Figure 11.1.

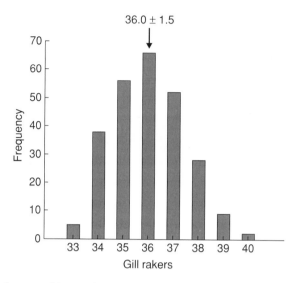

Figure 11.2 Distribution of the total number (left plus right) of gill rakers on the upper gill arch in the same population of pink salmon as Figure 11.1. There are no differences between males and females for this trait.

3 **Threshold characters**. These are characters in which individuals fall into a few discrete states, but there is an underlying continuously distributed genetic basis to the trait in question (e.g., alive or dead, sick or well).

The total amount of phenotypic variation for a quantitative trait within a population can be thought of as arising from two major sources: environmental differences between individuals and genetic differences between individuals. Writing this statement in the form of a mathematical model, we have:

$$V_P = V_E + V_G \tag{11.1}$$

The genetic differences between individuals can be attributed to three different sources:

V_A = additive effects (effects of gene substitution)

V_D = dominance effects (effects of interactions between alleles)

V_I = epistatic effects (effects of interactions between loci)

Therefore,

$$V_P = V_E + V_G$$
$$= V_E + V_A + V_D + V_I \tag{11.2}$$

11.1.1 Broad sense heritability

Heritability is a measure of the influence of genetics in determining phenotypic differences between individuals for a trait within a population. Heritability in the broad sense

(H_B) is the proportion of the phenotypic variability that results from genetic differences between individuals. That is,

$$H_B = \frac{V_G}{V_P} = \frac{V_A + V_D + V_I}{V_P} \tag{11.3}$$

For example, Sewall Wright removed virtually all of the genetic differences between guinea pigs within lines by continued sister-brother matings for many generations. The total variance (V_P) in the population as a whole (consisting of many separate inbred lines) for the amount of white spotting was 573. The average variance within the inbred lines was 340; this must be equal to V_E because genetic differences between individuals within the lines were removed from inbreeding. Thus,

$$V_P = V_E + V_G$$

$$573 = 340 + V_G$$

$$V_G = 573 - 340 = 233$$

and,

$$H_B = 233 / 573 = 0.409$$

11.1.2 Narrow sense heritability

Conservation biologists are often interested in predicting how a population will respond to selective differences among individuals in survival and reproductive success (Stockwell and Ashley 2004). Similarly, animal and plant breeders are often interested in improving the performance of agricultural species for specific traits of interest (e.g., growth rate or egg production). However, broad sense heritability may not provide a good prediction of the response. We will see shortly that a trait may not respond to selective differences even though variation for the trait is largely based upon genetic differences between individuals.

Another definition of heritability is commonly used because it provides a measure of the genetic resemblance between parents and offspring and, therefore, the predicted response of a trait to selection. Heritability in the narrow sense (H_N) is the proportion of the total phenotypic variation that is due to additive genetic differences between individuals:

$$H_N = \frac{V_A}{V_P} \tag{11.4}$$

If not specified, heritability generally (but not always) refers to narrow sense heritability, both in this chapter and in the published literature.

We can see the need for the distinction between broad and narrow sense heritability in the following hypothetical example. Assume that differences in length at sexual maturity are determined primarily by a single locus with two alleles in a fish species, and that heterozygotes at this locus are longer than either of the two homozygotes. In this example, the broad sense heritability for this trait is nearly 1.00 since almost all of the variation is due to genetic differences.

Let us do a thought selection experiment in a random mating population in which these two alleles are equally frequent:

Genotype	Length	Frequency
A_1A_1	10 cm	$p^2 = 0.25$
A_1A_2	12 cm	$2pq = 0.50$
A_2A_2	10 cm	$q^2 = 0.25$

Thus, one-half of the fish in this population are approximately 12 cm long and the other one-half are 10 cm long. What would happen if we selected only the longer 12 cm fish for breeding in an attempt to produce larger fish? All of the fish selected for breeding will be heterozygotes. The progeny from $A_1A_2 \times A_1A_2$ matings will segregate in Mendelian proportions of 25% A_1A_1 : 50% A_1A_2 : 25% A_2A_2. Therefore, the progeny generation is expected to have the same genotype and phenotype frequencies as the parental generation. That is, there will be no response to selection even though all of the phenotypic differences have a genetic basis ($H_B = 1.00$).

In this example, all of the phenotypic differences are due to dominance effects (V_D) resulting from to the interaction between the A_1 and A_2 alleles in the heterozygotes. Thus, there is no response to selection, and the narrow sense heritability is zero ($H_N = 0$). However, this population would have responded to this selection if the two alleles were not equally frequent. We will see in Section 11.2.1 that heritability will be different at different allele frequencies.

Heritability is another area where the nomenclature and the symbols used in publications can cause confusion to the reader. Narrow sense heritability is often represented by h^2 and broad sense heritability by H^2. The square in these symbols is in recognition of Wright's (1921b) original description of the resemblance between parents and offspring using his method of path analysis in which under the additive model of gene action, an individual's phenotypes is determined by $h^2 + e^2$, where e^2 represents environmental effects and h^2 is the proportion of the phenotypic variance due to the genotypic value (see Lynch and Walsh 1998, appendix 2). We have chosen not to use these symbols in hopes of reducing possible confusion.

11.1.3 Estimation of heritability

Heritability can be estimated by several different methods that depend upon comparing the relative phenotypic similarity of individuals with different degrees of genetic relationship. One of the most direct ways to estimate heritability is by regressing the progeny phenotypic values on the parental phenotypic values for a trait. The narrow sense heritability can be estimated by the slope of the regression of the offspring phenotypic value on the mean of the two parental values (called the mid-parent value).

Alatalo and Lundberg (1986) estimated the heritability of tarsus length in a natural population of the pied flycatcher using nest boxes in Sweden. The narrow sense heritability was estimated by twice the slope of the regression line of the progeny value on the maternal value. The slope is doubled in this case because only the influence of the maternal parent was considered. The male parents were not included in this analysis because previous

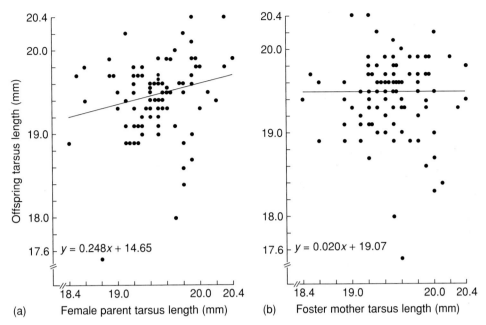

Figure 11.3 Mother–offspring regression estimation of heritability of tarsus length in the pied flycatcher ($H_N = 0.53$): (a) regression with mother and (b) regression with foster mother. Each point represents the mean tarsus length of progeny from one nest. From Alatalo and Lundberg (1986).

results had shown that nearly 25% of progeny were the result of extra-copulations rather than fathered by the social father. Heritability was estimated to be 0.53 in an examination of 331 nests (Figure 11.3a). In addition, 54 clutches were exchanged as eggs between parents to separate genetic and post-hatching environmental effects. There was no detectable resemblance at all between foster mothers and their progeny (Figure 11.3b).

The relatively high heritability in this example is somewhat typical of estimates for morphological traits in natural populations (Roff 1997). Our reanalysis in Chapter 2 of Punnett's (1904) data on vertebrae number in velvet belly sharks results in a heritability estimate of 0.63 ($P < 0.01$; see Figure 2.4). Similar high heritabilities for a variety of meristic traits (vertebrae number, fin rays, etc.) in many fish species have been reported (reviewed in Kirpichnikov 1981). Many morphological characters in bird species also have high narrow sense heritabilities (e.g., Table 11.1).

Quantitative genetic studies in natural populations have been rare because they require large breeding programs or known pedigrees. It is now possible to estimate relatedness in natural populations using a large number of molecular markers (Queller and Goodnight 1989; Wang 2002). These estimates of relatedness can now be used to provide pedigrees for species in the wild and to estimate heritabilities and genetic correlations (Ritland 2000; Moore and Kukuk 2002). Caution is necessary, however; even a small amount of pedigree uncertainty may make it difficult to obtain reliable heritability estimates from wild populations in this way (Coltman 2005).

Table 11.1 Heritability estimates from parent–offspring regression for morphological traits for three species of Darwin's finches in the wild (Grant 1986). Heritability ranges between zero and one. However, **estimates** of heritability can be greater than 1.0 (as here); for example, the slope of the regression of progeny on mid-parent values for weight and bill length was greater than 1.0 in *Geospiza conirostris*. *, $P < 0.05$; **, $P < 0.01$; ***, $P < 0.001$.

Character	G. fortis	G. scandens	G. conirostris
Weight	0.91***	0.58	1.09***
Wing cord length	0.84***	0.26	0.69*
Tarsus length	0.71***	0.92***	0.78**
Bill length	0.65***	0.58*	1.08***
Bill depth	0.79***	0.80*	0.69***
Bill width	0.90**	0.56*	0.77**

11.1.4 Genotype–environment interactions

We saw in Figure 2.9 that the environment can have a profound effect on the phenotypes of yarrow plants resulting from a particular genotype. We also saw evidence for local adaptation with reciprocal transplants in *Gilia capitata* (Figure 8.1) in which plants had greater fitness in their native habitat. In a statistical sense, these are examples of **interactions** between genotypes and environments. We can expand our basic model to include these important interactions as follows:

$$V_P = V_E + V_G + V_{G \times E} \tag{11.5}$$

Genotype–environment interactions are of major concern in conservation biology when translocating individuals to alleviate inbreeding depression (genetic rescue, Chapter 15) and when reintroducing captive populations into the wild (Chapter 18). We will see in Chapter 18 that many traits that are advantageous in captivity may greatly reduce the fitness of individuals in the wild.

11.2 Selection on quantitative traits

Evolutionary change by natural selection can be thought of as a two-step process. First, there must be phenotypic variation for the trait that results in differential survival or reproductive success (i.e., fitness). Second, there must be additive genetic variation for the trait ($H_N > 0$; fisher 1930). Heritability in the narrow sense can also be estimated by the response to selection. This is usually called the "realized" heritability (Figure 11.4). If a trait does not respond at all to selection then there is no additive genetic variation and $H_N = 0$. If the mean of the selected progeny is equal to the mean of the selected parents, then $H_N = 1$. Generally, the mean of the selected progeny will be somewhere in between these two extremes ($0 < H_N < 1.0$). Francis Galton coined the expression "regression" to describe the general tendency for progeny of selected parents to "regress" towards the mean of the

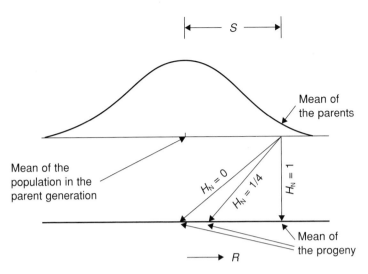

Figure 11.4 Illustration of the meaning of narrow sense heritability based upon a selection experiment and expression 11.5. If there is no response to selection then $H_N = 0$; if the mean of the progeny from selected parents is equal to the mean of the selected parents, then $H_N = 1$. Modified from Crow (1986).

unselected population. Galton lived in the 19th century and was a cousin of Charles Darwin (Provine 2001).

In this case, heritability is the response to selection divided by the total selection differential. Thus,

$$H_N = \frac{R}{S} \tag{11.6}$$

where S is the difference in the means between the selected parents and the whole population, and R is the difference in the means between the progeny generation and the whole population in the previous generation. Expression 11.6 is sometimes written in the form of the breeders' equation to allow prediction of the expected gain from artificial selection:

$$R = H_N S \tag{11.7}$$

Figure 11.5 illustrates two generations of artificial selection for a trait with a heritability of 0.33. The breeder selects by truncating the population and uses only individuals above a certain threshold as breeders. The mean of the progeny from these selected parents will regress two-thirds $(1 - 0.33)$ of the way toward the original population mean.

11.2.1 Heritabilities and allele frequencies

Heritability for a particular trait is not constant. It will generally vary in different populations that are genetically divergent. However, it may also vary within a single population under different environmental conditions. In addition, heritability within a population

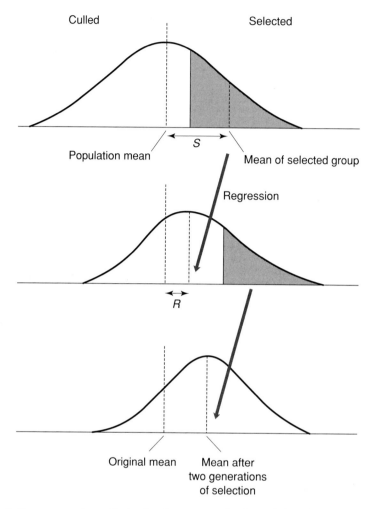

Figure 11.5 Two generations of selection for a trait with a heritability of 0.33. The progeny mean moves one-third (0.33) of the distance from the population mean toward the mean of the selected parents. From Crow (1986).

will not be constant even within the same environment because it is influenced by allele frequencies.

Let us examine this effect with a simple single locus model in which all phenotypic variation has a genetic basis. In a single locus case such as this, heritability can be calculated directly by calculating the appropriate variances:

$$V_G = p^2(w_{11} - \bar{w})^2 + 2pq(w_{12} - \bar{w})^2 + q^2(w_{22} - \bar{w})^2 \tag{11.8}$$

$$V_A = 2[p(w_{A1} - \bar{w})^2 + q(w_{A2} - \bar{w})^2] \tag{11.9}$$

where

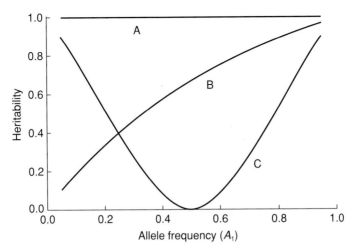

Figure 11.6 Hypothetical example of heritability (V_A/V_G) for plant height determined by a single locus with two alleles: (A) the additive case where heterozygotes are intermediate to both homozygotes, (B) complete dominance so that the heterozygotes have the same phenotype as the taller homozygotes (A_2A_2), and (C) the case of overdominance where the heterozygotes are taller than either of the homozygotes. It is assumed that all of the variability in height is genetically determined (i.e., $V_E = 0$).

$$w_{A1} = \frac{p^2(w_{11}) + pq(w_{12})}{p} \tag{11.10}$$

and

$$w_{A2} = \frac{pq(w_{12}) + q^2(w_{22})}{q} \tag{11.11}$$

V_A is very closely related to average heterozygosity, and will also be maximum at $p = q = 0.5$. Since we have assumed that V_E is zero, the narrow sense heritability is the additive genetic variance divided by the total genetic variance.

$$H_N = \frac{V_A}{V_G} \tag{11.12}$$

We can use this approach to estimate heritability in the example of height in a plant determined by a single locus where homozygotes are 10 cm and heterozygotes are 12 cm tall. We saw before that H_N was zero when the two alleles are equally frequent. We can use expressions 11.8–11.12 to estimate heritability over all possible allele frequencies (Figure 11.6). In addition, we can consider the case where the heterozygotes are intermediate to the homozygotes (case A) and the case of complete dominance when the heterozygotes have the same phenotype as the taller homozygotes (case B), as well as the overdominant case where the heterozygotes are taller than either homozygotes (case C):

Genotype	Additive (A)	Dominant (B)	Overdominant (C)
A_1A_1	10 cm	10 cm	10 cm
A_1A_2	11 cm	12 cm	12 cm
A_2A_2	12 cm	12 cm	10 cm

There are several important features of the relationship between allele frequency and heritability in this model (Figure 11.6). In case A, all of the genetic variance is additive so heritability is always 1.0. In the case of dominance (case B), heritability is high when the dominant allele (A_2) is rare but low when this allele is common. As we saw in Chapter 8, selection for a high frequency dominant allele is not effective so heritability will be low. Finally in the case of overdominance (case C), heritability is high when either of the alleles is rare because increasing the frequency of a rare allele will increase the frequency of het-erozygotes and thus the population will be taller the next generation. That is, the popula-tion will respond to selection when either allele is rare, and therefore heritability will be high. However, the frequency of heterozygotes is greatest when the two alleles are equally frequent; therefore, heritability will be zero at $p = q = 0.5$.

11.2.2 Genetic correlations

Genetic correlations are a measure of the genetic associations of different traits. Such cor-relations may arise from two primary mechanisms. First, many genes affect more than a single character so that they will cause simultaneous effects on different aspects of the phenotype; this is known as **pleiotropy**. For example, a gene that increases growth rate is likely to affect stature and weight. Pleiotropy will cause genetic correlations between different characters. Gametic disequilibrium between loci affecting different traits is another possible cause of genetic correlations between characters.

Selection (either natural of artificial) for a particular trait will often result in a secondary response in the value of another trait because of genetic correlations. Distinguishing between pleiotropy and gametic disequilibrium as causes of such secondary responses is difficult because many loci may affect the trait under selection. Therefore, the effects of associated alleles from other loci and pleiotropy of some of the selected alleles are both quite probable.

Figure 11.7 demonstrates genetic correlation between two meristic traits in the experi-mental population of pink salmon from Figure 11.1. The parental values for one trait, pelvic rays, were a good predictor of the progeny phenotypes for another trait, pectoral rays ($P < 0.05$). The reciprocal regression (progeny pelvic rays on mid-parent pectoral rays) is also significant ($P < 0.01$). The actual point estimate of the genetic correlation between these traits (0.64) takes into account both of these parental–progeny relationships (Funk et al. 2005):

$$r_A = \frac{cov_{XY}}{\sqrt{(cov_{XX} cov_{YY})}} \tag{11.13}$$

where cov_{XY} is the "cross-variance" that is obtained from the product of the value of trait X in parents and the value of trait Y in progeny, and cov_{XX} and cov_{YY} are the progeny–parent covariance for traits X and Y separately.

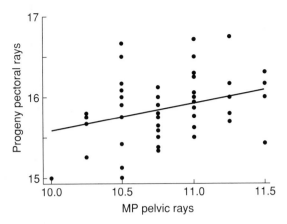

Figure 11.7 Genetic correlation between meristic traits in pink salmon. Regression of number of rays in the pectoral fins of progeny on the mid-parent (MP) values of number of rays in the pelvic fins (slope = 0.342; $P < 0.023$).

We can see with Figure 11.7 that if we selected by using parents with many pelvic rays the number of pectoral rays in the progeny would increase. In this case, the genetic correlation is almost certainly caused by pleiotropy. Genes that decrease developmental rate tend to increase counts for a suite of meristic traits in the closely related rainbow trout (Leary et al. 1984b).

11.3 Quantitative trait loci (QTLs)

The field of quantitative genetics developed to allow the genetic analysis of traits affected by multiple loci in which it was impossible to identify individual genes having major phenotypic effects. Formal genetic analysis could only be performed for traits in which discrete (qualitative) phenotypes (round or wrinkled, brown or albino, etc.) could be identified to test their mode of inheritance. Classic quantitative genetics treated the genome as a black box and employed estimates of a variety of statistical parameters (e.g., heritabilities, genetic correlations, and the response to selection) to describe the genetic basis of continuous (quantitative) traits. However, the actual genes affecting these traits could not be identified with this approach. Alan Robertson (1967) described this phenomenon as the "fog of quantitative variation".

The fog is now lifting. The current abundance of molecular markers now available makes it possible to identify those specific regions of the genome that are responsible for variation in continuous traits. Understanding the evolutionary effects of these so-called QTLs allows us to determine how specific genes influence phenotypic variation and understand the evolutionary importance of genetic variation. Perhaps even more important from a conservation perspective, this approach also allows us to determine how many genes affect phenotypic variation of particular interest.

Knowing whether a particular trait is affected primarily by a single gene or by many genes with small effects, may influence recommendations based upon genetic considerations. For example, captive golden lion tamarins suffer from a diaphragmatic defect that

Example 11.1 Should golden lion tamarins with diaphragmatic defects be released into the wild?

Golden lion tamarins held in captivity have a relatively high frequency of diaphragmatic defects detectable by radiography (Bush et al. 1996). No true hernias were observed, but 35% of captive animals had marked defects in the muscular diaphragm that provided a potential site for liver or gastrointestinal herniation. Only 2% of wild living animals had this defect. This difference between captive and wild animals could be the result of relaxation of selection against the presence of such defects in captivity or genetic drift in captivity (see Chapter 18).

All captive-born animals that are candidates for release in the wild are now screened for this defect before release. Individuals with a relatively severe defect are disqualified for reintroduction. Approximately 10% of all captive animals are expected to be disqualified using this criterion. This procedure is designed to protect the wild population against an increase in a potentially harmful defect that may have a genetic basis.

This screening procedure seems appropriate. However, its effectiveness will depend upon the genetic basis of this defect. If it is caused by a single gene with a major effect, then this selection is expected to be effective in limiting the increase of the genetic basis for this defect in the wild. If, however, this defect is caused by many genes with small effect, this selective removal of a few animals will have little effect.

In addition, there is a potential concern with such selection of animals for reintroduction. As we have seen, selection can reduce the amount of genetic variation. Disqualifying say 25 or 50% of candidates for reintroduction could potentially reduce genetic variation in the reintroduced animals. For example, Ralls et al. (2000) concluded that the selective removal of an allele responsible for chondrodystrophy in California condors would not be advisable because of the potential reduction of genetic variation in the captive population. Thus, there is no simple answer to the question of whether or not animals with defects should be released into the wild (see Example 18.4).

seems to be hereditary (Bush et al. 1996). The presence or absence of this condition is one criterion used in the selection of individuals for release into the wild (Example 11.1).

11.3.1 Genomic distribution of loci affecting quantitative traits

The basic problem with genetic analysis of a quantitative trait is that many different genotypes may result in the same phenotype. Therefore, genetic dissection of QTLs requires an experimental system in which the effects of individual genomic regions or marker genes can be identified.

DDT resistance in *Drosophila* is an example of this (Crow 1957). A DDT-resistant strain was produced by raising flies in a large experimental cage with the inner walls painted with DDT. The concentration of DDT was increased over successive generations as the flies became more and more resistant until over 60% of the flies survived doses of DDT that

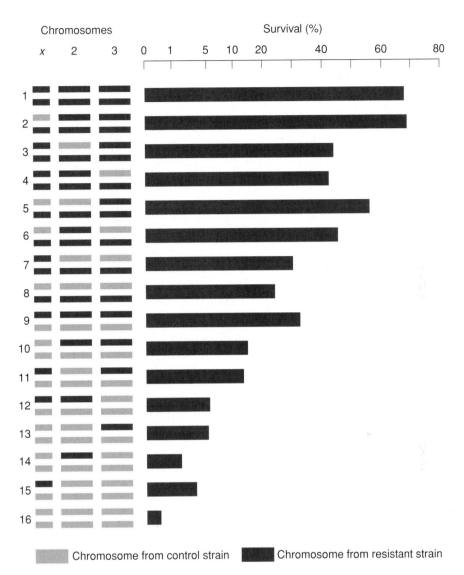

Chromosome from control strain   Chromosome from resistant strain

Figure 11.8 Results of an experiment demonstrating that genes affect DDT resistance in *Drosophila* (Crow 1957). The addition of any one of the chromosomes from the strain selected for DDT resistance increases DDT resistance. Redrawn from Crow (1986).

initially killed over 99% of all flies. flies from the resistant strain then were mated to flies from laboratory strains that had not been selected for DDT resistance. The F_1 flies were mated to produce an F_2 generation that had all possible combinations of the three major chromosomal pairs as identified by marker loci on each chromosome (Figure 11.8).

The results of this elegant experiment demonstrate that genes affecting DDT resistance occur on all three chromosomes. The addition of any one of the three chromosomes (Figure 11.8) increases DDT resistance. Furthermore, the effects of different chromosomes are cumulative. That is, the more copies of chromosomes from the resistant strain, the more resistant the F_2 flies were to DDT.

Table 11.2 Relationship between marker genotypes at a linked marker locus (*B*) and an unlinked locus (*A*) with a QTL (modified from Kearsey 1998). The Q^+ allele at the QTL causes an increase in the phenotype. The strong association between the phenotype of interest and genotypes at the *B* locus suggests that a QTL affecting the trait is linked to the *B* locus.

F_1

A_1	B_1	Q^+
A_2	B_2	Q^-

	Frequencies of F_2 genotypes			
Marker genotype	Q^+Q^+	Q^+Q^-	Q^-Q^-	**Mean phenotype**
A_1A_1	0.25	0.50	0.25	Intermediate
A_1A_2	0.25	0.50	0.25	Intermediate
A_2A_2	0.25	0.50	0.25	Intermediate
B_1B_1	Most	Few	Rare	High
B_1B_2	Few	Most	Few	Intermediate
B_2B_2	Rare	Few	Most	Low

11.3.2 Mapping QTLs

The availability of many genetic markers now makes it possible to determine more precisely the number and position of QTLs affecting phenotypes of interest by a variety of approaches. The most straightforward approach is similar to the DDT example: two parental lines that differ widely for a phenotype of interest are crossed (Table 11.2). If these lines are homozygous for alleles at many loci affecting the trait, the F_1 hybrids will be heterozygous at these loci. The average trait score or phenotype for any marker locus (*A*, *B*, *C*, etc.) can be calculated in the F_2 generation. If the marker is on a chromosome that does not contain a QTL, then the marker locus will segregate independently from the trait and all three genotypes at the marker will have the same phenotypic mean.

Table 11.2 shows the result of this approach on a chromosome that contains a QTL with two alleles, where the Q^+ allele causes a greater phenotypic value than the Q^- allele. If the marker locus is close to the QTL, then genotypes at the marker locus (*B* in this case) will have different mean phenotypes because of linkage. Some loci on the same chromosome will show independent segregation because of recombination. Such loci (*A* in this case) will segregate independently from the trait; therefore, all three genotypes at the marker will have the same phenotypic mean.

The size of the difference in mean phenotype effect will depend upon the magnitude of effect of the QTL and the amount of recombination between the marker and the QTL. The tighter the linkage, the greater will be the difference in mean phenotypes between

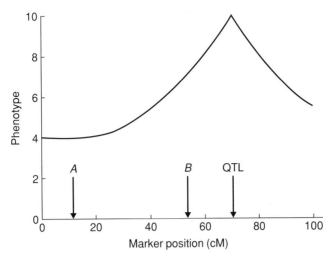

Figure 11.9 Hypothetical relationship between the mean phenotype of marker loci in the F_2 generation on a chromosome containing a QTL at map position 70 (see Table 11.2).

genotypes at the marker locus. Figure 11.9 shows the expected relationship between the mean phenotype of the marker locus and the map distance between the marker locus and the QTL. The genetic map distance is measured in centiMorgans (cM); 1 cM equals 1% recombination. Therefore, loci that are at least 50 cM apart will segregate independently and are said to be unlinked. Loci on the same chromosome are syntenic. In this example, the A locus and the QTL are syntenic but unlinked.

This approach has been used to determine the genetic basis of the derived life cycle mode of paedomorphosis in the Mexican axolotl (Voss 1995; Voss and Shaffer 1997). Mexican axolotls do not undergo metamorphosis and have a completely aquatic life cycle. In contrast, closely related tiger salamanders undergo metamorphosis in which their external gills are absorbed and other changes occur before they become terrestrial. Voss (1995) crossed these two species and found that the F_1 hybrids underwent metamorphosis. He then backcrossed the F_1 hybrids to a tiger salamander and found evidence for a major single gene controlling most of the variation in this life history difference, with additional loci having smaller effects.

Voss and Shaffer (1997) used a similar backcross approach in two crosses to search for QTLs affecting metamorphosis. They scored 262 AFLP marker loci in the backcross progeny. Only one region of the genome, which contained three AFLP markers, was associated with this life cycle difference. They hypothesized that a dominant allele (*MET*) at a QTL in this region was associated with the metamorphic phenotype in the tiger salamander, and the recessive *met* allele from the Mexican axolotl was associated with paedomorphosis. This result has been confirmed by additional analysis of nearly 1,000 segregating marker loci (Smith et al. 2005b). No other regions were found to contribute to this life history difference. It is interesting to note that this same QTL also contributes to continuous variation for the timing of metamorphosis in the tiger salamander (Voss and Smith 2005).

11.3.3 Candidate loci

Another way to identify genes associated with phenotypes of interest is the candidate locus approach in which knowledge of gene function is used to select a subset of genes that are likely to affect the trait of interest (Phillips 2005). The QTL approach is a top-down approach that begins with the phenotype in order to search for responsible genes. In contrast, the candidate gene approach begins with the genes and searches for phenotypic effects of interest (see Figure 2.2).

Voss et al. (2003) used the candidate locus approach to understand the genetic basis of paedomorphosis with the same species as described above. They tested if two thyroid hormone receptor loci affected the timing of metamorphosis. Thyroid hormone receptors are strong candidate genes for metamorphic timing because they mediate gene expression pathways that regulate developmental programs underlying the transformation of an aquatic larva into a terrestrial adult. They found a significant affect on metamorphic timing for one of these two loci. However, the effect was not straightforward as both the magnitude and direction of the phenotypic effect depended upon the genetic background.

11.4 Genetic drift and bottlenecks

The loss of genetic variation (heterozygosity and allelic diversity) via genetic drift will also affect quantitative variation. However, some surprises have been found in studies of the effects of population bottlenecks on quantitative genetic variation. In some cases, even extreme population bottlenecks have resulted in the **increase** of genetic variation as measured by quantitative genetics.

We saw in Chapter 6 that heterozygosity at neutral loci will be lost at a rate of $1/(2N_e)$ per generation. We can relate the effects of genetic drift on allele frequencies at a locus with two alleles to additive quantitative variation as follows (Falconer and Mackay 1996):

$$V_A = \sum 2pq[a + d(q - p)]^2 \tag{11.14}$$

where p and q are allele frequencies so that $2pq$ is the expected frequency of heterozygotes, a is half the phenotypic difference between the two homozygotes, and d is the dominance deviation.

This model results in the simple prediction that V_A, and therefore H_N, will be lost at the same rate as neutral heterozygosity in the case where all of the variation is additive ($d = 0$). Empirical results of the relationship between H_N, molecular genetic variation, and the effects of bottlenecks have been somewhat contradictory (see review in Gilligan et al. 2005). For example, Gilligan et al. (2005) found that quantitative genetic variation for two meristic characters in *Drosophila* (abdominal and sternopleural bristle numbers) declined at the same rate as heterozygosity at seven allozyme loci in laboratory populations maintained at a variety of effective population sizes. In contrast, van Oosterhout and Brakefield (1999) found no significant reduction in the heritability of wing pattern characters in the butterfly *Bicyclus anynana* reared at small population sizes in the laboratory.

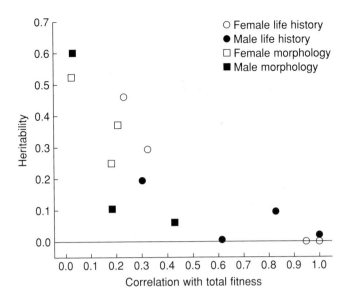

Figure 11.10 Relationship between the narrow sense heritability of traits and their correlation with total fitness in a population of red deer. These results demonstrate that morphometric traits tend to have greater heritabilities than life history traits, and traits with a greater effect on fitness tend to have lower heritabilities. From Kruuk et al. (2000).

Other studies have reported increases in quantitative genetic variation in populations following bottlenecks. In one of the earliest studies, Bryant et al. (1986) found that heritability of several morphological traits in the house fly increased after an experimental bottleneck.

The most important explanation for this apparent discrepancy is the assumption that all of the quantitative genetic variation is additive. We expect traits under strong natural selection to behave considerably differently to selectively neutral traits.

Consider the effect of natural selection on a phenotypic trait using the single locus selection models we examined in Chapter 8. We saw that all of the genetic variation for a trait under strong directional natural selection will eventually be depleted (see Figure 8.2). In general, natural selection will use up or remove additive genetic variation for a trait so alleles that increase fitness should increase in frequency until they reach fixation, while genes associated with reduced fitness should be reduced in frequency and eventually be lost from the population. Therefore, we expect that traits that are strongly associated with fitness should have lower heritabilites than traits that are under weak or no natural selection. A vast body of empirical information in many species supports this expectation (Mousseau and Roff 1987; Roff and Mousseau 1987). For example, Kruuk et al. (2000) found a strong negative association between the heritability of a trait and its association with fitness in a wild population of red deer (Figure 11.10).

We do not expect strong selection for a trait to remove all genetic variation. Consider our example in Section 11.1.2 of a single locus affecting length in a fish population in which the heterozygotes are longer than either of the homozygotes. This is a case of heterozygous advantage or overdominance as we saw in Section 8.2. Directional selection for length will eventually lead to an equilibrium allele frequency of 0.5 in which the population will no longer respond to selection for longer length ($H_N = 0$). However, the broad sense heritability may still be quite high if most of the phenotypic variation in this trait is caused by this locus. In this case, V_A will be zero but there will be substantial genetic variation affecting this trait due to dominance effects resulting from interactions between the two alleles (V_D, see expression 11.1).

What will happen in this case if the equilibrium population goes through a bottleneck which causes a change in allele frequency at this locus because of genetic drift? This bottleneck will change allele frequencies away from 0.5, and, therefore, narrow sense heritability (H_N) for this trait will increase!

The increase in heritability after a bottleneck can also result from simple dominance (Willis and Orr 1993). Selection against a rare deleterious recessive allele is ineffective when the allele is rare. Thus, these deleterious recessive alleles will remain in the population at low frequency. Most of these rare alleles will be lost during a bottleneck, but some will increase in frequency as we saw in Chapter 6. The increase in frequency of these alleles will decrease the average fitness of the population and will also increase the heritability of the traits affected by these alleles. Robertson (1952) found that the additive genetic variation for initially rare recessive alleles will increase in small populations until the expected reduction in heterozygosity is approximately 0.50, but declines to zero after that.

More complex models of quantitative genetic variation that he observed increase in additive genetic variation; this is primarily caused by the conversion of dominance to additive genetic variation (Wang et al. 1998). The conversion of epistatic effects (V_I) to additive genetic variation will also contribute to the phenomenon.

11.5 Divergence among populations (Q_{ST})

Quantitative genetics can also be used to understand genetic differentiation among local populations within species (Merilä and Crnokrak 2001). This approach has been especially interesting when used to understand the patterns of local adaptation caused by differential natural selection acting on heritable traits. We also expect local populations of species to differ for quantitative traits because of the effects of genetic drift. Understanding the relative importance of genetic drift and natural selection as determinants of population differentiation is an important goal when studying quantitative traits.

We saw in Chapter 9 that the amount of genetic differentiation among populations is often estimated by the fixation index F_{ST}, which is the proportion of the total genetic variation that is due to genetic differentiation among local populations. An analogous measure of population differentiation for quantitative traits has been termed Q_{ST} (Spitze 1993):

$$Q_{ST} = \frac{V_{GB}}{2V_{GW} + V_{GB}} \tag{11.15}$$

where V_{GB} is the additive genetic variation due to differences among populations and V_{GW} is the mean additive genetic variation within populations. Q_{ST} is expected to have the same value as F_{ST} if it is estimated from the allele frequencies at the loci affecting the quantitative

Table 11.3 Possible relationships in natural populations of divergence at neutral molecular markers (F_{ST}) and a quantitative trait (Q_{ST}).

Result	Interpretation
$Q_{ST} > F_{ST}$	Differential directional selection on the quantitative trait
$Q_{ST} = F_{ST}$	The amount of differentiation at the quantitative trait is similar to that expected by genetic drift alone
$Q_{ST} < F_{ST}$	Natural selection favoring the same phenotype in different populations

trait under investigation. This assumes that the local populations are in Hardy–Weinberg proportions ($F_{IS} = 0$) and the loci are in gametic equilibrium.

Q_{ST} is much more difficult to estimate than F_{ST} because it requires common garden experiments to distinguish genetic differences among populations from environmental influences on the trait. It also requires partitioning the within-population variation into genetic and environmental components, usually by measuring the variation among families. These requirements make it extremely difficult to estimate Q_{ST} except in plant species and animal species that can be readily raised in experimental conditions (e.g., *Drosphila* and *Daphnia*). Caution is needed in reading the literature because a number of papers that compare F_{ST} and Q_{ST} have not appropriately estimated Q_{ST} (e.g., Storz 2002). Simply partitioning the proportion of total phenotypic variation that is attributable to population differences is not comparable to F_{ST} because it includes nonadditive genetic effects, maternal effects, and nongenetic (i.e. environmental) effects.

The comparison of F_{ST} and Q_{ST} can provide valuable insight into the effects of natural selection on quantitative traits. F_{ST} is relatively easy to measure at a wide variety of molecular markers, which are generally assumed to not be strongly affected by natural selection. Thus, the value of F_{ST} depends only on local effective population sizes (genetic drift) and dispersal (gene flow) among local populations. Consequently, differences between F_{ST} and Q_{ST} can be attributed to the effects of natural selection on the quantitative trait.

There are three possible relationships between F_{ST} and Q_{ST} (Table 11.3). First, if $Q_{ST} > F_{ST}$, then the degree of differentiation in the quantitative trait exceeds that expected by genetic drift alone, and, consequently, directional natural selection favoring different phenotypes in different populations must have been involved to achieve this much differentiation. Second, if Q_{ST} and F_{ST} estimates are roughly equal, the observed degree of differentiation at the quantitative trait is the same as expected with genetic drift alone. Finally, if $Q_{ST} < F_{ST}$, the observed differentiation is less than that to be expected on the basis of genetic drift alone. This means that natural selection must be favoring the same mean phenotype in different populations.

Laboratory experiments with house mice have supported the validity of this approach to understand the effects of natural selection on quantitative traits (Morgan et al. 2005). Comparison of Q_{ST} and F_{ST} in laboratory lines with known evolutionary history generally produced the correct evolutionary inference in the interpretation of comparisons within and between lines. In addition, Q_{ST} was relatively greater than F_{ST} for those traits for which strong directional selection was applied between lines.

11.6 Quantitative genetics and conservation

How should we incorporate quantitative genetic information into conservation and management programs? Quantitative genetics has not played a major role in conservation genetics, which has been dominated by studies of molecular genetic variation at individual loci. Some have argued that quantitative genetic approaches may be more valuable than molecular genetics in conservation since quantitative genetics allows us to study traits associated with fitness rather than just markers that are neutral or nearly neutral with respect to natural selection (Storfer 1996). This is a compelling but somewhat misleading argument.

Quantitative genetics provides an invaluable conceptual basis for understanding genetic variation in populations. This approach is important for many crucial issues in conservation. For example, quantitative genetics is essential for understanding inbreeding depression (Chapter 13). If inbreeding depression is caused by a few loci with major effects, then the alleles responsible for inbreeding depression may be removed or "purged" from a population. However, inbreeding depression caused by many loci with small effects will be extremely difficult to purge (see discussion in Section 13.6). Similar considerations come into play with such important issues as the loss of evolutionary potential in small populations, and the effective population size required to maintain adequate genetic variation to increase the probability of long-term survival of populations and species.

Nevertheless, the empirical estimation of the primary parameters of quantitative genetics will generally not provide useful information for the management of particular species. For example, comparative estimates of quantitative genetic variation (i.e., heritability) in different populations are much less useful than molecular estimates of genetic variation. Part of the problem is statistical. The large standard errors generally associated with heritability estimates make it very difficult to detect significant differences (Storfer 1996). This problem is more limiting when working with rare species because hundreds to thousands of progeny from 25 or more families are needed for precise estimates of heritability (Falconer and Mackay 1996). We once heard a well-known evolutionary geneticist say that almost all heritability estimates are in the interval 0.50 ± 0.25.

The problem is also biological. As we saw in Section 11.4, small bottlenecks that cause a substantial loss of genetic variation and a reduction in fitness due to inbreeding depression may actually cause an increase in H_N because of the conversion of dominance and epistatic genetic variation into additive genetic variation. In addition, it is the estimates of additive genetic variation for those traits most closely associated with fitness that we expect to be most misleading. Finally, heritability estimates are strongly influenced by the environment and could change substantially between years within the same population (see Section 11.1.4). For these and other reasons, estimates of heritabilities are unlikely to be useful for making conservation decisions.

11.6.1 Response to selection in the wild

Quantitative genetics provides a framework to understand and predict the possible effects of selection acting on wild populations (see Guest Box 8). For example, many commercial fisheries target particular age or size classes within a population (Conover and Munch 2000). In general, larger individuals are more likely to be caught than smaller individuals. The expected genetic effect (i.e., response, R) of such selectivity on a particular trait depends upon the selection differential (S), and the narrow sense heritability (H_N), as we saw in expression 11.6. We saw in Figure 2.3 that the mean size of pink salmon caught off the coast of North America declined between 1950 and 1974 (12 generations) apparently because of size-selective harvest; approximately 80% of the returning adult pink salmon were harvested in this period (Ricker 1981). The mean reduction in body weight in 97 populations over these years was approximately was approximately 28%.

As we have seen, additive genetic variation is necessary for a response to natural selection. Empirical studies have found that virtually every trait that has been studied has some additive genetic variance (i.e., $H_N > 0$) (Roff 2003). Thus, evolutionary change is not expected to be constrained by the lack of genetic variation for a trait. An exception to this has been found in a rainforest species of *Drosophila* in Australia (Hoffmann et al. 2003). These authors found no response to selection for resistance to desiccation after 30 generations of selection! A parent-offspring regression analysis estimated a narrow sense heritability of zero with the upper 95% confidence value of 0.19. This result is especially puzzling since there are clinal differences between populations for this trait. This suggests there has been a history of selection and response for this trait. This is not the only report of lack of response to traits associated with the effects of global warming. Baer and Travis (2000) suggested that the lack of genetic variation was responsible for a lack of response to artificial selection for acute thermal stress tolerance in a livebearing fish.

Genetic correlations among traits can also constrain the response to selection in the wild when there is substantial additive genetic variation for a trait (see Section 11.2.2). For example, Etterson and Shaw (2001) studied that the evolutionary potential of three populations of a native annual legume in tallgrass prairie fragments in North America to respond to the warmer and more arid climates predicted by global climate models. Despite substantial heritabilities for the traits under selection, between-trait genetic correlations that were antagonistic to the direction of selection limited the adaptive evolution of these populations. The predicted rates of evolutionary response taking genetic correlations into account were much slower than the predicted rate of change with heritabilities alone.

11.6.2 Can molecular genetic variation provide a good estimate of quantitative variation?

As we saw in Section 11.4, we expect an equivalent loss of molecular heterozygosity and additive genetic variation during a bottleneck. Therefore, comparison of heterozygosity may provide a good estimate of the loss of quantitative variation. Briscoe et al. (1992) studied the loss of molecular and quantitative genetic variation in laboratory populations of *Drosophila*. They found that substantial molecular and quantitative genetic variation was lost during captivity even though census sizes were of the order of 5,000 individuals. More importantly for our question, they found that heterozygosity at nine allozyme loci provided an excellent estimate of the loss of quantitative genetic variation as measured by H_N for sternopleural bristle number (Figure 11.11). Some studies from natural populations have also found that molecular genetic variation support this result (Example 11.2).

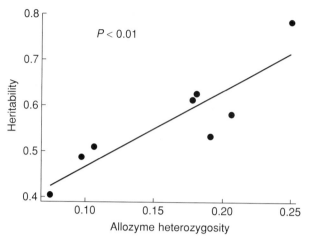

Figure 11.11 Relationship between quantitative genetic variation (H_N for sternopleural bristle number) and molecular genetic variation (heterozygosity at nine allozyme loci) in eight laboratory strains of *Drosophila*. From Briscoe et al. (1992).

Example 11.2 Lack of molecular and quantitative genetic variation in the red pine

The red pine is a common species with a broad range across the north–central and northeastern United States and southeastern Canada. Early studies detected very little genetic variation for quantitative characters in this species (Fowler and Lester 1970). This was a somewhat surprising result considering that red pine consists of millions of individuals occupying a vast range. The low level of quantitative genetic variation presumably results from one or a series of population size bottlenecks associated with the most recent glaciation, during which glaciers covered almost all of the present range of red pine, some 15,000–20,000 years ago.

A series of studies of molecular genetic variation in this species are concordant with this result. Fowler and Morris (1977) found no allozyme variability in samples from five widely separated geographic sources. Allendorf et al. (1982) found almost no variation at 27 allozyme loci in red pine; three of the four rare alleles detected in red pine were null alleles that produced no detectable enzyme. In contrast, these authors found an average heterozygosity of 18% using the same allozyme loci with ponderosa pine. Ponderosa pine also shows great quantitative genetic variation for all the traits that have been studied (Madsen and Blake 1977). Two additional allozyme studies found no variation throughout the range of red pine (Simon et al. 1986; Mosseler et al. 1991).

Direct examination of DNA has also been concordant with these results. Mosseler et al. (1992) found almost no variation for a number of randomly amplified polymorphic markers in red pine. Some genetic variation has been found in chloroplast microsatellites from red pine (Echt et al. 1998). For example, Walter and Epperson (2001) found genetic variation in 10 chloroplast microsatellite loci in individuals collected throughout the range of red pine. However, the amount of chloroplast DNA variation in red pine is substantially less than that found in other pine species. Only six haplotypes were found and 78% of all trees had the same haplotype. The other five haplotypes differed at a single locus, and all except one were caused by a single base pair change.

The red pine must have gone through a very small and prolonged bottleneck associated with glaciation within the last 20,000 years, less than 1,000 generations ago. There has not been enough time for this species to recover genetic variation for either quantitative genetic traits or a series of molecular genetic markers.

Nevertheless, as we saw in Section 11.4, the relationship between molecular and quantitative variation is not so simple. Bottlenecks have been found to increase the heritability for some traits, but are always expected to reduce molecular genetic variation. In addition, different types of genetic variation will recover at different rates from a bottleneck because of different mutation rates (see Section 12.5). The overall high mutation for quantitative traits because they are polygenic means that quantitative genetic variation may recover from a bottleneck more quickly than molecular quantitative variation (Lande 1996; Lynch 1996). In addition, even populations with fairly low effective population sizes can maintain enough additive genetic variation for substantial adaptive evolution (Lande 1996). We also expect a weak correlation between molecular and quantitative genetic variation because of statistical sampling. Substantial variation in additive genetic variation between small population is expected (Lynch 1996). There may also be large differences between quantitative traits because they will be differentially affected by natural selection.

Therefore, the amount of molecular genetic variation within a population should be used carefully to make inferences about quantitative genetic variation. The closer the relationship between the populations being compared, the more informative the comparison will be. The strongest case is the comparison of a single population at different times, as in the *Drosophila* example. Reduced genetic variation for molecular genetic variation over time is likely to reflect loss of genetic variation at the genes responsible for additive genetic variation. However, comparisons between species will not be very informative for the reasons described in the preceding paragraph. Therefore, lack of molecular genetic variation within a species should not be taken to mean that adaptive evolution is not possible because of the absence of additive genetic variation.

11.6.3 Does divergence for molecular genetic variation provide a good estimate of population divergence for quantitative traits?

Comparisons in natural populations have generally found that Q_{ST} and F_{ST} are highly correlated. Merilä and Crnokrak (2001), in a review of 18 studies, found a strong correlation between Q_{ST} and F_{ST} (mean $r = 0.75$), and concluded that differentiation at molecular markers is "closely predictive" of differentiation at loci coding quantitative traits. In addition, differentiation at quantitative traits (Q_{ST}) typically exceeds differentiation at molecular markers (F_{ST}). This suggests a prominent role for natural selection in determining patterns of differentiation at quantitative trait loci.

A comparison of quantitative and molecular markers is a useful approach for understanding the role of natural selection and drift in determining patterns of differentiation in natural populations. Nevertheless, we believe that there are too little data currently available to generalize these results and validate this approach. For example, it is somewhat surprising that Q_{ST} almost always exceeds F_{ST} given that so many quantitative traits seem to be under stabilizing selection. This result suggests that different local populations almost always have different optimum phenotypic values. Another interpretation is that the optimum mean value is similar in different populations but that environmental differences result in different combinations of genotypes producing a similar phenotype (see countergradient selection in Section 2.4).

Once again, the amount of molecular genetic variation between populations should be used carefully to make inferences about quantitative genetic variation. Substantial molecular genetic divergence between populations suggests some isolation between these populations and therefore provides strong evidence for the **opportunity** for adaptive divergence. And it is fair to say that some adaptive differences are **likely** to occur between populations that have been isolated long enough to accumulate substantial molecular genetic divergence. However, the reverse is not true. Lack of molecular genetic divergence should not be taken to suggest that adaptive differences do not exist. As we saw in Section 9.6, even fairly weak natural selection can have a profound effect on the amount of genetic divergence among populations.

Guest Box 11 Response to trophy hunting in bighorn sheep
David W. Coltman

Bighorn sheep populations are often managed to provide a source of large-horned rams for trophy hunting. In many places strict quotas of the number of rams that may be harvested each year are enforced through the use of a lottery system for hunting permits. However, in other parts of their endemic range, any ram that reaches a minimum legal horn size can be taken during the annual fall hunting season. In one population of bighorn sheep at Ram Mountain, Alberta, Canada, a total of 57 rams were harvested under such an unrestricted management regime over a 30-year period. This corresponded to an average harvest rate of about 40% of the legal-sized rams in a given year, with the average age of a ram at harvest of 6 years. Since rams in this population do not generally reach their peak reproductive years until 8 years of age (Coltman et al. 2002), hunters imposed an artificial selection pressure on horn size that had the potential to elicit an evolutionary response, provided that the horn size was heritable.

The heritability of horn size, or any other quantitative trait, can be estimated using pedigree information. Mother–offspring relationships in the Ram Mountain population were known through observation, and father–offspring relationships were determined using microsatellites for paternity (Marshall et al. 1998) and sibship analyses (Goodnight and Queller 1999). An "animal model" analysis (named as such because it estimates the expected genetic "breeding value" for each individual animal in the population) was conducted, which uses relatedness across the entire pedigree to estimate narrow sense heritability using maximum likelihood. Horn length was found to be highly heritable, with narrow sense $H_N = 0.69$ (Coltman et al. 2003).

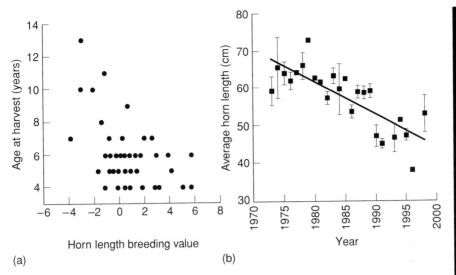

(a)   (b)

Figure 11.12 (a) Relationship between the age at harvest for trophy-harvested rams and their breeding value. (b) Relationship between mean ($\pm$ SE) horn length of 4-year-old rams and year ($N = 119$ rams).

The examination of individual "breeding values" (which is twice the expected deviation of each individual's offspring from the long-term population mean) revealed that rams with the highest breeding values were harvested earliest (Figure 11.12a) and therefore had lower fitness than rams of lower breeding value. As a consequence the average horn length observed in the population has declined steadily over time (Figure 11.12b). Unrestricted harvesting has therefore contributed to a decline in the trait that determines trophy quality by selectively targeting rams of high genetic quality before their reproductive peak.

Problem 11.1

Tarsus length in barn swallows is a polygenic trait. A population has an average length of 21 mm. The survivors of a severe winter storm from this population have an average tarsus length of 25 mm. The average tarsus length in the progeny of the swallows that survived the storm is 24 mm. What is the estimated narrow sense heritability (H_N) of tail length in this population?

Problem 11.2

The following two graphs show the relationship between parental and mean progeny phenotypes for two different traits (body length and number of eggs produced) from Morgan Creek, Resurrection Bay, Alaska. Which trait has the smaller narrow sense heritability (H_N) based on these relationships?

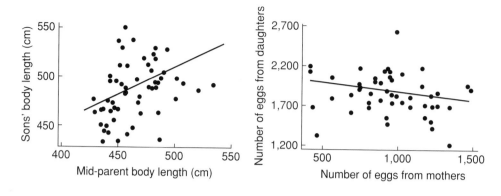

Problem 11.3

How much quantitative genetic variation would you expect to find in the Wollemi pine (see Example 4.2)?

Problem 11.4

The slope of the regression of progeny vertebrae number on maternal vertebrae number in velvet belly sharks is 0.42 (see Figure 2.4). What is the estimated narrow sense heritability for this trait?

Problem 11.5

We have seen that small population size will cause the loss of molecular genetic variation in populations (see Chapter 6). A number of studies have found that additive genetic variation (H_N) for traits associated with fitness often increases following a population bottleneck. Why?

Problem 11.6

Alan Robertson's classic 1967 paper on the nature of quantitative genetic variation contains the following quote from a colleague of his, "I can't help feeling that those people who introduced the notion of the selective advantage of a gene have completely confused the issue. After all, natural selection acts on phenotypes." Robertson states that this statement represents a legitimate point of view and would make a very good subject for a student essay. Why do geneticists insist on understanding the genetic basis for phenotypic differences associated with differences in fitness? Hint: see Section 11.2.

12

Mutation

Minke whale, Example 12.2

Mutation is the ultimate source of all the genetic variation necessary for evolution by natural selection; without mutation evolution would soon cease.
Michael Whitlock and Sarah Otto (1999)

Mutations can critically affect the viability of small populations by causing inbreeding depression, by maintaining potentially adaptive genetic variation in quantitative characters, and through the erosion of fitness by accumulation of mildly deleterious mutations.
Russell Lande (1995)

Mutations are errors in the transmission of genetic information from parents to progeny. The process of mutation is the ultimate source of all genetic variation in natural populations. Nevertheless, this variation comes at a cost because most mutations that have phenotypic effect are harmful (deleterious). Mutations occur both at the chromosomal level and the molecular level. As we will see, mutations may or may not have a detectable effect on the phenotype of individuals.

An understanding of the process of mutation is important for conservation for several reasons. The amount of standing genetic variation within populations is largely a balance between the gain of genetic variation from mutations and the loss of genetic variation from genetic drift. Thus, an understanding of mutation is needed to interpret patterns of genetic variation observed in natural populations.

Moreover, the accumulation of deleterious mutations in populations, which is dependent on mutation rates, is one source of inbreeding depression and may threaten the persistence of some populations. In addition, the rate of adaptive response to environmental change is proportional to the amount of standing genetic variation for fitness within populations. Thus, long-term persistence of populations may require large population sizes in order to maintain important adaptive genetic variation.

Unfortunately, there are few empirical data available about the process of mutation because mutation rates are rare. The data that are available generally come from a few model organisms (e.g., mice, *Drosophila*, or *Arabidopsis*) that are selected because of their short generation time and suitability for raising a large number of individuals in the laboratory. However, we must be careful in generalizing results from such model species; the very characteristics that make these organisms suitable for these experiments may make them less suitable for generalizing to other species. For example, the per generation mutation rate tends to be greater for species with longer generation length (Drake et al. 1998).

How common are mutations? On a per locus or per nucleotide level they are rare. For example, the rate of mutation for a single nucleotide is in the order a few per billion gametes per generation. However, from a genomic perspective, mutations are actually very common. The genome of most species consists of billions of base pairs. Therefore, it has been estimated (Lynch et al. 1999) that each individual may possess hundreds of new mutations! Fortunately, almost all of these mutations are in nonessential regions of the genome and have no phenotypic effect.

We will consider the processes resulting in mutations and examine the expected relationships between mutation rates and the amount of genetic variation within populations. We will examine evidence for both harmful and advantageous mutations in populations. Finally, we will examine the effects of mutation rates on the rate of recovery of genetic variation following a population bottleneck.

12.1 Process of mutation

Chromosomes and DNA sequences are normally copied exactly during the process of replication and are transmitted to progeny. Sometimes, however, errors occur that produce new chromosomes or new DNA sequences. Empirical information on the rates of mutation is hard to come by because mutations are so rare. Thousands of progeny must be examined to detect mutational events. Thus, estimating the rates of mutations or describing the types of changes brought about by mutation is generally incredibly difficult. Thus, most of our direct information about the process of mutation comes from model organisms (Example 12.1).

Example 12.1 Coat color mutation rate in mice

Schlager and Dickie (1971) presented the results of a direct and massive experiment to estimate the rate of mutation to five recessive coat color alleles in mice: albino, brown, nonagouti, dilute, and leaden. They examined more than seven million mice in 28 inbred strains. Overall, they detected 25 mutations in over two million gene transmissions for an average mutation rate of 1.1×10^{-5} per gene transmission. As expected, the reverse mutation rate (from the recessive to the dominant allele) was much lower, approximately 2.5×10^{-6} per gene transmission. The reverse mutation rate is expected to be lower because there are more ways to eliminate the function of a gene to reverse a defect. This assumes that the recessive coat color mutations are caused by mutations that cause loss of function (similar to the null alleles for allozyme loci).

Guest Box 12 presents a modern molecular study of the mutation process in pocket mice that produce a melanistic or dark morph.

Most mutations with phenotypic effects tend to reduce fitness. Thus, as we will see in Chapter 14, the accumulation of mutations can decrease the probability of survival of small populations. Nevertheless, rare beneficial mutations are important in adaptive evolutionary change (Elena et al. 1996). In addition, some recent experimental work with the plant *Arabidopsis thaliana* has suggested that roughly half of new spontaneous mutations increase fitness (Shaw et al. 2002); however, this result has been questioned in view of data from other experiments (Bataillon 2003; Keightley and Lynch 2003).

Mutations are commonly said to occur randomly, but there is some evidence that some aspects of the process of mutation may be an adaptive response to environmental conditions. There has been an ongoing controversy that mutations in prokaryotes may be directed toward particular environmental conditions (Lenski and Sniegowski 1995). In addition, there is accumulating evidence that the rate of mutations in eukaryotes may increase under stressful conditions and thus create new genetic variability that may be important in adaptation to changing environmental conditions (Capy et al. 2000).

12.1.1 Chromosomal mutations

We saw in Chapter 3 that rates of chromosomal evolution vary tremendously among different taxonomic groups. There are two primary factors that may be responsible for differences in the rate of chromosomal change: (1) the rate of chromosomal mutation; and (2) the rate of incorporation of such mutations into populations (Rieseberg 2001). Differences between taxa in rates of chromosomal change may result from differences in either of these two effects.

White (1978) estimated a general mutation rate for chromosomal rearrangements of the order of one per 1,000 gametes in a wide variety of species from lilies to grasshoppers to humans. Lande (1979) considered different forms of chromosomal rearrangements in animals and produced a range of estimates between 10^{-4} and 10^{-3} per gamete per generation. There is evidence in some groups that chromosomal mutation rates may be substantially higher than this. For example, Porter and Sites (1987) detected spontaneous chromosomal mutations in five of 31 males that were examined.

The apparent tremendous variation in chromosomal mutation rates suggests that some of the differences between taxa could result from differences in mutation rates. In addition, there is some evidence that chromosomal polymorphisms may contribute to increased chromosomal mutation rates. That is, chromosomal mutation rates may be greater in chromosomal heterozygotes than homozygotes (King 1993). In addition, we will see in Section 12.1.4 that genomes with more transposable elements may have higher chromosomal mutation rates.

12.1.2 Molecular mutations

There are several types of molecular mutations in DNA sequences: (1) substitutions, the replacement of one nucleotide with another; (2) recombinations, the exchange of a sequence from one homologous chromosome to the other; (3) deletions, the loss of one or more nucleotides; (4) insertions, the addition of one or more nucleotides; and (5) inversions, the rotation by $180°$ of a double-stranded DNA segment of two or more base pairs (see Grauer and Li 2000).

The rate of spontaneous mutation is very difficult to estimate directly because of their rarity. Mutation rates are sometimes estimated indirectly by an examination of rates of substitutions over evolutionary time in regions of the genome that are not affected by natural selection. The rate of substitution per generation will be equal to the mutation rate for selectively neutral mutations. The average rate of mutation in mammalian nuclear DNA has been estimated to be $3–5 \times 10^{-9}$ nucleotide substitutions per nucleotide site per year (Grauer and Li 2000). The mutation rate, however, varies enormously between different regions of the nuclear genome. The rate of mutation in mammalian mitochondrial DNA has been estimated to be at least 10 times higher than the average nuclear rate.

The mutation rate for protein coding loci (e.g., allozymes) is very low. Not all DNA mutations will result in a change in the amino acid sequence because of the inherent redundancy of the genetic code. Nei (1987, p. 30) reviewed the literature on direct and indirect estimates of mutation rates for allozyme loci. Most direct estimates of mutation rates in allozymes have failed to detect any mutant alleles; for example, Kahler et al. (1984) examined a total of 841,260 gene transmissions from parents to progeny at five loci and failed to detect any mutant alleles. General estimates of mutation rates for allozyme loci are in the order of 10^{-6} to 10^{-7} mutants per gene transmission (Nei 1987).

The rate of mutation at microsatellite loci is much greater than in other regions of the genome because of the presence of simple sequence repeats (Li et al. 2002). Two mechanisms are thought to be responsible for mutations at microsatellite loci: (1) mispairing of DNA strands during replication; and (2) recombination. Estimates of mutation rates at microsatellite loci have generally been approximately 10^{-3} mutants per gene transmission (Ellegren 2000a) (Table 12.1). Microsatellite mutations appear largely to follow the stepwise mutation model (SMM) where single repeat units are added or deleted with near equal frequency (Valdes et al. 1993) (Figure 12.1). However, the actual mechanisms of microsatellite mutation are much more complicated than this simple model (Estoup and Angers 1998; Li et al. 2002).

Recent evidence has suggested new mutations may occur in clusters because they occur early during gametogenesis (Woodruff et al. 1996). Woodruff and Thompson (1992) found

Table 12.1 Mutations observed at the *OGO1c* tetranucleotide repeat microsatellite locus in pink salmon (Steinberg et al. 2002). Approximately 1,300 parent–progeny transmissions were observed in 50 experimental matings. Mutations were found only in the four matings shown. The mutant allele is indicated by bold type and the most likely progenitor of the mutant allele is underlined. The overall mutation rate estimated from these data is $3.9 \times 10^{-3} (5/1,300)$.

Dam (a/b)	Sire (c/d)	Progeny genotypes				Mutant genotypes
		a/c	a/d	b/c	b/d	
342/350	408/<u>474</u>	1	1	3	3	342/**478**
295/366	303/<u>362</u>	1	2	4	2	295/**366**
269/420	346/<u>450</u>	8	16	10	8	420/**446** (2)
348/348	309/<u>448</u>	5	4	0	0	348/**444**

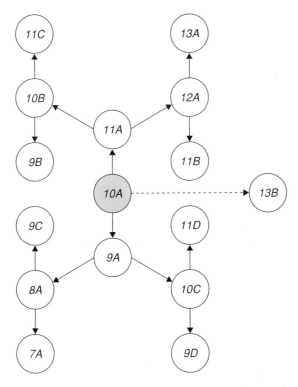

Figure 12.1 Pattern of mutation for microsatellites beginning with a single ancestral allele (shaded circle) with 10 repeats. Most mutations are a gain or loss of single repeat (stepwise mutation model, SMM). The dashed arrow shows a multiple step mutation (from 10 to 13 repeats). Alleles are designated by the number of repeats and a letter which distinguishes different alleles with the same number of repeats.

that as many as 20% of new mutations in *Drosophila* represented clusters of identical mutant alleles sharing a common premeiotic origin. Cluster mutations at microsatellite loci have been found in several other species (Steinberg et al. 2002). The occurrence of clustered mutations results in nonuniform distributions of novel alleles in a population that could influence interpretations of mutation rates and patterns as well as estimates of genetic population structure. For example, Woodruff et al. (1996) have shown that mutant alleles that are part of clusters are more likely to persist and be fixed in a population than mutant alleles entering the population independently.

12.1.3 Quantitative characters

As we saw in Chapter 11, the amount of genetic variation in quantitative characters for morphology, physiology, and behavior that can respond to natural selection is measured by the additive genetic variance (V_A). The rate of loss of additive genetic variance due to genetic drift in the absence of selection is the same for the loss of heterozygosity (i.e., $1/2N_e$). The effective mutation rate for quantitative traits is much higher than the rate for single gene traits because mutations at many possible loci can affect a quantitative trait. The input of additive genetic variance per generation by mutation is V_m. The expected genetic variance at equilibrium between these two factors is $V_A = 2N_e V_m$ (Lande 1995, 1996).

Estimates of mutation rates for quantitative characters are very rare and somewhat unreliable. It is thought that V_m is roughly on the order of $10^{-3} V_A$ (Lande 1995). However, some experiments suggest that the great majority of these mutations are highly detrimental and therefore are not likely to contribute to the amount of standing genetic variation within a population. Thus, the effective V_m responsible for much of the standing variation in quantitative traits in natural populations may be an order of magnitude lower, $10^{-4} V_A$ (Lande 1996).

12.1.4 Transposable elements, mutation rates, and stress

Much of the genome of eukaryotes consists of sequences associated with transposable elements that possess an intrinsic capability to make multiple copies and insert themselves throughout the genome. For example, approximately half of the human genome consists of DNA sequences associated with transposable elements (Lynch 2001). This activity is analogous to the "cut and paste" mechanism of a word processor. Transposable elements are potent agents of mutagenesis (Kidwell 2002). For example, Clegg and Durbin (2000) have studied mutations that affect flower color in the morning glory. Nine out of 10 mutations that they identified were the result of transposable elements. A consideration of the molecular basis of transposable elements is beyond our consideration (see chapter 7 of Grauer and Li 2000). Nevertheless, the mutagenic activity of these elements is of potential significance for conservation.

Transposable elements can cause a wide variety of mutations. They can induce chromosomal rearrangements such as deletions, duplications, inversions, and reciprocal translocations. Kidwell (2002) has suggested that "transposable elements are undoubtedly responsible for a significant proportion of the observed karyotypic variation among many groups". In addition, transposable elements are responsible for a wide variety of substitutions in DNA sequences, ranging from insertion of the transposable element sequence to substitutions, deletions, and insertions of a single nucleotide (Kidwell 2002).

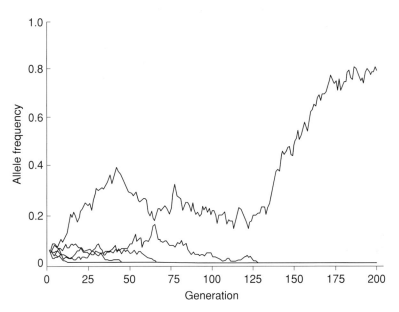

Figure 12.2 Simulations of genetic drift of neutral alleles introduced into a large population by mutation.

Stress has been defined as any environmental change that drastically reduces the fitness of an organism (Hoffmann and Parson 1997). McClintock (1984) first suggested that transposable element activity could be induced by stress. A number of transposable elements in plants have been shown to be activated by stress (Grandbastien 1998; Capy et al. 2000). Some transposable elements in *Drosophila* have been shown to be activated by heat stress, but other studies have not found an effect of heat shock (Capy et al. 2000). In addition, hybridization has also been found to activate transposable elements and cause mutations (Kidwell and Lisch 1998).

12.2 Selectively neutral mutations

Many mutations in the DNA sequence have no phenotypic effect so they are neutral with regard to natural selection (e.g., mutations in noncoding regions). In this case, the amount of genetic variation within a population will be a balance between the gain of variation by mutation and the loss by genetic drift (Figure 12.2). The distribution of neutral genetic variation among populations is primarily a balance between gene flow and genetic drift. Gene flow among subpopulations retards the process of differentiation until eventually a steady state may be reached between the opposing effects of gene flow and genetic drift. However, the process of mutation may also contribute to allele frequency divergence among populations in cases where the mutation rate approaches the migration rate.

12.2.1 Genetic variation within populations

The amount of genetic variation within a population at equilibrium will be a balance between the gain of variation as a function of the neutral mutation rate (μ) and the loss of genetic variation by genetic drift as a function of effective population size (N_e).

We will first consider the so-called infinite allele model in which we assume that every mutation creates a new allele that has never been present in the population. This model is appropriate if we consider variation in DNA sequences. A gene consists of a large number of nucleotide sites, each of which may be occupied by one of four bases (A, T, C, or G). Therefore, the total number of possible allelic states possible is truly a very large number! For example, there are over one million possible alleles if we just consider 10 base pairs $(10^4 = 1,048,576)$. In this case, the average expected heterozygosity (H) at a locus (or over many loci with the same mutation rate) is:

$$H = \frac{4N_e\mu}{(4N_e\mu + 1)} = \frac{\theta}{\theta + 1} \tag{12.1}$$

where μ is the neutral mutation rate and $\theta = 4N_e\mu$ (Kimura 1983).

The much greater variation at microsatellite loci compared to allozymes results from the differences in mutation rates that we discussed in Section 12.1.2. Figure 12.3 shows the equilibrium heterozygosity for microsatellites and allozymes using mutation rates of 10^{-4} and 10^{-6}, respectively. Thus, we expect a heterozygosity of 0.038 at allozyme loci and 0.80 at microsatellite loci with an effective population size of 10,000. However, we also expect a substantial amount of variation in heterozygosity between loci (Figure 12.4).

The heterozygosity values for microsatellite loci in Figure 12.3 are likely to be overestimates because of several important assumptions in this expectation. Microsatellite mutations tend to occur in steps of the number of repeat units. Therefore, each mutation will not be unique, but rather will be to an allelic state (say 11 copies of a repeat) that already occurs in the population. This is called **homoplasy** in which two alleles that are identical in state have different origins (e.g., alleles *11C* and *11D* in Figure 12.1. Therefore the actual expected heterozygosity is less than predicted by expression 12.1. Allozymes also tend to follow a stepwise model of mutation (Ohta and Kimura 1973), but this will have a smaller

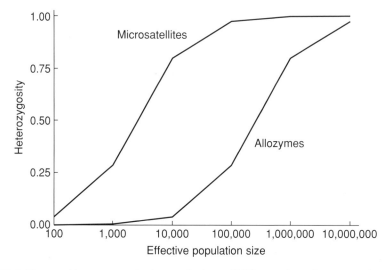

Figure 12.3 Expected heterozygosity in populations of different size using expression 12.1 for microsatellites ($\mu = 10^{-4}$) and allozymes ($\mu = 10^{-6}$).

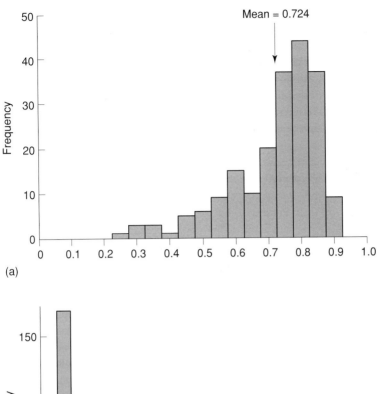

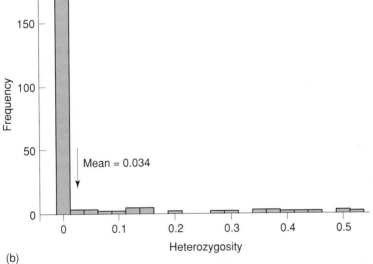

Figure 12.4 Simulated heterozygosities at 200 loci in a population with $N_e = 10,000$ and the infinite allele model of mutation produced with the program EASYPOP. (a) Microsatellite loci with $\mu = 10^{-4}$. (b) Allozyme with $\mu = 10^{-6}$. The expected heterozygosities are 0.800 and 0.038 (expression 12.1). From Balloux (2001).

effect because of the fewer number of alleles present in a population because of the smaller mutation rate.

It is also important to remember that the mutation rate used here (μ) is the neutral mutation rate. Mutations in DNA sequence within some regions of the genome are likely to not be selectively neutral. Therefore, different regions of the genome will have different effective neutral mutation rates, even though the actual rate of molecular mutations is the same. For example, mutations in protein coding regions may affect the amino acid

Example 12.2 How many whales are there in the ocean?

Expression 12.1 can also be used to estimate the effective population size of natural populations if we know the mutation rate (μ). For example, Roman and Palumbi (2003) estimated the historical (pre-whaling) number of humpback, fin, and minke whales in the North Atlantic Ocean by estimating θ for the control region of mtDNA. In the case of mtDNA, $\theta = 2N_{e(f)}\mu$ because of maternal inheritance and haploidy.

Their genetic estimates of historical population sizes for humpback, fin, and minke whales are far greater than those previously calculated, and are 6–20 times higher than the current population estimates for these species. This discrepancy is crucial for conservation because the International Whaling Commission management plan uses the estimated historical population sizes as guidelines for setting allowable harvest rates. We should be careful using estimates of N_e with this approach because there are a host of possible pitfalls (e.g., how reliable are our estimates of mutation rate?). Roman and Palumbi (2003) provide an excellent discussion of the limitations of this method for estimating N_e.

sequence of an essential protein and thereby reduce fitness. Such mutations will not be neutral and therefore will not contribute to the amount of variation maintained by our model of drift–mutation equilibrium considered here. In these regions, so-called **purging** selection will act to stop these mutations reaching high frequencies in a population. In contrast, mutations in the DNA sequence in regions of the genome that are not functional are much more likely to be neutral. This expectation is supported by empirical results; exons, which are the coding regions of protein loci, are much less variable than the introns, which do not encode amino acids (Grauer and Li 2000).

12.2.2 Population subdivision

The process of mutation may also contribute to allele frequency divergence among populations. The relative importance of mutation on divergence (e.g., F_{ST}) depends primarily upon the relative magnitude of the rates of migration and of mutation. Under the infinite alleles model (IAM) of mutation with the island model of migration (Crow and Aoki 1984), the expected value of F_{ST} is approximately:

$$F_{ST} = \frac{1}{(1 + 4Nm + 4N\mu)} \tag{12.2}$$

Greater mutation rates will increase F_{ST} when new mutations are not dispersed at sufficient rates to attain equilibrium between genetic drift and gene flow. Under these conditions, new mutations may drift to substantial frequencies in the population in which they occur before they are distributed among other populations via gene flow (Neigel and Avise 1993). Mutation rates for allozymes are generally thought to be less than 10^{-6} so that divergence at allozyme loci is unlikely to be affected by different mutation rates unless the subpopulations are completely isolated. Mutation rates for some microsatellite loci may be as high as 10^{-4} (Eisen 1999).

Mutation under the infinite alleles model may accelerate the rate of divergence at microsatellite loci among subpopulations that are very large and are connected by little gene flow. The actual expected effect of mutation is much more complicated than this, and it depends upon the rate of mechanism of mutation. For example, constraints on allele size at microsatellite loci under the stepwise mutation model may reverse the direction of this effect (i.e., decrease the rate of divergence) under some conditions (Nauta and Weissing 1996).

In general, mutations will have an important effect on population divergence only when the migration rates are very low (say 10^{-3} or less) and the mutation rates are unusually high (10^{-3} or greater) (Nichols and Freeman 2004; Epperson 2005). However, as we saw in Chapter 9, F_{ST} will underestimate genetic divergence at loci with very high within-deme heterozygosities (H_S) (Hedrick 1999). Large differences in H_S caused by differences in mutation rates among loci (e.g., Steinberg et al. 2002) can result in discordant estimates of F_{ST} among microsatellite loci. This may result in an underestimation of both the degree of genetic divergence among populations if all loci are pooled for analysis and the estimation of F_{ST} (see Olsen et al. 2004b).

We saw in the previous section that long-term N_e can be estimated using the amount of heterozygosity in a population if we know the mutation rate. However, we also know from Chapter 9 that the amount of gene flow affects the amount of genetic variation in a population. Therefore, estimates of N_e using expression 12.1 may be overestimates because they reflect the total N_e of a series of populations connected by gene flow rather than the N_e of the local population. Consider two extremes. In the first, a population on an island is completely isolated from the rest of the members of its species ($mN = 0$). In this case, estimates of N_e using expression 12.1 will reflect the local N_e. In the other extreme, a species consists of a number of local populations that are connected by substantial gene flow (say $mN = 100$); in this case the estimates of N_e using expression 12.1 will reflect the combined N_e of all populations (Table 12.2).

Table 12.2 Computer simulations of estimates of effective population size (N_e) using expression 12.1 in a series of 20 populations ($N_e = 200$) that are connected by different amounts of gene flow with an island model of migration (EASYPOP; Balloux 2001). A mutation rate of 10^{-4} was used to simulate the expected heterozygosities at 100 microsatellite loci. The simulations began with no genetic variation in the first generation and ran for 10,000 generations. F_{ST}^* is the expected F_{ST} with this amount of gene flow corrected for a finite number of populations (Mills and Allendorf 1996). $\hat{N}_e$ is the estimated effective population size based upon the mean local heterozygosity (H_S) using expression 12.1.

mN	H_T	H_S	F_{ST}	F_{ST}^*	$\hat{N}_e$
0	0.814	0.076	0.907	1.000	205
0.5	0.665	0.477	0.283	0.311	2,274
1.0	0.635	0.516	0.187	0.184	2,667
2.0	0.621	0.558	0.100	0.101	3,156
5.0	0.618	0.592	0.041	0.043	3,630
10.0	0.606	0.594	0.020	0.022	3,665

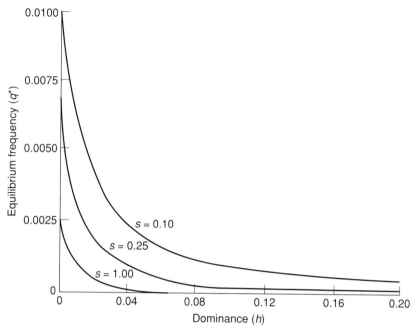

Figure 12.5 The expected equilibrium frequency of a deleterious allele ($q^\star$) with mutation–selection balance for different degrees of dominance (h) and intensity of selection (s). From Hedrick (1999).

12.3 Harmful mutations

Most mutations that affect fitness have a detrimental effect. Natural selection acts to keep these mutations from increasing in frequency. Consider the joint effects of mutation and selection at a single locus with a normal allele (A_1) and a mutant allele (A_2) that reduces fitness as shown below:

$$
\begin{array}{ccc}
A_1A_1 & A_1A_2 & A_2A_2 \\
1 & 1 - hs & 1 - s
\end{array}
$$

where s is the reduction in fitness of the homozygous mutant genotype and h is the degree of dominance of the A_2 allele. A_2 is recessive when $h = 0$, dominant when $h = 1$, and partially dominant when h is between 0 and 1.

If the mutation is recessive ($h = 0$), then at equilibrium:

$$q^\star = \sqrt{\frac{\mu}{s}} \tag{12.3}$$

When A_2 is partially dominant, q will generally be very small and the following approximation holds (Figure 12.5):

$$q^\star \approx \frac{\mu}{hs} \tag{12.4}$$

See Lynch et al. (1999) for a consideration of the importance of mildly deleterious mutations in evolution and conservation.

12.4 Advantageous mutations

Genetic drift plays a major role in the survival of advantageous mutations, even in extremely large populations. That is, most advantageous mutations will be lost during the first few generations because new mutations will always be rare. The initial frequency of a mutation will be one over the total number of gene copies at a locus (i.e., $q = 1/2N$). An advantageous allele that is recessive will have the same probability of initial survival in a population because the advantageous homozygotes will not occur in a population until the allele happens to drift to relatively high frequency. For example, a new mutation will have to drift to a frequency over 0.30 before even 10% of the population will be homozygotes with the selective advantage. Therefore, the great majority of advantageous mutations that are recessive will be lost.

Dominant advantageous mutations have a much greater chance of surviving the initial period because their fitness advantage will be effective in heterozygotes that carry the new mutation. However, even most dominant advantageous mutations will be lost within the first few generations because of genetic drift. For example, over 80% of dominant advantageous mutations with a selective advantage of 10% will be lost within the first 20 generations (Crow and Kimura 1970, p. 423). This effect can be seen in a simple example. Consider a new mutation that arises that increases the fitness of the individual that carries it by 50%. However, even if the individual that carries this mutation contributes three progeny to the new generation, there is a 0.125 probability that none of the progeny carry the mutation because of the vagaries of Mendelian segregation ($0.5 \times 0.5 \times 0.5 = 0.125$).

Gene flow and spread of advantageous mutations may be an important cohesive force in evolution (Rieseberg and Burke 2001). Ehrlich and Raven (1969) argued in a classic paper that the amounts of gene flow in many species are too low to prevent substantial differentiation among subpopulations by genetic drift or local adaptation, so that local populations are essentially independently evolving units in many species. We saw in Chapter 9 that even one migrant per generation among subpopulations can cause all alleles to be present in all subpopulations. However, even much lower amounts of gene flow can be sufficient to cause the spread of an advantageous allele (say $s > 0.05$) throughout the range of a species (Rieseberg and Burke 2001). The rapid spread of such advantageous alleles may play an important role in maintaining the genetic integration of subpopulations connected by very small amounts of genetic exchange.

12.5 Recovery from a bottleneck

The rate of recovery of genetic variation from the effects of a bottleneck will depend primarily on the mutation rate (Lynch 1996). The equilibrium amount of neutral

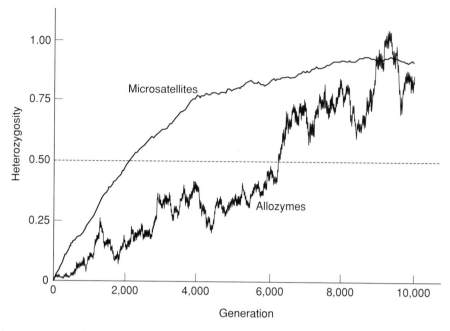

Figure 12.6 Simulated recovery of heterozygosity at 100 loci in a population of 5,000 individuals following an extreme bottleneck using EASYPOP. The initial heterozygosity was zero. The mutation rates are 10^{-4} for microsatellites and 10^{-6} for allozymes. Heterozygosity is standardized as the mean heterozygosity over all 100 loci divided by the expected equilibrium heterozygosity using expression 12.1 (0.670 and 0.020, respectively). From Balloux (2001).

heterozygosity in natural populations (see expression 12.1) will be approached at a time scale equal to the shorter of $2N_e$ or $1/(2\mu)$ generations (Kimura and Crow 1964).

We can see this expectation in Figure 12.6 for microsatellites and allozymes. In these simulations of 100 typical microsatellite and allozyme loci, the expected heterozygosity at microsatellite loci returned to 50% of that expected at equilibrium after 2,000 generations in populations of 5,000 individuals. It took approximately three times as long at the loci with mutation rates typical of allozymes. In this case, $1/(2\mu)$ is 5,000 generations for microsatellites and 100 times that for allozymes. However, $2N_e$ is 10,000 for both types of markers.

As we saw in Section 12.1.3, the estimated mutation rates (V_m) for phenotypic characters affected by many loci (quantitative characters) is similar to the rates of mutations at microsatellite loci. Therefore, we would expect quantitative genetic variance for quantitative characters to be restored at rates comparable to those of microsatellites (Lande 1996). Thus, recovery of microsatellite variation following a severe bottleneck may be a good measure of the recovery of polygenic variation for fitness traits.

Figure 12.7 provides a somewhat simplistic graphic representation of the effects of a severe population bottleneck on different sources of genetic variation. Microsatellites, allozymes, and quantitative traits are all expected to lose genetic variation at approximately the same rates. However, mtDNA will lose genetic variation more rapidly because of its smaller N_e. The rates of recovery of variation will depend upon the mutation rates for these different sources of genetic variation.

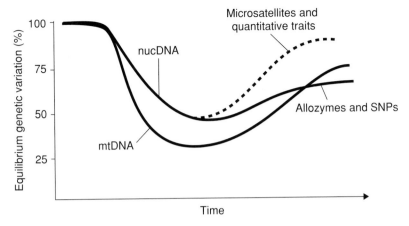

Figure 12.7 Diagram of relative expected effects of a severe population bottleneck on different types of genetic variation. The smaller N_e for mtDNA causes more genetic variation to be lost during a bottleneck. The rate of recovery following a bottleneck is largely determined by the mutation rate (see Section 12.1). SNP, single nucleotide polymorphism.

Guest Box 12 Color evolution via different mutations in pocket mice
Michael W. Nachman

Mutation is the ultimate source of genetic variation, yet the specific mutations responsible for evolutionary change have rarely been identified. We have been studying the genetic basis of color variation in pocket mice from the Sonoran Desert to try to find the mutations responsible for adaptive melanism. This research seeks to answer questions such as: Does adaptation result from a few mutations of major effect or many mutations of small effect? What kinds of genes and mutations underlie adaptation? Do these mutations change gene structure or gene regulation? Do similar phenotypes in different populations arise independently, and if so, do they arise from mutations at the same gene or from mutations at different genes?

Rock pocket mice are granivorous rodents well adapted to life in the desert. They remain hidden in burrows during the day, are active only at night, feed primarily on seeds, and do not drink water. In most places, these mice live on light-colored rocks and are correspondingly light in color. In several different places in the Sonoran Desert, these mice live on dark basalt of recent lava flows, and the mice in these populations are dark in color (see Figure 8.10). The close match between the color of the mice and the color of the rocks is presumed to be an adaptation to avoid predation. Owls are among the primary predators, and even though owls hunt at night, they are able to discriminate between mice that match and do not match their background under conditions of very low light.

The genetic basis of melanism is amenable to analysis because of the wealth of background information on the genetics of pigmentation in laboratory mice and other animals (Bennett and Lamoreux 2003). We developed markers in several candidate genes and we then looked for nonrandom associations between genotypes at these genes and color phenotypes in populations of pocket mice near the edge of lava flows, where both light and dark mice are found together.

This search revealed that mutations in the gene encoding the melanocortin-1 receptor (*Mc1r*) are responsible for color variation in one population in Arizona (Nachman et al. 2003). This receptor is part of a signaling pathway in melanocytes, the specialized cells in which pigment is produced. This work shows that an important adaptation – melanism – is caused by one gene of major effect in this case. Moreover, only four amino acid changes distinguish light and dark animals, demonstrating that relatively few mutations are involved. These mutations produce a hyperactive receptor that results in the production of more melanin in melanocytes and therefore in darker mice.

Surprisingly, nearly phenotypically identical dark mice have arisen independently in this species in several populations in New Mexico (Hoekstra and Nachman 2003). In these mice from New Mexico, however, *Mc1r* is not responsible for the differences in color. While the specific genes responsible for melanism in New Mexico have not yet been found, it is clear that the genetic basis of melanism in these populations is different from the genetic basis of melanism in the population from Arizona.

These results demonstrate that there may be different genetic solutions to a common evolutionary problem. Thus, the genetic basis of local adaptation may be different for isolated populations subjected to the same selective regime. Similarly, Cohan and Hoffmann (1986) found that isolated laboratory populations of *Drosophila* became adapted to high ethanol concentrations by different physiological mechanisms involving changes at different loci. From a conservation perspective, it is important to recognize that phenotypically similar populations may be quite distinct genetically. This has important implications, for example, for planning translocations of individuals between populations (e.g., to supplement declining populations). Translocations between phenotypically similar but isolated populations may result in reduced fitness when the translocated individuals mate with native individuals.

Problem 12.1

What is the expected equilibrium frequency of a lethal recessive mutation that has a mutation rate of 0.00001?

Problem 12.2

What is the expected equilibrium frequency of a partially dominant mutation that is lethal in homozygotes and reduces the fitness of heterozygotes by 0.05, and that has the same mutation rate as the recessive mutation in the previous problem (0.00001)?

Problem 12.3

What do you think the equilibrium frequency would be for a dominant lethal allele with a mutation rate of 0.00001?

Problem 12.4

We generally consider that genetic drift is only a concern in relatively small populations. Why are most advantageous mutations lost because of genetic drift even in populations with very large effective population sizes?

Problem 12.5

Why do you think most mutations are recessive or almost recessive ($h \approx 0$)?

Problem 12.6

An insect with a very large effective population size ($N_e \gg 1{,}000{,}000$) has a moderate amount of genetic variation for microsatellites, but virtually no variation at allozyme loci because of a very small historical bottleneck. Roughly how many generations ago do you think the bottleneck occurred? Assume the mutation rate for microsatellites is 10^{-3} and for allozymes 10^{-6} in this species.

Part III
Genetics and Conservation

13

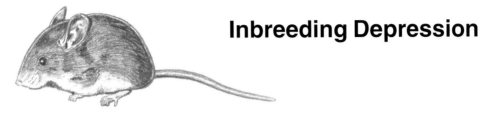

Inbreeding Depression

Peromyscus *mouse, Guest Box 13*

That any evil directly follows from the closest interbreeding has been denied by many persons; but rarely by any practical breeder; and never, as far as I know, by one who has largely bred animals which propagate their kind quickly. Many physiologists attribute the evil exclusively to the combination and consequent increase of morbid tendencies common to both parents: and that this is an active source of mischief there can be no doubt.

Charles Darwin (1896, p. 94)

Probably the oldest observation about population genetics is that individuals produced by matings between close relatives are often less healthy than those produced by mating between more distant relatives.

A. Ives and Michael Whitlock (2002)

The term "inbreeding" is used to mean many different things in population genetics (Question 13.1). Jacquard (1975) described five different effects of nonrandom mating that are measured by inbreeding coefficients. The multiple use of "inbreeding" can sometimes lead to confusion so it is important to be precise when using this term. Templeton and Read (1994) have described three different phenomena of special importance for conservation that are all measured by "inbreeding coefficients":

1   Genetic drift (F_{ST}, see Section 9.1).
2   Nonrandom mating within local populations (F_{IS}; see Section 9.1).
3   The increase in genome-wide homozygosity (pedigree F) caused by matings between related individuals (e.g., father–daughter mating in ungulates, or matings between cousins in birds).

Question 13.1   What is an "inbred population"?

We will focus on this last meaning in this chapter.

Inbreeding (mating between related individuals) will occur in both large and small populations. In large populations, inbreeding may occur by nonrandom mating because of self-fertilization or by a tendency for related individuals to mate with each other. For example, in many tree species nearby individuals tend to be related and are likely to mate with each other because of geographic proximity (Hall et al. 1994). However, substantial inbreeding will occur even in randomly mating small populations simply because all or most individuals within a small population will be related. In an extreme example of a population of two, after one generation, only brother–sister matings are possible. In a slightly larger population with 10 breeders, the most distant relatives will be cousins after only a few generations. This has been called the "inbreeding effect" of small populations (Crow and Kimura 1970, p. 101).

Inbred individuals generally have reduced fitness in comparison to noninbred individuals from the same population because of their increased homozygosity. **Inbreeding depression** is the reduction in fitness (or phenotype value) of progeny from matings between related individuals relative to the fitness of progeny between unrelated individuals (Example 13.1). Inbreeding depression in natural populations will contribute to the extinction of populations under some circumstances (Keller and Waller 2002) (see Chapter 14).

Example 13.1   Inbreeding depression in the monkeyflower

The monkeyflower is a self-compatible wildflower that occurs throughout western North America, from Alaska to Mexico. Willis (1993) studied two annual populations of this species on adjacent mountains about 2 km apart in the Cascade mountains of Oregon. Seeds were collected from both populations and germinated in a greenhouse. Hand pollinations produced self-pollinations and pollinations from another randomly chosen plant from the same population. Seeds resulting from these pollinations were germinated in the greenhouse, and randomly chosen seedlings were transplanted back into their original population. The transplanted seedlings were marked and followed throughout the course of their life (Figure 13.1). Cumulative inbreeding depression through several life history stages was estimated by the proportional reduction in fitness in selfed versus outcrossed progeny $(1 - w_s/w_o)$. A similar set of seedlings was maintained in the greenhouse. The amount of inbreeding depression in the greenhouse was similar to that found in the wild.

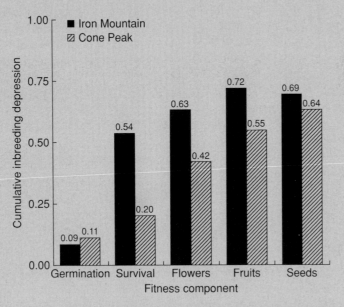

Figure 13.1 Cumulative inbreeding depression in two wild populations of the monkeyflower. Inbreeding depression was measured as the proportional reduction in fitness in selfed versus outcrossed progeny $(1 - w_s/w_o)$. From Willis (1993).

13.1 Pedigree analysis

"Inbred" individuals will have increased homozygosity and decreased heterozygosity over their entire genome. The pedigree inbreeding coefficient (F, the probability of identity by descent) is the expected increase in homozygosity for inbred individuals; it is also the expected decrease in heterozygosity throughout the genome. F ranges from zero (for noninbred individuals) to one (for totally inbred individuals).

An individual is inbred if its mother and father share a common ancestor. This definition must be put into perspective because any two individuals in a population are related if we trace their ancestries back far enough. We must, therefore, define inbreeding relative to some "base" population in which we assume all individuals are unrelated to one another. We usually define the base population operationally as those individuals in a pedigree beyond which no further information is available (Ballou 1983).

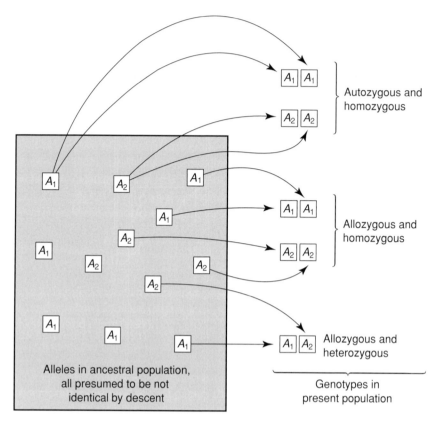

Figure 13.2 Patterns and definitions of genotypic relationships with pedigree inbreeding. Autozygous individuals in the present population contain two alleles that are identical by descent from a single gene in the ancestral population. In contrast, allozygous individuals contain two alleles derived from different genes in the ancestral population. Redrawn from Hartl and Clark (1997).

An inbred individual may receive two copies of the same allele that was present in a common ancestor of its parents. Such an individual is **identical by descent** at that locus (i.e., **autozygous**). The probability of an individual being autozygous is its pedigree inbreeding coefficient, F. All autozygous individuals will be homozygous unless a mutation has occurred in one of the two copies descended from the ancestral allele in the base population. The alternative to being autozygous is **allozygous**. Allozygous individuals possess two alleles descended from the different ancestral alleles in the base population. Figure 13.2 illustrates the relationship of the concepts of autozygosity, allozygosity, homozygosity, and heterozygosity.

Is it really necessary to introduce these two new terms? Yes. Autozygosity and allozygosity are related to homozygosity and heterozygosity, but they refer to the descent of alleles through Mendelian inheritance rather than the molecular state of the allele in question. This distinction is important when considering the effects of pedigree inbreeding on individuals. We assume that the founding individuals in a pedigree are allozygous for two unique alleles. However, these allozygotes may either be homozygous or heterozygous at a particular locus, depending upon whether the two alleles are identical in state or not. For example, an allozygote would be homozygous if it had two alleles that are identical in DNA sequence.

We can see this using Figure 12.1. An individual with one copy of the *10A* allele and one copy of the *10C* allele (i.e., *10A/10C*) would be homozygous in state for 10 repeats, but would be allozygous. In contrast, an individual with two copies of the *10B* allele (*10B/10B*) would be homozygous and autozygous.

13.1.1 Estimation of the pedigree inbreeding coefficient

Several methods are available for calculating the pedigree inbreeding coefficient. We will use the method of path analysis developed by Sewall Wright (1922). Figure 13.3 shows the pedigree of an inbred individual X. By convention, females are represented by circles, and males are represented by squares in pedigrees. Diamonds are used either to represent individuals whose sex is not unknown or to represent individuals whose sex is not of concern.

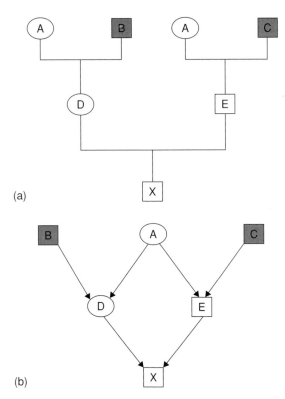

(a)

(b)

Figure 13.3 Calculation of the inbreeding coefficient for individual X using path analysis. (a) Conventional representation of pedigree for an individual whose mother and father had the same mother. (b) Path diagram to represent this pedigree to calculate the inbreeding coefficient. Shaded individuals in (a) need not be included in (b) because they are not part of the path through the common ancestor (individual A) and therefore do not contribute to the inbreeding of individual X.

What is the inbreeding coefficient of individual X in Figure 13.3a? The first step is to draw the pedigree as shown in Figure 13.3b so that each individual appears only once. Next we examine the pedigree for individuals who are ancestors of both the mother and father of X. If there are no such common ancestors, then X is not inbred, and $F_X = 0$. In this case, there is one common ancestor, individual A. Next, we trace all of the paths that lead from one of X's parents, through the common ancestor, and then back again to the other parent of X. There is only one such path in Figure 13.3 (D$\underline{A}$E); it is helpful to keep track of the common ancestor by underlining.

The inbreeding coefficient of an individual can be calculated by determining N, the number of individuals in the loop (not including the individual of concern) containing the common ancestor of the parents of an inbred individual. If there is a single loop then:

$$F = (1/2)^N(1 + F_{CA})$$

(13.1)

where F_{CA} is the inbreeding coefficient of the common ancestor. The term $(1 + F_{CA})$ is included because the probability of a common ancestor passing on the same allele to two offspring is increased if the common ancestor is inbred. For example, if the inbreeding coefficient of an individual is 1.0, then it will always pass on the same allele to two progeny. If there is more than one loop, then the inbreeding coefficient is the sum of the F values from the separate loops:

$$F = \sum [(1/2)^N(1 + F_{CA})]$$

(13.2)

In the present case (Figure 13.3), there is only one loop with $N = 3$ and the common ancestor (A) is not inbred, therefore:

$$F_X = (1/2)^3(1 + 0) = 0.125$$

This means that individual X is expected to be identical by descent (IBD) at 12.5% of his or her loci. Or, stated another way, the expected heterozygosity of individual X is expected to be reduced by 12.5%, compared to individuals in the base population. See Example 13.2 for calculating F when the common ancestor is inbred ($F_{CA} > 0$).

Figure 13.5 shows a complicated pedigree obtained from a long-term population study of the great tit in the Netherlands (van Noordwijk and Scharloo 1981). They have shown that the hatching of eggs is reduced by approximately 7.5% for every 10% increase in F. Ten different loops contribute to the inbreeding of the individual under investigation (Table 13.1). The total inbreeding coefficient of this individual is 0.1445.

13.2 Gene drop analysis

The pedigree analysis in the previous section provides an estimate of the increase in homozygosity and reduction in heterozygosity due to inbreeding. However, as we have seen in previous chapters, we are also interested in the loss of allelic diversity, as well as

Example 13.2 Calculating pedigree F

Figure 13.4 shows a pedigree in which a common ancestor of an inbred individual is inbred. What is the inbreeding coefficient of individual K in this figure?

There is one loop that contains a common ancestor of both parents of K (I $\underline{G}$ J). Therefore, using expression 13.2, $F_K = (1/2)^3(1 + F_G)$. The common ancestor in this loop, G, is also inbred; there is one loop with three individuals through a common ancestor for individual G (D $\underline{B}$ E). Therefore, $F_G = (1/2)^3(1 + F_B) = 0.125$. Individual B is not inbred ($F_B = 0$) since she is a founder in this pedigree. Therefore, $F_G = 0.125$, and $F_K = (1/2)^3(1.125) = 0.141$.

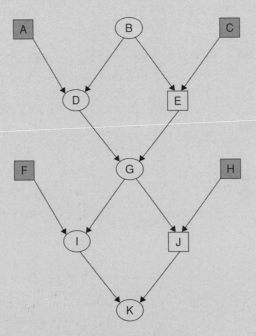

Figure 13.4 Hypothetical pedigree in which the common ancestor (G) of an inbred individual's (K) parents is also inbred. See Example 13.1. Shaded individuals are not part of the path through either of the common ancestors (individuals B and G) and therefore do not contribute to the inbreeding of individuals G and K.

heterozygosity. A simple computer simulation procedure called **gene drop** analysis has been developed for more detailed pedigree analysis (MacCluer et al. 1986).

In this procedure, two unique alleles are assigned to each individual in the base population. Monte Carlo simulation methods are used to assign a genotype to each progeny based upon its parents' genotypes and Mendelian inheritance (Figure 13.6). This procedure is followed throughout the pedigree until each individual is assigned a genotype. This

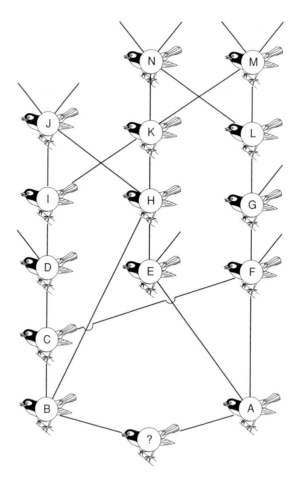

Figure 13.5 Complicated pedigree from a population of great tits in the Netherlands. The inbreeding coefficient of the bottom individual (?) is 0.1445 (see Table 13.1). From van Noordwijk and Scharloo (1981).

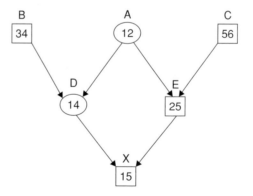

Figure 13.6 Pedigree from Figure 13.3 showing one possible outcome of gene dropping in which genotypes are assigned to descendants by Monte Carlo simulation of Mendelian segregation beginning with two unique alleles in each founder.

Table 13.1 Calculation of inbreeding coefficient of the individual (?) at the bottom of the pedigree shown in Figure 13.5.

Path	Length (N)	Common ancestor	F_{CA}	$(1/2)^N(1 + F_{CA})$
BC**F**A	4	F	0	0.0625
B**H**EA	4	H	0	0.0625
BCDI**K**HEA	8	K	0	0.0039
BCDI**J**HEA	8	J	0	0.0039
BHK**M**LGFA	8	M	0	0.0039
BHK**N**LGFA	8	N	0	0.0039
BCDIK**N**LGFA	10	N	0	0.0010
BCDIK**M**LGFA	10	M	0	0.0010
BCFGL**M**KHEA	10	M	0	0.0010
BCFGL**N**KHEA	10	N	0	0.0010
Total	–	–	–	0.1445

simulation is then repeated many times (say 10,000). Analysis of the genotypes in individuals of interest can provide information about the expected inbreeding coefficient, decline in heterozygosity, expected loss of allelic diversity, and many other characteristics that may be of interest (Example 13.3).

13.3 Estimation of *F* and relatedness with molecular markers

To understand the effect of inbreeding on fitness in natural populations it is necessary to know the inbreeding coefficient of individuals. Unfortunately, it is extremely difficult to estimate inbreeding coefficients in the field because pedigrees usually are not available. Pedigrees of wild populations are likely to be only a few generations deep, to have gaps, and be inaccurate. However, an individual's inbreeding coefficient can be estimated from the degree of homozygosity at molecular markers of its genome relative to the genomes of other individuals within the same population.

We saw in Section 13.1 that the pedigree inbreeding coefficient, *F*, is the expected increase in homozygosity due to identity by descent. For example, the offspring of a full-sib mating will have only 75% of the heterozygosity in their parents (Table 13.2). Offspring produced by half-sib matings will have only 87.5% of the heterozygosity observed in their parents. Therefore, we expect individual heterozygosity (*H*) at many loci to be reduced by a value of *F*. Pedigree *F* can be estimated by comparison of multilocus heterozygosity of individuals over many loci.

Individual inbreeding has been estimated using molecular markers in a wolf population from Scandinavia (Ellegren 1999; Hedrick et al. 2001). Twenty-nine microsatellite loci were examined in captive gray wolves for which the complete pedigree was known (Figure 13.8). The distribution of individual heterozygosity ranged from about 0.20 to 0.80. The pedigree inbreeding coefficient was significantly correlated with heterozygosity

Example 13.3 Gene drop analysis

Haig et al. (1990) have presented an example of gene drop analysis in their consideration of the effect of different captive breeding options on genetic variation in guam rails (Figure 13.7). The upper part of the figure shows a sample pedigree in which three founders (A, B, C) are given six unique alleles. The lower part of the figure shows one result of a gene drop simulation. Two of the birds (G and I) are heterozygous; the heterozygosity has thus declined 50%. Only three of the initial six alleles are present in the living population (1, 4, and 6); thus 50% of the alleles have been lost. The proportional representation of each of the founders can also be calculated from this result. Four of the eight genes are descended from A (50%); one of the eight from B (12.5%); and three of the eight genes from C (37.5%). This simulation would be repeated 10,000 times to get statistical estimates of these parameters.

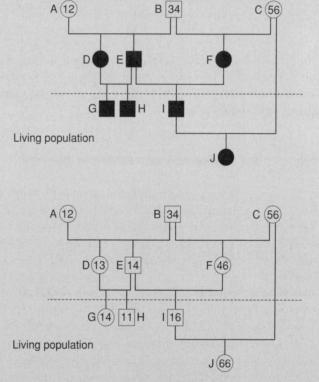

Figure 13.7 Example of a gene drop analysis in a hypothetical captive population with three founders (see Example 13.2). From Haig et al. (1990).

Table 13.2 Expected decline of genome-wide heterozygosity with different modes of inbreeding (modified from Dudash and Fenster 2000). The rate of loss per generation increases with the relatedness of parents. With selfing, 50% of the heterozygosity is lost per generation because one-half of the offspring from a heterozygote (*Aa*) will be homozygotes (following the 1 : 2 : 1 Mendelian ratios of 1 *AA* : 2 *Aa* : 1 *aa*).

	Mode of inbreeding		
Generation	Half-sib	Full-sib	Selfing
0	1.000	1.000	1.000
1	0.875	0.750	0.500
2	0.781	0.625	0.250
3	0.695	0.500	0.125
4	0.619	0.406	0.062
5	0.552	0.328	0.031
10	0.308	0.114	0.008

(Figure 13.9; $r^2 = 0.52, P < 0.001$). Thus, the 29 microsatellite loci are an accurate indicator of individual inbreeding level. Precise estimates of inbreeding coefficients using heterozygosity requires many loci because the variance in heterozygosity estimates is large (i.e., confidence intervals are wide; Pemberton 2004).

13.4 Causes of inbreeding depression

Inbreeding depression may result from either increased homozygosity or reduced heterozygosity (Crow 1948). Increased homozygosity leads to the expression of a greater number of deleterious recessive alleles in inbred individuals, thereby lowering their fitness. Reduced heterozygosity reduces the fitness of inbred individuals at loci where the heterozygotes have a selective advantage over homozygotes (heterozygous advantage or overdominance; see Section 8.2).

The probability of being homozygous for rare deleterious alleles increases surprisingly rapidly with inbreeding. Consider a recessive lethal allele at a frequency of 0.10 (q) and that the average inbreeding coefficient in a population is 10% (F). The proportion of heterozygotes will be reduced by 10% and each of the homozygotes will be increased by half of that amount (see Section 9.1):

	AA	*Aa*	*aa*
Expected	$p^2 + pqF$	$2pq - 2pqF$	$q^2 + pqF$
$F = 0$	0.810	0.180	0.010
$F = 0.10$	0.819	0.162	0.019

Thus, the expected proportion of individuals to be affected by this deleterious allele (*a*) will nearly double with just a 10% increase in inbreeding.

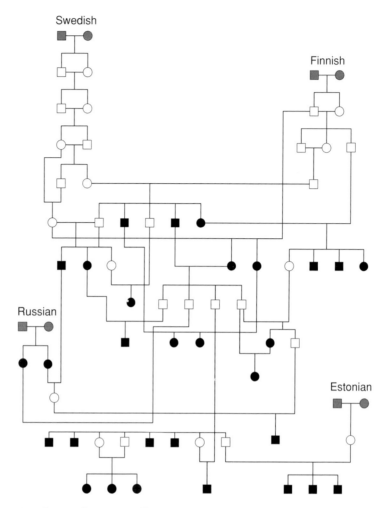

Figure 13.8 Pedigree of captive wolf population. The black symbols are those animals included in the study evaluating the use of 29 microsatellite loci to estimate inbreeding coefficient. The gray individuals are the four founder pairs (assumed $F = 0$) from four countries. From Hedrick et al. (2001).

It is crucial to know the mechanisms causing inbreeding depression because it affects the ability of a population to "adapt" to inbreeding. A population could adapt to inbreeding if inbreeding depression is caused by deleterious recessive alleles that potentially could be removed (purged by selection). However, inbreeding depression caused by heterozygous advantage cannot be purged because overdominant loci will always suffer reduced fitness as homozygosity increases due to increased inbreeding.

Both increased homozygosity and decreased heterozygosity are likely to contribute to inbreeding depression, but it is thought that increased expression of deleterious recessive alleles is the more important mechanism (Charlesworth and Charlesworth 1987; Ritland 1996; Carr and Dudash 2003). For example, Remington and O'Malley (2000) performed

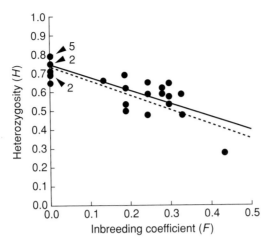

Figure 13.9 Relationship between individual heterozygosity (*H*) at 29 microsatellite loci and inbreeding coefficient (*F*) in a captive wolf population. The solid line represents the regression of *H* on *F*, the dashed line is the expected relationship between *H* and *F*, assuming an *H* of 0.75 in noninbred individuals. From Ellegren (1999).

a genome-wide evaluation of inbreeding depression caused by selfing during embryonic viability in loblolly pines. Nineteen loci were found that contributed to inbreeding depression. Sixteen loci showed predominantly recessive action. Evidence for heterozygous advantage was found at three loci.

13.5 Measurement of inbreeding depression

Some inbreeding depression is expected in all species (Hedrick and Kalinowski 2000; see Guest Box 13). Deleterious recessive alleles are present in the genome of all species because they are continually introduced by mutation, and natural selection is inefficient in removing them because most copies are "hidden" phenotypically in heterozygotes that do not have reduced fitness (see Chapter 12). We, therefore, expect all species to show some inbreeding depression due to the increase in homozygosity of recessive deleterious alleles. For example, Figure 13.10 shows inbreeding depression for infant survival in a captive population of callimico monkeys.

13.5.1 Lethal equivalents

The effects of inbreeding depression on survival are often measured by the mean number of "**lethal equivalents**" (**LEs**) per diploid genome. A lethal equivalent is a set of deleterious alleles that would cause death if homozygous. Thus, one lethal equivalent may either be a single allele that is lethal when homozygous, two alleles each with a probability of 0.5 of causing death when homozygous, or 10 alleles each with a probability of 0.10 of causing death when homozygous.

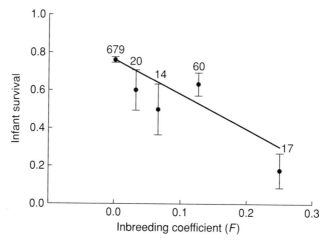

Figure 13.10 Relationship between inbreeding and infant survival in captive callimico monkeys. Callimico show a 33% reduction in survival resulting from each 10% increase in inbreeding ($P < 0.001$). Data are from 790 captive-born callimico, 111 of which are inbred. The numbers above the bars are the number of individuals studied at each inbreeding level. From Lacy et al. (1993, p. 367).

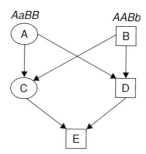

Figure 13.11 Effect of a single lethal equivalent per diploid genome on survival of inbred progeny produced by a mating between full-sibs ($F = 0.25$). The probability of E **not** being homozygous for a recessive allele inherited from either A or B is 0.879. Therefore, one LE per diploid genome will result in approximately a 12% reduction in survival of individuals with an F of 0.25. See text for explanation.

   We can see the effect of 1 LE in the example of a mating between full-sibs that will produce a progeny with an F of 0.25 (Figure 13.11). Individuals A and B each carry one lethal allele (a and b, respectively). The probability of individual E being homozygous for the a allele is $(1/2)^4 = 1/16$; similarly, there is a $1/16$ probability that individual E will be homozygous bb. Thus, the probability of E **not** being homozygous for a recessive allele at either of these two loci is $(15/16)(15/16) = 0.879$. Thus, 1 LE per diploid genome will result in approximately a 12% reduction ($1 - 0.879$) in survival of individuals with an F of 0.25.
   The number of LEs present in a species or population is generally estimated by regressing survival on the inbreeding coefficient (Figure 13.12). The effects of inbreeding on the probability of survival, S, can be expressed as a function of F (Morton et al. 1956):

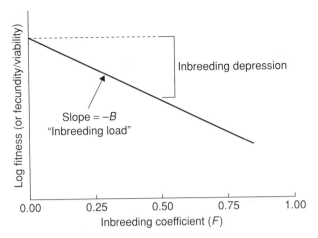

Figure 13.12 Relationship between inbreeding coefficient and reduction in fitness. Inbreeding depression is the reduction in fitness of inbred individuals and is measured by the number of lethal equivalents per gamete (B). Redrawn from Keller and Waller (2002).

$$S = e^{-(A+BF)}$$

$$\ln S = -A - BF$$

(13.3)

where e^{-A} is survival in an outbred population and B is the rate at which fitness declines with inbreeding (Hedrick and Miller 1992). B is the reduction in survival expected in a completely homozygous individual. Therefore, B is the number of LEs per gamete, and $2B$ is the number of LEs per diploid individual. B is estimated by the slope of the weighted regression of the natural log of survival on F. The callimico monkeys shown in Figure 13.10 have an estimated 7.90 LEs per individual ($B = 3.95$) (Lacy et al. 1993).

13.5.2 Estimates of inbreeding depression

The range of LEs per individual estimated for captive mammal populations varies from about 0 to 30. The median number of LEs per diploid individual for captive mammals was estimated to be 3.14 (see Figure 6.13; Ralls et al. 1988). This corresponds to about a 33% reduction of juvenile survival, on average, for offspring with an inbreeding coefficient of 0.25 (Figure 13.13). This value underestimates the magnitude of inbreeding depression in natural populations because it only includes the reduction of fitness for one life history stage (juvenile survival and not adult survival, embryonic survival, fertility, etc.), and because captive environments are less stressful than natural environments, and stress typically increases inbreeding depression (see below).

There are fewer estimates of the number of LEs using pedigree analysis in plant species. Most studies of inbreeding depression in plants compare selfed and outcrossed progeny from the same plants (see Example 13.1). In this situation, inbreeding depression is usually

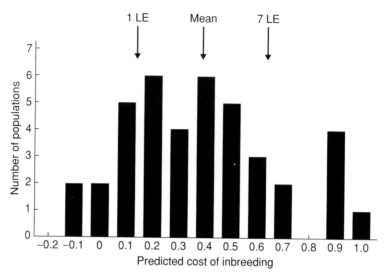

Figure 13.13 Distribution of the estimated cost of inbreeding in progeny with an inbreeding coefficient (*F*) of 0.25 in 40 captive mammal populations. Cost is the proportional reduction in juvenile survival; mean LE = 3.14. From Ralls et al. (1988).

measured as the proportional reduction in fitness in selfed versus outcrossed progeny ($\delta = 1 - (w_s/w_o)$). These can be converted by:

$$\delta = 1 - \frac{w_s}{w_o} = 1 - e^{-B/2} \tag{13.4}$$

and

$$B = -2\,\ln(1 - \delta) \tag{13.5}$$

Some plant species show a tremendous amount of inbreeding depression. For example, Figure 13.14 shows estimates of inbreeding depression for embryonic survival in 35 individual Douglas fir trees based upon comparison of the production of sound seed by selfing and crossing with pollen from unrelated trees. On average, each tree contained approximately 10 LEs. This is equivalent to over a 90% reduction in embryonic survival! Perhaps more interesting, however, is the wide range of LEs in different trees (Figure 13.14). Conifers in general seem to have high inbreeding depression, but inbreeding depression is especially great in Douglas fir (Sorensen 1999).

Most studies of inbreeding depression have been made in captivity or under controlled conditions, but experiments with both plants (e.g., Dudash 1990) and animals (e.g., Jiménez et al. 1994) have found that inbreeding depression is more severe in natural environments (Example 13.4). Estimates of inbreeding depression in captivity may be severe underestimates of the true effect of inbreeding in the wild (but not always, see Armbruster et al. 2000 and Example 13.1).

Crnokrak and Roff (1999) have recently reviewed the available empirical literature of inbreeding depression for wild species. They concluded that in general "the cost of inbreeding under natural conditions is much higher than under captive conditions". They tested this for mammals by comparing traits directly related to survival in wild mammals

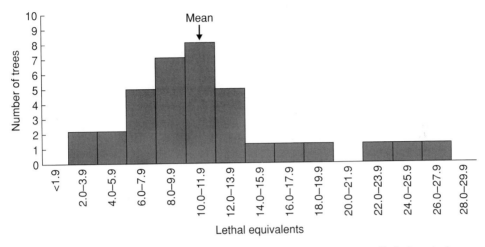

Figure 13.14 Inbreeding depression measured by the observed number of lethal equivalents for embryonic survival in 35 individual Douglas fir trees based upon comparison of the set of sound seed by selfing and crossing with pollen from unrelated trees. Redrawn from Sorensen (1969).

Example 13.4 Inbreeding depression in the wild

Meagher et al. (2000) compared inbreeding depression in house mice in the laboratory and in seminatural enclosures. Inbreeding ($F = 0.25$) caused an 11% reduction in litter size, but had no effect on birth-to-weaning survival in the laboratory. Inbred mice had a 52% reduction in mean fitness for adult mice in seminatural enclosures (81% for males and 22% for females). These two fitness components taken together resulted in a cumulative inbreeding depression (see Figure 13.1) of 57%:

$$1 - \left[(0.89)\left(\frac{(0.19 + 0.78)}{2} \right) \right] = 0.57$$

This fitness decline is over four times as great as previous estimates in the house mouse measured entirely in the laboratory. The primary difference was the greatly reduced fitness of inbred adult mice in the seminatural enclosures (Figure 13.15).

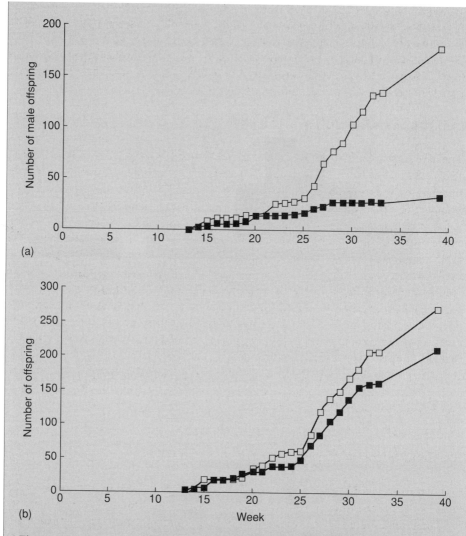

Figure 13.15 Total cumulative number of weaned progeny from inbred ($F = 0.25$; solid squares) and outbred mice ($F = 0$; open squares) in six seminatural enclosures. Equal numbers of inbred and outbred mice were placed in each of six seminatural enclosures. (a) Inbred males sired significantly fewer offspring in all six enclosures. (b) Inbred females produced significantly fewer offspring in three of six enclosures. From Meagher et al. (2000).

to the findings of Ralls et al. (1988) for captive species. They found that the cost of inbreeding on survival was much higher in wild than in captive mammals.

In addition, there is evidence that inbreeding depression is more severe under environmental stress and challenge-events (e.g., extreme weather, pollution, or disease) (Armbruster and Reed 2005). Bijlsma et al. (1997) found a synergistic interaction between stress and inbreeding with laboratory *Drosophila* so that the effect of environmental stress is greatly enhanced with greater inbreeding. These conditions may occur only occasionally, and thus it will be difficult to measure their effect on inbreeding depression. For example, Coltman et al. (1999) found that individual Soay sheep that were more heterozygous at 14 microsatellite loci had greater overwinter survival in harsh winters, apparently due to greater resistance to nematode parasites. In addition, this effect disappeared when the sheep were treated with antihelminthics (Figure 13.16).

13.5.3 Are there species without inbreeding depression?

Some have suggested that some species or populations are unaffected by inbreeding (e.g., Shields 1993). However, lack of statistical evidence for inbreeding depression does not demonstrate the absence of inbreeding depression. This is especially true because of the low power to detect even a substantial effect of inbreeding in many studies because of small sample sizes and confounding factors (Kalinowski and Hedrick 1999). In a comprehensive review of the evidence for inbreeding depression in mammals, Lacy (1997) was unable to find "statistically defensible evidence showing that any mammal species is unaffected by inbreeding".

Measuring inbreeding depression in the wild is extremely difficult for several reasons. First, pedigrees are generally not available and the alternative molecular-based estimates might not be precise (see Section 13.3). Thus power to detect inbreeding depression is low in most studies. Second, fitness is difficult to measure in the wild. Inbreeding depression might occur only in certain life history stages (e.g., zygotic survival or life time reproductive success) or under certain stressful conditions (e.g., severe winters or high predator density). Most studies in wild populations do not span enough years or life history stages to reliably measure inbreeding depression.

13.6 Genetic load and purging

Some have suggested that small populations may be "**purged**" of deleterious recessive alleles by natural selection (Templeton and Read 1984). Deleterious recessive alleles may reach substantial frequencies in large random mating populations because most copies are present in heterozygotes and are therefore not affected by natural selection. For example, over 5% of the individuals in a population in Hardy–Weinberg proportions will be heterozygous for an allele that is homozygous in only one out of 1,000 individuals. Such alleles will be exposed to natural selection in inbred or small populations and will thereby be reduced in frequency or eliminated. Thus, populations with a history of inbreeding because of nonrandom mating (e.g., selfing) or small N_e (e.g., a population bottleneck) may be less affected by inbreeding depression because of the purging of deleterious recessive alleles.

Reduced differences in fitness between inbred and noninbred individuals within a population that has gone through a bottleneck is not evidence for purging (Figure 13.17). Many nonlethal deleterious alleles may become fixed in such populations. The fixation of these alleles will cause a reduction in fitness of all individuals following the bottleneck relative to the individuals in the population before the bottleneck (so-called "**genetic load**"). However, inbreeding depression will appear to be reduced following fixation of deleterious

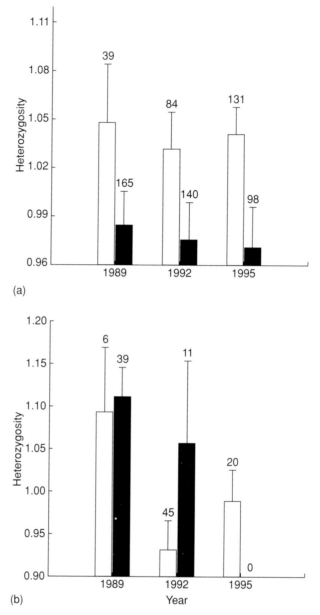

(a)

(b)

Year

Figure 13.16 Relative observed individual heterozygosity of Soay sheep during three severe winters with high mortality. (a) Sheep that died (black bars) had far lower heterozygosity than sheep that lived (open bars), when not treated with antihelminthics. (b) However, when treated with antihelminthics (to remove intestinal parasites), no difference in survival was detected between inbred (black bars) and outbred (open bars) individuals. Parasite load was higher in inbred individuals (with low heterozygosity), leading to their increased mortality during the stress of severe winters. A standardized relative heterozygosity was used because not all individuals were genotyped at all loci such that $H =$ (proportion heterozygous typed loci/mean heterozygosity at typed loci). Numbers above bars indicate sample size. From Coltman et al. (1999).

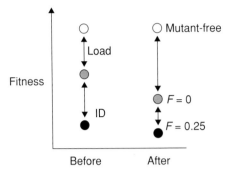

Figure 13.17 Diagram showing decreased inbreeding depression (ID) in a population before and after a bottleneck in which all of the inbreeding depression is due to increased homozygosity for deleterious recessive mutant alleles. The open circles show the average fitness of hypothetical "mutant-free" individuals that have no deleterious alleles. The shaded circles show the average fitness of individuals produced by random mating. The dark circles show the average fitness of individuals with an F of 0.25. This illustrates that reduced inbreeding depression following a bottleneck can be caused by an increase in the fixed genetic load rather than by purging.

alleles because of depressed fitness of outbred individuals ($F = 0$) rather than by increased inbred fitness (**fitness rebound**) (Byers and Waller 1999). To test for purging, the fitness of individuals in the post-bottleneck population must be compared with the fitness of individuals in the pre-bottleneck population. Alternatively, we might test for purging by comparing the fitness of offspring from resident (relatively inbred) individuals versus outbred offspring of crosses between residents and individuals from the pre-bottleneck population.

A review found little evidence for purging in plant populations (Byers and Waller 1999). Only 38% of the 52 studies included found evidence of purging. And when purging was found, it removed only a small proportion of the total inbreeding depression (roughly 10%). These authors concluded that "purging appears neither consistent nor effective enough to reliably reduce inbreeding depression in small and inbred populations" (Byers and Waller 1999).

A review of evidence for purging in animals came to similar conclusions. Ballou (1997) found evidence for a slight decline in inbreeding depression in neonatal survival among descendants of inbred animals in a comparison of 17 captive mammal species. However, he found no indication for purging in weaning survival or litter size in these species. He concluded that the purging detected in these species is not likely to be strong enough to be of practical use in eliminating inbreeding depression in populations of conservation interest. In addition, inbreeding depression can be substantial in both Hymenoptera (Antolin 1999) and mites (Saito et al. 2000) in which males are haploid so that deleterious recessive alleles are exposed to natural selection in males every generation and therefore should be purged relatively efficiently.

Failure of purging to decrease inbreeding depression can be explained by several mechanisms. First, purging is expected to be most effective in the case of lethal or semilethal recessive alleles (Lande and Schemske 1985; Hedrick 1994); however, when inbreeding depression is very high (more than 10 lethal equivalents), even lethals may not be purged except under very close inbreeding (Lande 1994). Second, the lack of evidence for purging is consistent with the hypothesis that a substantial proportion of inbreeding depression is caused by many recessive alleles with minor deleterious effects. Alleles with minor effects are unlikely to be purged by selection, because selection cannot efficiently target harmful alleles when they are spread across many different loci and different individuals. Third, it is not possible to purge the genetic load at loci under heterozygote advantage (**heterotic** loci), as mentioned above. On the contrary, the loss of alleles at heterotic loci generally reduces heterozygosity and thus fitness.

Husband and Schemske (1996) found that inbreeding depression for survival after early development and reproduction and growth was similar in selfing and nonselfing plant species (also see Ritland 1996). They suggested that this inbreeding depression is due primarily to mildly deleterious mutations that are not purged, even over long periods of time. Willis (1999) found that most inbreeding depression in the monkeyflower (see Example 13.1) is due to alleles with small effect, and not to lethal or sterile alleles. Bijlsma et al. (1999) have found that purging in experimental populations of *Drosophila* is effective only in the environment in which the purging occurred because additional deleterious alleles were expressed when environmental conditions changed.

Ballou (1997) suggested that **associative overdominance** may also be instrumental in maintaining inbreeding depression. Associative overdominance occurs when heterozygous advantage or deleterious recessive alleles at a selected locus results in apparent heterozygous advantage at linked loci (see Section 10.3.2; Pamilo and Pálsson 1998). Kärkkäinen et al. (1999) have provided evidence that most of the inbreeding depression in the self-incompatible perennial herb *Arabis petraea* is due to overdominance or associative overdominance.

Inbreeding depression due to heterozygous advantage cannot be purged. However, it is unlikely that heterozygous advantage is a major mechanism for inbreeding depression (Charlesworth and Charlesworth 1987). Nevertheless, there is recent strong evidence for heterozygous advantage at the major histocompatibility complex (MHC) in humans. Black and Hedrick (1997) found evidence for strong heterozygous advantage (nearly 50%) at both *HLA-A* and *HLA-B* in South Amerindians. Carrington et al. (1999) found that heterozygosity at *HLA-A*, *-B*, and *-C* loci was associated with extended survival of patients infected with the human immunodeficiency virus (HIV). Strong evidence for the selective maintenance of MHC diversity in vertebrate species comes from other approaches as well (see references in Carrington et al. 1999).

Speke's gazelle has been cited as an example of the effectiveness of purging in reducing inbreeding depression in captivity (Templeton and Read 1984). Willis and Wiese (1997), however, have concluded that this apparent purging may have been due to the data analysis rather than purging itself; this interpretation has been disputed by Templeton and Read (1998). Ballou (1997), in his reanalysis of the Speke's gazelle data, found that purging effects were minimal and nonsignificant. Perhaps most importantly, he found that the inbreeding effects in Speke's gazelle were the greatest in any of the 17 mammal species he examined. Kalinowski et al. (2000) recently have concluded that the apparent purging in Speke's gazelle is the result of a temporal change in fitness and not a reduction in inbreeding depression (see response by Templeton 2002).

The degree of inbreeding depression resulting from inbreeding or bottlenecks is difficult to predict in any given species for several reasons. We saw in Figure 13.14 that there may be great differences in the numbers of LEs within individuals from the same population. Moreover, the alleles across the genome that survive a bottleneck will depend on random sampling effects. For example, Lacy and Ballou (1998) found different magnitudes of inbreeding depression in different sets of individuals sampled from the same population of beach mice (see Guest Box 13; Wade et al. 1996). Second, mice populations with a history of inbreeding and bottlenecks were expected to suffer less from inbreeding depression. But instead these populations still suffered substantial inbreeding depression relative to other mice populations. In addition, they experienced less purging (over 10 generations in captivity) than other populations, probably because historical purging had already removed the "purgable" lethal and semilethal alleles.

Even if a population's history of inbreeding is known, it can be difficult to predict the cost of inbreeding on fitness (see Guest Box 13). The difficulty of predicting the magnitude of inbreeding depression makes it difficult for managers to design specific management strategies to minimize inbreeding depression, even when a population's history and biology is well known. Managers must simply consider that substantial inbreeding depression is possible in any population, especially under changing environments or stressful conditions.

Guest Box 13 Understanding inbreeding depression: 20 years of experiments with *Peromyscus* mice
Robert C. Lacy

At first glance, inbreeding depression would seem to be well understood scientifically. It is a widespread consequence of matings between close relatives, and there is a simple mechanistic explanation – increased expression of recessive deleterious alleles in inbred individuals. However, when conservation geneticists began looking more closely at inbreeding depression, in order to make predictions about the vulnerability of populations and to provide sound management advice, the picture became much less clear.

For almost 20 years, my colleagues at the Brookfield Zoo and I have been studying the effects of inbreeding in some *Peromyscus* mice (white-footed mice, old-field mice, and beach mice). We started with simple hypotheses and simple expectations, but have been led to conclude that inbreeding depression is a complex phenomenon that defies easy prediction. We first sought to show that populations of mice that were long isolated on small islands had already been purged of their deleterious alleles, leaving them able to inbreed with minimal further impact. Instead, we found that the island mice had low reproductive performance prior to inbreeding in the lab, and their fitness declined as fast or faster than in mice from more genetically diverse mainland populations when we forced inbred matings (Brewer et al. 1990).

We then tested whether we would get the same results if we repeated our measures of inbreeding depression on replicate laboratory stocks, each derived from the same wild populations. We found that different breeding stocks of a species could show very different effects of inbreeding (Lacy et al. 1996). The damaging effects of inbreeding showed up in different traits (e.g., in litter size versus pup survival versus growth rates), and to different extents. We found that the inbreeding depression was due largely to effects in the inbred descendants of some founders, while descendants of other founder pairs seemed to have little problem with inbreeding. This suggests that the effects of inbreeding are due mostly to a few alleles, and which animals carry these alleles is largely a matter of chance. This situation might provide a good opportunity for selection to be effective at purging those deleterious alleles, when they are expressed in inbred homozygotes. When we examined our data to see if inbreeding depression did become less through the generations of experimental inbreeding (Lacy and Ballou 1998), we found purging in some subspecies but not in others (Figure 13.18). Thus, it may be that the cause of inbreeding depression (recessive alleles versus heterozygous advantage or associative overdominance, few versus many loci, unconditional effects versus environmentally dependent effects) varies among natural populations, even of the same species.

While we cannot predict the specific effects of inbreeding for any one or a few lineages, at a broader scale, inbreeding depression may be more predictable. When averaged across many lineages, and assessed as an impact on overall fitness (considering the cumulative effect on mating propensity, litter size, and survival), inbreeding depression is very consistent across three subspecies of mice, as measured in our laboratory environment. Unfortunately, many studies of inbreeding depression have measured only one or a few components of fitness, and on unreplicated populations, with small sample sizes. Thus, we still do not know which species are most susceptible to inbreeding depression.

Conservation geneticists are now in the unenviable position of knowing that inbreeding depression can be a serious problem in small populations, but not being able to make accurate predictions about how severe the effects will be, or if they can be reduced or managed through selection programs. And, this is not a problem that can be solved simply with more data because of potential differences in the effects of inbreeding between different populations of the same species. Therefore, having good data on the effects of inbreeding in one population of a species may not necessarily be a reliable predictor of the effects of inbreeding in other populations of the same species. In addition, the effects of inbreeding within one population are likely to differ greatly depending on which individuals become the founders of the inbred lineage.

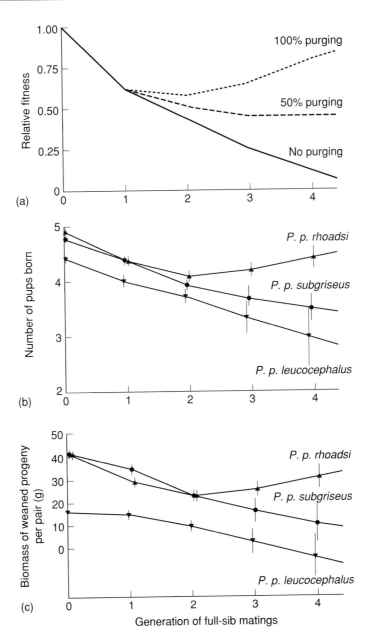

Figure 13.18 (a) Expected response of fitness to repeated generations of inbreeding, if the genetic load is due entirely to recessive lethal alleles (100% purging), over-dominance (no purging), or half of each. (b and c) Observed responses to inbreeding in three subspecies of *Peromyscus polionotus* mice, measured as the depression in initial litter size and biomass of progeny weaned within 63 days, as predicted from regressions on inbreeding levels of litters and of prior generations. From Lacy and Ballou (1998).

Problem 13.1

What is meant by the term "inbreeding" and why is it important to define the term "inbreeding" when using it?

Problem 13.2

In what sense are we all inbred?

Problem 13.3

Explain how inbreeding occurs even in random mating populations.

Problem 13.4

How can an individual that is autozygous (IBD) be heterozygous?

Problem 13.5

Identify all inbred individuals in Figure 13.5 and calculate their inbreeding coefficients.

Problem 13.6

How is inbreeding depression measured?

Problem 13.7

We saw in Example 13.1 that there was a 66.5% mean reduction in fitness as measured by seed production in two populations of monkeyflowers ($\delta = 0.665$). How many lethal equivalents are there per diploid individual in these two populations? How does this compare to the amount of inbreeding depression estimated in mammals?

Problem 13.8

What are the two primary genetic mechanisms that cause inbreeding depression? Which mechanism is thought to be the most common cause of inbreeding depression?

Problem 13.9

What is meant by the term "purging"?

Problem 13.10

Which of the two mechanisms that cause inbreeding depression does not allow for purging? Why is purging impossible at loci evolving under this mechanism?

Problem 13.11

Why are mildly deleterious alleles difficult to purge? Hint: why are lethal alleles easily purged?

Problem 13.12

The pedigree below is from a song sparrow population on Mandarte Island in western Canada that Keller (1998) has found to carry an average of 5 LEs per diploid genome. Identify all inbred individuals in the pedigree below and calculate their inbreeding coefficients (from Keller and Waller 2002).

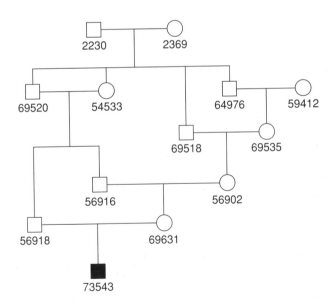

Problem 13.13

The following pedigree appeared in one of the original papers on inbreeding (from Wright 1923). Calculate the inbreeding coefficient for Favourite 252, the foundation bull for all Shorthorn cattle. Note: the upper parent in each case is the bull and the lower parent is the cow; and Studley Bull 626 is sometimes listed just as 626.

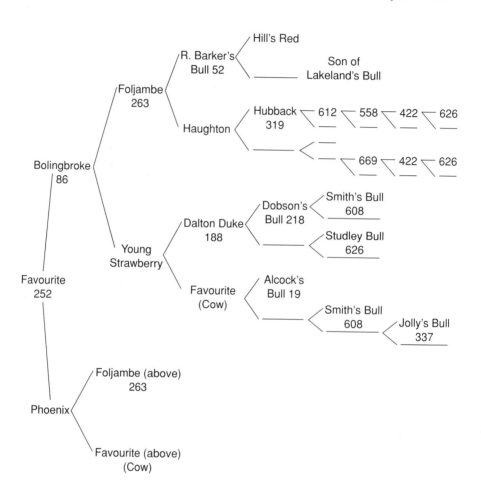

Problem 13.14

Assume that individuals F_1 to F_4 opposite are the founders of a captive breeding program for larlarlongs, an extraterrestrial species native to the planet Tralfalmadore. Use coin-flips to derive one possible outcome of a gene drop analysis for this pedigree. Assume that individuals A–D are the only surviving larlarlongs in the universe. What are the allele frequencies for individuals A–D based upon the outcome of your single gene drop? What are the observed and expected

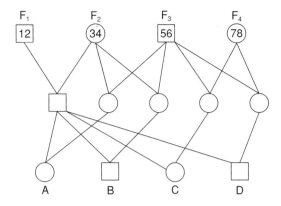

heterozygosities for individuals A–D based upon the outcome of your single gene drop? If you performed 10,000 such gene drops, what do you think the average observed heterozygosity would be for individuals A–D? How many of the eight original alleles remain in the captive population of larlarlongs based upon your single gene drop outcome?

Problem 13.15

A recent paper has come to the surprising conclusion that the most recent common ancestor (MRCA) of all humans present on the Earth today probably lived within the last few thousand years (Rohde et al. 2004). What are the crucial values that you would need to know in order to estimate the time since the MRCA for any species?

14

Demography and Extinction

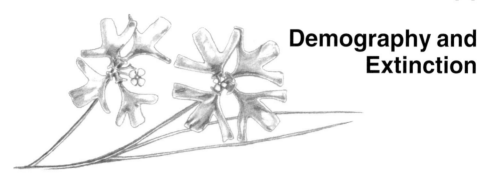

Clarkia pulchella, *Section 14.2*

*As some of our British parks are ancient, it occurred to me that there must have been long-continued close interbreeding with the fallow-deer (*Cervus dama*) kept in them; but on inquiry I find that it is a common practice to infuse new blood by procuring bucks from other parks.*

Charles Darwin (1896, p. 99)

What are the minimum conditions for the long-term persistence and adaptation of a species or a population in a given place? This is one of the most difficult and challenging intellectual problems in conservation biology. Arguably, it is the quintessential issue in population biology, because it requires a prediction based on a synthesis of all the biotic and abiotic factors in the spatial–temporal continuum.

Michael E. Soulé (1987)

The quote from Darwin above shows that both evolutionary biologists and wildlife managers have recognized for over 100 years that the persistence of small isolated populations may be threatened by inbreeding. Nevertheless, the potential harmful effects of inbreeding and the importance of genetics in the persistence of populations have been somewhat controversial and remain so to this day (Soulé and Mills 1992; Caughley 1994; Frankham and Ralls 1998; Mann and Plummer 1999).

There are a variety of reasons for this controversy. Some have suggested that inbreeding is unlikely to have significant harmful effects on individual fitness in wild populations. Others have suggested that inbreeding may affect individual fitness, but is not likely to affect **population viability** (Caro and Laurenson 1994). Still others have argued that genetic concerns can be ignored when estimating the viability of small populations because they are in much greater danger of extinction due to stochastic demographic effects (Lande 1988; Pimm et al. 1988). Finally, some have suggested that it may be best not to incorporate genetics into demographic models because genetic and demographic "currencies" are difficult to combine, and we have insufficient information about the effects of inbreeding in most wild populations (Beissinger and Westphal 1998).

The disagreement over whether or not genetics should be considered in demographic predictions of population persistence has been unfortunate and misleading. It is extremely difficult to separate genetic and environmental factors when assessing the causes of population extinction. This is because inbreeding depression initially usually causes subtle reductions in birth and death rates that interact with other factors to increase extinction probability (Mills et al. 1996). Obvious indications of inbreeding depression (severe congenital birth defects, monstrous abnormalities, or otherwise easily visible fitness deficiencies) are not likely to be detectable until after severe inbreeding depression has accumulated in a population.

Extinction is a demographic process that will be influenced by genetic effects under some circumstances. The key issue is to determine under what conditions genetic concerns are likely to influence population persistence (Nunney and Campbell 1993). There have been important recent advances in our understanding of the interaction between demography and genetics in order to improve the effectiveness of our attempts to conserve endangered species (e.g., Landweber and Dobson 1999; Oostermeijer et al. 2003).

Perhaps most importantly, we need to recognize when management recommendations based upon strict demographics or genetics may actually be in conflict with each other. For example, Ryman and Laikre (1991) have considered supportive breeding in which a portion of wild parents are brought into captivity for reproduction and their offspring are released back into the natural habitat where they mix with wild individuals. Programs like this are carried out in a number of species to increase population size and thereby temper stochastic demographic fluctuations. Under some circumstances, however, supportive breeding may reduce effective population size and cause a reduction in heterozygosity that may have harmful effects on the population (Ryman and Laikre 1991). This example demonstrates a conflict in that supplemental breeding can provide demographic benefits yet be genetically detrimental.

The primary causes of species extinction today are deterministic and result from human-caused habitat loss, habitat modification, and overexploitation (Caughley 1994; Lande 1999). Reduced genetic diversity in plants and animals is generally a symptom of endangerment, rather than its cause (Holsinger et al. 1999). Nevertheless, genetic effects of small populations have an important role to play in the management of many threatened species. For example, Ellstrand and Elam (1993) examined the population sizes of 743 sensitive plant taxa in California. Over 50% of the occurrences contained less than 100 individuals. In general, those populations that are the object of management schemes are often small and therefore are likely to be susceptible to the genetic effects of small populations. Many parks and nature reserves around the world are small and becoming so isolated that they are more like "megazoos" than healthy functioning ecosystems. Consequently many populations will require management (including genetic management) to insure their persistence (Ballou et al. 1994).

It is also important to be aware that genetic problems associated with small populations go beyond inbreeding depression and the associated loss of heterozygosity. In this chapter, we will consider the effects of inbreeding depression and several other genetic factors that can potentially act to reduce the probability of persistence of small populations.

14.1 Estimation of census population size

The number of individuals in a population is the most fundamental demographic characteristic of a population. Accurate estimates of abundance or **census population size** (N_c) are essential for effective conservation and management (Sutherland 1996). Moreover, the rate of loss of genetic variation in an isolated population will be primarily affected by the number of breeding individuals in the population. And, it seems that it should be relatively easy to estimate population size compared to the obvious difficulties of estimating other demographic characteristics, such as the gender-specific age distribution of individuals in a population. However, estimating the number of individuals in a population is usually difficult even under what may appear to be straightforward situations. For example, estimating the number of grizzly bears in the Yellowstone Eeosystem has been an especially contentious issue (Eberhardt and Knight 1996). This is perhaps surprising for a large mammal that is fairly easy to observe and occurs within a relatively small geographic area.

Genetic analyses can provide help in estimating the number of individuals in a population (see Guest Box 14). A variety of creative methods have been applied to this problem (Schwartz et al. 1998). For example, we saw in Section 12.2.1 that the amount of variation within a population can be used to estimate effective population sizes, which can be modified to estimate historical census population sizes (Roman and Palumbi 2003). Here we consider the two primary genetic methods for estimating population size. Bellemain et al. (2005) provide an excellent comparison of these methods.

14.1.1 Rarefaction methods

The simplest method for estimating the minimum size of a population is from the number of unique genotypes observed. Kohn et al. (1999) used feces from coyotes to genotype three hypervariable microsatellite loci in a 15 km² area in California near the Santa Monica Mountains. They detected 30 unique multilocus genotypes in 115 feces samples. Thus, their estimate of the minimum N_c was 30.

The actual N_c of a population may be much greater than the number of genotypes detected depending on what proportion of the population was sampled. For example, it is likely that not all the coyotes in this population were sampled in the collection of 115 feces. However, the estimate of total population size can be modified to take into account the probability of not sampling individuals. The cumulative number of unique multilocus genotypes (y) can be expressed as a function of the number of feces sampled (x), and the asymptote of this curve (a) can be estimated with iterative nonlinear regression with a computer to provide an estimate of local population size:

$$y = \frac{(ax)}{b + x} \tag{14.1}$$

where b is the rate of decline in value of the slope (Kohn et al. 1999). In this case, the estimate was 38 individuals with a 95% confidence interval of 36–40 coyotes. Eggert et al. (2003) have provided an alternative estimator that behaves similarly to expression 14.1 (Bellemain et al. 2005).

14.1.2 Capture–mark–recapture methods

A mark–recapture approach can also be used with genetic data to estimate population size (Bellemain et al. 2005) (Example 14.1). The multilocus genotypes of individuals can be considered as unique "tags" that exist in all individuals and are permanent.

The simplest mark–recapture method to estimate population size is the Lincoln-Peterson index (Lincoln 1930):

$$N_c = \frac{(N_1)(N_2)}{R} \tag{14.2}$$

where N_1 is the number of individuals in the first sample, N_2 is the total number of individuals in the second sample, and R is the number of individuals recaptured in the second sample. For example, suppose that 10 (N_1) animals were captured in the first sample, and one-half of 10 (N_2) animals captured in the second sample were marked ($R = 5$). This would suggest that the 10 animals in the first sample represented one-half of the population so that the total population size would be 20:

$$N_c = \frac{(10)(10)}{5} = 20$$

Example 14.1 Genetic tagging of humpback whales

Palsbøll et al. (1997) used a genetic capture–mark–recapture approach to estimate the number of humpback whales in the North Atlantic Ocean. Six microsatellite loci were analyzed in samples collected on the breeding grounds by skin biopsy or from sloughed skin in 1992 and 1993. A total of 52 whales sampled in 1992 were "recaptured" in 1993 as shown below:

	Females	Males
1992	231	382
1993	265	408
Recaptures	21	31

Substitution into expression 14.1 provides estimates of 2,915 female and 5,028 male humpback whales. Palsbøll et al. (1997) used a more complex estimator that has better statistical properties and estimated the North Atlantic humpback whale population to be 2,804 females (95% CI of 1,776–4,463) and 4,894 males (95% CI of 3,374–7,123). The total of 7,698 whales was in the upper range of previous estimates based on photographic identification.

Genetic capture–mark–recapture is potentially a very powerful method for estimating population size over large areas (Bellemain et al. 2005). It also is noninvasive (that is, does not require handling or manipulating animals). However, there are a variety of potential pitfalls (Taberlet et al. 1999). One potential problem is the failure to distinguish individuals due to using too few or insufficiently variable loci. For example, a new capture might be erroneously recorded as a recapture if the genotype is not unique (due to low marker polymorphism). This has been termed the **shadow effect** (Mills et al. 2000). The shadow effect can result in an underestimation of population size and will also affect confidence intervals (Waits and Leberg 2000). A second problem is that genotyping errors could generate false unique genotypes and thereby cause an overestimate of population size (Waits and Leberg 2000).

14.2 Inbreeding depression and extinction

We saw in Chapter 13 that inbreeding depression is a universal phenomenon. In this section we will examine when inbreeding depression is likely to affect population viability. Three conditions must hold for inbreeding depression to reduce the viability of populations:

1 Inbreeding must occur.
2 Inbreeding depression must occur.
3 The traits affected by inbreeding depression must reduce population viability.

Conditions 1 and 2 will hold to some extent in all small populations. As discussed earlier and below, matings between relatives must occur in small populations, and some deleterious recessive alleles will be present in all populations. However, condition 3 is the crux of the controversy. There is little empirical evidence that tells us when inbreeding depression will affect population viability and how important that effect will be.

For inbreeding depression to affect population viability it must affect traits that influence population viability. For example, Leberg (1990) found that eastern mosquitofish populations founded by two siblings had a slower growth rate than populations founded by two unrelated founders (see Guest Box 6). However, it has been difficult to isolate genetic effects in the web of interactions that affect viability in wild populations (Soulé and Mills 1998) (Figure 14.1). Laikre (1999) noted that many factors interact when a population is driven to extinction, and it is generally impossible to single out "the" cause.

Some authors have asserted that there is no evidence for genetics affecting population viability (Caro and Laurenson 1994):

> Although inbreeding results in demonstrable costs in captive and wild situations, it has yet to be shown that inbreeding depression has caused any wild population to decline. Similarly, although loss of heterozygosity has detrimental impact on individual fitness, no population has gone extinct as a result.

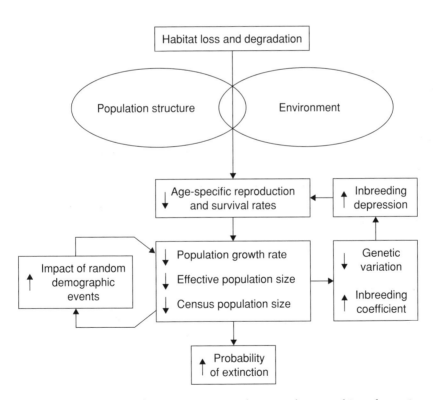

Figure 14.1 Extinction vortex showing interactions between demographic and genetic effects of habitat loss and isolation that can cause increased probability of extinction. Redrawn from Soulé and Mills (1998).

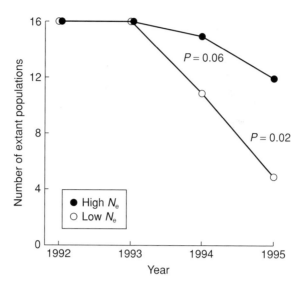

Figure 14.2 Population survival curves for populations of *Clarkia pulchella* founded by related (low N_e) and unrelated founders (high N_e). All populations were founded with 12 individuals. From Newman and Pilson (1997).

This observation prompted several papers that tested for evidence of the importance of genetics in population declines and extinction.

Newman and Pilson (1997) founded a number of small populations of the annual plant *Clarkia pulchella* by planting individuals in a natural environment. All populations were founded by the same number of individuals (12); however, in some populations the founders were unrelated (high N_e treatment) and in some they were related (low N_e treatment). All populations were demographically equivalent (that is, the same N_c) but differed in the effective population size (N_e) of the founding population. A significantly greater proportion of the populations founded by unrelated individuals persisted throughout the course of the experiment (Figure 14.2).

Saccheri et al. (1998) found that extinction risk of local populations of the Glanville fritillary butterfly increased significantly with decreasing heterozygosity at seven allozyme loci and one microsatellite locus after accounting for the effects of environmental factors. Larval survival, adult longevity, and hatching rates of eggs were all reduced by inbreeding, and were thought to be the fitness components responsible for the relationship between heterozygosity and extinction.

Westemeier et al. (1998) monitored greater prairie chickens for 35 years and found that egg fertility and hatching rates of eggs declined in Illinois populations after these birds became isolated from adjacent populations during the 1970s. These same characteristics did not decline in adjacent populations that remained large and widespread. These results suggested that the decline of birds in Illinois was at least partially due to inbreeding depression. This conclusion was supported by the observation that fertility and hatching success recovered following translocations of birds from the large adjacent populations (Figure 14.3).

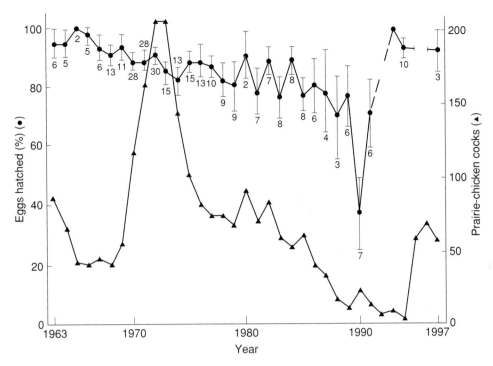

Figure 14.3 Annual means for success of greater prairie chicken eggs in 304 fully incubated clutches (circles) and counts of males (triangles) on booming grounds. Translocations of nonresident birds began in August 1992. Numbers on top line represent the number of fully incubated clutches. From Westemeier et al. (1998).

Madsen et al. (1999) studied an isolated population of adders in Sweden that declined dramatically some 35 years ago and has since suffered from severe inbreeding depression. The introduction of 20 males from a large and genetically variable population of adders resulted in a dramatic demographic recovery of this population. This recovery was brought about by increased survival rates, even though the number of litters produced by females per year actually declined during the initial phase of recovery.

Some have argued that the existence of species and populations that have survived bottlenecks is evidence that inbreeding is not necessarily harmful (Simberloff 1988; Caro and Laurenson 1994). However, we need to know how many similar populations went extinct following such bottlenecks to interpret the significance of such observations. For example, the creation of inbred lines of mice usually results in the loss of many of the lines (Bowman and Falconer 1960; Lynch 1977). This argument is similar to using the existence of 80-year-old smokers as evidence that cigarette smoking is not harmful. Only populations that have survived a bottleneck can be observed after the fact. Soulé (1987) termed this the "fallacy of the accident".

Nevertheless, this does not mean that populations that have lost substantial genetic variation because of a bottleneck are somehow "doomed" or are not capable of recovery. An increase in the frequency of some deleterious alleles and the loss of genome-wide heterozygosity is inevitable following a bottleneck. However, the magnitude of these effects on fitness-related traits (survival, fertility, etc.) may not be large enough to constrain recovery. For example, the tule elk of the Central Valley of California has gone through a series of bottlenecks since the 1849 gold rush (McCullough et al. 1996). Simulation analysis was used to estimate that tule elk have lost approximately 60% of their original heterozygosity (McCullough et al. 1996). Analyses of allozymes (Kucera 1991) and microsatellites (D. R. McCullough, personal communication) have confirmed relatively low genetic variation in the tule elk. Nevertheless, the tule elk has shown a remarkable capacity for population growth, and today there are 22 herds totaling over 3,000 animals (McCullough et al. 1996). Tule elk still may be affected by the genetic effects of the bottleneck in the future if they face some sort of challenge-event (e.g., disease).

14.3 Population viability analysis

Predictive demographic models are essential for determining whether or not populations are likely to persist in the future. Such risk assessment is essential for identifying species of concern, setting priorities for conservation action, and developing effective recovery plans. For example, one of the criteria for being included on the IUCN Red List (IUCN 2001) is the probability of extinction within a specified period of time (see Table 14.1). Some quantitative analysis is needed to estimate the extinction probability of a taxon based on known life history, habitat requirements, threats, and any specified management options. This approach has come to play an important role in developing conservation policy (Shaffer et al. 2002).

Population viability analysis (**PVA**) is the general term for models that take into account a number of processes affecting population persistence to simulate the demography of populations in order to calculate the risk of extinction or some other measure of population viability (Ralls et al. 2002). The first use of this approach was by Craighead et al. (1973) who used a computer model of grizzly bears in Yellowstone National Park. They demonstrated that closing of the park dumps to bears and the park's approach to problem bears was driving the population to extinction. McCullough (1978) developed an alternative model that came to different conclusions about the Yellowstone grizzly bear population. Both of these models were deterministic models in which the same outcome will always result with the same initial conditions and parameter values (e.g., stage-specific survival rates).

Mark Shaffer (1981) developed the first PVA model that incorporated chance events (stochasticity) into population persistence while a graduate student at Duke University. Shaffer described four sources of uncertainty: **demographic stochasticity**, **environmental stochasticity**, **natural catastrophes**, and **genetic stochasticity** (also see Shaffer 1987). Small populations with a positive growth rate will always increase in size and persist. However, chance fluctuations in demographic events may result in the extinction of small populations with a positive growth rate.

Incorporation of stochasticity into PVA was a crucial step in attempts to understand and predict the population dynamics of small populations. Many aspects of population dynamics are processes of sampling rather than completely deterministic events (e.g., stage-specific survival, sex determination, and transmission of alleles in heterozygotes, etc.). The predictability of an outcome decreases in a sampling process as the sample size is reduced. For example, in a large population the sex ratio will be near 50 : 50. However, this might not be true in a small population in which a large excess of males may significantly reduce the population growth rate (Leberg 1998).

In addition, there will be synergistic interactions between demographic processes and genetic effects. fluctuations in population size may result in genetic bottlenecks during which inbreeding may occur and substantial genetic variation may be lost. Even if the population grows and recovers from the bottleneck it will carry the legacy of this event in its genes. The loss of genetic variation during a bottleneck may have a variety of effects on demographic parameters (survival, reproductive rate, etc.). This may lead to large fluctuations in population size, increasing the probability of extinction (see Figure 14.1). These interactions have been called "extinction vortices" (Gilpin and Soulé 1986) and consideration of these interactions is a central part of PVA (Lacy 2000b).

14.3.1 The VORTEX simulation model

The complexity of the factors affecting population persistence means the useful PVA models must also be complex. It is possible for individuals to develop their own computer program to model population viability. The development of such a model, however, requires a lot of time and there is always a good probability that such a model will contain some programming errors. The usual alternative is to use an available software package for PVA that serve the same role as commercially available statistical packages. A number of such packages are available (Brook et al. 2000). We have chosen to present results using VORTEX because of its power, user friendliness, and widespread use (Lacy et al. 2003; Miller and Lacy 2003).

PVA requires information on birth and survival rates, reproductive rates, habitat capacity, and many other factors. It is important to understand the basic structure of the model being used in order to interpret the results. Figure 14.4 shows the relationships among the primary life history, environmental, and habitat components used by VORTEX.

As we saw in Section 6.5, a population growing exponentially increases according to the equation:

$$N_t = N_0 e^r \tag{14.3}$$

where N_0 is the initial population size ($t = 0$), N_t is the number of individuals in the population after t units of time (years in VORTEX), r is the exponential growth rate, and the constant e is the base of natural logarithms (approximately 2.72). A population is growing if $r > 0$ and is declining if $r < 0$. Population size is stable if $r = 0$. Lambda (λ) is the factor by which the population increases during each time unit. That is,

$$N_{t+1} = N_t \lambda \tag{14.4}$$

Let us use VORTEX to consider a PVA of grizzly bears from the Rocky Mountains of the USA. Figure 14.5 shows a summary of the VORTEX input values used. The actual values are taken from Harris and Allendorf (1989), but have been modified for use here.

These life history values result in a deterministic intrinsic growth rate (r) of 0.005 ($\lambda = 1.005$). Therefore, our simulated grizzly bear population is expected to increase by a factor of 1.005 each year (Figure 14.6). That is, if there are 1,000 bears in year $t = 0$ there will be 1,005 bears in year $t = 1$ (1,000 * 1.005) and 1,010 bears in year $t = 2$, etc. The

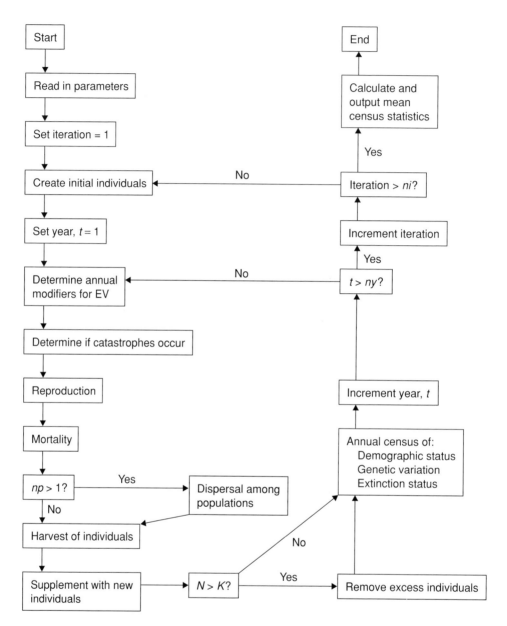

Figure 14.4 Flow chart of the primary components that occur within each subpopulation with the VORTEX simulation model. EV, environmental variation; K, carrying capacity; N, subpopulation size; ni, number of iterations; np, number of subpopulations; ny, years simulated; t, year. From Lacy (2000b).

generation interval for grizzly bears is approximately 10 years. Therefore, this growth rate will result in just over a 5% increase in population size after one generation. It is important to look at the deterministic projections of population growth in any analysis with VORTEX. If r is negative, then λ will be less than 1, and the population is in deterministic decline (the

```
VORTEX 9.42 -- simulation of population dynamics

   1 population(s) simulated for 200 years, 1 iterations
   Extinction is defined as no animals of one or both sexes.
   No inbreeding depression
   EV in reproduction and mortality will be concordant.

   First age of reproduction for females: 5    for males: 5
   Maximum breeding age (senescence): 30
   Sex ratio at birth (percent males): 50

Population 1: Population 1

  Polygynous mating;
     % of adult males in the breeding pool = 100

  % adult females breeding = 33
     EV in % adult females breeding: SD = 9

     Of those females producing progeny, ...
       28.00 percent of females produce 1 progeny in an average year
       44.00 percent of females produce 2 progeny in an average year
       28.00 percent of females produce 3 progeny in an average year

     % mortality of females between ages 0 and 1 = 20
       EV in % mortality: SD = 4
     % mortality of females between ages 1 and 2 = 18
       EV in % mortality: SD = 4
     % mortality of females between ages 2 and 3 = 15
       EV in % mortality: SD = 4
     % mortality of females between ages 3 and 4 = 15
       EV in % mortality: SD = 4
     % mortality of females between ages 4 and 5 = 15
       EV in % mortality: SD = 4
     % mortality of adult females (5<=age<=30) = 12
       EV in % mortality: SD = 4

     (Same mortality values for males)

   Initial size of Population 1: 100
     (set to reflect stable age distribution)
   Carrying capacity = 1000
     EV in Carrying capacity = 0

   Animals harvested from Population 1, year 1 to year 1 at 1 year
 intervals: 0

   Animals added to Population 1, year 1 through year 1 at 1 year
 intervals: 0
```

Figure 14.5 VORTEX input summary for population viability analysis of grizzly bears. This output has been slightly modified from that produced by the program. EV is the environmental variation for the parameter. Values used are modified from Harris and Allendorf (1989).

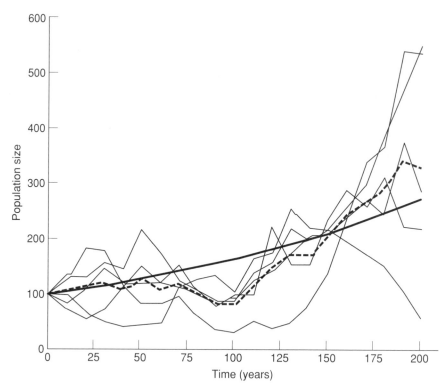

Figure 14.6 Stochastic variability in the growth of a grizzly bear population in five VORTEX simulations using input from Figure 14.5. The dark solid line is the expected growth rate with $r = 0.005$ ($\lambda = 1.005$) and an initial population size (N_0) of 100, using expression 14.4. The dark dashed line is the mean of the five simulated populations.

number of deaths outpaces the number of births) and will become extinct even in the absence of any stochastic fluctuations.

We can use VORTEX to examine how much stochastic variability in population growth we may expect (Figure 14.6). On the average, expression 14.3 does a good job of predicting growth rate. However, there is a wide range of results from each simulation even though the same input values were used. The differences among runs results from VORTEX using random numbers to mimic the life history of each individual.

What will happen if we incorporate genetic effects (inbreeding depression) into this model? Genetic effects due to inbreeding and loss of variation will come into play when the population size becomes small. For example, one of the runs reached an N of 34 after 100 years. This population than proceeded to grow very quickly and exceed 500 bears 90 years later. However, what would have happened if we had kept track of pedigrees within this populations and then reduced juvenile survival as a function of the inbreeding coefficient (F)? Remember that the effective population size of grizzly bears is approximately one-quarter of the population size (see Guest Box 7). Therefore, the N_e was much smaller than 34 during this period. The increased juvenile mortality of progeny produced by the mating of related individuals would have hindered this population's recovery.

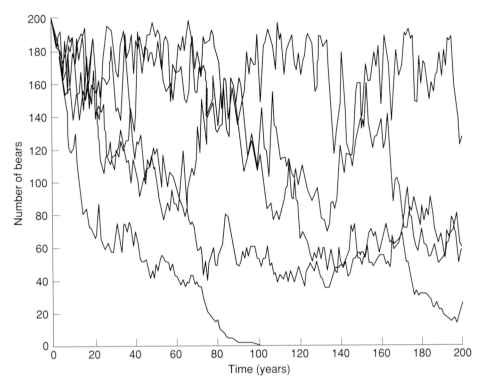

Figure 14.7 Results of five different VORTEX simulations of grizzly bears using input from Figure 14.5 except for the carrying capacity and initial population size. Each simulated population began with 200 bears and had a carrying capacity of 200. Inbreeding depression was incorporated as 3 LEs that increased mortality in the first year of life.

We can incorporate inbreeding depression with VORTEX by assigning a number of lethal equivalents (LEs) associated with decreased survival during the first year of life (Figure 14.7). Figure 14.8 shows the effects of inbreeding depression with these life history values on population persistence in 1,000 simulation runs for 0, 3, and 6 LEs per diploid genome. In the absence of any inbreeding depression (0 LE), the persistence probability is similar in the first and second hundred years of the simulations. However, even moderate inbreeding depression (3 LEs) reduces the probability of population persistence by approximately 25% in the second hundred years.

Note that even strong inbreeding depression has no effect on population persistence until after 100 years (approximately 10 generations) because it will take many generations for inbreeding relationships to develop within a population. This is an important point to recognize when considering the management of real populations. As we saw in Guest Box 7, Yellowstone grizzly bears have now been completely isolated for nearly 100 years. Some have argued that there is no reason to be concerned about the possible harmful genetic effects of this isolation because the population has persisted and seems to be doing well. However, it would be very difficult to detect inbreeding depression in a wild population of grizzly bears because we do not have good estimates of vital rates. In addition, Figure 14.7

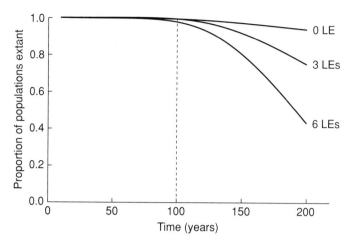

Figure 14.8 Effects of inbreeding depression on the persistence of a grizzly bear population based on VORTEX simulations using the values in Figure 14.5 except for the carrying capacity and initial population size. Each point represents the proportion of 1,000 simulated populations that did not go extinct during the specified time period. Simulated populations began with 200 bears and had a carrying capacity of 200. Inbreeding depression was incorporated as a different number of lethal equivalents (0, 3, and 6) that increased mortality in the first year of life.

shows that some populations that have accumulated substantial inbreeding depression may show positive growth rates. The effects of inbreeding depression are expected to be seen more quickly in species with shorter generation intervals.

14.3.2 What is a viable population?

An evaluation of the viability of a population requires identifying the time horizon of concern and the required probability of persistence or remaining above some minimum population size. There is no generally accepted time horizon or level of risk with regard to species extinctions (Shaffer et al. 2002). The World Conservation Union (IUCN) have offered standard criteria for placing taxa into categories of risk (Table 14.1). These criteria include predictions of the probability of extinction, as well as a variety of other alternative criteria (e.g., reduction in population size, current geographic range, or current population size). For example, a species may be considered to be facing an extremely high risk of extinction in the wild (i.e., "critically endangered") if its current population size is less than 50 mature individuals, without performing a PVA.

Early applications of PVA often set out to determine the minimum population size at which a population was likely to persist over some timeframe. The minimum viable population (MVP) concept was used to identify a goal or target for recovery actions. For example, one of the early grizzly bear recovery plans used the results of Shaffer's early work to set recovery targets for four of the six populations of between 70 and 90 bears (Allendorf and Servheen 1986). The term MVP has fallen out of favor for a variety of reasons. Many feel that the goal of conservation should not be to set a minimum number of individuals or a minimal distribution of a species. However, the concept of MVP is reasonable if we build

Table 14.1 Examples of demographic criteria for evaluating the results of population viability analyses.

Source	Category	Minimum persistence probability (%)	Timeframe
Shaffer (1978)	Minimum viable population (MVP)	95	100 years
Shaffer (1981)	MVP	99	1,000 years
Thompson (1991)	Threatened	50	10 years
	Endangered	95	100 years
Rieman et al. (1993)	Low risk	95	100–200 years
	High risk	50	100–200 years
AEPDC 1999*	Vulnerable	90	Medium-term future
	Endangered	80	Near future
	Critically endangered	50	Immediate future
IUCN (2001)	Vulnerable	90	100 years
	Endangered	80	20 years or 5 generations
	Critically endangered	50	10 years or 3 generations

*Australian Environment Protection and Biodiversity Act 1999.

in an appropriate margin of safety. Nevertheless, the term is not needed as long as we define the timeframe and probability of persistence that we are willing to accept.

Demographic criteria

Table 14.1 lists a variety of demographic criteria that have been used or suggested in the literature. There is no correct set of universal criteria to be used. Setting the timeframe and minimum probability of persistence are policy decisions that need to be specific for the situation at hand. Nevertheless, biological considerations should be used to set these criteria. Shorter periods have been recommended because errors are propagated in each time step in longer time periods (Beissinger and Westphal 1998). However, we should also be concerned with more than just the immediate future with which we can provide reliable predictions of persistence. The analogy of the distance we can see into the "future" using headlights while driving at night is appropriate here. We can only see as far as our headlights reach, but we need to be concerned about what lies beyond the reach of them (Shaffer et al. 2002). Population viability should be predicted on both short (say 10 generations) and long (more than 20 generations) timeframes.

There are a wide range of values presented in Table 14.1. The most stringent is the 99% probability of persistence for 1,000 years used by Shaffer in 1981. The IUCN values are the closest thing to generally accepted standards, and are fundamentally sound. They incorporate the concept of both short-term urgency (10 or 20 years) and long-term concerns. They also take into account that the appropriate timeframe will differ depending on the generation interval of the species under concern. For example, tuatara (see Example 1.1) do not become sexually mature until after 20 years, and their generation interval is approximately 50 years. Therefore, 10 years is just one-fifth of a tuatara generation, but it would represent 10 generations for an annual plant.

Genetic criteria

Persistence over a defined time period is not enough. We are also concerned that the loss of genetic variation over the time period does not threaten the long-term persistence of the population or species under consideration. A variety of authors have suggested genetic criteria to be used in evaluating the viability of populations. Soulé et al. (1986) suggested that the goal of captive breeding programs should be to retain 90% of the heterozygosity in a population for 200 years. By necessity, these kinds of guidelines are somewhat arbitrary. Nevertheless, the genetic goal of retaining at least 90–95% of heterozygosity over 100–200 years seems reasonable for a PVA (Allendorf and Ryman 2002). A loss of heterozygosity of 10% is equivalent to a mean inbreeding coefficient of 0.10 in the population.

14.3.3 Beyond viability

Probability viability analyses have great value beyond simply predicting the probability of extinction. Perhaps more importantly, PVA can be used to identify threats facing populations and identify management actions to increase the probability of persistence. This can be done by **sensitivity testing** in which a range of possible values for uncertain parameters are tested to determine what effects those uncertainties might have on the results. In addition, such sensitivity testing reveals which components of the data, model, and interpretation have the largest effect on population projections. This will indicate which aspects of the biology of the population and its situation contribute most to its vulnerability and, therefore, which aspects might be most effectively targeted for management. In addition, uncertain parameters that have a strong impact on results are those which might be the focus for future research efforts, to better specify the dynamics of the population. Close monitoring of such parameters might also be important for testing the assumptions behind the selected management options and for assessing the success of conservation efforts (Example 14.2).

14.4 Loss of phenotypic variation

Inbreeding depression is not necessary for the loss of genetic variation to affect population viability. Reduction in variability itself, even without a reduction in individual fitness, may reduce population viability (e.g., Conner and White 1999).

Honey bees present a fascinating example of the potential importance of genetic variation itself (Jones et al. 2004). Honey bee colonies have different amounts of genetic variation depending on how many males the queen mates with. Brood nest temperatures tend to be more stable in colonies in which the queen has mated with multiple males

Example 14.2   PVA of the Sonoran pronghorn

The Sonoran pronghorn is one of five subspecies and was listed as endangered under the US Endangered Species Act in 1967 (Hosack et al. 2002). The pronghorn is endemic to western North America, and it has received high conservation priority because it is the only species in the family Antilocapridae. The pronghorn resembles an antelope in superficial physical characteristics, but it has a variety of unusual morphological, physiological, and behavioral traits (Byers 1997).

The Sonoran subspecies is restricted to approximately 44,000 ha in southwestern Arizona. There were approximately 200 individuals in this population based on census estimates in the 1990s. A group of 22 biologists from a variety of federal, state, tribal, university, and environmental organizations convened a PVA workshop in September 1996. Nine primary questions and issues were identified as key to pronghorn recovery. All of these questions were explored with PVA simulation modeling during the workshop. The final three of those questions are presented below:

7   Can we identify a population size below which the population is vulnerable, but above which it could be considered for downlisting to a less threatened category?

8   Which factors have the greatest influence on the projected population performance?

9   How would the population respond (in numbers and in probability of persistence) to the following possible management actions: increase in available habitat; cessation of any research that subjects animals to the dangers of handling; exchange of some pronghorn with populations in Mexico; and supplementation of the wild population from a captive population?

Estimates of the life history parameters used by VORTEX were provided by participants of the workshop. Some of the values were available from field data, but there were no quantitative data available for many parameters. For these parameters, the field biologists provided "best guesses". The participants performed a sensitivity analysis to evaluate the response of the simulated populations to uncertainty by varying eight parameters: inbreeding depression, fecundity, fawn survival, adult survival, effects of catastrophes, harvest for research purposes, carrying capacity, and size and sex/age structure of the initial population.

Results indicated that the Sonoran pronghorn population had a 23% probability of extinction within 100 years using the best parameter estimates. This probability increased markedly if the population fell below some 100 individuals. Sensitivity analysis indicated that fawn survival rates had the greatest effects on population persistence (Figure 14.9). Sensitivity analysis also indicated that short-term emergency provisioning of water and food during droughts would substantially increase the probability of population persistence. The workshop concluded that this population is at serious risk of extinction, but that a few key management actions could greatly increase the probability of the population persisting for 100 years.

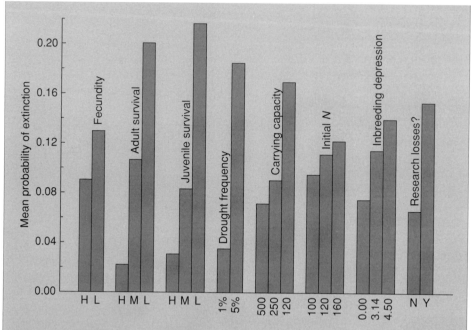

Figure 14.9 Results of PVA on the Sonoran pronghorn. The bars indicate the probability of extinction within 100 years for various values of eight parameters that were varied during sensitivity testing (H, high; M, medium; L, low; N, no; Y, yes). Extinction probabilities were fairly insensitive to some parameters (e.g., initial population size), but were greatly affected by others (e.g., adult and juvenile survival rates). From Hosack et al. (2002).

(Figure 14.10). Honey bee workers regulate temperature by their behavior; they fan out hot air when the temperature is perceived as being too hot and cluster together and generate metabolic heat when the temperature is perceived as being too low. Increased genetic variation for response thresholds produces a more graded response to temperature and results in greater temperature regulation within the hive.

14.4.1 Life history variation

Individual differences in life history (age at first sexual maturity, clutch size, etc.) that have at least a partial genetic basis occur in virtually all populations of plants and animals. Many of these differences may have little effect on individual fitness because of a balance or trade-off between advantages and disadvantages. Nevertheless, the loss of this life history variability among individuals may reduce the likelihood of persistence of a population.

For example, Pacific salmon return to fresh water from the ocean to spawn and then die (Groot and Margolis 1991). In most species, there are individual differences in age at reproduction that often have a substantial genetic basis (Hankin et al. 1993). For example, Chinook salmon usually become sexually mature at age 3, 4, or 5 years. The greater

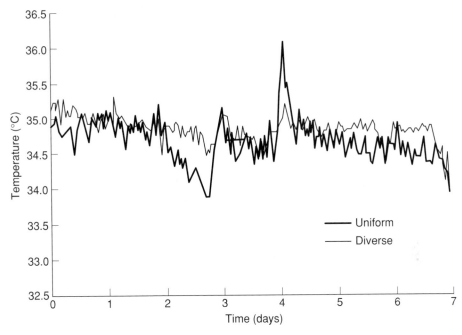

Figure 14.10 Hourly temperature variation in a genetically diverse and uniform honey bee colony. This graph shows the average hourly temperature for a representative pair of experimental colonies that differed only in the number of males with which the queen mated. The uniform colony queen mated with a single male; the diverse colony queen mated with multiple males.

fecundity of older females (because of their greater body size) is balanced by their lower probability of survival to maturity. These different life history types have similar fitnesses. Pink salmon are exceptional in that all individuals become sexually mature and return from the ocean to spawn in fresh water at 2 years of age (Heard 1991). Therefore, pink salmon within a particular stream comprise separate odd- and even-year populations that are reproductively isolated (Aspinwall 1974).

Consider a hypothetical comparison of two streams for purposes of illustration. The first stream has separate odd- and even-year populations, as is typical for pink salmon. In the second stream, there is phenotypic (and genetic) variation for the time of sexual maturity so that approximately 25% of the fish become sexually mature at age 1 year and 25% of the fish become sexually mature at age 3; the remaining 50% of the population becomes mature at age 2.

All else being equal, we would expect the population with variability in age of return to persist longer than the two reproductively isolated populations. The effective population size (N_e) of the odd- and even-year populations would be one-half the N_e of the single reproductive population with life history variability (Waples 1990). Thus, inbreeding depression would accumulate twice as rapidly in the two reproductively isolated populations than in the single variable population. The two smaller populations would also each be more susceptible to extinction from demographic, environmental, and catastrophic stochasticity. For example, a catastrophe that resulted in complete reproductive failure for 1 year would cause the extinction of one of the populations without variability.

14.4.2 Mating types and sex determination

The occurrence of separate genders or mating types is another case where the loss of phenotypic variation can cause a reduction in population viability without a reduction in the fitness of inbred individuals. Approximately 50% of flowering plant species have genetic incompatibility mechanisms (see Guest Box 3; Nettancourt 1977). In one of these self-incompatibility systems, individuals are of different mating types which possess different genotypes at the self-incompatibility (*S*) locus (Richards 1986). Pollen grains can only fertilize plants that do not have the same *S* allele as that carried by the pollen. Homozygotes cannot be produced at this locus, and the minimum number of alleles at this locus in a sexually reproducing population is three. Smaller populations are expected to maintain many fewer *S* alleles than larger populations at equilibrium (Wright 1960).

Les et al. (1991) have considered the demographic importance of maintaining a large number of *S* alleles in plant populations. A reduction in the number of *S* alleles because of a population bottleneck will reduce the frequency of compatible matings and may result in reduced levels of seed set. Demauro (1993) reported that the last Illinois population of the lakeside daisy was effectively extinct even though it consisted of approximately 30 individuals because all the plants apparently belonged to the same mating type. Reinartz and Les (1994) concluded that some one-third of the remaining 14 natural populations of *Aster furactus* in Wisconsin had reduced seed sets because of a diminished number of *S* alleles.

A similar effect can occur in the nearly 15% of animal species that are haplodiploid in which sex is determined by genotypes at one or more hypervariable loci (ants, bees, wasps, thrips, whitefly, certain beetles, etc.) (Crozier 1971). Heterozygotes at the sex-determining locus or loci are female, and the hemizygous haploids or homozygous diploid individuals are male (Packer and Owen 2001). Diploid males have been detected in over 30 species of Hymenoptera, and evidence suggests that single locus sex determination is common. Most natural populations have been found to have 10–20 alleles at this locus. Therefore, loss of allelic variation caused by a population bottleneck will increase the number of diploid males produced by increasing homozygosity at the sex-determining locus or loci.

Diploid males are often inviable, infertile, or give rise to triploid female offspring (Packer and Owen 2001). Thus, diploid males are effectively sterile, and will reduce a population's long-term probability of persisting both demographically and genetically. The decreased numbers of females will reduce the foraging productivity of the nest in social species or reduce the population size of other species. In addition, the skewed sex ratio will reduce effective population size and lead to further loss of genetic variation throughout the genome because of genetic drift.

A much weaker gender effect may occur in animal species in which sex is determined by three or more genetic factors. Leberg (1998) found that species with multiple factor sex determination (MSD) can experience large decreases in viability relative to species with simple sex determination systems in the case of very small bottlenecks. This effect results from increased demographic stochasticity because of greater deviations from a 1 : 1 sex ratio, not because of any reduction in fitness. MSD is rare, but it has been described in fish, insects, and rodents.

14.5 Loss of evolutionary potential

The loss in genetic variation caused by a population bottleneck may cause a reduction in a population's ability to respond by natural selection to future environmental changes. Bürger and Lynch (1995) predicted, on the basis of theoretical considerations, that small populations ($N_e < 1,000$) are more likely to go extinct due to environmental change because they are less able to adapt than are large populations.

The ability of a population to evolve is affected both by heterozygosity and the number of alleles present. Heterozygosity is relatively insensitive to bottlenecks in comparison to allelic diversity (Allendorf 1986). Heterozygosity is proportional to the amount of genetic variance at loci affecting quantitative variation (James 1971). Thus, heterozygosity is a good predictor of the potential of a population to evolve immediately following a bottleneck. Nevertheless, the long-term response of a population to selection is determined by the allelic diversity either remaining following the bottleneck or introduced by new mutations (Robertson 1960; James 1971).

The effect of small population size on allelic diversity is especially important at loci associated with disease resistance. Small populations are vulnerable to extinction by epidemics, and loci associated with disease resistance often have an exceptionally large number of alleles. For example, Gibbs et al. (1991) described 37 alleles at the major histocompatibility complex (MHC) in a sample of 77 adult blackbirds. Allelic variability at the MHC is thought to be especially important for disease resistance (Edwards and Potts 1996; Black and Hedrick 1997). For example, Paterson et al. (1998) found that certain microsatellite alleles within the MHC of Soay sheep are associated with parasite resistance and greater survival.

This effect of loss of variation due to inbreeding on response to natural selection has been demonstrated in laboratory populations of *Drosophila* by Frankham et al. (1999). They subjected several different lines of *Drosophila* to increasing environmental stress by increasing the salt (NaCl) content of the rearing medium until the line went extinct. Outbred lines performed the best; they did not go extinct until the NaCl concentration reached an average of 5.5% (Figure 14.11). Highly inbred lines went extinct at a salt concentration of 3.5%. Lines that experienced an expected 50–75% loss of heterozygosity due to inbreeding went extinct at a mean of roughly 5% NaCl. Thus, loss of genetic variation due to inbreeding made these lines less able to adapt to continuing environmental change.

14.6 Mitochondrial DNA

Recent results have suggested that mutations in mitochondrial DNA (mtDNA) may decrease the viability of small populations (Gemmell and Allendorf 2001). Mitochondria are generally transmitted maternally so that deleterious mutations that affect only males will not be subject to natural selection. Sperm are powered by a group of mitochondria at the base of the flagellum, and even a modest reduction in power output may reduce male fertility yet have little effect on females. A recent study of human fertility has found that mtDNA haplogroups are associated with sperm function and male fertility (Ruiz-Pesini et al. 2000). In addition, the mitochondrial genome has been found to be responsible for cytoplasmic male sterility, which is widespread in plants (Schnable and Wise 1998).

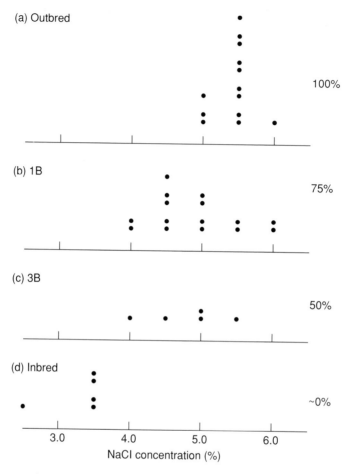

Figure 14.11 Results of an experiment demonstrating that loss of genetic variation can reduce a population's ability to respond by natural selection to environmental change. Lines of *Drosophila* with different relative amounts of expected heterozygosity (as indicated on the right of the figure) were exposed to increasing NaCl concentrations. Each dot represents the NaCl concentration at which lines went extinct. Modified from Frankham et al. (1999).

The viability of small populations may be reduced by an increase in the frequency of mtDNA genotypes that lower the fitness of males. Since females and males are haploid for mtDNA, it has not been recognized that mtDNA may contribute to the increased genetic load of small populations. The effective population size of the mitochondrial genome is generally only one-quarter that of the nuclear genome, so that mtDNA mutations are much more sensitive to genetic drift and population bottlenecks than nuclear loci.

Whether or not an increase in mtDNA haplotypes that reduce male fertility will affect population viability will depend on the mating system and reproductive biology of the particular population. However, it seems likely that reduced male fertility may decrease the number of progeny produced under a wide array of circumstances. At a minimum, the presence of mtDNA genotypes that reduce the fertility of some males would increase the variability in male reproductive success and thereby decrease effective population size.

This would increase the rate of loss of heterozygosity and other effects of inbreeding depression that can reduce population viability.

14.7 Mutational meltdown

Wright (1931, p. 157) first suggested that small populations would continue to decline in vigor slowly over time because of the accumulation of deleterious mutations that natural selection would not be effective in removing because of the overpowering effects of genetic drift (recall Section 8.5). Recent papers have considered the expected rate and importance of this effect for population persistence (Lynch and Gabriel 1990; Gabriel and Bürger 1994; Lande 1995). As deleterious mutations accumulate, population size may decrease further and thereby accelerate the rate of accumulation of deleterious mutations. This feedback process has been termed "mutational meltdown".

Lande (1994) concluded that the risk of extinction through this process "may be comparable in importance to environmental stochasticity and could substantially decrease the long-term viability of populations with effective sizes as a large as a few thousand". The expected timeframe of this process is hundreds or thousands of generations. Experiments designed to detect empirical evidence for this effect have had mixed results (e.g., Lynch et al. 1999).

14.8 Long-term persistence

When considering longer periods than those of a typical PVA, avoiding the loss of genetic variation is not enough for persistence. Environmental conditions are likely to change over time, and a viable population must be large enough to maintain sufficient genetic variation for adaptation to such changes. Evolutionary response to natural selection is generally thought to involve a gradual change of quantitative characters through allele frequency changes at the underlying loci, and discussions on the population sizes necessary to uphold "evolutionary potential" have focused on retention of additive genetic variation of such traits.

There is current disagreement among geneticists regarding how large a population must be to maintain "normal" amounts of additive genetic variation for quantitative traits (Franklin and Frankham 1998; Lynch and Lande 1998). The suggestions for the effective sizes needed to retain evolutionary potential range from 500 to 5,000. The logic underlying these contrasting recommendations is somewhat arcane and confusing. We, therefore, review some of the mathematical arguments used to support the conflicting views.

Franklin (1980) was the first to make a serious attempt to provide a direct estimate of the effective size necessary for the retention of additive genetic variation (V_A) of a quantitative character. He argued that for evolutionary potential to be maintained in a small population, the loss of V_A per generation must be balanced by new variation due to mutations (V_m). V_A will be lost at the same rate as heterozygosity ($1/2N_e$) at selectively neutral loci, so the expected loss of additive genetic variation per generation is $V_A/2N_e$. Therefore,

$$\Delta V_A = V_m - \frac{V_A}{2N_e} \tag{14.5}$$

(see also Lande and Barrowclough 1987; Franklin and Frankham 1998). At the equilibrium between loss and gain, ΔV_A is zero, and:

$$N_e = \frac{V_A}{2V_m} \tag{14.6}$$

Using abdominal bristle number in *Drosophila* as an example, Franklin (1980) also noted that $V_m \approx 10^{-3}V_E$, where V_E is the environmental variance (i.e., the variation in bristle number contributed from environmental factors). Furthermore, assuming that V_A and V_E are the only major sources of variation, the heritability (H_N, the proportion of the total phenotypic variation that is due to additive genetic effects; see Chapter 11) of this trait is $H_N = V_A/(V_A + V_E)$, and $V_E/(V_A + V_E) = 1 - H_N$. Thus, expression 14.6 becomes (cf. Franklin and Frankham 1998):

$$N_e = \frac{V_A}{2 \times 10^{-3}V_E} = 500\frac{V_A}{V_E} = 500\frac{H_N}{1 - H_N} \tag{14.7}$$

The heritability of abdominal bristle number in *Drosophila* is about 0.5. Therefore, the approximate effective size at which loss and gain of V_A are balanced (i.e., where evolutionary potential is retained) would be 500.

Lande (1995) reviewed the recent literature on spontaneous mutation and its role in population viability. He concluded that the approximate relation between mutational input and environmental variance observed for bristle count in *Drosophila* ($V_m \approx 10^{-3}V_E$) appears to hold for a variety of quantitative traits in several animal and plant species. He also noted, however, that a large portion of new mutations seem to be detrimental, and that only about 10% are likely to be selectively neutral (or nearly neutral), contributing to the potentially adaptive additive variation of quantitative traits. Consequently, he suggested that a more appropriate value of V_m is $V_m \approx 10^{-4}V_E$, and that Franklin's (1980) estimated minimum N_e of 500 necessary for the retention of evolutionary potential should be raised to 5,000.

In response, Franklin and Frankham (1998) suggest that Lande (1995) overemphasized the effects of deleterious mutations and that the original estimate of $V_m \approx 10^{-3}V_E$ is more appropriate. They argued that empirical estimates of V_m typically have been obtained from long-term experiments where a large fraction of the harmful mutations have had the opportunity of being eliminated, such that a sizeable portion of those mutations have already been accounted for. They also pointed out that in most organisms, heritabilities of quantitative traits are typically smaller than 0.5, and that this is particularly true for fitness-related characters. As a result, the quotient $H_N/(1 - H_N)$ in expression 14.7 is typically expected to be considerably smaller than unity, which reduces the necessary effective size. Franklin and Frankham (1998) concluded that an N_e of the order of 500–1,000 should be generally appropriate.

Lynch and Lande (1998) criticized the conclusions of Franklin and Frankham (1998) and argued that much larger effective sizes are justified for the maintenance of long-term genetic security. They maintain that the problems with harmful mutations must be taken seriously. An important point is that a considerable fraction of new mutations are expected to be only mildly deleterious with a selective disadvantage of less than 1%. Such mildly deleterious mutations behave largely as selectively neutral ones and are not expected to be "cleansed" from the population by selective forces even at effective sizes of several hundred individuals. In the long run, the continued fixation of mildly deleterious alleles may reduce population fitness to the extent that it enters an extinction vortex (i.e., mutational meltdown; Lynch et al. 1995).

According to Lynch and Lande (1998) there are several reasons why the minimum N_e for long-term conservation should be at least 1,000. At this size, at least the **expected** (average) amount of additive genetic variation of quantitative traits is of the same magnitude as for an infinitely large population, although genetic drift may result in considerably lower levels over extended periods of time. Furthermore, Lynch and Lande (1998) considered populations with $N_e > 1,000$ highly unlikely to succumb to the accumulation of unconditionally deleterious alleles (i.e., alleles that are harmful under all environmental conditions) except on extremely long time scales. They also stressed, however, that many single locus traits, such as disease resistance, require much larger populations for the maintenance of adequate allele frequencies (Lande and Barrowclough 1987), and suggest that effective target sizes for conservation should be of the order of 1,000–5,000.

Discussion of the population sizes adequate for long-term persistence of populations from a genetics perspective will continue. Regardless of the precise value of this figure, there is agreement that the long-term goal for **actual** population sizes to insure viability should be thousands of individuals, rather than hundreds.

14.9 The 50/500 rule

The 50/500 rule was introduced by Franklin (1980). He suggested that as a general rule-of-thumb, in the short term the effective population size should not be less than 50, and in the long term the effective population size should not be less than 500. The short-term rule was based upon the experience of animal breeders who have observed that natural selection for performance and fertility can balance inbreeding depression if ΔF is less than 1%; this corresponds to an effective population size using $\Delta F = 1/2N_e$ (see expression 6.2). The basis of the long-term rule was discussed in detail in the previous section.

There are many problems with the use of simple rules such as this in a complicated world. There are no real thresholds (such as 50 or 500) in this process; the loss of genetic variation is a continuous process. The theoretical and empirical basis for this rule is not strong and has been questioned repeatedly in the literature. In addition, such simple rules can and have been misapplied. We once heard a biologist for a management agency use this rule to argue that genetics need not be considered in developing a habitat management plan that affected many species. After all, if N_e is less than 50, then the population is doomed so that we don't need to be concerned with genetics, and if N_e is greater than 50, then the population is safe so we don't need to be concerned with genetics.

Nevertheless, we believe that the 50/500 rule is a useful guideline for the management of populations. Its function is analogous to a warning light on the dashboard of a car. If the N_e of an isolated population is less than 50, we should be concerned about a possible increased probability of extinction because of genetic effects. There is experimental evidence with house flies, however, that suggests that the N_e may have to be greater than 50 to escape extinction even in the short term (Reed and Bryant 2000). These numbers should, however, be used as targets. When the low fuel light comes on in your car, you do not stop filling the fuel tank once the light goes off. It is also important to remember that 50/500 is based only on genetic considerations. Some populations may face substantial risk of extinction because of demographic stochasticity before they are likely to be threatened by genetic concerns (Lande 1988; Pimm et al. 1988).

Guest Box 14 Noninvasive population size estimation in wombats
Andrea C. Taylor

Animals that are rare, cryptic, or endangered are notoriously difficult to study by traditional methods such as trapping and observation. One of the more intractable parameters for such species is population size. Two advances in molecular genetics have together paved the way for alternative approaches to "counting" animals. The first is the use of optimized methods for extracting the typically low quality and quantity DNA present in remotely collected field samples such as hair, feces, and saliva. The second is the development of polymerase chain reaction (PCR) based genetic assays (e.g., microsatellite analysis, Chapter 4) that allow researchers to assign individual-specific genotypes to such samples.

Amongst the world's most endangered mammals is the Australian northern hairy-nosed wombat. All known members of the species reside in a single small colony within Epping Forest National Park in central Queensland, thought to be as small as 25 individuals upon its "discovery" in the 1970s. Our ability to actively manage the species is crucially dependent on knowledge of its population size. However, methods available to date have been wholly inadequate: regular burrow activity assessments provide data only on long-term population trends, and estimates of population size based on trapping data are extremely imprecise (Figure 14.12). Being shy, nocturnal, and burrowing (individuals spend only 2–6 hours per night above ground) as well as highly endangered, this species is well qualified to receive the benefits of some creative alternative noninvasive approaches to abundance estimation.

The wombat's total reliance on burrows for daytime refuge means all individuals could theoretically be sampled by hair-collection devices, in this case double-sided carpet tape suspended between two posts on either side of burrow entrances. Over seven consecutive nights in September 2000, all active burrows (those showing signs of recent excavation, footprints, etc.) were taped in this manner, and 60 randomly selected hair-containing tapes chosen per night for further analysis. DNA (from single hairs, because of the possibility that multiple wombats may visit a burrow on any given night) was extracted in the field to minimize loss during hair storage, and later genotyped. Although northern hairy-nosed wombats have very low levels of genetic variation, use of the 10 most variable microsatellite markers provides sufficient individual specificity (Banks et al. 2003).

A total of 81 distinct genotypes were observed, presumably representing 81 different wombats. A mark–recapture analysis assuming heterogeneity in detection probability among individuals and implemented in the program CAPTURE (Otis et al. 1978), suggested a most likely population size of 113. This analysis uses information from the number of wombats detected only once, twice, three times, and so on to estimate the number of others that were present but not sampled at all, hence

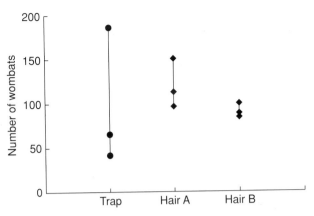

Figure 14.12 Population size estimates for hairy-nosed wombats based on trapping data (1993) and mark–recapture genetic analysis using noninvasive sampling (2000) and the programs CAPTURE (Hair A) and *capwire* (Hair B). The bars show the 95% confidence intervals of the estimates.

the disparity between the estimated population size and the number of distinct genotypes detected. Because CAPTURE was designed for trapping studies in which animals are typically only captured once per nightly session, it cannot incorporate multiple detections of individuals per sampling session, as occurs frequently in noninvasive studies. Craig Miller and colleagues have recently produced a modified mark–recapture model to incorporate multiple detections of individuals per session, i.e. *"capture with replacement"* (*capwire*). Analysis of the northern hairy-nosed wombat data with this new model gave an estimate of 88 individuals (Miller et al. 2005). The 95% confidence intervals for both hair-based estimates are a substantial improvement over those based on trap samples (Figure 14.12).

Problem 14.1

We saw in Section 14.1 that the multilocus genotypes of individuals can be considered as "tags" that exist in all individuals and are permanent. Palsbøll et al. (1997) used a mark–recapture approach with genotypes at six microsatellite loci to estimate the number of humpback whales in the North Atlantic Ocean. Per Palsbøll has generously provided a subset of their data, which is available in an Excel file (*humpback genotypes*) on the course web page. The first two columns give the identification number and sex of each sample; identification numbers that begin with 92XXXX were sampled in 1992, and numbers that begin with 93XXXX were sampled in 1993. The two alleles present at the six loci (A–F) are listed in the next 12 columns. For example, the two alleles present in each whale at each locus are in the A_1 and A_2 columns. Use these data, along with expression 14.1, to estimate the number of female and male humpback whales in the North Atlantic Ocean.

Problem 14.2

We saw in Section 14.4.2 that sex is determined in Hymenoptera (bees, ants, and wasps) by genotypes at one or more hypervariable loci. Heterozygotes at the sex-determining locus or loci are female, and the hemizygous haploid or homozygous diploid individuals are male. At equilibrium, all alleles at this type of sex-determining locus are expected to be equally frequent; that is, if there are A alleles, then each allele is expected to be at a frequency of $1/A$. What proportion of diploids is expected to be male in a population with 20 alleles at equilibrium? What increase in diploid males is expected if a bottleneck reduces the number of alleles to 10 or 5?

Problem 14.3

Wolves were hunted and poisoned to extinction in Sweden and Norway by the mid-20th century. Three wolves migrated from Russia and founded a new Scandinavian wolf population in the early 1980s. That population now consists of approximately 100 animals. How large do you think the Scandinavian wolf population should be to maintain genetic variation that may be important for the viability (continued persistence) and continued evolution of this population? There is some possibility of gene flow into this population from Russian wolves. Do you think that it is important that gene flow between the Scandinavian and the Russian populations be maintained in the future? Why?

Problem 14.4

What do you think is an appropriate timeframe for evaluating population viability? 100 years? 1,000 years? (See Table 14.1.)

Problem 14.5

Do you think the timeframe for the evaluation of population viability should be in units of years or generations? Why? (See Table 14.1.)

15

Metapopulations and Fragmentation

White campion, Section 15.5

An important case arises where local populations are liable to frequent extinction, with restoration from the progeny of a few stray immigrants. In such regions the line of continuity of large populations may have passed repeatedly through extremely small numbers even though the species has at all times included countless millions of individuals in its range as a whole.

Sewall Wright (1940)

Theoretical results have shown that a pattern of local extinction and recolonization can have significant consequences for the genetic structure of subdivided populations; consequences that are relevant to issues in both evolutionary and conservation biology.

David McCauley (1991)

The models of genetic population structure that we have examined to this point have assumed a connected series of equal size populations in which the population size is constant. However, the real world is much more complicated than this. Local populations differ in size, and local populations of some species may go through local extinction events and then be recolonized by migrants from other populations. These events will have complex, and sometimes surprising, effects on the genetic population structure and evolution of species.

These considerations are becomingly increasingly important because of ongoing loss of habitat and fragmentation. Many species that historically were nearly continuously distributed across broad geographic areas are now restricted to increasingly smaller and more isolated patches of habitat. In this chapter, we will combine genetic and demographic models to understand the distribution of genetic variation in species. We will also consider how these processes affect the viability of populations.

15.1 The metapopulation concept

Sewall Wright (1940) was the first to consider the effects of extinction of local populations on the genetics of species. Wright was interested in the effect that such local extinctions would have on the genetic structure and evolution of species. He considered the case where local populations are liable to frequent extinction and are restored with the "progeny of a few stray immigrants" (Figure 15.1).

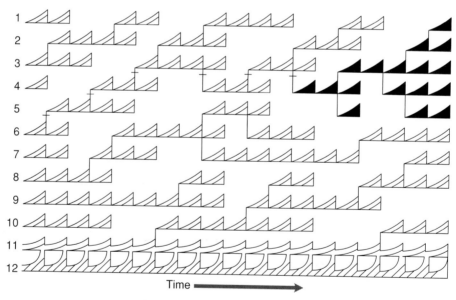

Figure 15.1 Diagram of a species in which local populations are liable to frequent extinction and recolonization. Time proceeds from left to right. Twelve different local patches are represented by a horizontal row (numbered 1 through 12). Note that the bottom two local populations never go extinct whereas all others go extinct every 2–9 time steps. For example, the subpopulation in patch 7 went extinct at the end of time steps 2, 6, and 15. The darkly shaded subpopulations in the upper right corner have passed through small groups of migrants six times. Modified from Wright (1940).

Wright pointed out that such local extinctions and recolonization events would act as bottlenecks that would make the effective population size of a group of local subpopulations much smaller than expected based on the number of individuals present within the subpopulations. Therefore, many of the subpopulations would be derived from a few local subpopulations that persist for long time periods. In modern terms, the genes in many of the subpopulations would "coalesce" to a single gene that was present in a "source" subpopulation in the relatively recent past.

He also discussed that genetic drift during such periodic bottlenecks would provide a mechanism for fixation of chromosomal rearrangements that are favorable when homozygous. Such arrangements are selected against in heterozygotes (see Section 8.2) and therefore will be selectively removed in large populations where natural selection is effective. Wright felt that such "nonadaptive inbreeding effects" in local populations might create a greater diversity of multilocus genotypes and thus make natural selection more effective within the species as a whole. He later suggested that differential rates of extinction and recolonization among local populations could result in intergroup selection that could lead to the increase in frequency of traits that were "socially advantageous" but individually disadvantageous (Wright 1945).

The term **metapopulation** was introduced by Richard Levins (1970) to describe a "population of populations", a collection of subpopulations that occupy separate patches of a subdivided habitat (Dobson 2003). In Levins's model, a metapopulation is a group of small populations that occupy a series of similar habitat patches isolated by unsuit-able habitat. The small local populations have some probability of extinction (e) during a particular time interval. Empty habitat patches are subject to recolonization with probability (c) by individuals from other patches that are occupied. Meta-population dynamics are a balance between extinction and recolonization so that at any particular time some proportion of patches are occupied (p) and some are extinct. At equilibrium,

$$p^{\star} = \frac{c}{c + e} \qquad (15.1)$$

The concept of metapopulations has become a valuable framework for understanding the conservation of populations and species (Hanski and Gilpin 1997). The general definition of a metapopulation is a group of local populations that are connected by dispersing individuals (Hanski and Gilpin 1991). More realistic models have incorporated differences in local population size and differential rates of exchange among populations as well as differential rates of extinction and colonization (Figure 15.2). In general, larger patches are less likely to go extinct because they will support larger populations. Patches that are near other occupied patches are more likely to be recolonized. In addition, immigration into a patch that decreases the extinction rate for either demographic or genetic reasons has been called the **rescue effect** (Brown and Kodric-Brown 1977; Ingvarsson 2001).

15.2 Genetic variation in metapopulations

It is important to consider both spatial and temporal scales in considering the genetic size of metapopulations. Slatkin (1977) described the first metapopulation genetic

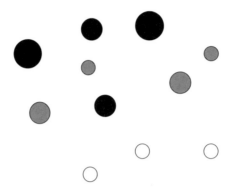

Figure 15.2 The pattern of occupancy of habitat patches of different sizes and isolation in a metapopulation. Darker shading indicates a higher probability that a patch will be occupied. Large patches that are close to other populations have the greatest probability of being occupied.

models. Hanski and Gilpin (1991) have described three spatial scales for consideration (Figure 15.3):

1   The **local scale** is the scale at which individuals move and interact with one another in their course of routine feeding and breeding activities.
2   The **metapopulation scale** is the scale at which individuals infrequently move from one local population to another, typically across habitat that is unsuitable for their feeding and breeding activities.
3   The **species scale** is the entire geographic range of a species; individuals typically have no possibility of moving to most parts of the range. Metapopulations on opposite ends of the range of a species do not exchange individuals, but they remain part of the same genetic species because of movement among intermediate metapopulations.

The effect of metapopulation structure on the pattern of genetic variation within a species depends upon the spatial and temporal scale under consideration. For example, the effective population size in the short term is the amount of genetic drift from generation to generation within local populations. It can be measured by allele frequency changes from generation to generation (see Chapter 7). The long-term effective population size is the rate at which genetic variation is lost over long periods of time (tens or hundreds of generations) within metapopulations.

We can use our models of genetic subdivision introduced in Section 9.1 to see this relationship (Waples 2002). Effective population size is a measure of the rate of loss of heterozygosity over time. The short-term effective population size is related to the decline of the expected average heterozygosity within subpopulations (H_S). The long-term effective population size is related to the decline of the expected heterozygosity if the entire metapopulation were panmictic (H_T).

Consider a metapopulation consisting of six subpopulations of 25 individuals each, that are "ideal" as defined in Section 7.1 so that $N_e = N = 25$. The total population size of this metapopulation is $6 \times 25 = 150 = N_T$. The subpopulations are connected by migration under the island model of population structure so that each subpopulation contributes a proportion m of its individuals to a global migrant pool every generation, and each subpopulation receives the same proportion of migrants drawn randomly from this migrant

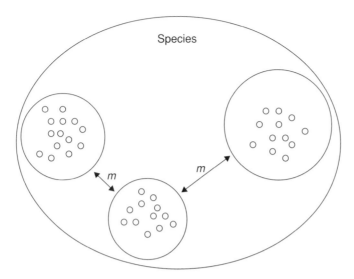

Figure 15.3 Hierarchical spatial organization of an entire species consisting of three metapopulations each consisting of a cluster of local populations that each exchange individuals. A small amount of gene flow between the three metapopulations (m) maintains the genetic integrity of the entire species. The two metapopulations on opposite ends of the range do not exchange individuals, but they remain part of the same genetic species because of movement between the intermediate metapopulation.

pool (see Section 9.4). The rate of decline of both H_S and H_T will depend upon the amount of migration among subpopulations (Figure 15.4).

In the case of complete isolation, the local effective population size is $N = 25$ and heterozygosities within subpopulations decline at a rate of $1/2N = 1/50 = 2.0\%$ per generation. However, different alleles will be fixed by chance in different subpopulations. Therefore, heterozygosity in the global metapopulation (H_T) will become "frozen" and will not decline. This can be seen in Figure 15.4. Five of the six isolated subpopulations went to fixation within the first 100 generations.

In the other extreme of effective panmixia among the subpopulations, the local effective population size will be N_T so that heterozygosities within subpopulations decline at a rate of $1/2N_T = 1/300 = 0.3\%$ per generation. In this case, the heterozygosity in the global metapopulation (H_T) will decline at the same rate as the local subpopulations. Eventually all subpopulations will go to fixation for the same allele so that H_T will become zero (Figure 15.4).

Thus, complete isolation will result in a small short-term effective population size, but greater long-term effective population size. The case of effective panmixia has the extreme opposite effect. That is, greater short-term effective population size, but smaller long-term effective population size (Figure 15.5).

The case of an intermediate amount of gene flow has the best of both worlds. The introduction of new genes by migration will maintain greater heterozygosities within local populations than the case of complete isolation. However, a small amount of migration will not be enough to restrain the subpopulations from drifting to near fixation of different alleles. Therefore, a small amount of gene flow will maintain nearly the same amount of

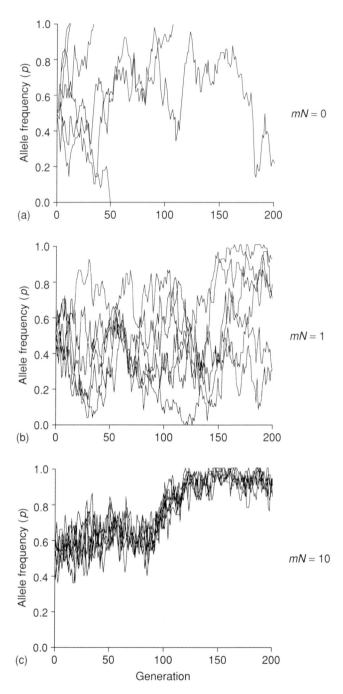

Figure 15.4 Changes in allele frequency in six subpopulations each of $N = 25$ connected by varying amount of migration under the island model. (a) The case of complete isolation ($m = 0$). (b) The case of one migrant per generation ($mN = 1$; $m = 0.04$). (c) The case of effective panmixia among subpopulations ($mN = 10$; $m = 0.4$). The graphs are from the *Populus* simulation program. From Alstad (2001).

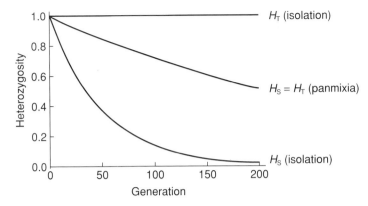

Figure 15.5 Expected decline in local (H_S) and total (H_T) heterozygosity in a population with six subpopulations of $N = 25$ each. In the case of effective panmixia ($mN = 10$) the decline in both local and global heterozygosities are equivalent and are equal to ($1/2N_T = 1/300$ per generation). In the case of complete isolation, local heterozygosity declines at an rate of $1/2N = 0.02$ per generation, but global heterozygosity is constant because random drift within local subpopulations causes the fixation of different alleles.

heterozygosity within local subpopulations as the case of effective panmixia and will maintain long-term heterozygosity at nearly the same rate as the case of complete isolation (Figure 15.4).

15.3 Effective population size

The effective population size of a metapopulation is extremely complex. Wright (1943) has shown that under these conditions with a large number of subpopulations that:

$$N_{eT} \approx \frac{N_T}{1 - F_{ST}} \tag{15.2}$$

where N_{eT} is the long-term effective population size of the metapopulation. Thus, increasing population subdivision (as measured by F_{ST}) will increase the long-term effective population size of the metapopulation (Nunney 2000). Expression 15.2 also indicates that that the effective size of the metapopulation will be greater than the sum of the N_es of the subpopulations when there is divergence among the subpopulations.

The validity of expression 15.2 and our conclusions for natural populations depend upon the validity of our assumptions of no local extinction ($e = 0$) and N within subpopulations being constant and equal. However, in the classic metapopulation of Levins (1970), extinction and recolonization of patches (subpopulations) is common. Wright (1940) pointed out that in the case of frequent local extinctions, the long-term N_e may be much smaller than the short-term N_e because of the effects of bottlenecks associated with recolonization:

$$N_e \text{ (long)} \ll N_e \text{ (short)}$$

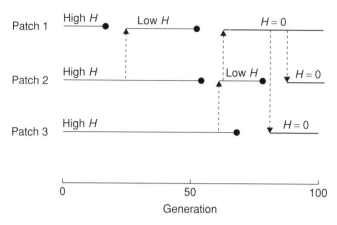

Figure 15.6 Effect of local extinction and recolonization by a few founders on the heterozygosity (*H*) in a metapopulation consisting of three habitat patches. Redrawn from Hedrick and Gilpin (1997).

For example, the entire ancestry of the darkly shaded group of related populations in Figure 15.1 have "passed through small groups of migrants six times in the period shown" (Wright 1940). Thus, these populations are expected to have low amounts of genetic variation even thought their current size may be very large.

This effect can be seen in Figure 15.6, which shows a metapopulation consisting of three habitat patches, from Hedrick and Gilpin (1997). The local populations in all three patches initially have high heterozygosity. The population in patch 1 goes extinct and is recolonized by a few individuals from patch 2 in generation 20, resulting in low heterozygosity. The population in patch 1 goes extinct and is recolonized again from patch 2. However, the few colonists from patch 2 have low heterozygosity because of an earlier extinction and recolonization in patch 2; this results in near zero heterozygosity in patch 1. Patches 2 and 3 are later recolonized by migrants from patch 1 so that heterozygosity is zero in the entire metapopulation.

Hedrick and Gilpin (1997) explored a variety of conditions with computer simulations to estimate long-term N_{eT} as a function of decline in H_T. They found that the rate of patch extinction (*e*) and the characteristics of the founders were particularly important. Slatkin (1977) described two extreme possibilities regarding founders. In the "**propagule** pool" model, all founders come from the same founding local population. In the "migrant pool" model, founders are chosen at random from the entire metapopulation. As expected, high rates of patch extinction greatly reduce N_{eT}. In addition, if vacant patches were colonized by a few founders, or if the founders came from the same subpopulation, rather than the entire metapopulation, H_T and N_{eT} are greatly reduced.

Relaxing the assumption that all subpopulations contribute an equal number of migrants also affects long-term N_{eT} as a function of decline in H_T. Nunney (1999) considered the case where differential productivity of the subpopulations brings about differential contributions to the migrant pool due to the accumulation of random differences among individuals in reproductive success. In this case, the effective size of a metapopulation (N_{eT}) is reduced by increasing F_{ST} by what Nunney (1997) has called "interdemic genetic drift".

As we have seen, most species have a large amount of heterozygosity at allozyme and nuclear DNA loci. On this basis, Hedrick and Gilpin (1997) concluded that most species have not functioned as a classic Levins-type metapopulation during their evolutionary history (see Figure 15.7).

We expect metapopulation dynamics to have a greater effect on variation at mtDNA than nuclear DNA because of the smaller effective population size of mtDNA (see Chapter 7). Grant and Leslie (1993) found that a variety of vertebrate species (mammals, birds, and fish) in southern Africa show greatly reduced amounts of variation at mtDNA compared to nuclear variation, relative to vertebrate species in the northern hemisphere. For example, a cichlid fish (*Pseudocrenilabrus philander*) had unusually high amounts of genetic variation both within ($H_S = 6.2\%$) and between populations ($F_{ST} = 0.30$) at allozyme loci. However, this species has virtually no genetic variation at mtDNA. Grant and Leslie (1993) suggested that the absence of genetic variation at mtDNA results from cycles of drought and rainfall in the semiarid regions of Africa, which have caused relatively frequent local extinctions and recolonizations that have not been severe enough to cause the loss of nuclear variability.

It is clear that the effect of metapopulation structure on the effective population size of natural populations is complex. The long-term effective population size may either be greater or less than the sum of the local N_es depending on a variety of circumstances: rates of extinction and recolonization, patterns of migration, and the variability in size and productivity of subpopulations. It is especially important to distinguish between local and global effective population size because these two parameters often respond very differently to the same conditions. All of these factors should be considered in evaluating conservation programs for endangered species (Waples 2002).

15.4 Population divergence and fragmentation

The effects of extinction and recolonization on the amount of divergence among populations (F_{ST}) is extremely complex (McCauley 1991). Gilpin (1991) has considered the effects of the relative rates of extinctions and recolonization on genetic divergence (Figure 15.7). If $e > c$, then the metapopulation is not viable. If both extinction and recolonization occur regularly and $c > e$, then patch coalescence will occur in which all patches descend from a single patch. For example, patches 1–5 (filled black) in Figure 15.1 coalesce to a single ancestral patch in seven steps back in time. If the rate of colonization is much greater than local extinctions ($c \gg e$) then all patches will have similar allele frequencies (panmixia). Allele frequency divergence ($F_{ST} > 0$) among local populations is expected only in a fairly narrow range of rates of colonization and extinction.

Perhaps most importantly, patterns of extinction and recolonization in nature may invalidate many inferences resulting from models that assume equilibrium (e.g., $F_{ST}^{\star} = 1/(4mN + 1)$ with the island model of migration). The effects of metapopulation dynamics (local extinctions and recolonizations) depend largely on the number and origin of founders that recolonize patches. We began this chapter with a model by Wright (1940) in which the genetic differentiation of local populations was enhanced because patches are founded by a few individuals so that genetic differentiation was enhanced by bottlenecks. In contrast, extinctions and recolonizations may act as a form of gene flow and reduce genetic differentiation if patches are founded by several individuals drawn from different patches (Slatkin 1987):

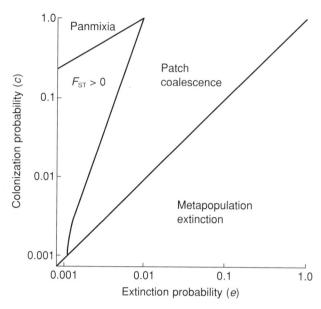

Figure 15.7 Effects of metapopulation dynamics on genetic divergence among local populations. From Gilpin (1991).

15.5 Genetic rescue

The term **genetic rescue** was coined to describe the increase in fitness of small populations resulting from the alleviation of inbreeding depression by immigrants (Thrall et al. 1998). Genetic rescue is generally considered to occur when the population fitness increases by more than can be attributed to the demographic contribution of migrant individuals (Ingvarsson 2001). Genetic rescue may play a critical role in the persistence of small natural populations and may, under some circumstances, be an effective conservation tool (Tallmon et al. 2004) (see Guest Box 15). Increasingly widespread evidence that genes from a pulse of immigrants into a local population often results in heterosis that increases population growth rate, also has important implications for the study of evolution and metapopulation dynamics. However, the occurrence of outbreeding depression following heterosis in the first generation indicates that care is needed in considering the source of populations for rescue (see Chapter 17).

Recent studies report positive fitness responses to low levels of migration (gene flow) into populations that have suffered recent demographic declines and suggest that natural selection can favor the offspring of immigrants (Example 15.1). Madsen et al. (1999) studied an isolated population of adders in Sweden that declined dramatically some 35 years ago, and that has since suffered from severe inbreeding depression. The introduction of 20 males from a large and genetically variable population of adders resulted in a dramatic demographic recovery of this population. This recovery was brought about by increased

Example 15.1 Genetic rescue of an isolated population of bighorn sheep

Hogg et al. (in press) documented genetic rescue in a natural population of bighorn sheep on the National Bison Range, an isolated wildlife refuge, in Montana. The population was founded in 1922 with 12 individuals from Alberta, Canada. The mean population size from 1922 to 1985 was approximately 40 individuals. Starting in 1985, 15 individuals (mostly from Alberta, Canada) were introduced over a 10-year period.

The restored gene flow caused an increase in the expected heterozygosity at eight microsatellite loci from 0.44 to more than 0.60. The gene flow also erased the genetic bottleneck signature consisting of a severe deficit of rare alleles (Hogg et al., in press).

Survival and reproductive success were remarkably higher in the outbred than in the inbred individuals. For example, the average annual reproductive success (number of lambs weaned) for females was 2.2-fold higher in outbred individuals compared to inbred resident animals. Average male annual reproductive success (number of lambs fathered) was 2.6-fold higher in outbred individuals. Survival for both females and males was higher in outbred individuals; average life span was 2 years longer in outbred animals compared to inbred ones with only resident genes.

This study was exceptional in that individual-based measures of fitness were available through a long-term (25-year) study. Most studies of genetic rescue report only a correlative increase in population size with genetic variation (e.g., mean heterozygosity), with no direct evidence that the increased genetic variation causes the increase in population size. In correlative studies, environmental factors could be the cause of increase in population size. In this study the inbred and outbred individuals coexisted in the same environment. This allowed for control of or removal of environmental effects. Finally, individual-based measures of fitness rescue are generally better than population-based measures (e.g., increased population growth rate) because it is possible to recover individual fitness without actually increasing population size, for example, if an environmental challenge or disease outbreak prevents a population size increase even following increased individual fitness.

survival rates, even though the number of litters produced by females per year actually declined during the initial phase of recovery. A genetic rescue effect has been uncovered in experimental populations of house flies, but only following many generations in which it was not detected (Bryant et al. 1999).

Two recent studies provide evidence that genetic rescue may be an important phenomenon. In experimentally inbred populations of a mustard (*Brassica campestris*), one immigrant per generation significantly increased the fitness of four of six fitness traits in treatment populations compared to (no immigrant) control populations (Newman and Tallmon 2001). Interestingly, there was no fitness difference between one-immigrant and 2.5-immigrant treatments after six generations, but there was greater phenotypic divergence among populations in the one-migrant treatment, which could facilitate local adaptation in spatially structured populations subject to divergent selection pressures. In small, inbred white campion populations, Richards (2000) found that gene flow increased germination success and that the success of immigrant pollen correlated positively with the amount of inbreeding in recipient populations.

Genetic rescue may be of critical importance to entire metapopulations by reducing local inbreeding depression and increasing the probability of local population persistence; in turn, this maintains a broad geographic range that buffers overall metapopulation extinction and provides future immigrants for other populations. Genetic rescue might also play a vital role in the spread of invasive species along the leading edge of invasion by supplying established, small propagules with adequate genetic variation to respond to selection and adapt to the new environment. In long-established plant populations, it is conceivable that plants emerging from long-dormant seed banks could also provide an intergenerational genetic rescue.

15.6 Long-term population viability

There is sometimes confusion regarding when to apply short- or long-term genetic goals, and how they relate to the conservation of local populations versus entire species. The short-term goals are appropriate for the conservation of local populations. As indicated above, those goals are aimed at keeping the rate of inbreeding at a tolerable level. The effective population sizes at which this may be achieved, however, are typically not large enough for new mutations to compensate for the loss of genetic variation through genetic drift. Some gene flow from neighboring populations is necessary to provide reasonable levels of genetic variation for quantitative traits to insure long-term persistence.

The long-term goal, where the loss of variation is balanced by new mutations, refers primarily to a global population, which may coincide with a species or subspecies that cannot rely on the input of novel genetic variation from neighboring populations. This global population may consist of one more-or-less panmictic unit, or it may be composed of multiple subpopulations that are connected by some gene flow, either naturally or through translocations (Mills and Allendorf 1996). It is the total assemblage of interconnected subpopulations that form a global population that must have an effective size meeting the criteria for long-term conservation (e.g., $N_e \geq 500-1,000$). The actual size of this global population will vary considerably from species to species depending on the number and size of the constituent subpopulations and on the pattern of gene flow between them (Waples 2002).

The accumulation of mildly deleterious mutations as considered in Section 14.6 may also affect the long-term viability of metapopulations (Higgins and Lynch 2001). Under some circumstances, metapopulation dynamics can reduce the effective population size so that even mutations with a selection coefficient as high as $s = 0.2$ can behave as nearly neutral and cause the erosion of metapopulation viability.

The long-term viability of a metapopulation or species is influenced by the number and complexity of the subpopulations. Metapopulation viability can be increased by the maintenance of a number of populations across multiple, diverse, and semi-independent environments as illustrated in Example 15.2, describing the study of Hilborn et al. (2003).

Example 15.2 Metapopulation structure and long-term productivity and persistence of sockeye salmon

Complex genetic population structure can play an important role in the long-term viability of populations and species. Sockeye salmon within major regions generally consist of hundreds of discrete or semi-isolated individual local demes (Hilborn et al. 2003). The amazing ability of sockeye salmon to return and spawn in their natal spawning sites results in substantial reproductive isolation among local demes. Local demes of sockeye salmon with major lake systems generally show pairwise F_{ST} values of 0.10–0.20 at allozyme and microsatellite loci, indicating relatively little gene flow.

These local demes occur in a variety of different habitats, which, combined with the low amount of gene flow, results in a complex of locally adapted populations. Sockeye salmon spawning in tributaries to Bristol Bay, Alaska, display a wide variety of life history types associated with different breeding and rearing habitats. Bristol Bay sockeye salmon spawn in streams and rivers from 10 cm to several meters deep in substrate ranging from small gravel to cobble. Some streams have extremely clear water while others spawn in sediment-laden streams just downstream from melting glaciers. Sockeye salmon also spawn on beaches in lakes with substantial ground water. Different demes spawn at different times of the year. The date of spawning is associated with the long-term average thermal regime experienced by incubating eggs so that fry emerge in the spring in time to feed on zooplankton and aquatic insects. Fish from different demes have a variety of morphological, behavior, and life history differences associated with this habitat complexity.

Up to 40 million fish are caught each year in the Bristol Bay sockeye fishery in several fishing areas associated with different major tributaries. There is a large year-to-year variability in overall productivity but the range of the productivity of this fishery has been generally consistent for nearly 100 years (Figure 15.8). However, the productivity of different demes and major drainages has changed dramatically over the years. The relative productivity of local demes has changed as the marine and freshwater climates change. Local reproductive units that are minor components of a mixed stock fishery during one climatic regime may dominate during others. Therefore, maintaining productivity over long time scales requires protecting against the loss of local populations during certain environmental regimes.

The long-term stability of this complex system stands in stark contrast to the dramatic collapse and extirpation of a highly productive population of an introduced population of this species in the Flathead River drainage of Montana (Spencer et al. 1991). The life history form of this species that spends its entire life in fresh water is known as kokanee. Sockeye salmon were introduced into Flathead Lake in the early 20th century, and by the 1970s some 50,000–100,000 fish returned to spawn in one primary local population and supported a large recreational fishery. Opossum shrimp were introduced into Flathead Lake in 1983 and had a major effect on the food web in this ecosystem. A primary effect was the

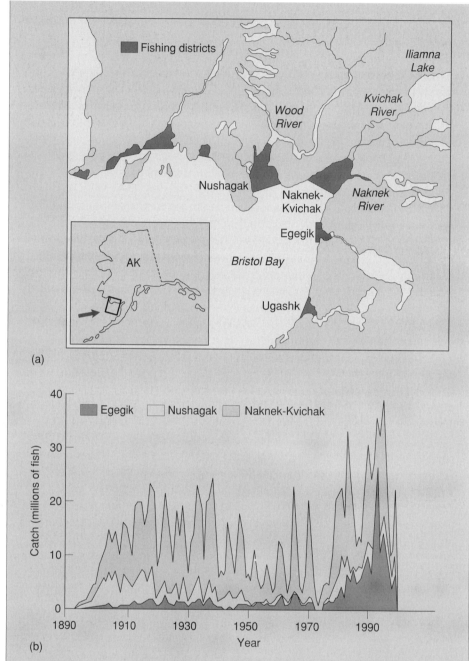

Figure 15.8 (a) Map of fishing districts (dark areas) around Bristol Bay, Alaska.
(b) Catch history of the three major sockeye salmon fishing areas within Bristol Bay.
The overall productivity of the system has been generally stable, but the relative
contributions of the three major areas have changed greatly. For example, the Egegik
district (see map) generally contributed less than 5% to the fishery until 1975, but has
been a major contributor since then.

predation of opossum shrimp on the zooplankton that was the major food resource of the kokanee. This productive single deme of kokanee went from over 100,000 spawners in 1985 to extirpation just 3 years later.

This example illustrates the importance of maintaining multiple semi-isolated subpopulations with different life histories to help insure long-term population and species persistence.

Guest Box 15   Fitness loss and genetic rescue in stream-dwelling topminnows
Robert C. Vrijenhoek

The "guppy-sized" livebearing topminnow *Poeciliopsis monacha* inhabits rocky arroyos in northwestern Mexico. The upstream portion of a small stream, the Arroyo de los Platanos, dried completely during a severe drought in 1976, but within 2 years fish recolonized this area from permanent springs that exist downstream. The founder population was homozygous at loci that were polymorphic in the source population, a loss of variation that corresponded with manifestations of inbreeding depression.

Fitness of the source and founder populations was compared with that of coexisting asexual forms of *Poeciliopsis* that experienced the same extinction/recolonization event. The reproductive mode of cloning preserves the heterozygosity and limits inbreeding depression in the asexual fish. Compared to local clones, the inbred founder population of *P. monacha* exhibited poor developmental stability (e.g., asymmetry in bilateral morphological traits) and an increased parasite load (Vrijenhoek and Lerman 1982; Lively et al. 1990). Genetic variation at allozyme loci in this species is associated with the ability to survive seasonal stresses in these desert streams – cold temperatures, extreme heat, and hypoxia (Vrijenhoek et al. 1992).

Prior to the extinction event in 1976, the sexual *P. monacha* constituted 76% of the fish population in the upper Platanos and 24% of the fish were asexual. After recolonization the sexual fish constituted no more than 10% for the next 5 years (10–15 generations). Corresponding frequency shifts did not occur downstream in permanent springs where levels of heterozygosity remained stable in *P. monacha*.

By 1983, *P. monacha* had been eliminated from several small pools in the upper Platanos while the clones flourished there. We rescued the founder population by transplanting 30 genetically variable females from a downstream location where *P. monacha* was genetically variable. By the following spring (2–3 generations), *P. monacha* regained numerical dominance over the clones (Vrijenhoek 1989), and its parasite loads dropped to levels that were typical of the permanent localities downstream (Lively et al. 1990). Restoration of genetic variability reversed inbreeding depression in *P. monacha* and restored its fitness relative to that of the competing fish clones (Figure 15.9).

Setting aside the special reproductive features of *Poeciliopsis*, it is easy to imagine similar interactions between a rare endangered species and its competitors and

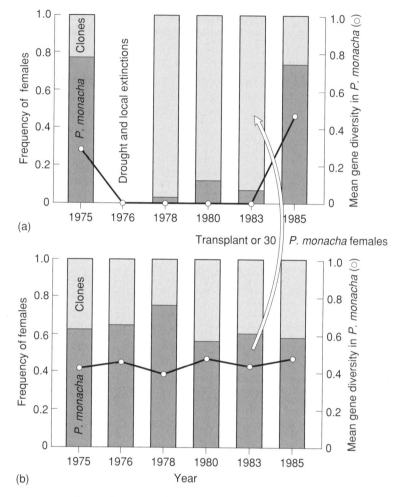

Figure 15.9 Population dynamics of *Poeciliopsis* topminnows. (a) The upper portion of the Arroyo de los Platanos. (b) The main stream of the Arroyo de Jaguari. Histogram bars, arranged by year, record the frequencies of *P. monacha* females (dark gray) and the triploid clones (light gray) in each sample. The mean gene diversity in *P. monacha* across four polymorphic allozyme loci is traced by the black line. The open arrow indicates a single transplant of 30 *P. monacha* females into the upper Platanos population in 1983. Modified from Vrijenhoek (1989).

parasites. Furthermore, loss of heterozygosity in small populations and inbreeding depression can have manifold effects on fitness that might reduce a population's capacity to resist displacement by alien competitors and combat novel diseases. The genotypic differences among individuals of a sexually reproducing species help to reduce intraspecific competition and provide the variability needed to persist in an evolutionary arms race with rapidly evolving parasites and pathogens (Van Valen 1973).

Problem 15.1

We saw in this chapter that we expect metapopulation dynamics of relatively high rates of local extinction (e) and recolonization (c) to have a greater effect on genes coded in the mitochondria than in the nucleus because of the smaller N_e of mtDNA. Do you think it would be possible to identify populations with high rates of local extinction by a comparison of the patterns of genetic variation at mtDNA versus nuclear genes in a species?

Problem 15.2

Explain how the process of extinction and recolonization in local populations could reduce genetic variation in an entire population or species.

Problem 15.3

In the extreme case of complete isolation of subpopulations within a metapopulation, how will heterozygosity in the global metapopulation (H_T) change? Why? How will heterozygosity within local subpopulations (H_S) change?

Problem 15.4

In a model by Wright (1940), the genetic differentiation among local populations was enhanced by metapopulation structure. In contrast, work by Slatkin (1987) suggested that metapopulation structure can reduce genetic differentiation among populations. How can we reconcile these two views? Explain how both can be true depending on the characteristics of the metapopulation.

Problem 15.5

Does demographic or genetic connectivity require more exchange of individuals between local populations? How would your answer to this question change if the local populations were very small (say $N_e < 100$) or very large (say $N_e \gg 1,000$)?

16

Units of Conservation

Grand skink, Section 16.4.2

The zoo directors, curators, geneticists and population biologists who attempt to pursue the elusive goal of preservation of adaptive genetic variation are now considering the question of which gene pools they should strive to preserve.

Oliver A. Ryder (1986)

The choices of what to conserve must often be made with regard to populations that are not separate completely from others, or when information regarding the relationships and degrees of distinction among populations is very incomplete.

Jody Hey et al. (2003)

The identification of appropriate taxonomic and population units for protection and management is essential for the conservation of biological diversity. For species identification and classification, genetic principles and methods are relatively well developed; nonetheless species identification can be controversial. Within species, the identification and protection of genetically distinct local populations should be a major focus in conservation because the conservation of many distinct populations helps maximize evolutionary potential and minimize extinction risks (Hughes et al. 1997; Hilborn et al. 2003; Luck et al. 2003). Furthermore, the local population is often considered the functional unit in ecosystems.

Identification of population units is necessary so that management and monitoring programs can be efficiently targeted toward distinct or independent populations. Biologists and managers must be able to identify populations and geographic boundaries between populations in order to effectively plan harvesting quotas (e.g., to avoid overharvesting) or to devise translocations and reintroductions of individuals (e.g., to avoid mixing of adaptively differentiated populations). In addition, it is sometimes necessary to prioritize which population units (or taxa) to conserve because limited financial resources preclude conservation of all units.

Finally, many governments and agencies have established legislation and policies to protect intraspecific population units. This requires the identification of population units. For example, the ESA (Endangered Species Act of the USA) allows listing and full protection of **distinct population segments** (DPS) of vertebrate species (Example 16.1). Species and subspecies identification is based upon traditional, established taxonomic criteria as well as genetic criteria (although criteria for species identification are sometimes controversial). The choice of criteria to use to delineate intraspecific units for conservation has been highly controversial. Other countries, for example in Europe and Australia, also have laws that depend on the identification of distinct taxa and populations for the protection of species and habits (Example 16.1).

Example 16.1 The US Endangered Species Act (ESA) and conservation units

The ESA of the United States is one of the most powerful pieces of conservation legislation in the world. The ESA has been a major stimulus motivating biologists to develop criteria for identifying population units for conservation. This is because the ESA provides legal protection for subspecies and "distinct population segments" (DPSs) of vertebrates, as if they were full species. According to the ESA:

> The term "species" includes any subspecies of fish or wildlife and plants, and any distinct population segment of any species of vertebrate fish or wildlife which interbreeds when mature.

However, the ESA does not provide criteria or guidelines for delineating DPSs. The identification of intraspecific units for conservation is controversial. This is not surprising given that the definition of a "good species" is controversial (see Section 16.5). Biologists have vigorously debated the criteria for identifying DPSs and other conservation units ever since the US Congress extended full protection of the ESA to "distinct" populations, but did not provide guidelines.

Legislations in other countries around the world have provisions that recognize and protect intraspecific units of conservation. For example, Canada passed the Species at Risk Act (SARA) in 2003. The SARA aims to "prevent wildlife species from becoming extinct, and to secure the necessary actions for their recovery". Under the SARA, "wildlife species" means a species, subspecies, variety, or geographically or genetically distinct population of animal, plant, or other organism, other than a bacterium or virus, which is wild by nature.

In Australia, the Endangered Species Protection Act (ESPA) also allows protection for subspecies and distinct populations. But, like the ESA in the United States, there are problems with defining and identifying intraspecific units (Woinarski and Fisher 1999).

In this chapter, we examine the components of biodiversity and then consider methods to assess taxonomic and population relationships. We discuss the criteria, difficulties, and controversies in the identification of conservation units. We also consider the identification of appropriate population units for legal protection and for management actions (e.g., supplemental translocation of individuals between geographic regions). Recall that in the previous chapter, we considered three spatial scales of genetic population structure for conservation: local, metapopulation, and species.

16.1 What should we try to protect?

Genes, species, and ecosystems are three primary levels of biodiversity (Figure 16.1) recognized by the IUCN. There has been some controversy as to which level should receive priority for conservation efforts (e.g., Bowen 1999). However, it is clear that all three levels must be conserved for successful biodiversity conservation. For example, it is

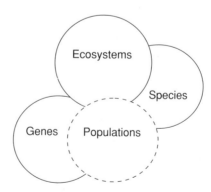

Figure 16.1 Primary levels of biodiversity recognized by the IUCN (solid circles), and a fourth level – populations – recognized as perhaps most crucial for species' long-term persistence (Hughes et al. 1997; Luck et al. 2003). In reality, biodiversity exists across a continuum of many hierarchical levels of organization including genes, genomes (i.e., multilocus genotypes), local populations, communities, ecosystems, and biomes. Additional levels of diversity include metapopulations, subspecies, genera, families, and so on.

as futile to conserve ecosystems without species, as it is to save species without large, healthy ecosystems.

An example of this kind of futility is that of the African rhinoceros, which are being protected mainly in zoos and small nature reserves, but for which little habitat (free from poachers) is currently available. Without conserving vast habitats for future rhino populations, it seems pointless to protect rhinos in small nature reserves surrounded by armed guards and fences.

It is not too late for rhinos. Vast habitats do exist, and rhinos could be successful in these habitats if poaching is eliminated. In addition to conserving rhino species and their habitats, it is also important to conserve genetic variation within rhino species because variation is a prerequisite for long-term adaptive change and the avoidance of fitness decline through inbreeding depression (see Chapter 14). Clearly, it is important to recognize and conserve all levels of biodiversity: ecosystems, species, and genes.

The debate over whether to protect genes, species, or ecosystems is, in a way, a false trichotomy because each level is an important component of biodiversity as a whole. Nonetheless, considering each level separately can help us appreciate the interacting components of biodiversity, and the different ways that genetics can facilitate conservation at different levels. Appreciation of each level also can promote understanding and multidisciplinary collaborations across research domains. Finally, a fourth level of biodiversity – that of genetically distinct local populations – is arguably the most important level for focusing conservation efforts (Figure 16.1). The conservation of multiple, genetically distinct populations is necessary to insure long-term species survival and the functioning of ecosystems, as mentioned above (Luck et al. 2003).

We can also debate which temporal component of biodiversity to prioritize for conservation: past, present, or future biodiversity. All three components are important, although future biodiversity often warrants special concern (Example 16.2).

Example 16.2 Temporal considerations in conservation: past, present, and future

What temporal components of biodiversity do we wish to preserve? Do we want to conserve ancient isolated lineages, current patterns of diversity (ecological and genetic), or the diversity required for future adaptation and for novel diversity to evolve? Most would agree "all of the above". All three temporal components are interrelated and complementary (Figure 16.2). For example, conserving current diversity helps insure future adaptive potential. Similarly, conserving and studying ancient lineages ("living fossils") can help us understand factors important for long-term persistence. Nonetheless, one can argue that the most important temporal component to consider is future biodiversity, i.e., the ability of species and populations to adapt to future environments (e.g., global climate change). If populations do not adapt to future environments then biodiversity will decline – leading to loss of ecosystem functioning and services. Figure 16.2 illustrates how different temporal components of biodiversity (past, present, and future) can be related to different scientific disciplines (systematics, ecology, and evolutionary biology, respectively). These components also are often related to different hierarchical levels of biodiversity: species, ecosystems, and genes, respectively.

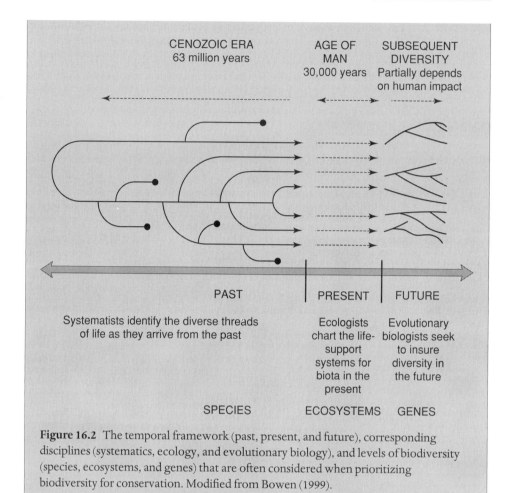

Figure 16.2 The temporal framework (past, present, and future), corresponding disciplines (systematics, ecology, and evolutionary biology), and levels of biodiversity (species, ecosystems, and genes) that are often considered when prioritizing biodiversity for conservation. Modified from Bowen (1999).

Another choice that is often debated is whether we should emphasize protecting the existing **patterns** of diversity or the **processes** that generate diversity (e.g., ecological and evolutionary processes themselves)? Again the answer is, in general, both. It is clear that we should prioritize the preservation of the process of adaptation so that populations and species can continually adapt to future environmental changes. However, one important step toward preserving natural processes is to quantify, monitor, and maintain natural patterns of population subdivision and connectivity (e.g., identify intraspecific population units and boundaries). This, for example, would prevent extreme fragmentation and promote continued natural patterns of gene flow among populations.

How do we conserve the "processes" of evolution, including adaptive evolutionary change? We must first maintain healthy habitats and large wild populations because only in large populations can natural selection proceed efficiently (see Section 8.5). In small populations, genetic drift leads to random genetic change, which is generally nonadaptive. Drift can preclude selection from maintaining beneficial alleles and eliminating deleterious ones. To maintain evolutionary process we also must preserve multiple populations –

ideally from different environments so that selection pressures remain diverse and multilocus genotype diversity remains high. In this scenario, a wide range of local adaptations are preserved within species, as well as some possibility of adaptation to different future environmental challenges.

16.2 Systematics and taxonomy

The description and naming of distinct taxa is essential for most disciplines in biology. In conservation biology, the identification of taxa (taxonomy) and assessing their evolutionary relationships (systematics) is crucial for the design of efficient strategies for biodiversity management and conservation. For example, failing to recognize the existence of a distinct and threatened taxon can lead to insufficient protection and subsequent extinction. Identification of too many taxa (oversplitting) can waste limited conservation resources. The misidentification of a sister taxon could lead to nonideal choice of source populations for supplementing endangered populations.

There are two fundamental aspects of evolution that we must consider: phenotypic change through time (**anagenesis**) and the branching pattern of reproductive relationships among taxa (**cladogenesis**). The two primary taxonomic approaches are based on these two aspects.

Historically, taxonomic classification was based primarily upon phenotypic similarity (**phenetics**), which reflects evolution via anagenesis. That is, groups of organisms that were phenotypically similar were grouped together. This classification is conducted using clustering algorithms (described below) that group organisms based exclusively on "overall similarity" or outward appearance. For example, populations that share similar allele frequencies are grouped together into one species. In this example, the clustering by overall similarity of allele frequencies is phenetic. The resulting diagram (or tree) used to illustrate classification is called a **phenogram**, even if based upon genetic data (e.g., allele frequencies).

A second approach is to classify organisms on the basis of their phylogenetic relationships (**cladistics**). Cladistic methods group together organisms that share **derived** traits (originating in a common ancestor), reflecting cladogenesis. Under cladistic classification, only **monophyletic** groups can be recognized, and only genealogical information is considered. The resulting diagram (or tree) used to illustrate relationships is called a **cladogram** (or sometimes, a **phylogeny**). Phylogenetics is discussed below (see Section 16.3).

Our current system of taxonomy combines cladistics and phenetics, and it is sometimes referred to as evolutionary classification (Mayr 1981). Under evolutionary classification, taxonomic groups are usually classified on the basis of phylogeny. However, groups that are extremely phenotypically divergent are sometimes recognized as separate taxa even though they are phylogenetically related. A good example of this is birds (Figure 16.3). Birds were derived from a dinosaur ancestor, as evidenced from the fossil record showing reptiles with feathers (bird–reptile intermediates). Therefore, birds and dinosaurs are sister groups that should be classified together under a strictly cladistic classification scheme. However, birds underwent rapid evolutionary divergence associated with their development of flight. Therefore, birds are classified as a separate class while dinosaurs are classified as a reptile (class Reptilia).

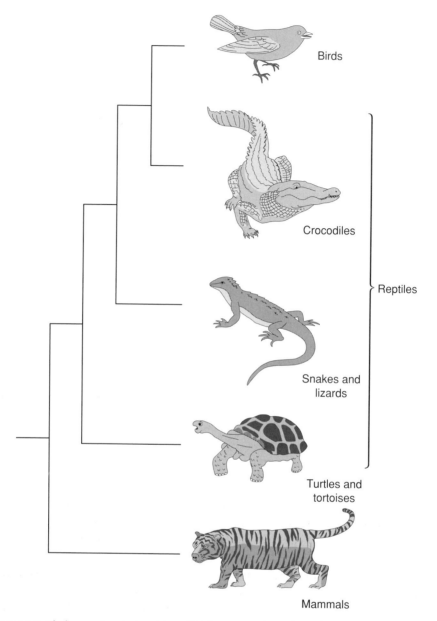

Figure 16.3 Phylogenetic relationships of birds, mammals, and reptiles. Note that crocodiles and birds are more closely related to each other than either is to other reptiles. That is, crocodiles share a more recent common ancestor with birds than they do with snakes, lizards, turtles, and tortoises. Therefore, the classification of the class Reptilia is not monophyletic.

There is a great deal of controversy associated with the "correct" method of classification. We should use all kinds of information available (morphology, physiology, behavior, life history, geography, and genetics) and the strengths of different schools (phenetic and cladistic) when classifying organisms (Mayr 1981; see also Section 16.6).

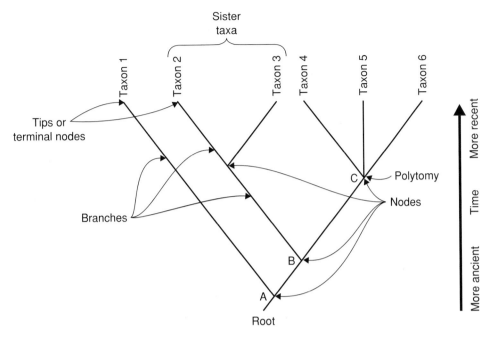

Figure 16.4 A phylogenetic tree (phylogeny). A **polytomy** (node 'C') is when more than two taxa are joined at the same node because data cannot resolve which (two) of the three taxa are most closely related. A widely controversial polytomy 10–20 years ago was that of chimpanzees, gorillas, and humans. However, extensive genetic data now show that chimps and humans are more closely related (i.e., sister taxa). From Freeman and Herron (1998).

16.3 Phylogeny reconstruction

A phylogenetic tree is a pictorial summary that illustrates the pattern and timing of branching events in the evolutionary history of taxa (Figure 16.4). A phylogenetic tree consists of **nodes** for the taxa being considered and branches that connect taxa and show their relationships. Nodes are at the tips of branches and at branching points (representing extinct ancestral taxa). A phylogenetic tree represents a hypothesis about relationships that is open to change as more taxa or characters are added. The same phylogeny can be drawn many different ways. Branches can be rotated at any node without changing the relationship between the taxa, as illustrated in Figure 16.5.

Branch lengths are often proportional to the amount of genetic divergence between taxa. If the amount of divergence is proportional to time, a phylogeny can show time since divergence between taxa. Molecular divergence (through mutation and drift) will be proportional to time if mutation accumulation is stochastically constant (like radioactive decay). The idea that molecular divergence can be constant is called the **molecular clock** concept. In conservation biology, the molecular clock and divergence estimates can help identify distinct populations and prioritize them based on their distinctiveness or divergence times. One serious problem with estimating divergence times is that extreme genetic drift (e.g., bottlenecks and founder events) can greatly inflate estimates of

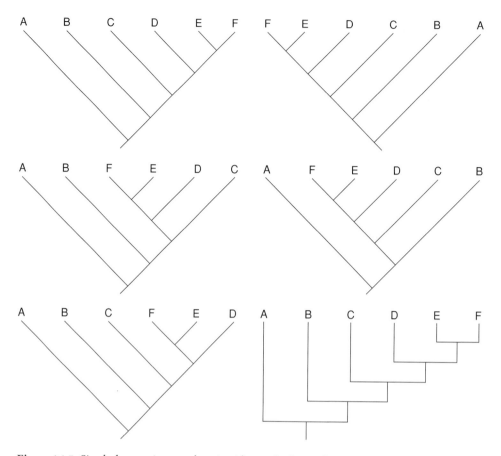

Figure 16.5 Six phylogenetic trees showing identical relationships among taxa. Note that branches can be rotated at the nodes without changing relationships represented on the trees (e.g., E vs. F in the top two trees). From Freeman and Herron (1998).

divergence times leading to long branch lengths and misleading estimates of phylogenetic distinctiveness (see Section 9.7).

16.3.1 Methods

There are two basic steps in phylogeny reconstruction: (1) generate a matrix of character states (e.g., derived versus ancestral states); and (2) build a tree from the matrix. Cladistic methods use only shared derived traits, **synapomorphies**, to infer evolutionary relationships. Phenogram construction is based on overall similarity. Therefore, a phylogenetic tree may have a different topology from a phenogram using the same character state matrix (Example 16.3).

   The actual construction of phylogenies is much more complicated than this simple example. It is sometimes difficult to determine the ancestral state of characters. Moreover, the number of possible evolutionary trees to compare rises at an alarming rate. For example, there are nearly 35 million possible rooted, bifurcating trees with just 10 taxa and over 8×10^{21} possible trees with 20 taxa! In addition, there are a variety of other methods

Example 16.3 Phenogram and cladogram of birds, crocodiles, and lizards

As we have seen, birds and crocodiles are sister taxa based upon phylogenetic analysis, but crocodiles are taxonomically classified as reptiles because of their phenetic similarity with snakes, lizards, and turtles. These conclusions are based on a large number of traits. Here we will consider five traits (Table 16.1) to demonstrate how a different phenogram and cladogram can result from the matrix of character states.

Lizards and crocodiles are more phenotypically similar to each other than either is to birds because they share three out of five traits (0.6), while crocodile and birds share just two out of five traits (0.4) (Table 16.2). We can construct a phenogram based upon clustering together the most phenotypically similar groups (Figure 16.6a). The phenotypic similarity of lizards and crocodiles results from their sharing ancestral character states because of the rapid phenotypic changes that occurred in birds associated with adaptation to flight.

Parsimony methods were among the first to be used to infer phylogenies, and they are perhaps the easiest phylogenetic method to explain and understand (Felsenstein 2004, p. 1). There are many possible phylogenies for any group of taxa. Parsimony is the principle that the phylogeny to be preferred is the one that

Table 16.1 Character states for five traits used to construct a phenogram and cladogram of lizards, crocodiles, and birds. Traits: 1, heart (three or four chambered); 2, inner ear bones (present or absent); 3, feathers (present or absent); 4, wings (present or absent); and 5, hollow bones (present or absent).

	Traits*				
Taxon	1	2	3	4	5
A Lizards	0	0	0	0	0
B Crocodiles	1	1	0	0	0
C Birds	1	1	1	1	1

* 0, ancestral; 1, derived.

Table 16.2 Phenotypic similarity matrix for lizards, crocodiles, and birds based upon the proportion of shared characters states in Table 16.1.

	Lizards	Crocodiles	Birds
Lizards	1.0		
Crocodiles	0.6	1.0	
Birds	0.0	0.4	1.0

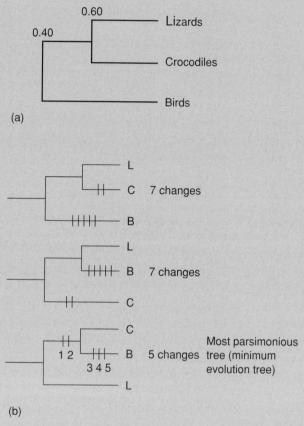

(a)

(b)

Figure 16.6 (a) Phenogram and (b) cladograms showing phenotypic and evolutionary relationships, respectively, among lizards, crocodile, and birds. Numbers in (a) are genetic distance estimates (e.g., 0.60 distance units between lizards and crocodiles). Vertical slashes in (b) on branches represent changes. Numbers below slashes on the bottom (most parsimonious) tree correspond to the traits (i.e., evolutionary change in traits) listed in Table 16.1.

requires the minimum amount of evolution. To use parsimony, we must search all possible phylogenies and identify the one or ones that minimize the number of evolutionary changes.

There are only three possible bifurcating phylogenies for lizards, crocodiles, and birds. Figure 16.6b shows these trees and the number of evolutionary changes from the ancestral to the derived trait to explain the character state matrix. The upper two phylogenies both require seven changes because certain evolutionary changes had to occur independently in the crocodile and bird branches. The bottom tree requires only five evolutionary changes to explain the character state matrix. Thus, the bottom tree is the most parsimonious. Birds and crocodiles form a monophyletic group because they share two synapomorphies (traits 1 and 2).

besides parsimony for inferring phylogenies (Hall 2004). The field of inferring phylogenies has been marked by more heated controversy than perhaps any other area of evolutionary biology (see Felsenstein 2004).

16.3.2 Gene trees and species trees

It is important to realize that different genes can result in different phylogenies of species, and that gene trees are often different from the true species phylogeny (Nichols 2001). Different gene phylogenies can arise due to four main phenomena: lineage sorting and associated genome sampling error, sampling error of individuals or populations, natural selection, or introgression (following hybridization).

Lineage sorting and sampling error

Ancestral **lineage sorting** occurs when different DNA sequences from a mother taxon are sorted into different daughter species such that lineage divergence times do not reflect population divergence times. For example, two divergent lineages can be sorted into two recently isolated populations, where less-divergent lineages might become fixed in different ancient daughter populations. Lineage sorting makes it important to study many different genes (or independent DNA sequences) to avoid sampling error associated with sampling too few (or an unrepresentative set of) genetic characters (loci).

Sampling error of individuals occurs when too few individuals or nonrepresentative sets of individuals are sampled from a species such that the inferred gene tree differs from the true species tree. For example, many early studies using mtDNA analysis included only a few individuals per geographic location, which could lead to erroneous phylogeny inference. Limited sampling is likely to detect only a subset of local lineages (i.e., alleles), especially for lineages at low frequency.

We can use simple probability to estimate the sample size that we need to detect rare genotypes. For example, how many individuals must we sample to have a greater than 95% chance of detecting an allele with frequency of 0.10 ($p = 0.1$)? Each time we examine one sample, we have a 0.90 chance ($1 - p$) of not detecting the allele in question and a 0.10 chance (p) of detecting it. Using the product rule (see Appendix Section A1), the probability of not detecting an allele at $p = 0.1$ in a sample of size x is $(1 - p)^x$. Therefore, the sample size required to have a 95% chance of sampling an allele with frequency of 0.10 is 29 haploid individuals or 15 diploids for nuclear markers: $(1 - 0.1)^{29} = 0.047$.

Natural selection

Directional selection can cause gene trees to differ from species trees if a rare allele increases rapidly to fixation because of natural selection (**selective sweep**, see Section 10.3.1). For example, a highly divergent (ancient) lineage may be swept to fixation in a recently derived species. Here the ancient age of the lineages would not match the recent age of the newly derived species. In another example, balancing selection could maintain the same lineages in each of two long-isolated species, and lead to erroneous estimation of species divergence, as well as a phylogeny discordant with the actual species phylogeny (and with neutral genes). To avoid selection-induced errors in phylogeny reconstruction, many loci should be used. Analysis of many loci can help identify a locus with unusual (deviant) phylogenetic patterns due to selection (as in Section 9.6.3). For example, selection might cause rapid divergence at one locus that is not representative of the rest of the genome (or of the true species tree).

Introgression

Introgression also causes gene trees to differ from species trees. For example, hybridization and subsequent backcrossing can cause an allele from species X to introgress into species Y. This has happened between wolves and coyotes that hybridize in northeastern United States, where coyote mtDNA has introgressed into wolf populations. Here, female coyotes hybridize with male wolves, followed by the F_1 hybrids mating with wolves, such that coyote mtDNA introgresses into wolf populations (Roy et al. 1994). This kind of unidirectional introgression of mtDNA (maternally inherited) has been detected in deer, mice, fish, and many other species.

MtDNA gene tree versus species tree

An example of a gene tree not equaling the species tree is illustrated in a study of mallard ducks and black ducks (Avise 1990). The black duck apparently recently originated (perhaps via rapid phenotypic evolution) from the more widely distributed mallard duck. This likely occurred when a peripheral mallard population became isolated, evolved into the black duck and became fixed for a single mtDNA lineage (e.g., via lineage sorting or selection). The mallard population is much larger and maintains several divergent mtDNA lineages, including the lineage fixed in the black duck (Figure 16.7). Thus, while the black duck is monophyletic, the mallard is paraphyletic relative to the black duck for mtDNA. Because the black duck mtDNA is common in the mallard, the black duck appears to be part of the mallard species when considering only mtDNA data. However, the black duck has important phenotypic, adaptive, and behavioral differences meriting recognition as a separate species.

   This duck example illustrates a problem that is likely to occur when identifying species from molecular data alone (and from only one locus). It shows the importance of considering nonmolecular characteristics (such as life history, morphology, and geography) along with the molecular data (see Section 16.6). This example is analogous to the widely cited example of brown bears that are paraphyletic to polar bears for mtDNA lineages. Despite the lack of **monophyly**, brown bears have important phenotypic, adaptive, and behavioral differences meriting recognition as a separate species apart from polar bears (Paetkau 1999).

16.4 Description of genetic relationships within species

Identifying populations and describing population relationships is crucial for conservation and management (e.g., monitoring population status, measuring gene flow, and planning translocation strategies). Population relationships are generally assessed using multilocus allele frequency data and statistical approaches for clustering individuals or populations with a dendrogram or tree in order to identify genetically similar groups.

   Population trees and phylogenetic trees look similar to each other, but they display fundamentally different types of information. Phylogenies show the time since the most recent

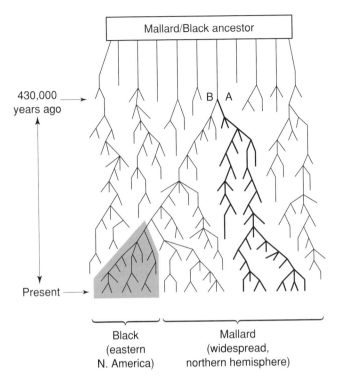

Figure 16.7 Simplified diagrammatic representation of the possible matriarchal ancestry of mallard and black ducks. The mtDNA lineage A is shown in dark lines, and the black duck portion of the phylogeny is shaded. From Avise (1990).

common ancestor (TMRCA) between taxa. Phylogenies represent relationships among taxa that have been reproductively isolated for many generations. A phylogeny identifies monophyletic groups – isolated groups that shared a common ancestor. Phylogenetic trees can be used both for species and for genes (e.g., mtDNA) (Nichols 2001). In the case of species, the branch points represent speciation events; in the case of genes, branch points represent common ancestral genes.

Population trees, in contrast, generally identify groups that have similar allele frequencies because of ongoing genetic exchange (i.e., gene flow). The concept of TMRCA is not meaningful for populations with ongoing gene flow. Populations with high gene flow will have similar allele frequencies and cluster together in population trees.

The differences between population and phylogenetic trees, as described here, are somewhat oversimplified to help explain the differences. In reality there is a continuum in the degree of differentiation among populations in nature. Some populations within the same species may have been reproductively isolated for many generations. In this case, genealogical information and the phylogenetic approach can be used to infer population relationships (see Section 16.4.3).

The description of genetic population structure is the most common topic for a conservation genetics paper in the literature. Individuals from several different geographic locations are genotyped at a number of loci to determine the patterns and amounts of gene flow among populations. This population-based approach assumes that all individuals sampled from one area were born there and represent a local breeding population. However, new powerful approaches have been developed that allow the description of population structure using an individual-based approach. That is, many individuals are sampled, generally over a wide geographic range, and then placed in population units on the basis of genotypic similarity.

16.4.1 Population-based approaches

A bewildering variety of approaches have been used to describe the genetic relationships among a series of populations. We will discuss several representative approaches.

The initial step in assessing population relationships (after genotyping many individuals) often is to conduct statistical tests for differences in allele frequencies between sampling locations. For example, a chi-square test is used to test for allele frequency differences between samples (e.g., Roff and Bentzen 1989). If two samples are not significantly different, they often are pooled together to represent one population. It can be important to resample from the same geographic location (in different years or seasons) to test for sampling error and for stability of genetic composition through time. After distinct population samples have been identified, the genetic relationships (i.e., genetic similarity) among populations can be inferred.

Population dendrograms

Population relationships are often assessed by constructing a dendrogram based on the genetic similarity of populations. The first step in dendrogram construction is to compute a genetic differentiation statistic (e.g., F_{ST} or Nei's D; see Sections 9.1 and 9.7) between each pair of populations. A genetic distance can be computed using any kind of molecular marker (e.g., allozyme frequencies, DNA haplotypes) and a vast number of metrics (e.g., Cavalli-Sforza's chord distance, Slatkin's R_{ST}, and Wright's F_{ST}; see Section 9.7). This yields a **genetic distance matrix** (Table 16.3).

The second step is to use a clustering algorithm to group populations with similar allele frequencies (e.g., low F_{ST}). The most widely used cluster algorithms are UPGMA

Table 16.3 Genetic distance (D; Nei 1972) matrix based upon allele frequencies at 15 allozyme loci for five populations of a perennial lily. Data from Godt et al. (1997).

	Population				
	FL1	**FL2**	**FL3**	**SC**	**NC**
FL1	–				
FL2	0.001	–			
FL3	0.003*	0.002*	–		
SC	0.029	0.032	0.030	–	
NC	0.059	0.055	0.060	0.062	–

Asterisks and underlining are explained in Example 16.4.

Example 16.4 Dendrogram construction via UPGMA clustering of lily populations

UPGMA (unweighted pair group method with arithmetic averages) clustering was used to assess relationships among five populations of a perennial lily (*Tofieldia racemosa*) from northern Florida (Godt et al. 1997). Allele frequencies from 15 polymorphic allozyme loci were used to construct a genetic distance matrix (Table 16.3) and subsequently a dendrogram using the UPGMA algorithm.

The UPGMA algorithm starts by finding the two populations with the smallest interpopulation distance in the matrix. It then joins the two populations together at an internal node. In our lily example here, population FL1 and FL2 are grouped together first because the distance (0.001) is the smallest (underlined in Table 16.3). Next, the mean distance from FL1 and from FL2 to each other population is used to cluster taxa. The next shortest distance is the mean of FL3 to FL1 and FL3 to FL2 (i.e., the mean of 0.002 and 0.003; see asterisks in Table 16.3); thus FL3 is clustered as the sister group of FL1 and FL2. Next SC is clustered followed by NC (Figure 16.8).

In this example, the genetic distance is correlated with the geographic distance in that SC is geographically and genetically closer to the Florida populations than it is to the North Carolina one. See Guest Box 16 for another example application of dendrogram construction using UPGMA.

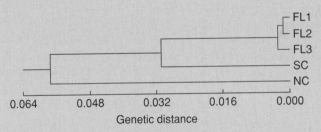

Figure 16.8 Dendrogram generated using the UPGMA clustering algorithm and the genetic distance matrix from Table 16.3. FL is Florida; SC and NC are South Carolina and North Carolina, respectively. From Godt et al. (1997).

(unweighted pair group method with arithmetic averages) and neighbor-joining (Salemi and Vandamme 2003). UPGMA clustering for dendrogram construction is illustrated by a study assessing population relationships of a perennial lily from Florida (Example 16.4).

Neighbor-joining is one of the most widely used algorithms for constructing dendrograms from a distance matrix (Salemi and Vandamme 2003). Neighbor-joining is different from UPGMA in that the branch lengths for sister taxa (e.g., FL1 and FL2, Table 16.3) can be different, and thus can provide additional information on relationships between populations. For example, FL1 is more distant from FL3 than FL2 is from FL3 (Table 16.3). This is not evident in the UPGMA dendrogram (Figure 16.8), but would be in a neighbor-joining tree. It follows that neighbor-joining is especially useful for data sets with lineages evolving at substantially different rates. Other advantages include that neighbor-joining is fast and thus useful for large data sets and for bootstrap analysis (see next paragraph), which involves the construction of hundreds of replicate trees. It also permits correction for multiple character changes when computing distances between taxa. Disadvantages are it gives only one possible tree and it depends on the model of evolution used.

Bootstrap analysis is a widely used sampling technique for assessing the statistical error when the underlying sampling distribution is unknown. In dendrogram construction, we can bootstrap resample across loci from the original data set, meaning that we sample with replacement from our set of loci until we obtain a new set of loci, called a "bootstrap replicate". For example, if we have genotyped 12 loci, we randomly draw 12 numbers from 1 and 12 and these numbers (loci) become our bootstrap replicate data set. We repeat this procedure 100 times to obtain 100 data sets (and 100 dendrograms). The proportion of the random dendrograms with the same cluster (i.e., branch group) will be the bootstrap support for the cluster (see Figure 16.9a).

Multidimensional representation of relationships among populations

Dendrograms cannot illustrate complex relationships among multiple populations because they consist of a one-dimensional branching diagram. Thus dendrograms can oversimplify and obscure relationships among populations. Note that this is not a limitation in using dendrograms to represent phylogenic relationships, as these can be represented by a one-dimensional branching diagram as long as there has not been secondary contact following speciation.

There are a variety of multivariate statistical techniques (e.g., principal component analysis, PCA) that summarize and can be used to visualize complex data sets with multiple dimensions (e.g., many loci and alleles) so that most of the variability in allele frequencies can be extracted and visualized on a two- or three-dimensional plot (Example 16.5). Related multivariate statistical techniques include PCoA (principal coordinates analysis), FCA (frequency correspondence analysis), and MDS (multidimensional scaling).

Example 16.5 How many species of tuatara are there?

We saw in Example 1.1 that tuatara on North Brother Island in Cook Strait, New Zealand, were described as a separate species primarily on the basis of variation at allozyme loci (Daugherty et al. 1990). A neighbor-joining dendrogram based on allele frequencies at 23 allozyme loci suggested that the North Brother tuatara population is highly distinct because it is separated on a long branch (Figure 16.9a).

More recent molecular genetic data, however, have raised some important questions about this conclusion. Analysis of mtDNA sequence data indicates that tuatara from North Brother and three other islands in Cook Strait are similar to each other, and that they are all distinct form the northern tuatara populations (Hay et al. 2003; Hay et al., in preparation). Allele frequencies at microsatellite loci also support the grouping of tuatara from Cook Strait (Hay et al., in preparation).

Principal component analysis (PCA) of the allozyme data supports the similarity of the Cook Strait tuatara populations. Three major population groupings are

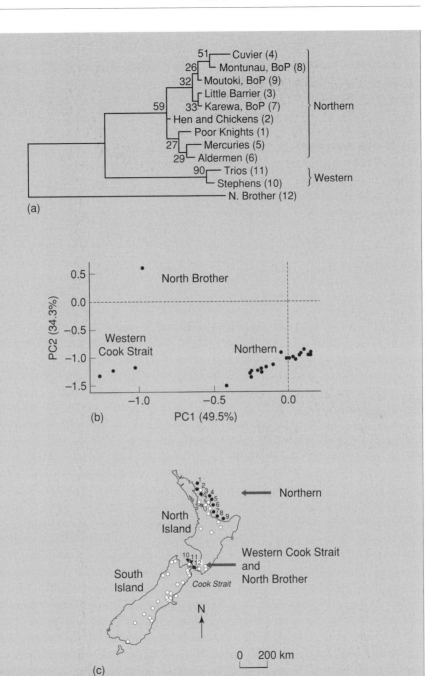

Figure 16.9 (a) Neighbor-joining dendrogram (the numbers on the branches are the bootstrap values) and (b) principal component analysis based on allele frequencies at 23 allozyme loci. (c) The map of New Zealand shows the geographic locations of populations sampled. Open circles indicate where fossil remains have been found.

apparent in the plot of the first two components of the PCA analysis (Figure 16.9b). The first component distinguishes between the northern group and the Cook Strait populations; the North Brother population clusters closely with the Western Cook Strait populations on this axis. PC2 separates the North Brother population from the other populations. The North Brother population clusters with the other Cook Strait populations on PC1, which explains nearly 50% of the variation. The North Brother population is distinct only for the second main variance component (PC2), which explains 34% of the variation. These results suggest that the Cook Strait populations are much more genetically similar to each other than they are to the northern populations.

North Brother Island is very small and the tuatara on this island have substantially less genetic variation at microsatellite loci. Thus, genetic distinctiveness of the North Brother tuatara is likely due to a small population and rapid genetic drift rather than long-term isolation that might warrant species status.

This example illustrates the limitations of one-dimensional tree diagrams and the possible loss (or oversimplification) of information when data are collapsed into one dimension.

16.4.2 Individual-based methods

Individual-based approaches are used to assess population relationships through first identifying populations by delineating genetically similar clusters of individuals. Clusters of genetically similar individuals are often identified by building a dendrogram in which each branch tip is an individual. Second, we quantify genetic relationships among the clusters (putative demes).

Individual-based methods for assessing population relationships make no a priori assumptions about how many populations exist or where boundaries between populations occur on the landscape. If individual-based methods are not used, we risk wrongly grouping individuals into populations based on somewhat arbitrary traits (e.g., color) or an assumed geographic barrier (a river) identified by humans subjectively.

One example of erroneous a priori grouping would be migratory birds that we sample on migration routes or on overwintering grounds. Here, we might wrongly group together individuals from different breeding populations, because we sampled them together at the same geographic location. A similar potential problem could exist in migratory butterflies, salmon, and whales, for example, if we sample mixtures containing individuals from different breeding groups with different geographic origins.

An individual-based approach was used by Pritchard et al. (2000) to assess relationships among populations of the endangered Taita thrush in Africa. The authors built a tree of individuals based on pairwise genetic distance between individuals. Each individual was genotyped at seven microsatellite loci (Galbusera et al. 2000). The genetic similarity index (Nei's genetic distance) between each pair of individuals was computed, and then a clustering algorithm (neighbor-joining) was used to group similar individuals together on branches. The geographic location of origin of individuals was also plotted on the branch

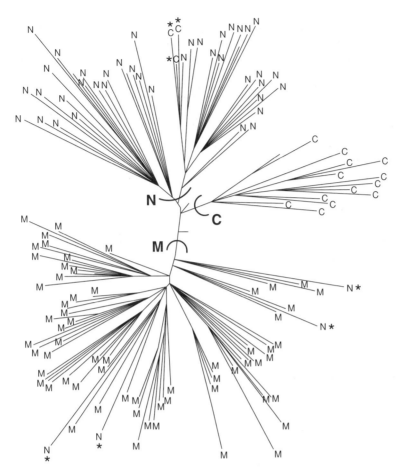

Figure 16.10 Tree of individuals (Taita thrush) constructed using the genetic distance between individuals and the neighbor-joining tree building algorithm (Chawia, 17 individuals; Ngangao, 54 individuals; Mbololo, 80 individuals). The three curved slashes (N, M, and C) across the branches identify the three population clusters. The letters on branch tips are sampling locations (i.e., population names); asterisks on branch tips represent putative immigrants (e.g., three migrants from Ngangao into Chawia (top of figure). Modified from Pritchard et al. (2000).

tips to help identify population units. The analysis revealed three distinct populations represented by three discrete clusters of individuals (Figure 16.10).

This example illustrates a strength of the individual-based approach: the ability to identify migrants. Individuals (i.e., branches) labeled with "N" and an asterisk (bottom of tree, Figure 16.10) were sampled from the "N" location (Ngangao) but cluster genetically with Mbololo (labeled "M"). This suggests these individuals are migrants from Mbololo into the Ngangao population.

An individual-based and model-based approach to identify populations as clusters of individuals was introduced by Pritchard et al. (2000). "Model based" refers to the use of a model with k populations (demes) that are assumed to be in Hardy–Weinberg (HW) and gametic equilibrium. This approach first tests if our data fit a model with $k = 1, 2, 3$, or more populations. The method uses a computer algorithm to search for the set (k) of individuals that minimizes the amount of HW and gametic disequilibrium in the data. Many or all possible sets of individuals are tested. Once k is inferred (step 1), the algorithm estimates, for each individual, the (posterior) probability (Q) of the individual's genotype originating from each population (step 2). If an individual is equally likely to have originated from population X and Y, then Q will be 0.50 for each population.

For example, Berry et al. (2005) used 15 microsatellites and Pritchard's model-based approach to study dispersal and the affects of agricultural land conversion on the connectivity of insular populations of the grand skink from New Zealand. The skink lives in small populations (approximately 20 individuals) on rock outcrops separated by 50–150 m of inhospitable vegetation (native tussocks grassland or exotic pasture). A total of 261 skinks were genotyped from 12 rock outcrops. The number of dispersers inferred from Pritchard's cluster analysis was lower for the exotic pasture than for the native grassland habitat. For example, nine known dispersers were detected among rock outcrops within the native grassland site T1, versus only one disperser within the exotic pasture site P1 (Figure 16.11; see open squares above bar graphs representing dispersers). This study suggests that exotic pasture fragments populations; this likely increases population extinction risks by increasing genetic and demographic stochasticity (see Chapter 14).

Individual-based analyses can also be conducted with many multivariate statistical methods (e.g., PCA) if individuals are used as the operational unit (instead of populations). These multivariate approaches make no prior assumptions about the population structure model, e.g., HW and gametic equilibrium are not assumed.

Individual-based methods are useful to identify cryptic subpopulations and localize population boundaries on the landscape. Once genetic boundaries are located, we can test if the boundaries are concordant with some environmental gradient or some ecological or landscape feature (e.g., a river or temperature gradient). This approach of associating population genetic "boundaries" with landscape or environmental features has been called **landscape genetics** (Manel et al. 2003).

A final strength of individual-based methods is that they can help evaluate data quality by detecting human errors in sampling; for example, a sample with the wrong population label. Such mislabeled samples would show up as outliers (or candidate "migrants") from a different population (Figure 16.11).

A disadvantage of individual-based methods is that they often require the analysis of many individuals (hundreds), sampled across relatively evenly spaced locations. In a continuous population, we might wrongly infer a genetic discontinuity (barrier) between sampling locations if clusters of individuals are sampled from locations far apart. For example, in an isolation by distance scenario (see Figure 9.8), we could infer different (discrete) populations by sampling distant locations with no individuals in between the locations. However, this problem could arise even with the classic population-based methods (see Section 16.4.1).

A potential problem with individual-based methods is that they still can yield uncertain results if genetic differentiation among populations is not substantial. Plus the performance and reliability of individual-based methods has not been thoroughly evaluated (but see Evanno et al. 2005 for a performance evaluation of the individual-based clustering method of Pritchard et al. 2000). Thus it seems useful and prudent to use both individual-based and population-based methods.

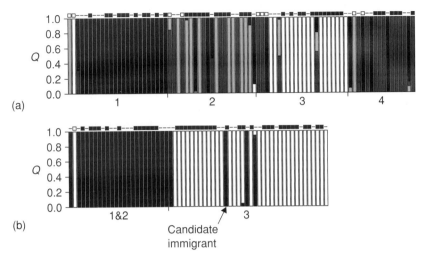

(a)

(b)

Candidate
immigrant

Figure 16.11 Bayesian clustering of individual skink genotypes. Each site is shown separately: (a) site T1 (native tussocks) and (b) site P1 (exotic pasture). Individuals are represented across the x-axis by a vertical bar that may be divided into shaded segments that represent the individual's probability of originating (Q) from each of the rocks at a study site (computed using STRUCTURE 2.0; Pritchard et al. 2000). Skinks are also grouped across the x-axis according to the rock they were captured on (e.g., 1, 2, 3, or 4). Filled squares above an individual indicate that the natal rock was known and the individual did not disperse, open squares indicate that the individual was a known disperser (from mark–recapture data), and dashes indicate that the natal rock was not known for that individual. The arrow points toward one (of several) putative immigrants. From Berry et al. (2005).

16.4.3 Phylogeography

Phylogeography is the assessment of the correspondence between phylogeny and geography (Avise 2000b). We expect to find phylogeographic structuring among populations with long-term isolation. Isolation for hundreds of generations is generally required for new mutations to arise locally, and to preclude their spread across populations. Phylogeographic structure is expected in species with limited dispersal capabilities, with **philopatry**, or with distributions that span strong barriers to gene flow (e.g., mountains, rivers, roads, and human development). In conservation biology, detecting phylogeographic structuring is important because it helps identify long-isolated populations that might have distinct gene pools and local adaptations. Long-term reproductive isolation is one major criterion widely used to identify population units for conservation (see Section 16.5.2).

Intraspecific phylogeography was pioneered initially by J. C. Avise and colleagues (Avise et al. 1987). In a classic example, Avise et al. (1979a) analyzed mtDNA from 87 pocket gophers from across their range in southeastern United States. The study revealed 23 different mtDNA genotypes, most of which were localized geographically (Figure 16.12). A major discontinuity in the maternal phylogeny clearly distinguished eastern and western populations. A potential conservation application of such results is that eastern and

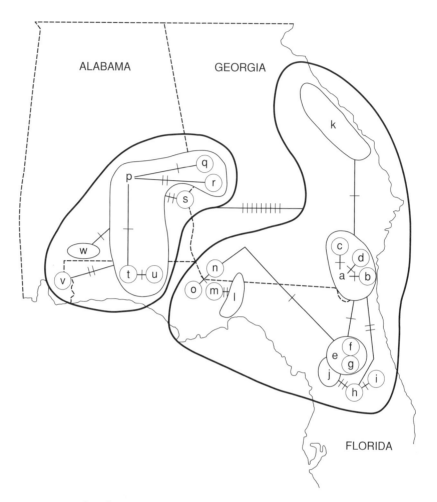

Figure 16.12 Mitochondrial DNA phylogenetic network for 87 pocket gophers; mtDNA genotypes are represented by lower case letters and are connected by branches in a parsimony network. Slashes across branches are substitutions. Nine substitutions separate the two major mtDNA clades encircled by heavy lines. From Avise (1994).

western populations of pocket gopher appear to be highly divergent with long-term isolation and thus potentially adaptive differences; this could warrant management as separate units. However additional data (including nuclear loci and nongenetic information) should be considered before making conservation management decisions (e.g., Section 16.6 and Guest Box 4).

Phylogeographic studies can help identify **biogeographic** provinces containing distinct flora and fauna worth conserving as separate geographic units in nature reserves. For example, multispecies phylogeographic studies in the southwest United States (Avise 1992) and northwest Australia (Moritz and Faith 1998) have revealed remarkably concordant phylogeographic patterns across multiple different species. Such multispecies concordance can be used to identify major biogeographic areas that can be prioritized as separate conservation units and to identify locations to create nature reserves (Figure 16.13).

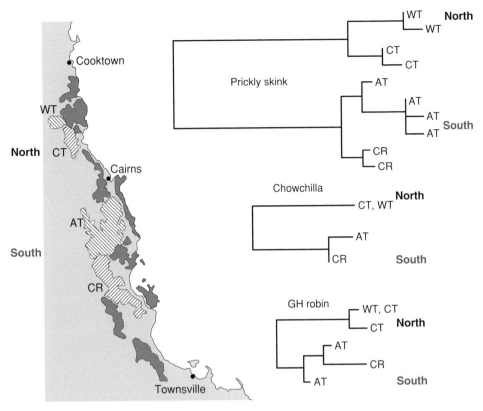

Figure 16.13 Phylogeographic analysis for three species sampled from each of four geographic areas from the tropical rain forests of northeastern Australia between Cooktown and Townsville: Windsor Tableland (WT), Carbine Tableland (CT), Atherton Tableland (AT), and Cardwell Ranges (CR). Note the deep phylogenetic break (long branches) separating the WT/CT populations in the north from the CR/AT populations in the south for all three species (prickly skink, chowchilla, and gray-headed robin). These results suggest long-term isolation for numerous species between the northern and southern rain forests. These regions merit conservation as separate systems. From Moritz and Faith (1998).

A promising phylogeographic approach is nested clade analysis (NCA) (Templeton 1998). NCA involves three steps: (1) building a parsimony network of alleles such that hierarchically nested clades are identified (i.e., recent derived clades versus ancestral clades); (2) testing for statistically significant geographic structuring of alleles within clades; and (3) interpreting the biological cause of structuring (e.g., isolation by distance, recent fragmentation, or range expansion). Step 3 uses an inference key that lists expectations of each cause of a given structuring pattern. For example, NCA predicts that under isolation by distance (i.e., restricted gene flow), the derived alleles will be localized geographically whereas ancestral alleles will be less localized. This is because under restricted gene flow new alleles will not have had time to spread geographically. This pattern is not expected under range expansion (Templeton 1998).

There has been substantial debate over the usefulness of NCA (e.g., Knowles and Maddison 2002). The main shortfall of NCA is that it does not incorporate error or uncertainty. This is the same problem with most phylogenetic approaches. For example, NCA does not consider interlocus variation (as do coalescent-based population genetic models; e.g., Appendix A9 and Figure A12). Thus NCA might provide the correct inference about phylogeographic history, but we cannot quantify the probability of it being correct. Another occasionally cited shortfall is that NCA is somewhat ad hoc in using an inference key in order to distinguish among many different historical processes.

Fortunately, the emerging field of "statistical phylogeography" promises to combine the strengths of NCA with formal modeling and statistical tests to allow for more objective testing of alternative hypotheses that could explain phylogeographic patterns (Knowles and Maddison 2002). Until formal modeling and validated statistical phylogeography approaches are available, it seems prudent to use NCA in combination with other approaches such as AMOVA (analysis of molecular variation) that use genealogical information in ways similar to or complementary to NCA (e.g., see Turner et al. 2000).

16.5 Units of conservation

It is critical to identify species and units within species to help guide management, monitoring, and other conservation efforts, and to facilitate application of laws to conserve taxa and their habitats. In this section we consider the issues of identifying species and intraspecific conservation units.

16.5.1 Species

Identification of species is often problematic, even for some well-known taxa. One problem is that biologists cannot even agree on the appropriate criteria to define a species. In fact, more than two dozen species concepts have been proposed over the last decades. Darwin (1859) wrote that species are simply highly differentiated varieties. He observed that there is often a continuum in the degree of divergence from between populations, to between varieties, species, and higher taxonomic classifications. In this view, the magnitude of differentiation that is required to merit species status can be somewhat arbitrary.

The biological species concept (BSC) of Mayr (1942, 1963) is the most widely used species definition, at least for animals. This concept emphasizes reproductive isolation and isolating mechanisms (e.g., pre- and postzygotic). Criticisms of this concept are that: (1) it can be difficult to apply to allopatric organisms (because we cannot observe or test for natural reproductive barriers in nonoverlapping populations); (2) it cannot easily accommodate asexual species (that may not interbreed only because they are selfing); and (3) it has difficulties dealing with introgression between highly distinct forms. Further, an emphasis on "isolating mechanisms" implies that selection counteracts gene flow. However, the BSC generally does not allow for interspecific gene flow, even at a few segments of the genome (i.e., limited introgression) (Wu 2001).

The phylogenetic species concept (PSC; Cracraft 1989) relies largely on monophyly, such that all members of a species must share a single common ancestor. This concept has fewer problems dealing with asexual organisms (e.g., many plants, fish, etc.) and with allopatric forms. However, it does not work well under hybridization and it can lead to oversplitting, for example as more and more characters are used (e.g., using powerful DNA sequencing techniques) more "taxa" might be identified.

A problem using the PSC can arise if biologists interpret fixed DNA differences (monophyly) between populations as evidence for species status. For example, around the world many species are becoming fragmented, and population fragments are becoming fixed (monophyletic) for different DNA polymorphisms. Under the PSC, this could cause the proliferation of "new" species if biologists strictly apply the PSC criterion of monophlyly for species identification. This could result in oversplitting and the waste of limited conservation resources. This potential problem of fragment-ation-induced oversplitting is described in a paper titled "Cladists in wonderland" (Avise 2000a). To avoid such oversplitting, multiple independent DNA sequences (e.g., not mtDNA alone) should be used, along with many nongenetic characters when possible.

Other species definitions include the ecological species concept based on a distinct ecological niche (Van Valen 1976), and the evolutionary species concept often used by paleontologists to identify species based on change within lineages through time but without splitting (anagenesis) (Simpson 1961). The different concepts overlap, but emphasize different types of information. Generally, it is important to consider many kinds of information or criteria when identifying and naming species. If most criteria (or species concepts) give the same conclusion (e.g., species status is warranted), than we can be more confident in the conclusion.

African cichlid fishes illustrate some of the difficulties with the different species concepts. Approximately 1,500 species of cichlids have recently evolved a diverse array of morphological differences (e.g., mouth structure, body color) and ecological differences (e.g., feeding and behaviors such as courtship). Morphological differences are relatively pronounced among cichlids. However, the degree of genetic differentiation among cichlids is relatively low compared to other species, due to the recent radiation of African cichlid species (less than 1–2 million years ago!). Further complicating species identification using molecular markers, is that reproductive isolation can be transient. For example, some cichlid species are reproductively isolated due to mate choice based on fixed color differences between species. However, this isolation breaks down during years when murky water prevents visual color recognition and leads to temporary interspecific gene flow (Seehausen et al. 1997)!

Molecular genetic data can help identify species, especially cryptic species that have similar phenotypes (see also Section 20.1). For example, the neotropical skipper butterfly was recently identified as a complex of at least 10 species, in part by the sequencing of a standard gene region (**DNA "barcoding"**). The 10 species have only subtle differences in adults and are largely sympatric (Hebert et al. 2004). However, they have distinctive caterpillars, different caterpillar food plants, and a relatively high genetic divergence (3%) in the mitochondrial gene cytochrome *c* oxidase I (COI) gene.

Molecular data can also help identify taxa that are relatively well studied. For example, a recent study of African elephants used molecular genetic data to detect previously unrecognized species. Elephants from tropical forests are morphologically distinct from savannah elephants. Roca et al. (2001) biopsy-dart sampled 195 free-ranging elephants from 21 populations. Three populations were forest elephants in central Africa, 15 were savannah elephants (located north, east, and south of the forest populations), and three were unstudied and thus unclassified populations. DNA sequencing of 1,732 base pairs from four nuclear genes revealed 52 nucleotide sites that were phylogenetically informative (i.e., at least two individuals shared a variant nucleotide).

All savannah elephant populations were closer genetically to every other savannah population than to any of the forest populations, even in cases where the forest population was geographically closer (Roca et al. 2001). Phylogenetic analyses revealed five fixed site differences between the forest and savannah elephants (Figure 16.14). By comparison, nine fixed differences exist between Asian and African elephants. Hybridization was considered

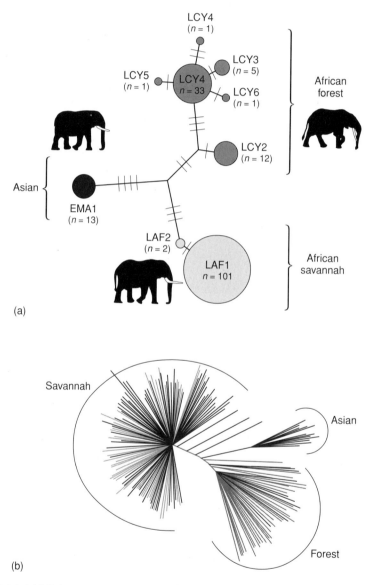

Figure 16.14 (a) Minimum spanning network showing relationships among nine haplotypes observed for the X-linked *BNG* gene for African forest and savannah elephants and the Asian elephant. Each slash mark along branches separating each haplotypes represents one nucleotide difference. From Roca et al. (2001). (b) Neighbor-joining cluster analysis of 189 African elephants and 14 Asian elephants based on proportion of shared alleles (Dps) at 16 microsatellite loci. From Comstock et al. (2002).

to be "extremely limited", although the number of individuals sampled was only moderate, and one savannah individual apparently contained one nucleotide diagnostic for the forest elephants. The genetic data (see also Comstock et al. 2002; Figure 16.14b), combined with the morphological and habitat differences, suggests that species-level status is warranted. This study represents a nice example of combining genetic and nongenetic data. The results could influence conservation strategies, making it more urgent to protect and manage these increasingly endangered taxa separately.

Sometimes genetic data may show that recognized species are not supported by reproductive relationships. Some authors have recognized the black sea turtle (*Chelonia* spp.) as a distinct species on the basis of skull shape, body size, and color (Pritchard 1999). However, molecular analyses of mtDNA and three independent nuclear DNA fragments suggest that reproductive isolation does not exist between the black and green forms (Karl and Bowen 1999). Over the years, taxonomists have proposed more than a dozen species for different *Chelonia* populations, with oversplitting occurring in many other taxa as well. Nonetheless, for conservation purposes, it is clear that black turtles are distinct and could merit recognition as an intraspecific conservation unit (see Section 16.5.2) that posses potential local adaptations. Unfortunately, populations are declining, and additional data on potentially adaptive differences are needed (e.g., food sources and feeding behavior, etc.).

16.5.2 Evolutionary significant units

An **evolutionary significant unit** (**ESU**) can be defined broadly as a population or group of populations that merit separate management or priority for conservation because of high distinctiveness (both genetic and ecological). The first use of the term ESU was by Ryder (1986). He used the example that five (extant) subspecies of tigers exist, but there is not space in zoos or captive breeding programs to maintain viable populations of all five. Thus sometimes we must choose which subspecies to prioritize for conservation action, and perhaps maintain only one or two global breeding populations (each perhaps consisting of more than one named subspecies). Since Ryder, the term ESU has been used in a variety of frameworks for identifying conservation units (Example 16.6).

There is considerable confusion and controversy in the literature associated with the term ESU. For example, the US Endangered Species Act (ESA) lacks any definition of a distinct population segment (DPS; see Example 16.1). Waples (1991) suggested that a population or group of populations (of salmon) would be a DPS if it is an ESU. This has lead to some confusion because some biologists equate a DPS and an ESU. We will use the term DPS when referring to officially recognized "species" under the ESA, and the term ESU in the more generally accepted sense.

It can be difficult to provide a single concise, detailed definition of the term ESU because of the controversy and different uses and definitions of the term in the literature. This ESU controversy is analogous to that surrounding the different species concepts mentioned above. The controversy is not surprising considering the problems surrounding the definition of species, and the fact that identifying intraspecific units is generally more difficult than identifying species (Waples 1991). It is also not surprising considering the different rates of evolution that often occur for different molecular markers and phenotypic traits used in ESU identification. Different evolutionary rates lead to problems analogous to that in the classification of birds (Aves) as a separate taxonomic class (due to their rapid evolution), when in fact birds are monophyletic within the class Reptilia (see Figure 16.3).

Example 16.6 Proposed definitions of evolutionary significant units

1 **Ryder (1986)**: populations that actually represent significant adaptive variation based on concordance between sets of data derived by different techniques. Ryder (1986) clearly argues that this subspecies problem is "considerably more than taxonomic esoterica". (Main focus: zoos for potential *ex situ* conservation of gene pools of threatened species.)

2 **Waples (1991)**: populations that are reproductively separate from other populations (e.g., as inferred from molecular markers) and that have distinct or different adaptations and that represent an important evolutionary legacy of a species. (Main focus: integrating different data types, and providing guidelines for identifying "distinct population segments" or DPSs (of salmon) which are given "species" status for protection under the United States Endangered Species Act.)

3 **Dizon et al. (1992)**: populations that are distinctive based on morphology, geographic distribution, population parameters, and genetic data. (Main focus: concordance across some different data types, but always requiring some degree of genetic differentiation.)

4 **Moritz (1994)**: populations that are reciprocally monophyletic (see Figure 16.15) for mtDNA alleles and that show significant divergence of allele frequencies at nuclear loci. (Main focus: defining practical criteria for recognizing ESUs based on population genetics theory, while considering that variants providing adaptation to recent or past environments may not be adaptive (or might even retard the response to natural selection) in future environments.)

5 **USFWS and NOAA (1996b)** (US policy for all vertebrates): (1) discreteness of the population segment (DPS) in relation to the remainder of the species to which it belongs; and (2) the significance of the population segment to the species to which it belongs. This DPS policy does not use the term ESU, but has a framework similar to that of Waples' (1991) salmon ESU policy.

6 **Crandall et al. (2000)**: populations that lack: (1) "ecological exchangeability" (i.e., they have different adaptations or selection pressures – e.g., life histories, morphology, quantitative trait locus variation, habitat, predators, etc. – and different ecological roles within a community); and (2) "genetic exchangeability" (e.g., they have had no recent gene flow, and show concordance between phylogenetic and geographic discontinuities). (Main focus: emphasizing adaptive variation and combining molecular and ecological criteria in a historical timeframe. Suggests returning to the more holistic or balanced and two-part approach of Waples.)

In practice, an understanding of the underlying principles and the criteria used in the different ESU frameworks will help when identifying ESUs. The main criteria in several different ESU concepts are listed in Example 16.6, and synthesized at the end of this section (see also Fraser and Bernatchez 2001). Here we discuss some details about three widely used ESU frameworks, each with somewhat different criteria as follows: (1) reproductive isolation and adaptation (Waples 1991); (2) **reciprocal monophyly** (Moritz 1994); and (3) "exchangeability" of populations (Crandall et al. 2000). This will provide a background on principles and concepts, as well as a historical perspective of the controversy surrounding the different frameworks for identifying units of conservation.

Isolation and adaptation

Waples (1991) was the first to provide a detailed framework for ESU identification. His framework included the following two main requirements for an ESU: (1) long-term reproductive isolation (generally hundreds of generations) so that an ESU represents a product of unique past evolutionary events that is unlikely to re-evolve (at least on an ecological time scale); and (2) ecological or adaptive uniqueness such that the unit represents a reservoir of genetic and phenotypic variation likely important for future evolutionary potential. This second part requiring ecological and adaptive uniqueness was termed the "evolutionary legacy" of a species by Waples (1991, 1998). This framework has become the official policy of the United States Fish and Wildlife Service and the National Marine Fisheries Service (USFWS and NOAA 1996b).

Waples (2005) recently argued that ESU identification is often most helpful if an intermediate number of ESUs are recognized within each species, e.g., when the goal is to preserve a number of genetically distinct populations within a species. Waples (2005) reviewed many of the published ESU concepts and criteria (e.g., see Example 16.6) and concluded that they could often identify only a single ESU or a large number (hundreds) of ESUs in Pacific salmon species. This is a tentative conclusion based on the published criteria for other ESU concepts, many of which are subjective or qualitative. There is a need for more empirical examples in which multiple ESU concepts are applied to a common problem (as in Waples 2005).

Reciprocal monophyly

Moritz (1994) offered simple and thus readily applicable molecular criteria for recognizing an ESU: "ESUs should be reciprocally monophyletic for mtDNA (in animals) and show significant divergence of allele frequencies at nuclear loci". Mitochondrial DNA is widely used in animals because it has a rapid rate of evolution and lacks recombination, thus facilitating phylogeny reconstruction. Cytoplasmic markers are often used in plants as they also lack recombination. "Reciprocally monophytic" means that all DNA lineages within an ESU must share a more recent common ancestor with each other than with lineages from other ESUs (Figure 16.15). These molecular criteria are relatively quick and easy to apply in most taxa because the necessary molecular markers (e.g., "universal" PCR primers) and data analysis software have become widely available. Further, speed is often important in conservation where management decisions may have to be made quickly, and before thorough ecological studies of a species can be conducted.

An occasionally cited advantage of the Moritz (1994) monophyly criterion is that it can employ population genetics theory to infer the time since population divergence. For example, it takes a mean of $4N_e$ generations for a newly isolated population to coalesce to a single gene copy and therefore become reciprocally monophyletic (through drift and

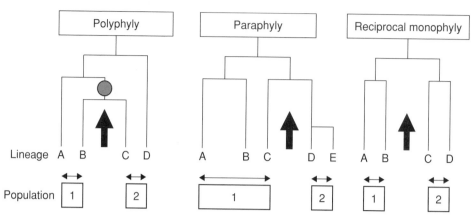

Figure 16.15 Development of phylogenetic structure of alleles between populations. After a population splits into two, the phylogenetic relationship of the alleles in the two daughter populations usually proceeds from polyphyly through paraphyletic conditions to reciprocal monophyly. The alleles (or lineages) are labeled A, B, C, D, and E. When two populations (1 and 2) become isolated, **both** will initially have some alleles that are more closely related to alleles in the other population (**poly**phyly). The filled circle at the root of the B and C branches represents the most recent common ancestor between B and C. After many generations of isolation, one population might become monophyletic, e.g., for alleles for D and E in the paraphyly example (see also the black duck, Figure 16.7). But the other population might maintain an allele that is more related to an allele in the other population (e.g., the mallard duck, Figure 16.7). After approximately four N_e generations, both daughter populations will usually be monophyletic with respect to each other (reciprocal monophyly). Modified from Moritz (1994).

mutation) at a nuclear locus (Neigel and Avise 1986). This means that if a population splits into two daughter populations of size $N_e = 1,000$, it would take an expected 1,000 generations to become reciprocally monophyletic for mtDNA. For mtDNA to become monophyletic it requires fewer generations because the effective population size is approximately four times smaller for mtDNA than for nuclear DNA; thus lineage sorting is faster (see Section 9.5). Here it is important to recall that adaptive differentiation can occur in a much shorter time period than does monophyly (see, for example, Guest Box 8).

A disadvantage of the Moritz ESU concept is it generally ignores adaptive variation, unlike the two-step approach that also incorporates the "evolutionary legacy" of a species (Waples 1991). The framework of Moritz is based on a cladistic phylogenetic approach (Section 16.2) using neutral loci. Thus, unfortunately, the Moritz approach makes it more likely that smaller populations (e.g., bottlenecked populations) will be identified as ESUs. Small populations quickly become monophyletic due to drift or lineage sorting. Worse, natural selection is most efficient in large populations. Consequently, the strict Moritz framework is unlikely to identify ESUs with substantial adaptive differences.

One limitation of using only molecular information is that a phylogenetic tree might not equal the true population tree. This is analogous to the "gene tree versus species tree" problem discussed in Section 16.3.2 (Example 16.7). This problem of population trees not equaling gene trees is worse at the intraspecific level because there is generally less time

Example 16.7 Lack of concordance between mtDNA and nuclear genes in white-eyes

Degnan (1993) compared dendrograms from mtDNA and nuclear DNA for two species of white-eye from Australia. The mtDNA data yielded a single gene tree that does not reflect the organismal tree based on phenotypic characters. In contrast, the two nuclear DNA loci revealed phylogeographic patterns consistent with the traditional classification of the two species (Figure 16.16). The author concluded that the discordance between the mtDNA and nuclear DNA (and phenotype) likely results from past hybridization between the two species of white-eye and mtDNA introgression. Evidence for hybridization might have been lost in nuclear genes through recombination.

This study provides a clear empirical demonstration that single gene genealogies cannot be assumed to accurately represent the true organismal phylogeny. Further, it emphasizes the need for analyses of multiple, independent DNA sequences when inferring phylogeny and identifying conservation units. This is especially true for populations within species where relatively few generations have passed. For example, if we were trying to identify ESUs (or species) in this study by using mtDNA alone, we might identify three ESUs (corresponding to the three mtDNA haplogroups in Figure 16.16); However these three are not concordant with the two groups identified by phenotype, nuclear DNA, and geography.

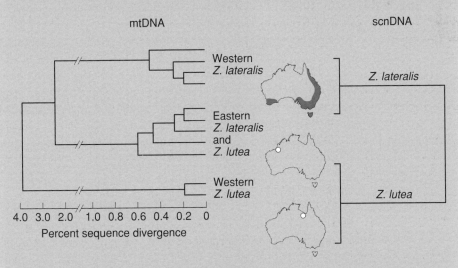

Figure 16.16 UPGMA dendrograms for mtDNA haplotypes (left) and scnDNA (single copy nuclear DNA) genotypes (right) in silver-eyes (*Zosterops lateralis*) and yellow white-eyes (*Z. lutea*). The distribution of silver-eye haplotypes is shown in solid black (top map), and for yellow white-eyes in white circles (bottom two maps). Note the middle map shows that the yellow white-eye samples from northwestern Australia group in the mtDNA tree with silver-eye samples from eastern Australia, but group with the yellow white-eyes in the scnDNA tree. Modified from Degnan (1993).

since reproductive isolation at the intraspecific level, and thus more problems with lineage sorting and paraphyly. Consequently, problems of gene trees not matching population trees (and different genes giving different trees) will be relatively common at the intra-specific level. Unfortunately, in the conservation literature, mtDNA data are often used alone to attempt to identify ESUs. This should occur less often as nuclear DNA markers become more readily available.

Exchangeability

Crandall et al. (2000) suggested that ESU identification be based on the concepts of ecological and genetic "exchangeability". The idea of exchangeability is that individuals can be moved between populations and can occupy the same niche, and can perform the same ecological role as resident individuals, without any fitness reduction due to genetic mechanisms (e.g., outbreeding depression). If we can reject the hypothesis of exchangeability between populations, then those populations represent ESUs. Ideally, exchangeability assessment would be based on heritable adaptive quantitative traits. Strengths of this approach are that it integrates genetic and ecological (adaptive) information and that it is hypothesis based.

Exchangeability can be tested using common-garden experiments and reciprocal transplant experiments. For example, if two plant populations from different locations have no reduced fitness when transplanted between locations, they might be exchangeable and would not warrant separate ESU status (see also Section 2.4, Figure 2.9, and Figure 8.1).

The main problem with this approach is it is not generally practical – i.e., it is difficult to test the hypothesis of exchangeability in many species. For example, it is difficult to move a rhinoceros (or most any endangered species) from one population to another and then to measure its fitness and the fitness of its offspring. Such studies are especially problematic in endangered species where experiments are often not feasible. Although difficult to test, exchangeability is a still worthy concept to consider when identifying ESUs. Even when we cannot directly test for exchangeability, we might consider surrogate measures of exchangeability, such as life history differences, the degree of environmental differentiation, or the number of functional genes showing signatures of adaptive differentiation (see Section 16.6). Surrogates are often used when applying Waples' ESU definition (see Guest Box 16).

Synthesis

Substantial overlap in criteria exists among different ESU concepts. Several concepts promote a two-step approach involving isolation and adaptive divergence. The main principles and criteria are the following: reproductive isolation (no gene flow), adaptive differentiation, and concordance across multiple data types (e.g., genetic, morphological, behavioral, life history, and geographic). The **longer the isolation** and the **more different the environment** (i.e., selection pressures), the more likely are populations to represent distinct units worthy of preservation and separate management. We should not rely on any single criterion, such as reciprocal monophyly of mtDNA. In fact, the greater the number of different data types showing concordant differentiation between populations, the stronger the evidence for ESU status.

16.5.3 Management units

Management units (MUs) usually are defined as populations that are demographically independent. That is, their population dynamics (growth rate) depend mostly on local birth and death rates rather than on immigration. The identification of these units, similar to "stocks" recognized in fisheries biology, would be useful for short-term management – such as delineating hunting or fishing areas, setting local harvest quotas, and monitoring habitat and population status.

MUs, unlike ESUs, generally do not show long-term independent evolution or strong adaptive differentiation. MUs should represent populations that are important for the long-term persistence of an entire ESU (and/or species). The conservation of multiple populations, not just one or two, is critical for insuring the long-term persistence of species (Hughes et al. 1997; Hobbs and Mooney 1998).

MUs are generally smaller than ESUs, such that an ESU might contain several MUs. MUs often are divergent subpopulations within a major metapopulation that represents an ESU. For example, fish populations are often structured on hierarchical levels such as small streams (as MUs) that are nested within a major river drainage (an ESU, e.g., Guest Box 16). Moritz (1994) defined the term "management unit" as a population that has substantially divergent allele frequencies at many loci.

One potential limitation of using allele frequency differentiation (e.g., F_{ST}) to identify MUs is that F_{ST} cannot directly be interpreted as evidence for demographic independence. For example, large populations experience little drift (and little allele frequency differentiation) and thus can be demographically independent even if allele frequencies are similar. The same Nm (and hence F_{ST}) can result in different migration rates (m) for different population sizes (N). As N goes up, m goes down for the same F_{ST} (Table 16.4). So in a large population, the number of migrants can be very small, and the population could be demographically independent yet have a relatively low F_{ST}.

A related difficulty is determining if migration rates would be sufficient for recolonization on an ecological time scale, e.g., if a MU became extinct or overharvested. Allele frequency data can be used to estimate migration rates (Nm), but at moderate to high rates of migration ($Nm > 5$–10) genetic estimators are notoriously imprecise, such that confidence intervals on the Nm estimate might include infinity (Waples 1998). Unfortunately, the

Table 16.4 Inferring demographic independence of populations by using genetic differentiation data (F_{ST}) requires knowledge of the effective population size (N_e). Here, the island model of migration was assumed to compute $N_e m$ and m (proportion of migrants) from the F_{ST} (as in Figure 9.9). Recall that the effective population size is generally far less than the census size in natural populations (see Section 7.10).

F_{ST}	N_e	m	$N_e m$	Demographic independence
0.10	50	0.040	2	Unlikely
0.10	100	0.020	2	Likely
0.10	1,000	0.002	2	Yes

range of *Nm* we are most interested in for MU identification is often moderate to high (5–50). Additional problems with *Nm* estimation exist (e.g., Whitlock and McCauley 1999; and Section 9.8.1).

The identification of conservation units can be difficult when population differences are subtle or if hierarchical structure is complex. For example, in green turtles (mentioned in Section 16.5.1), two ESUs have been proposed: the Atlantic Ocean and the Indo-Pacific region (e.g., Karl and Bowen 1999). Within each ESU, more than 10 MUs have been recognized; however population differences and demographic independence is difficult to delineate. In the humpback whale, extensive molecular and demographic studies have suggested the presence of one ESU containing numerous MUs, most of which correspond to major stocks identified by migration routes (Baker et al. 1998). A similar scenario was proposed for koalas in Australia, which was suggested to contain only one ESU but many MUs, based on mtDNA, microsatellite DNA variation, and biogeography (e.g., Houlden et al. 1999).

Two general errors can occur in MU diagnosis (as with ESU identification): identifying too few or too many units. Recognizing too few units could lead to underprotection and then to the reduction or loss of local populations. This problem could arise, for example, if statistical power is too low to detect genetic differentiation when differentiation is biologically significant.

For example, too few MUs (and underprotection) may be established if only one MU is identified when the species is actually divided into five demographically independent units. Consider that the sustainable harvest rate is 2% per year on the basis of total population, but that all the harvest comes from only one of the five MUs. Then the actual harvest rate for the single harvested MU is 10% (assuming equal size of the five MUs). This high harvest rate could result in overexploitation and perhaps extinction of the one harvested MU population. For example, if the harvested population's growth rate is only 4% per year and the harvest rate is 10%, overexploitation would be a problem (Taylor and Dizon 1999).

Here, undersplitting could result from either a lack of statistical power (e.g., due to too few data) or to the misidentification of population boundaries (e.g., due to cryptic population substructure). To help avoid misplacement of boundaries, researchers should sample many individuals that are widely distributed spatially, and use recently developed, individual-based statistical methods (see Section 16.4.2 and caveats therein; see also Manel et al. 2003).

Diagnosing too many MUs (oversplitting) could lead to unnecessary waste of conservation management resources. This error could occur if, for example, populations are designated as MUs because they have statistically significant differences in allele frequencies, but this differentiation is not associated with important biological differences. This becomes a potential problem as more and more molecular markers are used that are highly polymorphic (and thus statistically powerful).

For example, if many highly polymorphic microsatellites are genotyped, and different populations have "significantly" different allele frequencies ($P < 0.01$), they might not necessarily warrant recognition as different MUs. This is because the magnitude of differentiation could be low (e.g., $F_{ST} \ll 0.01$) even though this relatively small difference is significantly different ($P < 0.01$). Note that an F_{ST} of 0.01 suggests that populations probably exchange numerous migrants, on average, per generation (recall that if $F_{ST} = 0.1$, then approximately two migrants per generation would occur, assuming the island model, expression 9.3). In this case, the populations might not be demographically independent and not merit separate MU status. Researchers and managers must be careful to understand the difference between the biological significance and the statistical significance of genetic differentiation measures (e.g., see Waples 1998; Hedrick 1999). Also recall that the island model has numerous assumptions unlikely to be met in natural populations (see Section 9.8.1; Whitlock and McCauley 1999).

16.6 Integrating genetic, phenotypic, and environmental information

Many kinds of information should be integrated, including life history traits, environmental characteristics, phenotypic divergence, and patterns of gene flow (isolation/phylogeography), for the identification of conservation units (Figure 16.17; Guest Box 16). For example, if two geographically distant populations (or sets of populations) show large molecular differences that are concordant with life history (e.g., flowering time) and morphological (e.g., flower shape) differences, we would be relatively confident in designating them as two geographic units important for conservation (perhaps ESUs).

Researchers should always consider if the environment or habitat type of different populations has been different for many generations, because this could lead to adaptations (even in the face of high gene flow) that are important for the long-term persistence of species. The more kinds of independent information that are concordant, the more sure one can be that a population merits recognition as a conservation unit. The principle of considering multiple data types and testing for concordance is critical for identifying conservation units.

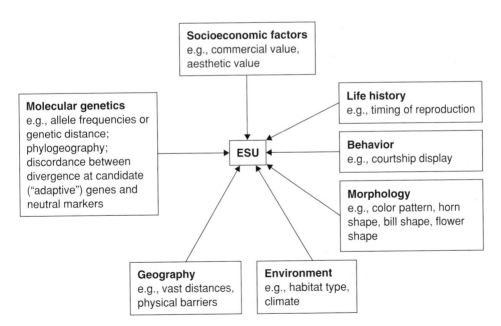

Figure 16.17 Sources of information that can help diagnose an evolutionary significant unit. Modified from Moritz et al. (1995).

When concordance is lacking among data types, difficulties arise. For example, imagine that two populations show morphological differences in size or color, but show evidence of extensive recent gene flow. This scenario has arisen occasionally in studies that measure phenotypic traits from only small samples or nonrepresentative samples of individuals from each population (e.g., only 5–10 individuals of different sexes or ages). In this example taxonomic "oversplitting" results from biased or limited sampling, and status is not warranted. This hypothetical example relates to the green/black turtle "species" dilemma described above, where more extensive sampling and study (including life history and adaptive trait information) is needed.

It is prudent to not use only molecular or only morphological information to identify ESUs, because adaptive differences can exist between populations even when little molecular or morphological differentiation is detectable (e.g., see Merilä and Crnokrak 2001). It is especially important to not base gene flow estimates on only one type of molecular marker (e.g., mtDNA). Rather, researchers should combine many loci along with ecological information. When ecological information is scarce (for example life history information can be difficult to collect), researchers could at least consider climate, habitat type, adaptive gene markers, etc., when identifying an ESU.

One example of how to integrate adaptive and "neutral" molecular variation is to consider them on two separate axes in order to identify populations with high distinctiveness for both adaptive and neutral diversity (Figure 16.18). For many species (e.g., mammals, salmon, and agricultural plants and their relatives), it is becoming feasible to detect adaptive molecular variation by genotyping numerous mapped markers including candidate genes (Luikart et al. 2003; Morin et al. 2004).

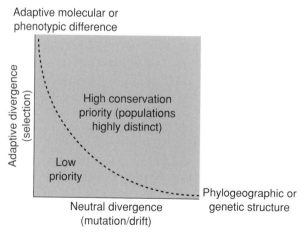

Figure 16.18 Adaptive information, including adaptive genes under selection, could be integrated with information from neutral markers and information on long-term isolation. Such an approach could help identify the most appropriate source population (i.e., non-adaptively differentiated population) from which to translocate individuals into small, declining populations that require supplementation. This approach could also help rank or prioritize populations for conservation management. From Luikart et al. (2003); modified from Moritz (2002).

Guest Box 16 Identifying conservation units in Pacific salmon
Robin S. Waples

Pacific salmon populations considered for listing as threatened or endangered "species" under the US Endangered Species Act (ESA) have been evaluated by the National Marine Fisheries Service (NMFS) using a concept of evolutionary significant units (ESUs) developed by Waples (1991, 1995). Under this framework, a population or (more often) group of populations is considered an ESU if it is substantially isolated reproductively and contributes substantially to ecological and genetic diversity of the species as a whole. Molecular genetic data are particularly informative for the first criterion. The second criterion emphasizes adaptive differences, but direct information about adaptations is generally lacking, so life history and ecological data are typically used as proxies.

In one example application, the NMFS evaluated a petition to list steelhead in the Illinois River in southern Oregon under the ESA. The Biological Review Team (BRT; Busby et al. 1993) found some support for the petitioners' claims of local differences in phenotypic and life history traits between Illinois River steelhead and nearby Rogue River steelhead, but in a broader geographic survey the traits of Illinois River fish were found to be shared by many other populations in southern Oregon and northern California. Furthermore, three of four genetic samples collected from within the Illinois River drainage were more similar to a population from outside the basin than to any of the other Illinois River samples. As a consequence of these findings, which illustrate the importance of an appropriate geographic context for evaluating distinctiveness, the BRT concluded that Illinois River steelhead do not, by themselves, constitute an ESU.

The BRT again expanded the geographic scope of its evaluations to determine the boundaries of the ESU to which Illinois River steelhead belong. Several lines of evidence suggest that Cape Blanco, which forms the northern boundary for the Klamath Mountains Geological Province (KMP), is also the northern boundary for this ESU. The KMP is distinctive geologically and ecologically (interior valleys receive less precipitation than any other location in the Pacific Northwest west of the Cascade Range) and supports a large number of endemic species. In the marine environment, the strength and consistency of coastal upwelling south of Cape Blanco yields high productivity in nearshore waters utilized by salmon. Tagging studies suggest that coho salmon and steelhead from south of Cape Blanco may not be strongly migratory, remaining instead in these productive oceanic waters.

Identifying the southern extent of this ESU was more problematical. The KMP and the distinctive Klamath–Rogue freshwater zoogeographic zone include the Klamath River basin but not areas further south. However, Cape Mendocino (well to the south of the Klamath River) is a natural landmark associated with changes in ocean currents and represents the approximate southern limit of two important life history traits for steelhead: adult fish that return to fresh water in the summer, and subadults that spend only a few months at sea before returning to fresh water on a false spawning run at a size that inspired their name, "half-pounders". Finally, the area of increased upwelling extends well into central coastal California.

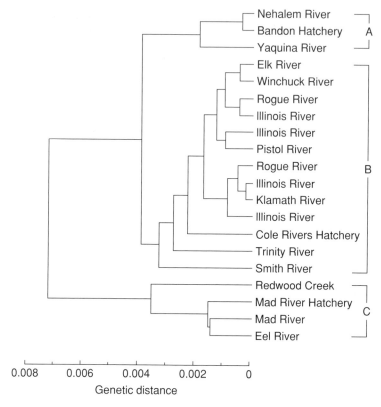

Figure 16.19 Dendogram constructed using UPGMA and pairwise genetic distances between populations (Nei's 1978 unbiased distance) computed from allele frequencies at 39 polymorphic allozyme loci. The population groups A, B, and C are in different ESUs. From Busby et al. (1994).

This issue was resolved by additional genetic sampling from the northern California coast, which showed a sharp genetic transition south of the Klamath River. At several genetic loci, alleles that were rare or absent north of the Klamath River suddenly appeared at appreciable frequencies (Figure 16.19). These results suggest considerable reproductive isolation between steelhead from the Klamath River and populations to the south (those in cluster C), and as a result the BRT concluded that the Klamath River forms the southern boundary for this ESU (Busby et al. 1994).

This example illustrates how combining different kinds of information can help identify intraspecific units for conservation. Here, the kinds of information included life history (migratory behavior), geology (river drainage system), ecology and environment (precipitation, ocean currents, and productivity), and demography (tagging and movement studies), as well as genetics (allele frequency differences).

Problem 16.1

What are the three hierarchical levels of biodiversity recognized by the IUCN and some other organizations? Name two additional organizational levels. Is any one level most important to focus on for conservation efforts? Why or why not?

Problem 16.2

Name three temporal aspects of biodiversity conservation. Which is most important and should be prioritized?

Problem 16.3

Describe several scenarios, mechanisms, and evolutionary processes that can lead to isolated populations failing to show reciprocal monophyly of DNA lineages.

Problem 16.4

What are the three main schools of taxonomic classification? Which school is a combination of the other two? Which is most widely used today? Which is most appropriate for studies of evolutionary history?

Problem 16.5

Define paraphyly and polyphyly. Does our currently accepted classification of birds (relative to reptiles) represent an example of paraphyly or polyphyly? Are reptiles monophyletic in our currently accepted classification? (Consider Figure 16.3.)

Problem 16.6

The figure below (a) shows a hypothetical phylogenetic tree with three derived alleles x, y, and z, that arose from the ancestral allele w. Figure (b) shows circles representing geographic areas in which each allele from (a) is distributed (modified from Moritz and Faith 1998). Conduct a phylogeographic analysis and overlay the phylogeny of alleles onto the geographic distribution (b) of alleles. Is there evidence for phylogeographic structuring? Why or why not? Now, imagine another geographic distribution of alleles (c); does figure (c) reveal phylogeographic structuring?

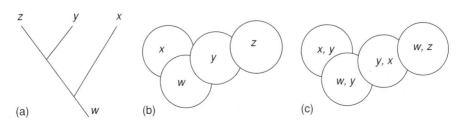

Problem 16.7

What is an ESU? What are the main principles and criteria used for ESU identification?

Problem 16.8

How does a management unit differ from an ESU? How might molecular markers be used to help identify management units?

Problem 16.9

What kinds of information are most useful for identifying units for conservation?

Problem 16.10

Below are four allelic DNA sequences from a species of domestic ungulate (family Bovidae). Nucleotide positions are numbered 1 to 40 (numbers are written vertically) above the sequences. For example, nucleotide position 7 has a substitution (**G**) in sequence 2 and 4, which is different from the "t" in reference sequence 1.

(a) Build a parsimony network (as in Figures 16.12 and 16.14a) by hand by connecting the sequences based on their similarity at polymorphic nucleotide sites (in bold and capitals). The two circles (connected by a line) below the sequences are to get you started drawing the network. Circles represent sequences 1 and 3 and the line connecting the circles show the one substitution at base pair position 23 that exists between the two sequences (haplotypes) 1 and 3.

```
             Base pair position (1-40)

   1234567891 1111111112 2222222223 3333333334
            0 1234567890 1234567890 1234567890

1) gagtattata agggcgagtg tcatttcttc aacgggaccg
2) gagtatGCta agAgcgagtg tcatttcttc aaccggacgg
3) gagtattata agggcgagtg tcTtttcttc aacgggaccg
4) gagtatGCta agAgcgagtg tcatttGttc aacgggacgg
```

(b) Conduct a BLAST search (at http://www.ncbi.nlm.nih.gov/BLAST/) with sequence number 1, and identify the gene and species of origin of the sequence.

17

Hybridization

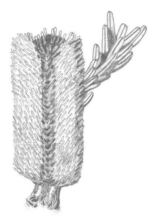

Banksia, Section 17.2

Hybridization, with or without introgression, frequently threatens populations in a wide variety of plant and animal taxa because of various human activities.
Judith M. Rhymer and Daniel Simberloff (1996)

We conclude by suggesting that when viewed over the long-term of millennia, introgressive hybridization may have contributed importantly to the generation of species diversity in birds.
Barbara R. Grant and Peter R. Grant (1998)

Rates of **hybridization** and **introgression** have increased dramatically worldwide because of widespread intentional and incidental translocations of organisms and habitat modifications by humans (see Guest Box 17). Hybridization has contributed to the extinction of many species through direct and indirect means (Levin et al. 1996; Allendorf et al. 2001).

The severity of this problem has been underestimated by conservation biologists (Rhymer and Simberloff 1996). The increasing pace of the three interacting human activities that contribute most to increased rates of hybridization (introductions of plants and animals, fragmentation, and habitat modification) suggests that this problem will become even more serious in the future. For example, increased turbidity in Lake Victoria, Africa, has reduced color perception of cichlid fishes and has interfered with the mate choice that produced reproductive isolation among species (Seehausen et al. 1997). Increased turbidity because of land development and forest harvesting has led to increased hybridization among stickleback species in British Columbia, Canada (Wood 2003).

On the other hand, hybridization is also part of the evolutionary process. Hybridization has long been recognized as playing an important role in the evolution of plants (Arnold 1997) (Figure 17.1). In addition, recent studies have found that hybridization also has played an important role in the evolution of animals (Arnold 1997; Dowling and Secor 1997; Grant and Grant 1998). Several reviews have emphasized the creative role that

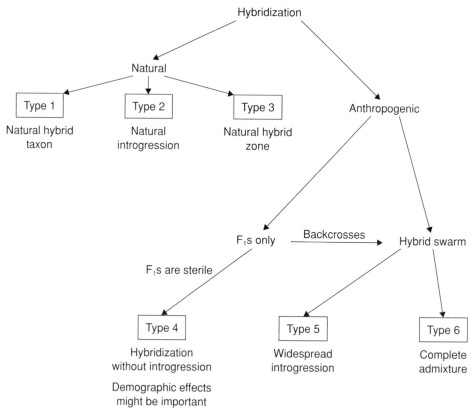

Figure 17.1 Framework to categorize hybridization. Each type should be viewed as a general descriptive classification used to facilitate discussion rather than a series of strict, all-encompassing divisions. Types 1–3 represent hybridization events that are a natural part of the evolutionary legacy of taxa; these taxa should be eligible for protection. Types 4–6 divide anthropogenic hybridization into three categories that have different consequences from a conservation perspective. From Allendorf et al. (2001).

hybridization may play in adaptive evolution and speciation (e.g. Rieseberg 1997; Grant and Grant 1998; Seehausen 2004). Many early conservation policies generally did not allow protection of hybrids. However, increased appreciation of the important role of hybridization as an evolutionary process has caused a re-evaluation of these policies. Determining whether hybridization is natural or anthropogenic is crucial for conservation, but it is often difficult (Allendorf et al. 2001).

Hybridization provides an exceptionally tough set of problems for conservation biologists. The issues are complex and controversial, beginning with the seemingly simple task of even defining hybridization (Harrison 1993). Hybridization has sometimes been used to refer to the interbreeding of species (e.g., Grant and Grant 1992b). However, we believe that this taxonomically restrictive use of hybridization can be problematic (especially since it is sometimes difficult to agree on what is a species!). We have adopted the more general definition of Harrison (1990) that includes matings between "individuals from two populations, or groups of populations, which are distinguishable on the basis of one or more heritable characters".

The term "hybrid" itself sometimes has a negative connotation, especially when used in conjunction with its opposite "purebred". In the United States, a proposed policy to treat hybrids and hybridization under the Endangered Species Act (ESA) used the term "intercross" (suggested by John Avise) and "intercross progeny" rather than hybrid to avoid the connotations of the term hybrid (USFWS and NOAA 1996a).

Detection of hybridization can also be difficult, although it is becoming much easier through the application of various molecular techniques over the last two decades. Despite improved molecular data that can be collected with relative ease, interpreting the evolutionary significance of hybridization and determining the role of hybrid populations in developing conservation plans is more difficult than often appreciated. According to one review: "It is an understatement to say that hybridization is a complex business!" (Stone 2000).

In this chapter, we first consider the role that natural hybridization has played in the process of evolution. We next consider the possible harmful effects of anthropogenic hybridization and the fitness of hybrid individuals and populations. We also present and discuss genetic methods for detecting and evaluating hybridization. Finally, we consider the possible use of hybridization as a tool in conservation.

17.1 Natural hybridization

Consideration of the role of hybridization in systematics and evolution goes back to Linnaeus and Darwin (see discussion in Arnold 1997, p. 6). Botanical and zoological workers have tended to focus on the two opposing aspects of hybridization. Botanists have generally accepted hybridization as a pervasive and important aspect of evolution (e.g., Stebbins 1959; Grant 1963). They demonstrated that many plant taxa have hybrid origins and demonstrated that hybridization is an important mechanism for the production of new species and novel adaptations. In contrast, early evolutionary biologists working with animals were very interested in the evolution of reproductive isolation leading to speciation (Mayr 1942; Dobzhansky 1951). They emphasized that hybrid offspring were often relatively unfit, and that this led to the development of reproductive isolation and eventually speciation.

17.1.1 Intraspecific hybridization

Intraspecific hybridization in the form of gene flow among populations has several important effects. It has traditionally been seen as the cohesive force that holds species together as units of evolution (Mayr 1963). This view was challenged by Ehrlich and Raven (1969) who argued that the amount of gene flow observed in many species is too low to prevent differentiation thorough genetic drift or local adaptation.

The resolution to this conflict is the recognition that even very small amounts of gene flow can have a major cohesive effect. We saw in Chapter 9 that an average of one migrant individual per generation with the island model of migration is sufficient to make it likely that all alleles will be found in all populations. That is, populations may diverge quantitatively in allele frequencies, but qualitatively the same alleles will still be present. We saw in Chapters 9 and 12 that just one migrant per generation can greatly increase the local effective population size.

Rieseberg and Burke (2001) have presented a model of species integration that considers the effects of the spread of selectively advantageous alleles. They have shown that new mutations that have a selective advantage will spread across the range of a species much faster than selectively neutral mutations with low amounts of gene flow. They have proposed that it is the relatively rapid spread of highly advantageous alleles that holds a species together as an integrated unit of evolution.

High amounts of gene flow can restrict the ability of populations to adapt to local conditions. **Genetic swamping** occurs when gene flow causes the loss of locally adapted alleles or genotypes (Lenormand 2002). This effect may be greatest in sparsely populated populations in which gene flow tends to be from densely populated areas (García-Ramos and Kirkpatrick 1997). In such cases, the continued immigration of locally unfit genotypes reduces the mean fitness of a population and potentially could lead to what has been called a **hybrid sink** effect. This is a self-reinforcing process in which immigration produces hybrids that are unfit, which reduces local density and increases the immigration rate (Lenormand 2002).

Riechert et al. (2001) have provided evidence that gene flow in a desert spider has caused genetic swamping and the reduction in fitness of local populations. Riparian habitats favor spiders with a genetically determined nonaggressive phenotype in comparison to adjacent arid habitats in which a competitive, aggressive phenotype is favored. Nearly 10% of the matings of riparian spiders are with an arid-land partner. The resulting offspring have reduced survival in the riparian habitat compared to matings between riparian spiders. Modeling has shown that cessation of gene flow between spiders in different habitat types is expected to quickly result in the divergence in frequency of aggressive and nonaggressive phenotypes in the two habitats (Figure 17.2).

17.1.2 Interspecific hybridization

Hybridization and introgression between species may occur more often than usually recognized (see Guest Box 17). For example, Grant and Grant (1992b) estimated that approximately 10% of all bird species have bred with another species and produced hybrid progeny. Bush (1994) defined speciation as a process of divergence of lineages that are sufficiently distinct from one another to follow independent evolutionary paths. Many

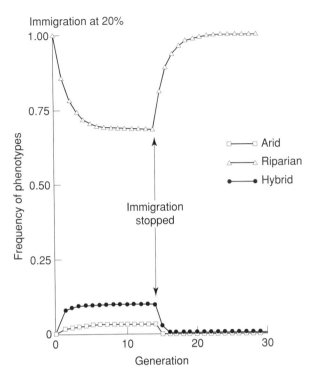

Figure 17.2 Predicted response to cessation of gene flow between desert spiders living in riparian and arid habitat patches as measured by the frequency of the nonaggressive phenotype that has higher fitness in riparian but not arid environments. From Riechert et al. (2001).

independent lineages are capable of hybridizing and exchanging genes (introgression) for quite long times without losing their phenotypic identities.

Interspecific hybridization can be an important source of genetic variation for some species. Grant and Grant (1998) have studied two species of Galápagos finches on the volcanic island of Daphne Major for over 30 years. The have found that hybridization between the two species (medium ground finch and cactus finch) has been an important source of genetic variation for the rarer cactus finch species. And they have suggested that their results may apply to many bird species. Interspecific introgression, or leakage between species, may cause a major shift in the way we think about species (Zimmer 2002).

Such introgression is especially important for island populations in which the effective population size is restricted because of isolation and the amount of available habitat. Two species of land snails (*Partula*) occur sympatrically on the island of Moorea in French Polynesia. In spite of being markedly different both phenotypically and ecologically, estimates of genetic distance based on molecular markers between some sympatric populations of these species are lower than is typical for conspecific comparisons for these taxa on different islands. Clarke et al. (1998) concluded that this apparent paradox was best explained by "molecular leakage, the convergence of neutral and mutually advantageous genes in two species through occasional hybridization".

Mitochondrial DNA seems particularly prone to introgression and molecular leakage (Ballard and Whitlock 2004). There are many examples of cases where the mtDNA molecule of one species has completely replaced the mtDNA of another species, in some populations without any evidence of nuclear introgression. For example, the mtDNA in a population of brook trout in Lake Alain in Québec is identical to the Québec Arctic char genotype, yet the brook trout are morphologically indistinguishable from normal brook trout and have diagnostic brook trout alleles at nuclear loci (Bernatchez et al. 1995).

Hybrid zones

An interspecific **hybrid zone** is a region where two species are sympatric and hybridize to form at least partially fertile progeny. Hybrids zones usually result from secondary contact between species that have diverged in allopatry. Recent molecular analysis of plants and animals has revealed that hybrid zones occur widely in many taxa (Harrison 1993). Barton and Hewitt (1985) reviewed 170 reported hybrid zones and concluded that hybrids were selected against in most hybrid zones that have been studied. Nevertheless, some hybrid zones appear to be stable and persist over long periods of time through a balance between dispersal of parental types and selection against hybrids (Harrison 1993). Hybrid zones may act as selective filters that allow introgression of only selectively advantageous alleles between species (Martinsen et al. 2001).

Arnold (1997) has proposed three models to explain the existence of a stable hybrid zone without genetic swamping of one or both of the parental species. In the **Tension Zone Model**, first- and second-generation hybrids are less fit than the parental types, but a balance between dispersal into the hybrid zone and selection against hybrids produces an equilibrium, with a persistent, narrow hybrid zone containing F_1 individuals but few or no F_2 or beyond hybrids. This model does not depend upon ecological differences between the habitats of the two parental types. In the **Bounded Hybrid Superiority Model**, hybrids are fitter than either parental species in environments that are intermediate to the parental habitats, but are less fit than the parental species in their respective native habitats (Example 17.1). The **Mosaic Model** is similar to the Bounded Hybrid Superiority Model, but the parental habitats are patchy rather than there being an environmental gradient between two spatially separated parental habitats. Under both models, theory predicts that hybridization and backcrossing would occur for many generations, creating introgressed populations containing individuals varying in their proportions of genetic material from the parental species.

17.1.3 Hybrid taxa

Approximately one-half of all plant species have been derived from **polyploid** ancestors, and many of these polyploid events involved hybridization between species or populations within the same sp ecies (Stebbins 1950). Recent evidence suggests that all vertebrates went through an ancient polyploid event that might have involved hybridization (Lynch and Conery 2000). Other major vertebrate taxa have gone through additional polyploid events. For example, all salmonid fishes (trout, salmon, char, whitefish, and grayling) went through an ancestral polyploid event some 25–50 million years ago (Allendorf and Waples 1996).

Example 17.1 Genetic analysis of a hybrid zone between Sitka and white spruce

Bennuah et al. (2004) have used genetic analysis of nuclear DNA markers to study a hybrid zone between the more coastal Sitka spruce and the more inland white spruce in northwestern British Columbia. Their results suggested that the stable, narrow hybrid zone is most likely maintained by hybrid superiority limited to environments that are intermediate to the ecological niches of the parental species (bounded hybrid superiority).

Genotypes of individuals in the hybrid zone suggested that they were all later generation hybrids (beyond the F_1 and F_2 generations). A hybrid index based upon the genotypes was estimated to reflect the relative contribution of Sitka spruce and white spruce genomes. There is also a steep and wide geographic cline across the hybrid zone of parental contributions that was concordant with the climatic gradient from maritime to continental climate of the two parental species. A patchy distribution of hybrid index estimates corresponding to environmental variation, as expected with the Mosaic Model, was not observed.

The hybrid superiority in transitional habitats could result from a combination of higher cold and drought hardiness of the white spruce and the higher growth potential of the Sitka spruce. The authors recommended that the relatively steep cline observed in the hybrid index across the maritime/continental climate **ecotone** should be managed by limiting longitudinal seed transfer for reforestation.

Some hybrid taxa of vertebrates are unisexual. For example, unisexual hybrids between the northern redbelly dace and the finescale dace occur across the northern USA. Reproduction of such unisexual species is generally asexual or semisexual, and they are often regarded as evolutionary dead ends. However, it appears that some tetraploid bisexual taxa had their origins in a unisexual hybrid (e.g., all salmonid fish).

Asexual hybrid taxa may provide some interesting challenges to conservation. Some recognized species of corals have been found to be long-lived first-generation hybrids that primarily reproduce asexually (Vollmer and Palumbi 2002). However, this interpretation has been controversial. Others have argued that interbreeding does occur between hybrid corals and their parental species (Miller and van Oppen 2003). Regardless, understanding which corals are reproducing asexually and which are reproducing sexually is essential for setting conservation priorities for corals.

17.1.4 Transgressive segregation

Hybridization sometimes produces phenotypes that are extreme or outside the range of either parental type. This has been called **transgressive segregation**. Rieseberg et al. (2003) have shown that sunflower species that are found in extreme habitats tend to be ancient interspecific hybrids. They argue that new genotypic combinations resulting from hybridization have led to the ecological divergence and success of these species. A review of many hybrid species concluded that transgressive phenotypes are generally common in plant populations of hybrid origin (Rieseberg et al. 1999).

17.2 Anthropogenic hybridization

The increasing pace of introductions of plants and animals and habitat modifications have caused increased rates of hybridization among plant and animal species. The introduction of plants and animals outside their native range clearly provides the opportunity for hybridization among taxa that were reproductively isolated. However, it is sometimes not appreciated just how much habitat modifications have increased rates of hybridization.

In many cases, it is difficult to determine whether hybridization is "natural" or the direct or indirect result of human activities. In some cases, authors have referred to hybridization events resulting from habitat modifications as natural since they do not involve the introduction of species outside their native range. Decline in abundance itself because of anthropogenic changes also promotes hybridization among species because of the greater difficulty in finding mates. In both of these cases, however, we believe that hybridization should be considered to be the indirect result of human activities.

Wiegand (1935) was perhaps the first to suggest that introgressive hybridization is observed most frequently in habitats modified by humans. The creation of extensive areas of new habitats around the world has the effect of breaking down mechanisms of isolation between species (Rhymer and Simberloff 1996). For example, two native *Banksia* species in western Australia hybridize only in disturbed habitats where more vigorous growth has extended the flowering seasons of both species and removed asynchronous flowering as a major barrier to hybridization (Lamont et al. 2003). In addition, taxa that can adapt quickly to new habitats may undergo adaptive genetic change very quickly. It now appears that many of the most problematic invasive plant species have resulted from hybridization events (Ellstrand and Schierenbeck 2000; Gaskin and Schaal 2002). This topic is considered in more detail in Chapter 19.

Increased turbidity in aquatic systems because of deforestation, agricultural practices, and other habitat modifications has increased hybridization among aquatic species that use visual clues to reinforce reproductive isolation (Wood 2003). This has threatened sympatric species on the western coast of Canada (Kraak et al. 2001) and cichlid fish species in Lake Victoria (Seehausen et al. 1997). It is estimated that nearly half of the hundreds of species in Lake Victoria have gone extinct in the last 50 years primarily because of the introduction of the Nile perch in the 1950s (Goldman 2003). The waters of this lake have grown steadily murkier, in part due to algal blooms resulting from the decline of cichlids. Mating between species now appears to be widespread and the loss of this classic example of adaptive radiation is now threatened (Goldman 2003).

Many other forms of habitat modification can lead to hybridization (Rhymer and Simberloff 1996). For example, the modification of patterns of water flows may bring species into contact that have been previously geographically isolated. It is likely that hybridization will continue to be more and more of a problem in conservation. Global environmental change may further increase the rate of hybridization between species in cases where it allows geographic range expansion.

Hybridization can contribute to the decline and eventual extinction of species in two general ways. In the case of sterile or partially sterile hybrids, hybridization results in a loss of reproductive potential and may reduce the population growth rate below that needed for replacement (demographic swamping). In the case of fertile hybrids, genetically distinct populations may be lost through genetic mixing.

17.2.1 Hybridization without introgression

Many interspecific hybrids are sterile so that introgression (i.e. gene flow between populations whose individuals hybridize) does not occur. For example, matings between horses and donkeys produce mules, which are sterile because of chromosomal pairing problems during meiosis. Sterile hybrids are evolutionary dead-ends. Nevertheless, the production of these hybrids reduces the reproductive potential of populations and can contribute to the extinction of species (see Section 17.5.3).

17.2.2 Hybridization with introgression

In many cases, hybrids are fertile and may displace one or both parental taxa through the production of **hybrid swarms** (populations in which all individuals are hybrids by varying numbers of generations of backcrossing with parental types and mating among hybrids). This phenomenon has been referred to by many names (genetic assimilation, genetic extinction, or genomic extinction). The term "**genetic assimilation**", which has been used in the literature (e.g., Cade 1983), should not be used in order to avoid confusion. Waddington (1961) used this phrase to mean a process in which phenotypically plastic characters that were originally "acquired" become converted into inherited characters by natural selection (Pigliucci and Murren 2003).

   Genomic extinction is a more appropriate term than the phrase genetic extinction as it is not genes or single locus genotypes that that are lost by hybridization; it is combinations of genotypes over the entire genome that are irretrievably lost. Genomic extinction results in the loss of the legacy of an evolutionary lineage. That is, the genome-wide combination of alleles and genotypes that have evolved over evolutionary time will be lost by genetic swamping through introgression with another lineage.

17.3 Fitness consequences of hybridization

Hybridization may have a wide variety of effects on fitness. In the case of **heterosis**, or **hybrid vigor**, hybrids have enhanced performance or fitness relative to either parental taxa. In the case of **outbreeding depression**, the hybrid progeny have lower performance or fitness than either parent (Lynch and Walsh 1998). Both heterosis and outbreeding have many possible causes, and the overall fitness of hybrids results from an interaction among these different effects. To further complicate matters, much of the heterosis that is often detected in F_1 hybrids is lost in subsequent generations so that a particular cross may result in heterosis in the first generation and outbreeding depression in subsequent generations (Figure 17.3).

   There are two primary mechanisms that may reduce the fitness of hybrids. The first mechanism is genetic incompatibilities between the hybridizing taxa; this has been referred to as both intrinsic outbreeding depression and endogenous selection. Outbreeding depression may also result from reduced adaptation to environmental conditions by hybrids; this has been referred to as extrinsic outbreeding depression and also as exogenous selection.

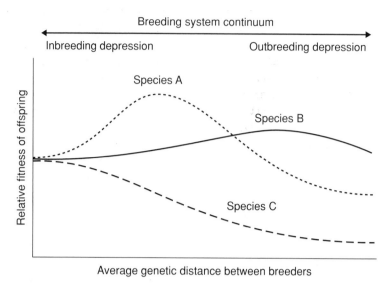

Figure 17.3 Heuristic model for visualizing the balance between inbreeding depression, hybrid vigor (heterosis), and outbreeding depression. Individual species exhibit different optimal levels of outcrossing, as illustrated by the plot of fitness relative to average genetic distance among breeders. For example, species A shows considerable inbreeding depression and also outbreeding depression. Species B exhibits little inbreeding depression and hybrid vigor. Redrawn from Waples (1995).

17.3.1 Hybrid superiority (heterosis)

In many regards, heterosis is the opposite of inbreeding depression. Therefore, the underlying causes of heterosis are the same as the causes of inbreeding depression: increased homozygosity and reduced heterozygosity (Crow 1993). The primary cause of heterosis is the sheltering of deleterious recessive alleles in hybrids. In addition, increased heterozygosity will increase the fitness of hybrid individuals for loci where the heterozygotes have a selective advantage over homozygote types. There is evidence that hybridization can serve as a stimulus for the evolution of invasiveness in plants (Ellstrand and Schierenbeck 2000).

The best example of heterosis has come from crossing inbred lines of corn to produce high-yielding hybrid corn. Virtually all agricultural corn grown today in the United States is hybrid, as compared to less than 1% of the corn planted in 1933 (Sprague 1978). A large number of self-fertilized lines of corn have been established from highly polymorphic populations. The yield of each inbred line decreases as homozygosity increases. Many lines are discontinued because their performance is so low. Inbred lines are expected to become homozygous for different combinations of deleterious recessive alleles. These deleterious recessive alleles will be sheltered in hybrids by heterozygosity. In addition, many alleles resulting in increased yield are dominant. The hybrids between two inbred lines are superior to both the inbred lines as well as corn from the original highly polymorphic populations. Many combinations of inbred lines are tested so that the combinations that produce the most desirable hybrids can be used.

Population subdivision of natural populations (see Chapter 9) can provide the appropriate conditions for heterosis. Different deleterious recessive alleles will drift to relatively high frequencies in different populations. Therefore, progeny produced by matings between immigrant individuals are expected to have greater fitness than resident individuals. This effect is expected to result in a higher effective migration rate because immigrant alleles will be present at much higher frequencies than predicted by neutral expectations (Ingvarsson and Whitlock 2000; Whitlock et al. 2000; Morgan 2002).

Experiments involving immigration into inbred, laboratory populations of African satyrne butterflies have revealed surprisingly strong heterosis (Saccheri and Brakefield 2002). Immigrants were, on average, over 20 times more successful in contributing descendants to the fourth generation than were inbred nonimmigrants. The mechanism underlying this rapid spread of immigrant alleles was found to be heterosis. The disproportionately large impact of some immigrants suggests that rare immigration events may be very important in evolution, and that heterosis may drive their fitness contribution.

Hybrids may also have a fitness advantage because they possess advantageous traits from both parental populations (see Example 17.1). As we saw in Section 17.1.4, Rieseberg et al. (2003) found that hybridization between sunflower species produced progeny that are adapted to environments very different from those occupied by the parental species. This was associated with the hybrids possessing new combinations of genetic traits. Choler et al. (2004) found hybrids between two subspecies of an alpine sedge in the Alps that occurred only in marginal habitats for the two parental subspecies.

17.3.2 Intrinsic outbreeding depression

Intrinsic outbreeding depression results from genetic incompatibilities between hybridizing taxa.

Chromosomal

Reduced fitness of hybrids can result from heterozygosity for chromosomal differences between populations or species (see Chapter 3). Differences in chromosomal number or structure may result in the production of aneuploid gametes that result in reduced survival of progeny. We saw in Table 3.4 that hybrids between races of house mice with different chromosomal arrangements produce smaller litters in captivity. Hybrids between chromosomal races of the threatened owl monkey from South America show reduced fertility in captivity (De Boer 1982).

Genic

Reduced fitness of hybrids can also result from genetic interactions between genes originating in different taxa (Whitlock et al. 1995). Dobzhansky (1948) first used the word "co-adaptation" to describe reduced fitness in hybrids between different geographic populations of the fruit fly Drosophila pseudoobscura. This term became controversial (and somewhat meaningless) following Mayr's (1963) argument that most genes in a species are co-adapted because of the integrated functioning of an individual.

Reduced fitness of hybrids can potentially occur because of the effects of genotypes at individual loci. Perhaps the best example is that of the direction of shell coiling in snails

(Johnson 1982). Shells of some species of snails coil either to the left (sinistral) or to the right (dextral). Variation in shell-coiling direction occurs within populations of snails of the genus *Partula* which are found on islands in the Pacific Ocean. Many species in this genus are now threatened with extinction because of the introduction of other snails (Mace et al. 1998). The variation in shell coiling in many snail species is caused by two alleles at a single locus (Sturtevant 1923; Johnson 1982). Snails that coil in different directions find mating difficult or impossible. Thus, the most common phenotype (sinistral or dextral) in a population will generally be favored leading to the fixation of one type or the other. Hybrids between sinistral and dextral coiling populations may have reduced fitness because of the difficulty of mating with snails of the other type (Johnson et al. 1990).

Outbreeding depression may result from genic interactions between alleles at multiple loci (**epistasis**) (Whitlock et al. 1995). That is, alleles that enhance fitness within their parental genetic backgrounds may reduce fitness in the novel genetic background produced by hybridization. Such interactions between alleles are known as Dobzhansky–Muller incompatibilities because they were first described by these two famous *Drosophila* geneticists (Johnson 2000). Dobzhansky referred to these interactions as co-adapted gene complexes. Such interactions are thought to be responsible for the evolution of reproductive isolation and eventually speciation.

There are few empirical examples of specific genes that show such **Dobzhansky–Muller incompatibilities**. Rawson and Burton (2002) have recently presented an elegant example of functional interactions between loci that code for proteins involved in the electron transport system of mitochondria in an intertidal copepod *Tigriopus californicus*. A nuclear gene encodes the enzyme cytochrome *c* (CYC) while two mtDNA genes encode subunits of cytochrome oxidase (COX). CYC proteins isolated from different geographic populations each had significantly higher activity in combination with the COX proteins from their own source population. These results demonstrate that proteins in the electron transport system form co-adapted combinations of alleles and that disruption of these co-adapted gene complexes lead to functional incompatibilities that may lower the fitness of hybrids.

Self-fertilization in plants has long been recognized as potentially facilitating the evolution of adaptive combinations of alleles at many loci. Many populations of primarily self-fertilizing plants are dominated by a few genetically divergent genotypes that differ at multiple loci. Parker (1992) has shown that hybrid progeny between genotypes of the highly self-fertilizing hog peanut have reduced fitness (Figure 17.4). These genotypes naturally co-occur in the same habitats and the hybrid progeny have reduced fitness in a common-garden. Thus, the reduced fitness of hybrids apparently results from Dobzhansky–Muller incompatibilities between genotypes.

17.3.3 Extrinsic outbreeding depression

Extrinsic outbreeding depression results from the reduced fitness of hybrids because of loss of local adaptation by ecologically mediated selection.

Mechanisms of escape from predation in northwestern garter snakes provide an excellent example of potential extrinsic outbreeding depression. Brodie (1992) has described a variety of color patterns and multiple behavioral strategies for escape from bird predators. Striped snakes are more visible than spotted snakes when still, but their stripes make it more difficult to detect their motion or judge their speed. Spots or blotches disrupt the

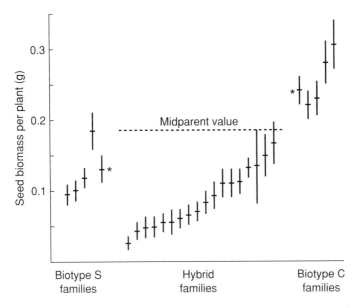

Figure 17.4 Fitness as measured by lifetime seed biomass of parental genotypes (biotypes S and C) and their hybrids in the highly self-fertilizing hog peanut. The two parental families marked with asterisks are the parents of the hybrids. From Parker (1992).

outline of the snake's body and make them more difficult to detect visually. The escape behaviors of garter snakes tend to match their color pattern. That is, striped snakes try to escape when threatened, while spotted snakes often use an evasive stop-and-go behavior.

This association between color patterns and behavior also occurs at the species level. In general, striped patterns in North American snakes are associated with diurnal activity and flight as a primary defense (Jackson et al. 1976). Blotched or broken patterns tend to occur in snakes with secretive habits or aggressive antipredator behavior. Hybrids between species or populations with different combinations of coloration and behavior are likely to have reduced fitness because of having the wrong combination of coloration and behavior.

Sage et al. (1986) studied a hybrid zone between two species of mice in Europe (*Mus musculus* and *M. domesticus*). Hybrids had significantly greater loads of pinworm (nematodes) and tapeworm (cestodes) parasites than either of the parental taxa (Figure 17.5). A total of 93 mice were examined within the hybrid zone. Fifteen of these mice had exceptionally high nematodes (>500) while 78 mice had "normal" numbers of nematodes (<250). Fourteen of the 15 mice with high nematode loads were hybrids while 37 of the 78 mice with normal loads were hybrids ($P < 0.005$). Cestode infections showed a similar pattern in hybrid and parental mice.

Increased susceptibility to diseases and parasites, as above, is an important potential source of outbreeding depression because of the importance of disease in conservation and the complexity of immune systems and their associated gene complexes. Currens et al. (1997) found that hybridization with introduced hatchery rainbow trout native to a different geographic region increased the susceptibility of wild native rainbow trout to myxosporean parasites. Similarly, Goldberg et al. (2005) found that hybrid largemouth bass from two genetically distinct subpopulations were more susceptible to largemouth

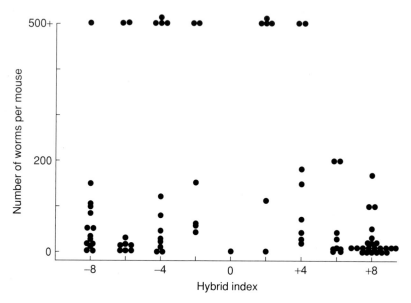

Figure 17.5 Nematode burdens (number of worms per mouse) in hybrid and parental mice. The hybrid index is based upon four diagnostic allozyme loci. Pure *Mus musculus* has an index of −8 and pure *M. domesticus* has an index of +8. From Sage et al. (1986).

bass virus. Parris (2004) found that hybrid frogs show increased susceptibility to emergent pathogens compared to the parental species.

Hybridization between hatchery and wild populations is a major conservation problem for many fish species. For example, up to 2 million Atlantic salmon are estimated to escape from salmon farms each year. Hybrids (F_1, F_2, and backcrosses) between farm and wild fish all show lower survival rates than wild salmon (McGinnity et al. 2003). Nevertheless, farm and hybrid salmon show faster growth rates as juveniles and therefore may displace juvenile wild salmon. The repeated escapes of farmed salmon present a substantial threat to the remaining wild populations of Atlantic wild salmon through accumulation of fitness depression by introgression. These issues are considered in more detail in the next chapter.

17.4 Detecting and describing hybridization

The detection of hybrid individuals relied upon morphological characteristics until the mid-1960s. However, not all morphological variation has a genetic basis, and the amount of morphological variation within and among populations is often greater than recognized (Campton 1987). The detection of hybrids using morphological characters generally assumes that hybrid individuals will be phenotypically intermediate to parental individuals (Smith 1992). This is often not the case because hybrids sometimes express a mosaic of parental phenotypes. Furthermore, individuals from hybrid swarms that contain most of their genes from one of the parental taxa are often morphologically indistinguishable from that parental taxon (Leary et al. 1996). Morphological characters do not allow one to determine whether an individual is a first-generation hybrid (F_1), a backcross, or a later generation hybrid. These distinctions are crucial because if a population has not become a hybrid swarm and still contains a reasonable number of parental individuals, it could potentially be recovered by removal of hybrids or by a captive breeding program.

The use of molecular genetic markers greatly simplifies the identification and description of hybridized populations. This procedure began with the development of protein electrophoresis (allozymes) in the mid-1960s. Recent advances in molecular techniques, especially the development of the polymerase chain reaction (PCR), have greatly increased the number of loci that can be used to detect hybridization (see Figure 4.10). In addition, these techniques are more applicable to small populations threatened with extinction because sampling can be noninvasive.

Genetic analysis of hybrids and hybridization is based upon loci at which the parental taxa have different allele frequencies. **Diagnostic loci** that are fixed or nearly fixed for different alleles in two hybridizing populations are the most useful although hybridization can also be detected using multiple loci at which the parental types differ in allele frequency (Cornuet et al. 1999) (Example 17.2).

Example 17.2 Hybridization between the threatened Canada lynx and bobcat

The Canada lynx is a wide ranging felid that occurs in the boreal forest of Canada and Alaska (Schwartz et al. 2004). The southern distribution of native lynx extends into the northern contiguous USA from Maine to Washington State. Lynx are also located in Colorado where a population was introduced in 1999. The Canada lynx is listed as "threatened" under the US ESA. Canada lynx are elusive animals and their presence routinely has been detected by genetic analysis of mtDNA from hair and fecal samples (Mills et al. 2001). Samples of hair and feces confirmed that Canada lynx were present in northern Minnesota, after a 10-year suspected absence from the state. In 2001, a trapper was prosecuted for trapping a lynx. The trapper thought it was a bobcat, while the biologist registering the pelt and the enforcement officer processing the case thought it was a lynx. Initial analysis based on mtDNA showed the sample was a lynx. However, upon hearing the controversy and recognizing that mtDNA could only determine the matriline of the cat, Schwartz et al. (2004) decided to design and test an assay that could detect hybridization between bobcats and lynx. Hybridization between these species had never before been confirmed in the wild.

The controversial sample was a hybrid. In addition, another sample from a carcass and a hair sample collected on a putative lynx backtrack were also identified as hybrids using microsatellite analysis. The hybrids were identified as having one lynx diagnostic allele and one bobcat diagnostic allele (Figure 17.6). A heterozygote with one allele from each parental species is expected in a F_1 hybrid (although some F_2 hybrids will also be heterozygous for species-diagnostic alleles at some loci). The species-diagnostic alleles were identified (at two loci) by analyzing microsatellites in 108 lynx and 79 bobcats across North America, far away from potential hybridization zones between the two species. In addition, mtDNA analysis revealed that all hybrids had lynx mothers (i.e., lynx mtDNA). All three hybrid samples had Canada lynx mtDNA and therefore were produced by matings

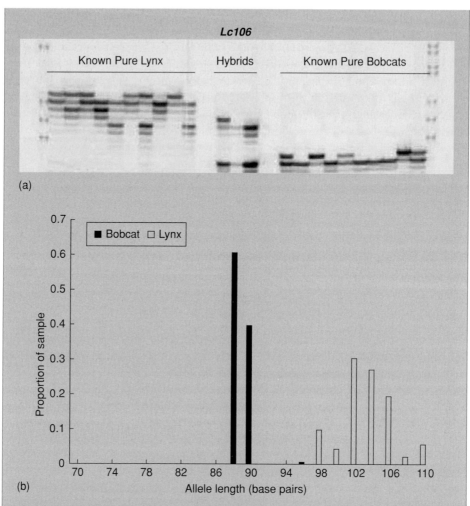

(a)

(b)

Figure 17.6 (a) Microsatellite gel image showing genotype profiles (locus *Lc106*) for 10 lynx, 10 bobcats, and three putative hybrids. Dark bands represent alleles; lighter bands are "stutter" bands. Outer lanes show size standards. (b) Allele frequencies for locus *Lc106* in bobcats and lynx. This locus is diagnostic because the allele size ranges do not overlap between species. From Schwartz et al. (2004).

between female Canada lynx and male bobcats. After these results were published, researchers from Maine and New Brunswick requested that some of their study animals be screened using the hybrid test. Four additional hybrids were discovered – two in Maine and two in New Brunswick. However, in a screening of hundreds of lynx samples from the Rocky Mountains (another area where the two species co-occur), no hybrids were discovered.

These data have important conservation implications. First, bobcat trapping is legal, while it is illegal to trap lynx anywhere in conterminous United States. The trapping of bobcats in areas where Canada lynx are also present could be problematic because both lynx and lynx–bobcat hybrids can be incidentally taken from extant populations. On the other hand, any factors that may favor bobcats in lynx habitat may lead to the production of hybrids and thus be potentially harmful to lynx recovery. Efforts need to be undertaken to describe the extent, rate, and nature of hybridization between these species, and to understand the ecological context in which hybridization occurs.

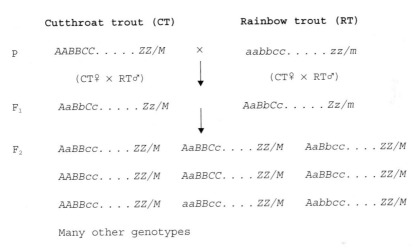

Figure 17.7 Outline of genetic analysis of hybridization between two species: cutthroat trout and rainbow trout. Alleles present in one species at diagnostic nuclear loci are designated by capital letters and the alleles in the other species by lower case letters. The parental (P) mtDNA haplotypes are designated by M and m, respectively.

Figure 17.7 outlines the use of diagnostic loci to analyze hybridization. First-generation (F_1) hybrids will be heterozygous for alleles from the parental taxa at all diagnostic loci. Later generation hybrids may result either from matings between hybrids or backcrosses between hybrids and one of the parental taxa (Example 17.3). The absence of such genotypes resulting in later generation hybrids suggests that the F_1 hybrids are sterile or have reduced fertility (Example 17.4). These two examples with different pairs of trout species demonstrate the contrasting results depending upon whether or not the F_1 hybrids are fertile (Example 17.3) or sterile (Example 17.4).

Example 17.3 Hybrid swarms of cutthroat trout and rainbow trout

The loss of native cutthroat trout by hybridization with introduced rainbow trout has been recognized as a major threat for over 75 years in the western USA (Allendorf and Leary 1988). The westslope cutthroat trout is one of four major subspecies of cutthroat trout. The geographic range of westslope cutthroat trout is the largest of all cutthroat trout subspecies and includes the Columbia, Fraser, Missouri, and Hudson Bay drainages of the United States and Canada. The west-slope cutthroat trout is genetically highly divergent at both nuclear and mitochondrial genes from the three other major subspecies of cutthroat trout: the coastal, Yellowstone, and Lahontan cutthroat trout. For example, 10 of 46 nuclear allozyme loci are fixed or nearly fixed for different alleles between westslope and Yellowstone cutthroat trout. This amount of divergence is far beyond that usually seen within a single species.

Introgressive hybridization with introduced rainbow trout and Yellowstone cutthroat trout occurs throughout the range of the westslope subspecies. Hybridization of cutthroat and rainbow trout generally results in the formation of random mating populations in which all individuals are hybrids by varying numbers of generations of backcrossing with parental types and mating among hybrids (i.e., hybrid swarms).

Table 17.1 shows genotypes at eight diagnostic nuclear loci between native westslope cutthroat trout and Yellowstone cutthroat trout introduced into Forest

Table 17.1 Genotypes at eight diagnostic nuclear allozyme loci in a sample of westslope cutthroat trout, Yellowstone cutthroat trout, and their hybrids from Forest Lake, Montana (Allendorf and Leary 1988). Heterozygotes are WY, while individuals homozygous for the westslope cutthroat trout allele are indicated as W and individuals homozygous for the Yellowstone cutthroat trout allele are indicated as Y. All individuals in this sample are later generation hybrids; thus, the fish in this lake are a hybrid swarm.

No.	mtDNA	Aat1	Gpi3	Idh1	Lgg	Me1	Me3	Me4	Sdh
				Nuclear encoded loci					
1	YS	W	W	WY	W	W	W	W	Y
2	YS	W	WY	WY	WY	Y	W	WY	Y
3	WS	WY	Y	Y	W	Y	WY	Y	WY
4	WS	Y	W	WY	WY	W	Y	W	WY
5	YS	Y	Y	Y	WY	WY	WY	Y	Y
6	YS	WY	Y	W	WY	W	W	W	Y
7	WS	WY	WY	Y	W	WY	W	W	W
8	WS	WY	Y	WY	WY	Y	W	Y	Y
9	WS	Y	Y	WY	WY	W	WY	WY	W
10	WS	WY	Y	WY	WY	WY	Y	W	Y
11	YS	Y	W	W	WY	W	Y	W	Y
12	WS	W	WY	Y	WY	W	WY	WY	Y
13	YS	W	Y	W	Y	W	WY	W	W
14	YS	Y	Y	WY	WY	WY	WY	WY	W
15	WS	WY	Y	WY	Y	W	Y	WY	W

Lake, Montana in a representative sample of 15 individuals taken from the lake. All but one of these 15 fish are homozygous for both westslope and Yellowstone alleles at different loci. Each individual in this sample appears to be a later generation hybrid (see Figure 17.7). Thus, the fish in this lake are a hybrid swarm.

Example 17.4 Near sterility in hybrids between bull and brook trout

Bull trout are legally protected as "threatened" in the United States under the US ESA. Hybridization with introduced brook trout is potentially one of the major threats to the persistence of bull trout. Bull trout and brook trout have no overlap in their natural distribution, but secondary contact between these species has occurred as a result of the introduction of brook trout into the bull trout's native range.

Leary et al. (1993a) described a rapid and almost complete displacement of bull trout by brook trout in which the initial phases were characterized by frequent hybridization. In the South Fork of Lolo Creek in the Bitterroot River drainage, Montana, brook trout first invaded in the late 1970s. In the initial sample collected in 1982, bull trout (44%) were the most abundant, followed by hybrids (36%) and brook trout (20%), and matings appeared to be occurring at random. By 1990, however, brook trout (65%) were more abundant than bull trout (24%) and hybrids (12%).

Table 17.2 shows genotypes at eight diagnostic nuclear loci between native bull trout and brook trout introduced into Mission Creek, Montana in a sample of 15 individuals that were selected to be genetically analyzed because they appeared to be hybrids. Eleven of the 15 fish in this sample contained alleles from both species indicating that they are hybrids. However, in striking contrast to Example 17.3, 10 of the 11 hybrids were heterozygous at all eight loci suggesting that they are F_1 hybrids. It is extremely unlikely that a later generation hybrid would be heterozygous at all loci. For example, there is a 0.50 probability that an F_2 hybrid will be heterozygous at a diagnostic locus (see Figure 17.7). Thus, thus there is a $(0.50)^8 = 0.004$ probability that an F_2 will be heterozygous at all eight loci. The F_1 hybrids in this sample have both bull and brook trout mtDNA indicating that both reciprocal crosses are resulting in hybrids.

This general pattern has been seen throughout the range of bull trout. Almost all hybrids appear to be F_1 hybrids with very little evidence of F_2 or backcross individuals (Kanda et al. 2002). The near absence of progeny from hybrids of bull and brook trout may result from either the sterility of the hybrids, their lack of mating success, the poor survival of their progeny, or combinations of these factors. Over 90% of the F_1 hybrids are male suggesting some genetic incompatibility between these two genomes.

Table 17.2 Genotypes at eight diagnostic nuclear loci in a sample of bull trout, brook trout, and their hybrids from Mission Creek, Montana (Kanda et al. 2002). Heterozygotes are *LR*, while individuals homozygous for the bull trout allele are indicated as *L* and individuals homozygous for the brook trout allele are indicated as *R*. Individuals are identified in the status column as either parental-type bull trout (BL) or brook trout (BR), or hybrids on the basis of their genotype.

| No. | mtDNA | Nuclear encoded loci | | | | | | | | Status |
		Aat1	Ck-A1	Iddh	sIdhp-2	Ldh-A1	Ldh-B2	Mdh-A2	sSod-1	
1	L	LR	LR	LR	LR	LR	LR	LR	LR	F_1
2	L	LR	LR	LR	LR	LR	LR	LR	LR	F_1
3	L	R	R	LR	LR	LR	LR	LR	R	$F_1 \times BR$
4	L	L	L	L	L	L	L	L	L	BL
5	L	LR	LR	LR	LR	LR	LR	LR	LR	F_1
6	R	LR	LR	LR	LR	LR	LR	LR	LR	F_1
7	L	LR	LR	LR	LR	LR	LR	LR	LR	F_1
8	R	LR	LR	LR	LR	LR	LR	LR	LR	F_1
9	R	LR	LR	LR	LR	LR	LR	LR	LR	F_1
10	L	LR	LR	LR	LR	LR	LR	LR	LR	F_1
11	R	LR	LR	LR	LR	LR	LR	LR	LR	F_1
12	R	LR	LR	LR	LR	LR	LR	LR	LR	F_1
13	R	R	R	R	R	R	R	R	R	BR
14	R	R	R	R	R	R	R	R	R	BR
15	R	R	R	R	R	R	R	R	R	BR

17.4.1 Multiple loci and gametic disequilibrium

The distribution of gametic disequilibria (D) between pairs of loci is helpful to describe the distribution of hybrid genotypes and to estimate the "age" of hybridized populations (see Table 10.2; Guest Box 10). Recently hybridized populations will have a high D because they will contain parental types and many F_1 hybrids. By contrast, genotypes will be randomly associated among loci in hybrid swarms that have existed for many generations. This will occur rather quickly for unlinked loci because D will decay by one-half each generation (see expression 10.3). However, nonrandom association of alleles at different loci might persist for many generations at pairs of loci that are closely linked. Barton (2000) and Agapow and Burt (2001) have provided single measures of gametic disequilibrium that can provide a meaningful measure to compare the amount of gametic disequilibrium at a number of unlinked loci in hybrid swarms.

Genetic data must be interpreted at both the individual and population level to understand the history of hybridization in populations (Barton and Gale 1993). Hybrid individuals can be first-generation (F_1) hybrids, second-generation hybrids (F_2s), backcrosses to one of the parental taxa, or later generation hybrids. Parental types and F_1 hybrids can be reliably identified if many loci are examined. However, it is very difficult to distinguish between F_2s, backcrosses, and later generation hybrids, even if many loci are examined (Boecklen and Howard 1997).

New statistical approaches for assigning individuals to their population of origin based upon many highly polymorphic loci are especially valuable for identifying hybrids (Hansen et al. 2000). These techniques may be useful even when putative "pure" populations are not available to provide baseline information. Example 17.5 presents genetic analysis of a particularly difficult hybridization situation in which known parental types were not available to determine the genetic composition of the parental taxa.

Example 17.5 Genetic mixture analysis of Scottish wildcats

The Scottish wildcat has had full legal protection since 1988 (Beaumont et al. 2001). The presence of feral domestic cats and the possibility of hybridization have, however, made this protection ineffective because it has been impossible to unambiguously distinguish wildcats from hybrids. The amount of hybridization between domestic cats and existing wildcats is unknown. Some believe that there has been little hybridization until recently. However, the behavioral similarity between cats in a study area containing morphologically domestic cat and morphologically wildcat individuals suggests that hybridization may have had a substantial impact on the genetic composition of wildcats in Scotland.

Beaumont et al. (2001) studied nine microsatellite loci in 230 wild-living Scottish cats (including 13 museum skins) and 74 house cats from England and Scotland. In addition, pelage characteristics of the wild-living cats were recorded (Figure 17.8). Hybridization between Scottish wildcats and domestic cats was tested in order to identify hybrid populations and to guide conservation management.

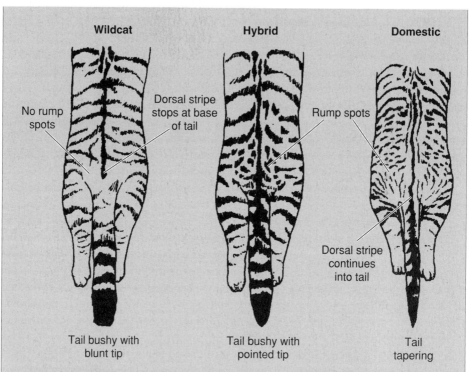

Figure 17.8 Diagram showing three (tail shape, dorsal stripe, and rump spots) of the five morphologically diagnostic characters used to distinguish wildcats and domestic cats. Tail tip color and paw color are not illustrated. From Beaumont et al. (2001).

The genetic mixture analysis method of Pritchard et al. (2000) using a Markov chain Monte Carlo approach (see Appendix Section A5) was used without specifying the allele frequencies of the source populations to estimate q (the proportion of an individual's genome that comes from the wildcat population). The distribution of q is integrated over all possible gene frequencies in the two parental populations weighted by their posterior density to obtain a posterior density for q that is independent of the parental frequencies.

This analysis revealed five main genetic groupings of individuals (Figure 17.9): two groups of wild-living cats (high $\hat{q}$ values), one group intermediate to the two wild-living cats, one group very similar to the domestics (perhaps introgressed), and the domestics themselves. The authors concluded that most of the wildcats have not experienced recent introgression from domestics. However, morphological and genetic data suggest that earlier introgression from domestic cats has occurred. The authors conclude there is strong evidence of a population of individuals that are different from domestic cats that may be worthy of legal protection. However, this will be difficult because there is no diagnostic test of a true wildcat that contains no domestic cat ancestry. In fact the evidence suggests that such cats may not exist.

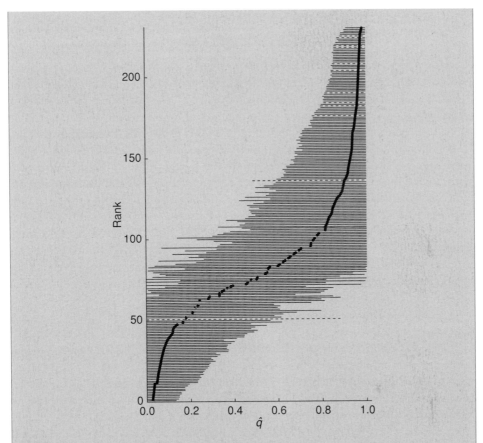

Figure 17.9 Genetic mixture analysis of hybridzation between domestic cats and Scottish wildcats. $\hat{q}$ is the proportion an individual's genome that comes from the wildcat population. This figure illustrates the ranked distribution of $\hat{q}$ among individuals. Also shown are lines giving the 95% equal-tail posterior probability intervals for each individual. The dashed lines are from museum specimens. Individuals with low $\hat{q}$ values are likely to be domestic cats, and individuals with high $\hat{q}$ values are likely to be wildcats.

17.5 Hybridization and conservation

Hybridization is a natural part of evolution. Taxa that have arisen through natural hybridization should be eligible for protection. Nevertheless, increased anthropogenic hybridization is causing the extinction of many taxa (species, subspecies, and locally adapted populations) by both replacement and genetic mixing. Conservation policies should be designed to reduce anthropogenic hybridization. Nevertheless, developing policies to deal with the complex issues associated with hybridization has been difficult.

17.5.1 Protection of hybrids

Protection of hybrids under the US ESA has had a controversial history (Haig and Allendorf, in press). In May 1977, the US Department of the Interior's Office of the Solicitor issued a statement that "because it defines 'fish or wildlife' to include any offspring without limitation, the Act's plain meaning dictates coverage of hybrids of listed animal species. The legislative history buttresses this conclusion for animals and also makes clear its applicability to plants." However, response from the US fish and Wildlife Service (July 1977) indicated ". . . since the Act was clearly passed to benefit endangered species, . . . it must have meant the offspring of two listed species and was not meant to protect a hybrid where that protection would in fact cause jeopardy to the continued existence of a species." The Solicitor responded (August 1977, and reaffirmed in 1983) stating that "hybrids of listed species are not protected under the ESA" because he had learned there was the potential for a listed species to be harmed by hybridization. Overall, the US fish and Wildlife Service's early position was to "discourage conservation efforts for hybrids between taxonomic species or subspecies and their progeny because they do not help and could hinder recovery of endangered taxon."

This series of correspondences and decisions that denied ESA protection for organisms with hybrid ancestry became known as the "Hybrid Policy" (O'Brien and Mayr 1991). O'Brien and Mayr pointed out that we would lose invaluable biological diversity if the ESA did not protect some subspecies or populations that interbreed (e.g., the Florida panther), or taxa derived from hybridization (e.g., the red wolf). Further, Grant and Grant (1992b) pointed out that few species would be protected by eliminating protection for any species interbreeding since so many plant and animal species interbreed to some extent. Discussions such as these and the Florida panther situation contributed to the US Fish and Wildlife Service suspending the Hybrid Policy in December 1990.

A proposed policy on hybrids was published in 1996 (USFWS and NOAA 1996a). This "Intercross Policy" was scheduled to be finalized 1 year later, but has still not been approved. Thus, no official policy provides guidelines for dealing with hybrids under the ESA. The absence of a final policy probably results from the difficulty in writing a hybrid policy that would be flexible enough to apply to all situations, but which would still provide helpful recommendations.

17.5.2 Intentional hybridization

Some populations of listed taxa are small or have gone through a recent bottleneck, and therefore they contain little genetic variation. In some cases, it might be advisable to increase genetic variation in these populations through intentional hybridization. Under what circumstances should genetic rescue (see Section 15.5) by purposeful hybridization be used as a tool in conservation?

In extreme cases, some taxa might only be recovered through the use of intentional hybridization. However, the very characteristics of the local populations that make them unusual or exceptionally valuable could be lost through this purposeful introgression. In addition, such introductions could cause the loss of local adaptations and lower the mean fitness of the target population. The most well-known example of this dilemma is the decision to bring in panthers from Texas (USA) to reduce the apparent effects of inbreeding depression in Florida panthers.

Intentional hybridization should be used only after careful consideration of potential harm. Intentional hybridization would be appropriate when the population has lost substantial genetic variation through genetic drift and the detrimental effects of inbreeding depression are apparent (e.g., reduced viability or an increased proportion of obviously deformed or asymmetric individuals). Populations from as similar an environment as possible (that is, the greatest ecological exchangeability) should be used as the donor population (Crandall et al. 2000). In these situations, even a small amount of introgression might sufficiently counteract the effects of reduced genetic variation and inbreeding depression without disrupting local adaptations (Ingvarsson 2001).

Hybridization is least likely to result in outbreeding depression when there is little genetic divergence between the populations. Thus, it is most appropriate in the case of intraspecific hybridization as with the Florida panther example. Intrinsic outbreeding depression is probably not a major concern in most circumstances of intraspecific hybridization. However, in some circumstances genetic exchange between intraspecific populations could result in extrinsic outbreeding depression through the loss of important local adaptations that are crucial for the viability of local populations. This is more probable as the amount of genetic divergence between populations increases at molecular markers. Thus, populations that are genetically similar at molecular makers and are similar for a wide range of adaptive traits are the best candidates for intentional hybridization.

17.5.3 Hybridization without introgression

Hybridization may present a demographic threat to native species even without the occurrence of genetic admixture through introgression (type 4). In this case, hybridization is not a threat through genetic mixing, but wasted reproductive effort could pose a demographic risk. For example, females of the European mink hybridize with males from the introduced North American mink. Embryos are aborted so that hybrid individuals are not detected, but wastage of eggs through hybridization has accelerated the decline of the European species (Rozhnov 1993).

The presence of primarily F_1 hybrids should not jeopardize protection of populations affected by type 4 hybridization. However, care should be taken to determine conditions that favor the native species to protect and improve its status and reduce the wasted reproductive effort of hybridization. In extreme (and expensive) cases, it may be possible to selectively remove all of the hybrids and the non-native species to recover a native population.

Bull trout in Crater Lake National Park, Oregon occur in a single stream, Sun Creek (Buktenica 1997). Introduced brook trout far outnumbered bull trout in this stream in the early 1990s and threatened to completely replace the native bull trout. The US National Park Service undertook a plan to recover this population by removing the brook trout. An effort was made to capture and identify every fish in this stream. Brook trout were identified by visual observation and removed. Putative bull trout and hybrids were sampled by fin clips for genetic analysis and held until genetic testing revealed the identity of each individual (Spruell et al. 2001). Pure bull trout were then held in a small fishless stream and a hatchery. Sun Creek was then chemically treated to remove all fish from the stream. After treatment, the pure bull trout were placed back into Sun Creek. This population is currently increasing in abundance and distribution.

17.5.4 Hybridization with introgression

This is a much more difficult conservation situation than the case of sterile hybrids. It is not possible to select nonhybridized individuals from a hybrid swarm to be used in recovery because every fish is a hybrid. Thus, once introgression spreads throughout the range of a species, that species is effectively extinct and cannot be recovered.

Perhaps surprisingly, introgression and admixture may spread even if hybrid individuals have reduced fitness. Heterosis is not necessary for introgression to spread and cause genomic extinction. In fact, population models (Epifanio and Philipp 2001) indicate that introgression may spread even when hybrids have severely reduced fitness (e.g., just 10% that of the parental taxa). This occurs because the production of hybrids is unidirectional, a sort of **genomic ratchet**. That is, all of the progeny of a hybrid will be hybrids. Thus, the frequency of hybrids within a local population may increase even when up to 90% of the hybrid progeny do not survive. The increase in the proportion of hybrid individuals in the population may occur even when the proportion of admixture in the population (i.e., the proportion of alleles in a hybrid swarm that come from each of the hybridizing taxa) is constant.

Guest Box 17 Hybridization and the conservation of plants
Loren H. Rieseberg

Hybridization is a common feature in vascular plants. Estimates from several well-studied floras suggest that approximately 11% of plant species hybridize (Ellstrand et al. 1996) and that close to a quarter of these are rare or endangered (Carney et al. 2000). In most instances, hybridization will not harm the rare taxon. Species with strong premating barriers, for example, or that have coexisted naturally for thousands of generations, are unlikely to be threatened by hybridization. Only when premating barriers are weak or when rare species come into contact with non-native species (or native species that have recently become aggressive due to human-induced habitat disturbance) is hybridization likely to cause genomic extinction. Because loss of rare populations may occur quickly, contact between native species and recently introduced or newly aggressive congeners requires swift assessment and action (Buerkle et al. 2003).

Plant species from islands or other isolated floras are particularly vulnerable to hybridization because premating barriers often are weak and geographic ranges are small. Perhaps the best-studied example is the Catalina Island mahogany, whose population size has dwindled to six pure adult trees (Rieseberg and Gerber 1995). This distinctive species is restricted to Wild Boar Gully on the southwest side of Santa Catalina Island off the coast of California. When the population was first discovered in 1897, it consisted of more than 40 trees, but it has declined rapidly over the past century. Two factors appear to have caused this decline: grazing and rooting by introduced herbivores, and interspecific hybridization with its more abundant congener, mountain mahogany. Although the mountain mahogany is not found in Wild Boar Gully, hybridization between the two mahogany species appears to be frequent. In addition to the six pure Catalina mahogany trees in the gully, five other adult trees and at least 7% of newly established seedlings are of hybrid origin. Presumably, wind pollination allows mountain mahogany trees from nearby canyons to sire hybrid plants in Wild Boar Gully.

Other vulnerable island species include the Haleakala greensword and several Canary Island species in the genus *Argyranthemum*. In fact, the Haleakala greensword is now extinct; two hybrids contain the only known remnants of its genome (Carr and Medeiros 1998).

The wild relatives of crops often are vulnerable to hybridization as well; indeed, 22 of the 25 most important crops are known to hybridize with wild relatives (Ellstrand 2003)! Examples of ongoing introgression include the California black walnut, which hybridizes with the cultivated walnut, and the common sunflower. Populations of common sunflowers along cultivated sunflower fields consist entirely of crop–wild hybrids (Figure 17.10; Linder et al. 1998), a finding consistent with computer simulations indicating that wild plants were likely to be replaced by crop–wild hybrids in less than 20 generations (Wolf et al. 2001).

Although most examples of genomic extinction by hybridization represent island endemics or crop relatives, even abundant mainland species may be at risk if faced with an aggressive congener. The native cordgrass (*Spartina foliosa*) in the San Francisco Bay is threatened by invading cordgrass (*S. alternifolia*) because the invader produces 21-fold more viable pollen than the native, and hybrids are strong and vigorous (Antilla et al. 1998). Simulations predict that native cordgrass could be extinct in 3–20 generations (Wolf et al. 2001).

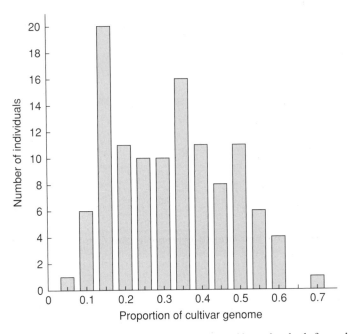

Figure 17.10 The proportion of cultivar genome carried by individuals from three "wild" sunflower populations that are sympatric with cultivated sunflowers. Note that all "wild" individuals are actually crop–wild hybrids. From Linder et al. (1998).

Problem 17.1

Do you expect allozyme or microsatellite loci to become diagnostic more quickly following the advent of reproductive isolation between two taxa? Why?

Problem 17.2

Why is the examination of mtDNA alone not adequate to detect hybridization between taxa?

Problem 17.3

What is a hybrid swarm?

Problem 17.4

Some authors have argued that hybridization is not harmful because it simply introduces new genetic variation. They further argue that this increased genetic variation may allow an increase in fitness and adaptation by natural selection. Furthermore, any introduced variation that is harmful will be removed by natural selection. What is wrong with this argument? Hint: consider the effect of hybridization on polygenic traits and genetic divergence among local populations.

Problem 17.5

The tree species hau kuahiwi (*Hibiscadelphus giffardianus*) is native to the Island of Hawai'i and was derived from a single tree in 1911. Closely related trees in the genus *Hibiscadelphus* occur on other nearby islands in the Hawai'ian archipelago. Some conservationists have suggested that this species should be hybridized with congeneric species to increase genetic variation and to increase the long-term viability of this species. What do you think? What information should be considered in making this recommendation?

Problem 17.6

The Haleakala greensword from Hawai'i is extinct (Guest Box 17). However, two hybrid individuals between this and the relatively common Haleakala silversword have been found in the wild. How would you go about regenerating the Haleakala greensword by selective breeding beginning with these two hybrid individuals?

Guam rail, Example 18.7

18

Conservation Breeding and Restoration

For nearly 3,000 taxa of birds and mammals, conservation breeding may be the only possible way to avoid extinction.

Torbjörn Ebenhard (1995)

A major challenge of ex situ conservation will be to ensure that sexually propagated samples of rare plants do not become museum specimens incapable of surviving under natural conditions.

Spencer Barrett and Joshua Kohn (1989)

Captive breeding represents the last chance of survival for many species faced with imminent extinction in the wild. The Guam rail, black-footed ferret, and the kakapo (Example 18.1) would all almost certainly be extinct if the last few remaining individuals in the wild were not captured and brought into captivity where they have been bred successfully. Less charismatic and well-known animal species have also avoided extinction by captive breeding programs. The white abalone became the first marine invertebrate to be listed under the Endangered Species Act (ESA) of the United States in 2001. A captive breeding program was begun in 1999 to bring this species back from the brink of extinction and establish a self-sustaining population in the wild (USGS 2002).

Example 18.1 The kakapo: a conservation breeding challenge

The kakapo (night parrot) is one of the most unusual and rarest birds in the world (Cresswell 1996). It is a flightless and large (1.5–4 kg) parrot that was widespread throughout New Zealand. Kakapo are solitary birds that breed once every 2–5 years and live for many decades. Kakapo are the only flightless bird, the only New Zealand bird, and the only parrot in which **lek** behavior has been observed. Males construct tracks that lead to a shallow bowl on a prominent high point. The low frequency booming of the males from their bowls, to attract females, travels up to 5 km and can go on every night for up to 4 months.

By the 1950s the only known kakapo consisted of a relic population in Fiordland on the South Island. The primary cause of decline was predation by introduced mammals (rats, cats, and stoats). Intensive investigation of this population in the 1970s revealed that it consisted only of a few males (Merton et al. 1984; Elliot et al. 2001). Another small population was discovered on Stewart Island in 1977. Some 61 kakapo were transferred to other islands because of high rates of predation by cats on Stewart Island. One male (named Richard Henry) from Fiordland, the last known surviving individual from mainland New Zealand, was transferred in 1975 and is part of the conservation breeding program.

The kakapo breeding program reached a low of 51 birds in 1995. There are currently 86 living kakapo, nearly one-half have been produced by the conservation breeding program. This program has faced a series of challenges associated with the unusual natural history of this bird. The lek breeding system has resulted in very high variance in male mating success. Approximately one-third of the birds born in the breeding program have been sired by a single male (named Felix). Supplemental feeding has been used in an attempt to increase the frequency of breeding. This did increase breeding success but not breeding frequency, but it also produced a significant excess of males (Clout et al. 2002). This effect is consistent with the general observation that polygynous birds produce an excess of the larger and more costly sex by females that are in good condition.

Little genetic variation has been found in the founding birds from Stewart Island (Triggs et al. 1989; Robertson et al. 2000; Miller et al. 2003). In contrast, the single bird from the mainland, Richard Henry, is more genetically variable; he is also substantially genetically divergent from the Stewart Island birds at these markers (see Figure 4.10). Thus, the Stewart Island birds may have naturally had a very small effective population size compared to birds on the South Island. Only 40% of eggs produced by females have hatched. This is an extremely low number for a bird species and may be caused by inbreeding depression associated with the low effective population size of birds on Stewart Island.

A rapid increase in population size is essential for kakapo recovery since it is vulnerable to extinction because of its small population size and low reproductive rate. However, the near absence of genetic variation in Stewart Island birds means that it is essential that Richard Henry contribute progeny. He has only sired three progeny to date. Thus, there is a conflict between the demographic needs of increasing the number of birds as soon as possible and increasing the genetic variation in kakapo by incorporating the lone remaining bird from the more genetically variable Fiordland population.

The management plan emphasizes the importance of increasing the contribution of the Fiordland population into the breeding population. Felix has been temporarily removed from the breeding population in the hope that Richard Henry will sire more progeny. Richard Henry is at least 50 years old and may no longer be able to sire progeny. An effort is currently underway to find new birds from Fiordland. However, it is not at all clear that any birds remain in this population – the last known individuals died in the late 1980s and the characteristic booming has not been heard since then. However, a hunter did report seeing what he thought to be a kakapo in 2004.

The kakapo has become an icon of conservation in New Zealand. Every animal is named, and births and deaths are national news. Six chicks hatched in March 2005, and four of these survived. A national contest was held to name them.

Several plant species have also been rescued from extinction by similar intervention. *Kokia cookei* is one of Hawai'i's most beautiful and endangered plants (Mehrhoff 1996). It is a medium-sized tree with very large red and somewhat curved flowers. This species was discovered on the island of Molokai in 1871, and became extinct in the wild in 1918. Extinction resulted from habitat loss and predation by introduced species. The species is apparently adapted to bird pollination, and the loss of native nectar-feeding birds may have contributed to the decline of the species (USFWS 1998). Four seeds were collected from the last remaining tree in 1915. Only one mature tree resulted from these four seeds. This tree produced hundreds of progeny, but none of the progeny survived reintroduction. In 1976, a branch from the last remaining *Kokia cookei* was successfully grafted onto a closely related species. Twenty-eight grafted *Kokia cookei* were transplanted back to Molokai in 1991. Most of these transplants survived, but none have yet flowered.

The World Conservation Union (IUCN) has defined ***ex situ* conservation** as "the conservation of components of biological diversity outside their natural habitats" (IUCN 2002). There are a variety of *ex situ* (or **offsite**) techniques that are potentially valuable tools in the conservation of a wide variety of taxa that are threatened with extinction (e.g., captive breeding and germplasm banking). The Russian N. I. Vavilov initiated systematic collection of plant germplasm samples that have long been used to conserve genetic resources associated with plants used by humans (Frankel 1974). Eberhart et al. (1991) have reviewed the long-term management of germplasm collections for the conservation of wild plant species. Others have considered the application of biotechnical advances (e.g., cloning and artificial insemination) in the conservation of wild animals (Bawa et al. 1997; Ryder and Benirschke 1997; Rennie 2000) and microbes (Gams 2002).

We use the more general term **conservation breeding** to include efforts to manage the breeding of plant and animals species that do not strictly involve captivity. For example, kakapo breeding is managed by moving groups of birds to predator-free islands, but they are not held in captivity (see Example 18.1).

Captive breeding has played a major role in the development of conservation biology. The first book on conservation biology (Soulé and Wilcox 1980) devoted five of 19 chapters to captive breeding. Modern conservation genetics had its beginnings in the use and application of genetic principles for the off-site preservation of plant genetic resources (Frankel 1974) and the development of genetically sound protocols for captive breeding programs in zoos (Ralls et al. 1979). Some conservationists even have equated conservation genetics with captive breeding. Caughley (1994) concluded that there was little application of genetics in conservation other than captive breeding programs in zoos.

The maintenance of genetic diversity and demographic security are the primary goals for management of conservation breeding programs. These two goals often are compatible. However, there are situations in which maintaining the genetic characteristics of a population may reduce the population growth rate so that a conflict arises (see Example 18.1). This is most likely to occur when a species with only a few remaining individuals is brought into captivity in a last-ditch effort for survival. Demographic security will best be achieved by rapidly increasing the census size of the captive population.

Maintenance of genetic diversity generally requires maximizing effective population size by reducing variation in reproductive success among individuals (see Chapter 7). However, some individuals or pairs of individuals may by much more successful in captivity than others. Thus, maximizing the growth rate of a captive population may actually reduce the effective population size and result in more rapid erosion of genetic variation. In addition, allowing just a few founders to produce most of the captive population is expected to accelerate the rate of adaptation to captive conditions. Thus, maintaining the genetic characteristics of a captive population may come at the cost of reduced population growth rate.

Our goal in this chapter is to consider the genetic issues involved in conservation breeding and the introduction of individuals into the wild (Doremus 1999). When should captive breeding be considered as a conservation option? What are the potential problems with a conservation breeding program? What criteria should be used when choosing populations and individuals to introduce or move between populations? We also provide an overview of the principles involved in actually genetically managing captive populations. Interested readers should consult other sources that provide detailed instructions for genetic management of conservation breeding programs (e.g., Ballou and Foose 1996).

18.1 The role of conservation breeding

There are three primary roles of offsite conservation breeding as part of a management or recovery program to conserve a particular species:

1 Provide demographic and genetic support for wild populations.
2 Establish sources for founding new populations in the wild.
3 Prevent extinction of species that have no immediate chance of survival in the wild.

The genetic objectives of these three roles are very different. Captive individuals used to provide demographic and genetic support for wild populations should be genetically matched to the wild population into which they will be introduced so that they do not reduce the fitness of the population by outbreeding depression. In contrast, introduced new populations should have enough genetic variation present so that they can become adapted to their new environment by natural selection. In the last case, the initial concern of a captive breeding program is to insure that the species can be maintained in captivity (Midgley 1987). This may involve preferentially propagating individuals capable of reproducing in captivity and may result in the adaptation to captivity.

Captive breeding has made many contributions to conservation other than just conservation breeding (e.g., public education, research, and professional training). The public display of species plays a very important role in conservation in providing opportunities for the public to come into contact with a wide variety of species that would otherwise just be names or pictures in books. The first author of this book became interested in biology because of visits to the Philadelphia zoo as a school child.

The goals of a display program are to establish an easily managed population that is well adapted to the captive environment (Frankham et al. 1986). These experiences provide an excellent opportunity for education and also provide the setting for the public to develop affection and appreciation of a wide variety of species. Most people around the world will never have the opportunity to see a tiger, elephant, or a great ape in the wild. Zoos provide an important role in allowing the public to develop a first-hand connection to these species. People are more likely to support conservation efforts if they have knowledge, understanding, and appreciation of the species involved.

There is also a danger in this. Seeing elephants or tigers in the zoo may encourage the public and politicians to believe that these "species" are now protected from extinction. However, a species is not just a collection of individuals that has been removed from the ecosystem in which they have survived and evolved for millions of years.

> A condor is 5 percent feathers, flesh, blood, and bone. All the rest is place. Condors are soaring manifestations of the place that built them and coded their genes. (Devall and Sessions 1984, p. 317.)

The ecologist David Barash (1973) has said this in a somewhat different fashion:

> Thus, the bison cannot be separated from the prairie, or the epiphyte from its tropical perch. Any attempt to draw a line between these is clearly arbitrary, so the ecologist studies the bison–prairie, acacia–bromeliad units.

Thus, the display of charismatic species to the public should be accompanied with educational efforts that emphasize that long-term species existence can only occur within the complex web of connections and interactions in their native ecosystems.

18.1.1 When is conservation breeding an appropriate tool for conservation?

This is an important and difficult question. Conservation breeding should be used sparingly because it is difficult and expensive, and worldwide resources are limited. In addition, directing resources to captive breeding and taking individuals into captivity may hamper efforts to recover species in the wild.

Captive breeding is perhaps too often promoted as a recovery technique. For example, Conservation Assessment and Management Plans under the Conservation Breeding Specialist Group of the IUCN have recommended captive breeding for 36% of the 3,314 taxa considered (Seal et al. 1993). In the USA, captive breeding has been recommended in 64% of 314 approved recovery plans for species listed under the ESA (Tear et al. 1993). The resources are not available to include captive breeding in the recovery plans of such a high proportion of species. It is important that it be used only for those species in which it can have the greatest effect.

Intensive field-based conservation may be an effective and cost-efficient alternative to captive propagation. Balmford et al. (1995) found that *in situ* management of well-protected reserves for large-bodied mammals resulted in comparable population growth rates and was consistently less expensive than captive propagation. These authors suggest that captive breeding is most cost-effective for smaller bodied taxa and will only remain the best option for large mammals that are restricted to one or two vulnerable wild populations.

18.1.2 Priorities for conservation breeding

It is clear that only a relatively small proportion of the thousands of animal species that are threatened in the wild can be maintained in captivity because of constraints on space and other resources (Balmford et al. 1996; Snyder et al. 1996). It is generally assumed that a maximum of roughly 500 animal species could be maintained offsite in conservation breeding programs (IUDZG/CBSG 1993). As we have seen, however, captive breeding programs are often recommended for many taxa. Given this situation, what criteria should be used to determine which species should be maintained in conservation breeding programs?

Zoos have historically focused on large and charismatic species in breeding programs. Balmford et al. (1996) have spelled out three general sets of criteria that should be considered in selecting candidate animal species for captive breeding:

1 Economic considerations. Which species can be conserved successfully in a captive breeding program most economically?
2 Biological suitability for captive breeding. Which species can be bred and raised successfully in captivity?
3 Likelihood of successful reintroduction. For which species is successful reintroduction to the wild a realistic option?

We suggest a fourth criterion: the potential effect on habit preservation. Will development of a captive breeding program increase or decrease the likelihood of habitat protect?

Invertebrates are generally better candidates for captive breeding than are large and charismatic vertebrates, for which enormous resources have been used (Pearce-Kelly et al. 1998). Invertebrates have a relatively high probability of success for both the rearing and release phases. They also have small size and require relatively little space and cost. They typically have high reproductive potential and population size increases relatively rapidly in captivity and after release. Finally, there is a wealth of knowledge and techniques for rearing numerous invertebrate species. For example, crickets, katydids, beetles, and butterflies have been widely and successfully raised in captivity.

Plants generally are better candidates for offsite breeding programs than animals for a variety of reasons (Templeton 1991). Many plants can be maintained for long periods as

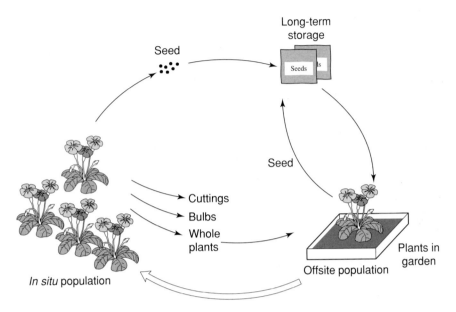

Figure 18.1 Possible modes of reproduction for offsite breeding of plants and possible interchange between offsite plants and *in situ* populations. Redrawn from Brown and Briggs (1991).

dormant seeds. This may be used to increase the generation interval and therefore reduce the rate of genetic change during offsite breeding. Other plants, such as trees, live a long time so that offsite breeding programs that may take hundreds of years may only represent a handful of generations. This again will minimize the rate of genetic change by genetic drift and selection. Other problems with offsite breeding can be reduced because of the variety of modes of reproduction that are possible for plants with short generation intervals (e.g., selfing, apomixis, and clonal reproduction; Figure 18.1).

Guidelines for selecting candidate plants for **conservation collections** have been presented by the Center for Plant Conservation (1991). The decision to protect (or abandon) a particular population or species must be made within a larger framework of conservation. In addition, these guidelines are based on a natural genetic hierarchy: species, populations (or ecotypes), individuals, and alleles. The goal is to address diversity at several levels of organization rather than sampling a particular species without regard to genetic variation and future long-term viability. This approach includes five sampling decisions:

1   Which species should be collected?
2   How many populations within a species should be sampled?
3   How many individuals should be sampled per population?
4   How many propagules should be collected from each individual?
5   When should collections be made from multiple year periods?

18.1.3 Potential dangers of captive propagation

The use of captive breeding has been controversial. It is expensive, is sometimes ineffective, and may harm wild populations both indirectly and directly if not done correctly

(Snyder et al. 1996). Perhaps the most serious criticism is that efforts directed toward captive breeding detract from grappling with the real problems (e.g., loss of habitat and protection). The dangers of captive breeding are clearly demonstrated by the use of fish hatcheries to maintain stocks of Pacific salmon on the west coast of North America (Example 18.2).

Example 18.2 Who needs protection? We have hatcheries

Fish hatcheries have a long and generally unsuccessful history in conservation efforts to protect populations of fish. Pacific salmon began a rapid decline on the west coast of the lower United States in the late 1800s with the advent of the salmon-canning industry (Lichatowich 1999). The State of Oregon sought advice from the newly created US Commission on Fish and Fisheries that was directed by Spencer Baird, a scientist with the Smithsonian Institution. In 1875 Baird (Lichatowich 1999, p. 112) recommended that:

> . . . instead of protective laws, which cannot be enforced except at very great expense and with much ill feeling, measures be taken, either by the joint efforts of the States and Territories interested or by the United States, for the immediate establishment of a hatching establishment on the Columbia River, and the initiation during the present year of the method of artificial hatching of these fish.

Unfortunately, this recommendation from the leading fisheries scientist of the United States set in motion a paradigm for the conservation of salmon through hatcheries rather than facing the real problems of excessive fishing, dams that blocked spawning migrations, and habitat changes in the spawning rivers and streams. These efforts have failed profoundly (Meffe 1992). Some 26 different groups of Pacific salmon and anadromous rainbow trout (steelhead) are listed as threatened or endangered under the US ESA at the time of writing this chapter. The role of hatcheries in salmon conservation continues to be controversial. There is current disagreement about whether hatchery populations should be considered part of the distinct population segments that are listed and protected under the ESA (Myers et al. 2004; see also Guest Box 16).

A number of recent studies have been performed to assess the possible genetic effects on wild populations of releasing hatchery fish into the wild. The consensus is clear: hybridization with hatchery fish has a dramatic, harmful effect on the fitness of wild populations of salmon (Reisenbichler and Rubin 1999; Waples 1999; McGinnity et al. 2003).

Reisenbichler and others have published several papers that compare the relative fitness of progeny of hatchery steelhead (the anadromous form of rainbow trout) to wild fish. Three primary results emerge from these studies. First, progeny from hatchery fish uniformly show reduced rates of survival. For example,

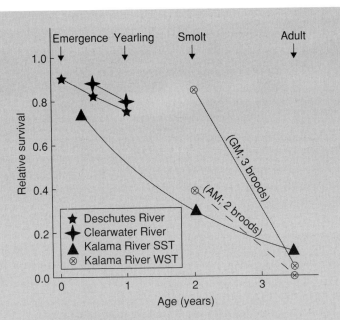

Figure 18.2 Results showing reduction in relative survival throughout the life cycle of progeny from hatchery steelhead spawning in the wild relative to the survival of progeny from wild fish. Data for the Kalama River winter steelhead (WST) are the geometric means (GM) for three year classes, or the arithmetic means (AM) for two year classes with an exceptional year class omitted. SST, summer steelhead. From Reisenbichler and Rubin (1999).

Leider et al. (1990) found that the reproductive success of hatchery fish spawning in the wild relative to wild fish ranged from 5 to 15% in four successive year classes. Second, progeny of hatchery fish have reduced survival at all life history stages between emergence from the gravel until returning from the ocean as adults (Reisenbichler and Rubin 1999 (Figure 18.2). Finally, the decline of fitness observed in hatchery fish is proportional to the number of generations that the hatchery stock has been maintained in captivity. Chilcote (2003) found that intrinsic measures of population productivity of 12 populations of steelhead in Oregon declined as a function of the number of hatchery spawners.

18.2 Reproductive technologies and genome banking

Reproductive technologies initially developed for agricultural species (e.g., cattle, sheep, and chickens) can be transferred to some related wild species to facilitate their conservation. These technologies include genome banking, cryopreservation, artificial insemination, and cloning.

Genome banking is the storage of sperm, ova, embryos, seeds, tissues, or DNA. Genome resource banking can help move genetic material without moving individuals. It might, for example, allow for managed gene flow into isolated populations without the risks of translocating individuals. Genome banking also serves as an insurance against population or species extinction. It lengthens generation intervals and thereby reduces random genetic drift. It increases efficiency of captive breeding and reduces the number of individuals kept in captivity. Finally, banks are a source of tissue and DNA for basic and applied research.

Genome banking is widely used for agricultural crop and farm animal preservation to help insure future agricultural productivity. Genome banking is also increasingly used for wild taxa. For example, for wild animals there exists a genome bank at the Smithsonian Institution's National Zoo where there are more than 1,500 samples of frozen sperm or embryos from 69 species (including ~2% of mammalian species worldwide). Similarly, the San Diego Zoo maintains a "frozen zoo" with samples (including cell lines and tissues) from more than 7,000 species of endangered mammals, birds, and reptiles.

For wild plants there is increasing interest to establish seed banks. Researchers suggest that many tropical and rain forest species seeds can be banked. However the ability to bank seeds is known for only about 4% of angiosperms (flowering plants). The FAO (Food and Agriculture Organization) reports that 6 million accessions exist in over 1,300 seed banks around the world. However less than about 10% are from wild plants.

Cryopreservation is the freezing and storage (often in liquid nitrogen at $-180°C$) of sperm, ova, embryos, seeds, or tissues to manage and safeguard against loss of genetic variation in agricultural or wild populations. It is the principal storage method for animal material. In plants, seeds are often preserved dry at room temperature and can remain viable for 50–200 years. But for some plant species longevity increases if seeds are frozen and some seed banks are kept at $-20°C$ or colder.

Artificial insemination (AI) is a widely used and important technique for captive breeding. AI allows animals to breed that would not breed naturally, perhaps due to behavior problems such as aggression towards mates. Further, a genetically important male can still be used within a breeding program long after his death. Finally, instead of moving animals, sperm can be collected and cryopreserved and shipped for AI (as mentioned above). AI has been used, for example, in breeding programs for the black footed ferret and killer whales in the USA, the cheetah in Namibia, koalas in Australia, and gazelles in Spain and Saudi Arabia. AI was recently used successfully with corn snakes at the Henry Doorly Zoo in Omaha, Nebraska.

Cloning for conservation is controversial and often debated (Example 18.3). Cloning is generally conducted by: (1) removing the nucleus from a donor egg cell of the animal that will carry the cloned embryo; and (2) injecting into the carrier's egg cell the nucleus from a cell of the animal to be cloned. For example, a nucleus from a tissue cell of a European wild sheep (mouflon) was injected into the nucleus-free cell of a close relative species, the domestic sheep. The resulting mouflon lamb was born and mothered by the domestic sheep. This is an example of cross-species cloning, which is more difficult than within-species cloning because of risks of incompatibility of mitochondrial genes from the donor egg and the nuclear genes from the animal to be cloned.

These technologies provide valuable opportunities for protecting species and increasing genetic variation within species on the brink of extinction. Nevertheless, it is essential that they be integrated so that they support ongoing conservation efforts rather than being used as alternatives.

Example 18.3 Is cloning a useful tool for animal conservation?

Cloning could potentially allow resurrection of a recently extinct species (Holt et al. 2004). For example, a subspecies of wild goat from the Pyrenees Mountains in Spain became extinct recently when the last individual was killed by a falling tree. Biologists had sampled tissue before the animal died, and a company called Advanced Cell Technology is currently trying to clone the individual.

Cloning is very expensive, and it is technologically feasible for only a few species that are related to model research organisms (e.g., mice) or important in agriculture (e.g., cattle and sheep). Further, the success rate is very low, less than 0.1 to 5% of renucleated embryos lead to a live birth (Holt et al. 2004). It is generally agreed that long-extinct species, such as the woolly mammoth from the frozen Siberian permafrost, cannot be cloned because their DNA is fragmented.

Another potential advantage of cloning is to help bolster populations and avoid the extinction of critically endangered species such as the panda. However, the benefits of cloning compared to that of more traditional captive breeding programs is questionable and the disadvantages are substantial.

Disadvantages are that cloned individuals and populations are genetically identical and thus would be highly susceptible to the same infectious diseases and to have low adaptive potential to environmental change. Further, the money spent on cloning would often be better spent preserving habitat and conducting less expensive breeding programs. Extensive healthy habitats are necessary to insure long-term persistence of any species anyway.

Cloning should never be viewed as an alternative to habitat preservation and breeding programs. In certain limited scenarios, cloning could be a last resort approach in combination with habitat conservation and breeding programs to help insure species persistence and even recover extinct taxa.

18.3 Founding populations for conservation breeding programs

Developing a captive breeding program begins with the selection of the founding individuals. In many situations when a species is on the brink of extinction, there is no choice involved because all remaining individuals are brought into captivity. In other cases, however, captive breeding programs are established when long-term survival of a species in the wild is unlikely even though many individuals currently occur in the wild (e.g., tigers). In such cases, there are a variety of questions to be decided. Which subspecies or populations should be the source of individuals to be brought into captivity? How many subspecies or populations should be maintained? Should subspecies and populations be maintained separately or mixed together? How many individuals should there be in the founder population?

18.3.1 Source populations

Selecting the founding individuals and populations for a captive breeding program is an important and difficult problem for many species. Source populations should be selected in order to maximize genetic and ecological (adaptive) diversity. For example, there are currently four remaining subspecies of tigers in the wild. Recent genetic results have indicted substantial genetic divergence among these subspecies (Luo et al. 2004). A strong argument can be made that each of these subspecies represents a separate evolutionary significant unit (ESU; see Chapter 16) and separate captive breeding programs should be established for each. However, space and other resources for captive breeding of tigers are limited. There are currently approximately 1,000 spaces for tigers in captive breeding programs throughout the world. We then face a dilemma. How should we partition available captive breeding spaces among the four subspecies to enhance survival and retention of genetic variation?

Maguire and Lacy (1990) have provided a very informative consideration of this problem. They identified three conservation goals: (1) to maximize the number of surviving subspecies; (2) to maximize genetic variation at the species level; and (3) to maximize genetic diversity at the subspecies level. They choose a timeframe of 200 years (32 tiger generations) to match recommendations for long-term conservation plans (Soulé et al. 1986). Their analysis also included consideration of the probabilities of persistence of the subspecies in the wild.

The two extreme options are to choose only one subspecies for captive breeding or to divide the 1,000 spaces equally among the four subspecies. They assume that the $N_e : N_c$ ratio in captive tigers is 0.4 (Ballou and Seidensticker 1987). In the latter case, each of the four subspecies would have an N_e of approximately 100 tigers (250 × 0.40). Using expression 6.7, we would expect to lose approximately 14% of the heterozygosity in each subspecies after 200 years ($t = 32$). General recommendations suggest a goal of retaining at least 90% of the heterozygosity after 200 years (Soulé et al. 1986). This would require an N_e of approximately 150, and an N_c of 375 for each subspecies. Maguire and Lacy (1990) recommend devoting half of the available captive spaces to the *tigris* subspecies and dividing the remainder equally among the other three subspecies.

18.3.2 Admixed founding populations

Another option is to establish a captive population by hybridizing genetically divergent populations. For example, the State of Montana established a captive population of westslope cutthroat trout in 1985 to be used in a variety of restoration projects. Geographic populations of westslope cutthroat trout show substantial genetic divergence among populations: $F_{ST} = 0.32$ (Allendorf and Leary 1988). Space limitations required that only a single captive population could be maintained. The choice was to use a single representative population to establish the captive population or to create a hybrid captive population by crossing individuals from a wide spectrum of native westslope cutthroat trout populations.

Do we choose one population to be brought into captivity or do we create a captive population by hybridizing individuals from different populations? The genetic choice that we face here is between genes and genotypes. We can maximize the allelic diversity of westslope cutthroat trout in the captive population by including fish from many streams in our founding population. However, hybridizing these populations will cause the loss of the unique combination of alleles (genotypes) that exist in each population. These genotypes may be important for local adaptations. These combinations of genes, and the resulting locally adapted phenotypes, will be lost through hybridization. In addition, the hybridization of different populations could result in outbreeding depression (see Section 17.3).

In some cases, genetically distinct populations have been brought into captivity and hybridized without realizing potential problems. For example, we saw in Table 3.3 that approximately 20% of orangutans born in captivity were hybrids between orangs captured in Borneo and Sumatra. These two populations are fixed for chromosomal differences and it has been proposed they are classified as separate species. Current conservation breeding plans avoid the production and use of hybrids between these taxa.

There are no simple prescriptive answers to the best strategy in establishing a captive population. In the case of westslope cutthroat trout, the captive population was established by mixing from some 20 natural populations. There was some concern in this case about possible outbreeding depression caused by mixing together so many local populations. However, the alternative of using just one local population, which would contain such a small proportion of the total overall genetic variation, was considered less desirable.

18.3.3 Number of founder individuals

The number of founders recommended for establishing a captive population depends substantially on the proportion of rare alleles desired to be captured, and on the population growth rate expected in captivity. Approximately 30 diploid founders are required to have a 95% probability of sampling an allele at frequency 0.05. However with 30 founders there is only approximately a 45% probability of including an allele of frequency 0.01 (see expression 6.8 and Figure 6.8). Thus we recommend a minimum of 30 founders and preferably at least 50. Thirty founders will maintain approximately 98% of the original heterozygosity (see expression 6.6). If the rate of population growth is low, additional founders or subsequent supplementation with additional individuals is recommended.

18.4 Genetic drift in captive populations

A primary genetic goal of captive breeding programs is to minimize genetic change in captivity. Genetic changes in captive populations may reduce the ability of captive populations to reproduce and survive when returned to the wild. There are two primary sources of genetic change in captivity: genetic drift and natural selection.

18.4.1 Minimizing genetic drift

Genetic drift may cause the loss of heterozygosity and allelic diversity. This reduced genetic diversity can have several consequences. First, inbreeding depression may limit population growth and lower the probability that the introduced population will persist. Second, reduced genetic diversity will limit the ability of introduced populations to evolve in their new or changing environments. In general, the effects of genetic drift can be minimized in captivity by managing the population to maximize the effective population size.

The primary method for minimizing genetic drift and maximizing effective population size is to equalize reproductive success among individuals. This is especially important for the founder individuals of a captive breeding program. We saw in Chapter 7 that the ideal population includes random variability in reproductive success. Under controlled captive conditions, it may be possible to reduce variance in reproductive success to near zero. In this case, the effective population may actually be nearly twice as great as the census population size (see expression 7.5). The most effective method to reduce variance in reproductive success depends upon the type of breeding scheme used in captivity (see Section 18.5).

18.4.2 Deleterious alleles and mutational meltdown

Deleterious alleles that are present at low frequencies in natural populations may drift to high frequencies in captive populations because of the founder effect combined with relaxed natural selection (Example 18.4). Joron and Brakefield (2003) have suggested that relaxed natural selection in captivity can mask reduced fitness due to inbreeding. For example, wolves bred for conservation purposes in Scandinavia were found to have a high frequency of hereditary blindness apparently caused by an autosomal recessive allele (Laikre et al. 1993). Only six founders were originally brought into captivity (Figure 18.3). At least one of these founders apparently was heterozygous for a recessive allele associated with blindness. It is also possible that partial blindness may actually have some advantage in captivity for a wild animal such as a wolf.

Some populations of salmon and trout have high frequencies of null alleles at enzyme coding loci that are enzymatically inactive or nonfunctional (see Section 5.4.2; Allendorf

---

Example 18.4 Chondrodystrophy in California condors (Ralls et al. 2000)

The captive population of California condors was founded with the last remaining 14 individuals in 1987. California condors have bred well in captivity and the first individuals were reintroduced into the wild in 1992. However, nearly 5% of birds born in captivity have suffered from chondrodystrophy, a lethal form of dwarfism. This defect is apparently caused by a recessive allele that occurs at a frequency of 0.09 in the captive population.

Such deleterious alleles are likely to occur in any captive population founded by a small number of founders (Laikre 1999). What should be done? Ralls et al. (2000) considered three management options for this allele: (1) reduce its frequency by selection; (2) minimize its phenotypic frequency by avoiding matings between possible heterozygotes; or (3) ignore it.

Selective removal of this allele would require not using possible heterozygotes in the breeding program. Under this scheme, over 50% of all birds would be eliminated from the breeding population. This is a very high cost to pay for elimination of a trait that affects less than 5% of all birds. In addition, it is likely that other traits caused by deleterious recessive alleles occur in this population. Selective removal of relatively low frequency alleles at multiple loci is generally not worth the cost of reducing the effective population size and further eroding genetic variation in the captive populations.

Ralls et al. (2000) recommend minimizing the phenotypic frequency of this trait by avoiding pairings between possible heterozygotes. They suggest that some selection would be feasible once the captive population has reached the carrying capacity in captivity. In addition, possible heterozygotes could be given a lower priority as candidates for introduction.

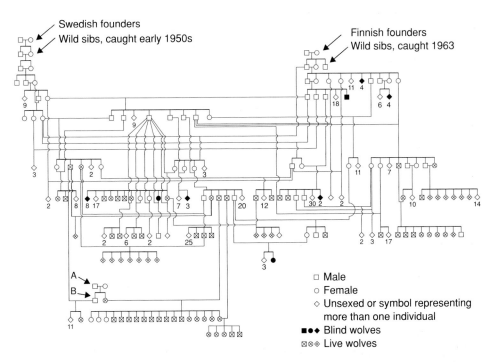

Figure 18.3 Pedigree of the captive population of 442 wolves bred in Scandinavian zoos as of January 1988. The numbers below the symbols indicate the number of individuals in a particular family. A, Russian founders; B, full-sibs, imported 1980. From Laikre et al. (1993).

et al. 1984). Such alleles are not as deleterious in these fishes because the gene duplication provided by their polyploid ancestry provides some redundancy. Nevertheless, developmental studies have found that these alleles do have harmful effects on developmental rate and developmental stability (Leary et al. 1993b).

Extensive surveys of natural and hatchery populations of trout and salmon indicate that enzymatically null alleles occur at high frequencies only in hatchery populations or natural populations that are restricted to lakes (Allendorf et al. 1984; Leary et al. 1993b, unpublished data). For example, a null allele at a lactate dehydrogenase (LDH) locus occurred at a frequency of 0.122 in a hatchery population of rainbow trout (Leary et al. 1993b). Homozygotes for this allele exhibited a 70% reduction in LDH activity in heart tissue. These hatchery populations usually have a large number of founders so that it is unlikely that the founder effect contributed to the high frequencies of these alleles.

In addition, new mildly deleterious mutations will occur in captive populations; these mutations may drift to high frequency in populations with a small N_e because natural selection is not effective in small populations (see Chapter 8). Many of these new mutations with mildly deleterious affects could accumulate in small populations and lead to so-called "mutational meltdown" (Lande 1995; see Chapter 14).

18.4.3 Inbreeding or genetic drift?

It is crucial to distinguish between the effects of inbreeding and genetic drift in captive populations. Some inbreeding (the mating of related individuals) will be unavoidable in small captive populations; this is the so-called inbreeding effect of small populations (see Chapter 6). In general, inbreeding should be avoided in captive populations because the reduced fitness associated with inbreeding depression may threaten short-term persistence of the captive population.

However, the loss of genetic variation by genetic drift is a more serious and lasting effect than inbreeding. The harmful effects of inbreeding last for a single generation. That is, a mating between an inbred individual and an unrelated mate will produce a noninbred progeny. The long-term genetic well being of a captive population is more affected by the unequal representation of founders and effective population size than by matings between related individuals.

Schemes of mating with maximum avoidance of inbreeding will minimize the initial rate of loss of heterozygosity. However, perhaps surprisingly, there are often systems of mating that do a better job of retaining heterozygosity in the long term (Kimura and Crow 1963; Robertson 1964; Wright 1965b).

18.5 Natural selection and adaptation to captivity

Natural selection will occur in captivity and bring about adaptation to captive conditions. Such changes will almost inevitably reduce the adaptiveness of the captive population to wild or natural conditions. For example, tameness in response to contact with humans is generally advantageous in captivity, but can have serious harmful effects in the wild.

The emphasis of captive breeding protocols has been primarily to reduce genetic drift by maximizing effective population size. This emphasis is appropriate for captive breeding programs of mammals and birds in zoos that have a relatively small number of individuals that are managed using pedigrees (Ballou and Foose 1996). However, increasing effective population size for some captive species (e.g., fish and plants) may increase the rate of adaptation to captive conditions.

18.5.1 Adaptation to captivity

Adaptation to captivity is probably the greatest threat in species that produce many offspring (e.g., insects, fish, amphibians, etc.). For example, females of many fish species produce thousands of eggs. Extremely strong natural selection can occur in the first few generations when founding a captive hatchery population of fish.

Darwin (1896) was very interested in the genetic changes brought about by selection during the process of domestication of animals bred in captivity. He attributed such changes to three mechanisms:

1 Systematic selection.
2 Incidental (unintentional) selection.
3 Natural selection.

Systematic selection occurs when purposeful selection occurs for some desirable characteristics. For example, many hatchery populations of fish are selected for rapid growth rate. Incidental selection occurs when captive management favors a particular phenotype without being aware of their preference. For example, hatchery personnel may unconsciously favor a particular phenotype (e.g., large, colorful, etc.) when choosing fish to be mated. Finally, natural selection will act to favor those individuals who have characteristics that are favored under captive conditions. For example, many wild fish will not feed when brought into captivity. Therefore, natural selection for behaviors that permit feeding and surviving in captivity will be very strong.

This issue was raised many years ago by A. Starker Leopold (1944) in his consideration of the effects of release of 14,000 hybrid (wild X domestic) turkeys on the wild population of turkeys in southern Missouri, USA. It was common practice throughout many parts of the USA to release such hybrid turkeys in order to enhance wild populations that were hunted. Hybrid stocks were used because of the great difficulty in raising wild turkeys in captivity. He found that the hybrid birds were unsuccessful in the wild because of their tranquility, early breeding, and inappropriate behavior of chicks in response to the warning note of the hen:

> Wild turkeys are wary and shy, which are advantageous characters in eluding natural and human enemies. They breed at a favorable time of the year. The hens and young automatically react to danger in ways that are self-protective. . . . Birds of the domestic strain, on the other hand, are differently adapted. Many of the physiological reactions and psychological characteristics are favorable to existence in the barnyard but many preclude success in the wild. (Leopold 1944.)

Systematic selection and incidental selection can be greatly reduced in captivity by intensive effort. However, genetic divergence between wild and captive populations because of natural selection cannot be eliminated. Efforts are currently underway to reduce these effects in fish hatcheries by mimicking the natural environment (Brannon et al. 2004). Nevertheless, it is impossible for a hatchery to simulate the complex and dynamic ecological heterogeneity of a natural habitat. In fact, any hatchery must create an environment that differs dramatically from the natural one to achieve its goal of producing more progeny per parent than occurs under natural circumstances. By definition then, a goal of reducing mortality while retaining natural environmental conditions cannot be achieved; it is impossible to synthetically create conditions that are both identical to the natural ones and at same time provide a basis for increased survival (Spruell et al., submitted).

18.5.2 Minimizing adaptation to captivity

Natural selection is most effective in large populations (see Chapter 8). Thus, rapid adaptation to captivity is expected to occur most rapidly in captive populations with a large N_e. Minimizing variance in reproductive success via pedigree management will also act to delay adaptation to captive conditions. However, pedigree management is probably not necessary or not practical for many species kept in captivity.

In contrast to this view, Bryant and Reed (1999) have suggested that the absence of any selection in captivity can lead to a deterioration in fitness and that captive programs should allow the alleles of less adapted individuals to be lost from the captive population. We agree with Lacy (2000a) that Bryant and Reed overestimate the likely deterioration of fitness in this case; they also overlook several other problems with the strategy that they propose.

In species with high fecundity (such as many fish, amphibians, and insects), rapid adaptation to captivity is most likely to occur because hundreds of progeny can be produced by single matings. Thus, natural selection may be very intense, especially in the first few generations after being brought into captivity.

For example, the Apache trout, which is native to the southwestern USA, is currently listed as "threatened" under the ESA. A single captive population, originating from individuals captured in the wild in 1983 and 1984, is the cornerstone of a recovery effort with an established goal of establishing 30 discrete populations within the native range of this species. Advances in culture techniques and the high fecundity of these fish have resulted in a program that spawns hundreds of mature fish and produces hundreds of thousands of fry per year for reintroduction.

The large number of spawners suggests that the effective population size of this population is very large so the loss of genetic variation due to drift is not a concern. Nevertheless, these circumstances are ideal for natural selection to bring about rapid changes to captive conditions that would reduce the probability of successful establishment of reintroduced populations.

18.5.3 Interaction of genetic drift and natural selection

In many regards, actions taken to reduce genetic drift will also reduce the potential for natural selection. For example, minimizing variability in reproductive success among individuals will both maximize N_e and reduce the effects of natural selection (Allendorf 1993). However, as we saw in Chapter 8, natural selection is most effective in very large populations. Therefore, intermediate size populations would be large enough to avoid rapid genetic drift, but not so large that even weak natural selection could bring about adaptation to captive conditions.

Woodworth et al. (2002) tested these predictions with experimental populations of *Drosophila* to mimic captive breeding. They evaluated adaptation to captivity under benign captive conditions for 50 generations using effective population sizes of 25, 50, 100, 250, and 500. The small populations demonstrated reduced fitness after 50 generations due to inbreeding depression. The large populations demonstrated the most rapid adaptation to captive conditions. The least genetic change in captivity was observed in intermediate size populations as measured by moving the populations to simulated wild conditions (Figure 18.4). These authors suggested that adaptation to captivity can be minimized by subdividing or fragmenting the captive population into a series of intermediate size populations. The effective population size of each population should be large enough to minimize the harmful effects of inbreeding and genetic drift, but slow enough to minimize rapid adaptation to captive conditions.

18.6 Genetic management of conservation breeding programs

A primary genetic goal of captive breeding programs is to minimize genetic change caused by genetic drift and natural selection. Specific actions to achieve this goal depend upon the biology of the species. We first consider captive populations that are managed by keeping track of individual pedigrees (e.g., large mammals and birds). We then consider species for which large groups of individuals are held, but it is difficult or impractical to keep track of

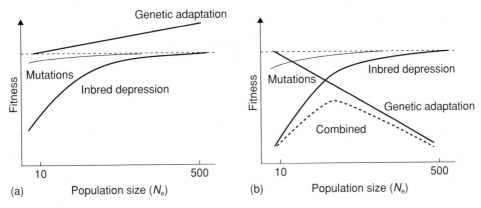

Figure 18.4 Expected relationship between fitness and population size (N_e) due to the inbreeding effect of small populations and genetic adaptation to captivity. The combined line represents the net effects of both factors. The effects are shown for populations maintained for approximately 50 generations under (a) benign captive conditions and (b) for these populations when introduced into the wild. Redrawn from Woodworth et al. (2002).

individuals (e.g., fishes and insects). Most of our examples here concern animals, but the same underlying genetic principles hold for plants. The wide variety of possible modes of reproduction in plants (see Figure 18.1) makes it harder to provide general guidelines to apply these genetic principles. Guerrant (1996) provides an excellent review of maintaining offsite populations of plants for reintroduction.

18.6.1 Pedigreed populations

Genetic management by individual pedigrees is extremely powerful. It provides both maximum genetic information about the captive population and also maximum power to control the reproductive success of individuals chosen for mating. This approach is most appropriate for large mammals and birds. Most of the genetics literature dealing with management of captive populations deals with this situation.

Simply maximizing N_e may not be the best strategy for maintaining genetic variation in pedigreed populations (Ballou and Lacy 1995). Remember that genetic variation can be measured by either heterozygosity or allelic diversity. Maximizing N_e will minimize the loss of heterozygosity (by definition), but it may not be the best approach to retain allelic diversity. A strategy that uses all of the information contained in a pedigree can be developed to minimize the loss of heterozygosity and allelic diversity.

Ballou and Lacy (1995) provide a lucid explanation of captive breeding strategies to maintain maximum genetic variation that is beyond the detail we will consider here. This problem is extremely difficult because the pedigrees of captive populations are often extremely complicated (see Figure 18.3) and genetic planning is often not initiated until after the first few generations of captivity.

Simple rules of thumb such as equalizing the genetic contributions of founders to the captive population are not valid. We can see this in the hypothetical example presented in Figure 18.5 in which there are four founders of a captive population. What would be the result of a breeding strategy that equalized the genetic contributions of the founders? We

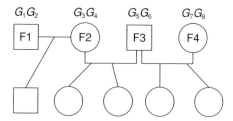

Figure 18.5 Hypothetical pedigree of a captive population founded by four individuals. We know that one allele at each locus has been lost from founder F1 because he left only one descendant in the captive population. Therefore, equalizing the contributions of these four founders in future generations would lead to an overrepresentation of genes from F1.

can be absolutely certain that we have lost one of the two alleles carried by individual F1 at every locus since this founder only contributed one offspring to the captive population. However, there is some possibility that both alleles from the three other founders have been retained because they have contributed multiple progeny. Thus, we can maximize the retention of allelic diversity in the captive population by weighting the desirable contribution of each founder by weighting it by the expected proportion of a founder's alleles retained (founder genome equivalents; Lacy 1989).

Accurate calculations of kin relationships, inbreeding coefficients, and retention of founder alleles require a complete knowledge of the pedigree. However, many pedigreed captive populations have some individuals with one or both parents unknown. Traditionally such individuals have been treated as founders unrelated to all nondescendant animals. In some circumstances, this can cause substantial errors in estimating genetic parameters (Ballou and Lacy 1995). Incorporation of molecular genetic information can often be used to resolve unknown relationships and may result in a substantially different view of a captive population (e.g., the whooping crane; Jones et al. 2002).

Similarly, the founders of a captive population brought into captivity are generally assumed to be unrelated for pedigree analysis. However, this may often not be the case. Incorrect assignment of founder relatedness will result in erroneous estimates of inbreeding coefficients, effective population size, and population viability. For example, the last remaining individuals in the wild may consist of a just a few groups of sibs. This information should be taken into account along with molecular genetic analysis of relationships in order to maximize the retention of genetic variation in the captive population. Thus, correct classification of kin structure among founders is important for a captive breeding program.

18.6.2 Nonpedigreed populations

For many species held in captivity it is difficult or impractical to keep track of individuals and pedigrees. For example, a single female of the endangered Colorado pikeminnow may produce as many as 20,000 eggs each year. Other procedures, therefore, need to be developed to achieve the goal of minimizing genetic change by genetic drift or selection.

The large census population sizes at which some species are maintained in captivity should not be taken to mean that genetic drift is not a concern. For example, Briscoe et al. (1992) studied the genetics of eight captive populations of *Drosophila* held in populations with approximately thousands of individuals. All eight populations lost substantial heterozygosity at nine allozyme loci. Values of N_e estimated by the decline in heterozygosity were less than 5% of the census population size. Delpuech et al. (1993) reported similar results in their review of five species of insects held in captivity. Populations of all of these species had retained approximately 20% or less of their original heterozygosity at allozyme loci.

These results demonstrate the importance of genetic monitoring of populations. Regular examination of allele frequencies at molecular genetic loci should be used to detect the effects of genetic drift in captive populations in which individual reproductive success is not being monitored.

Adaptation to captive conditions is an even greater concern for large populations held in captivity. For example, Frankham and Loebel (1992) found that the average fitness in captivity of *Drosophila* doubled after being maintained for eight generations in captivity. Many other studies have found evidence for rapid adaptation to captivity in a variety of organisms (see discussion in Gilligan and Frankham 2003). The genes selected for in captivity are almost certain to decrease the fitness of individuals when they are returned to wild conditions. In addition, the strong selection in captivity will reduce the effective population size of the captive population. In fact, strong variance in reproductive success associated with this adaptation is the likely explanation of the small $N_e : N_c$ ratios often found in captive populations.

A conceptual framework for minimizing the rate of adaptation to captivity (R) is provided by a modified form of the breeders' equation (see expression 11.7) (Frankham and Loebel 1992):

$$R = \frac{H_N S(1 - m)}{g} \tag{18.1}$$

where H_N is the narrow sense heritability, S is the selection differential, m is the proportion of genes contributed from wild individuals, and g is the generation interval.

Continued introduction of individuals from the wild (increased m) will slow the rate of adaptation to captivity. However, this will often not be possible.

The generation length (g) can be manipulated by increasing the average age of the parents. For example, doubling the mean age of parents will double the generation interval and halve the rate of adaptation to captivity. However, increasing the age of parents will also slow the rate of population growth so this approach is less feasible during the early stages of captivity before the population reaches carrying capacitiy.

Of course, reducing the intensity of selection (S) will slow adaptation to captivity. All efforts should be made to reduce differential survival and reproduction (fitness) in captivity. This can be done by minimizing mortality in captivity and by making the environmental conditions as close as possible to wild conditions.

Reducing differences in the number of progeny produced by individuals (family size) will also diminish the effects of selection in captivity (Allendorf 1993; Frankham et al. 2000). There will be no reproductive differences between individuals if all individuals produce the same number of progeny. In this situation, natural selection will only operate through differences in the relative survival of genotypes within families of full- or half-sibs. In a random mating population, approximately one-half of the additive genetic variance is within families and half is between families. Therefore, the rate of adaptation will be reduced by approximately 50% by equalizing family size. Equalizing family size will also increase N_e.

18.7 Supportive breeding

Supportive breeding is the practice of bringing in a fraction of individuals from a wild population into captivity for reproduction and then returning their offspring into their native habitat where they mix with their wild counterparts (Ryman and Laikre 1991). The goal of these programs generally is to increase survival during key life stages in order to support the recovery of a wild population that is threatened with eminent extirpation. These programs would seem to pose a relatively small risk of causing genetic problems. Nevertheless, the favoring of only a segment of the wild population may also bring about changes in the wild population due to genetic drift and selection (Example 18.5).

18.7.1 Genetic drift and supportive breeding

Supportive breeding acts to increase the reproductive rate of one segment of the population (those brought into captivity). This will increase the variance in reproductive success (family size) among individuals and therefore potentially reduce effective population size. Demographic increases in population size may reduce the overall N_e and accelerate the loss of genetic variation. This effect is most likely to occur for species with high reproductive rates where large differentials in reproductive success are possible (e.g., fishes, amphibians, reptiles, and insects).

Example 18.5 Supportive breeding of the world's largest freshwater fish

The Mekong giant catfish is a spectacular example of the potential problem with supportive breeding (Hogan et al. 2004). This is perhaps the largest species of fish found in fresh water. It grows up to 3 m long and weighs over 300 kg! A century ago, this species was found throughout the entire Mekong River from Vietnam to southern China. This species began disappearing from fish markets in the 1930s, and efforts to find individuals in fish markets have failed in the last few years. Very few fish remain in the wild and the species is currently listed as endangered on the IUCN "Red List" (critical world distribution).

   The Department of fisheries of Thailand began a captive breeding program in 1984. Over 300 adult fish have been captured in the wild and brought into captivity over the last 20 years. However, this program further threatens this species because of the removal of adult fish from the wild and the release of large numbers of young fish from very few parents. For example, over 20 wild adults were sacrificed in 1999 to supply eggs and milt for artificial propagation. More than 10,000 of these fingerlings were released back into the wild in 2001. However, genetic analysis of the progeny indicated that roughly 95% of these progeny were full-sibs produced by just two parents (Hogan et al. 2004).

Consider the situation where the breeding population consist of N_w effective parents that are reproducing in the wild and N_c effective parents that are breeding in captivity and their progeny are then released into the wild to supplement the wild population (Figure 18.6). Figure 18.7 presents the overall N_e as a function of the progeny that are produced in

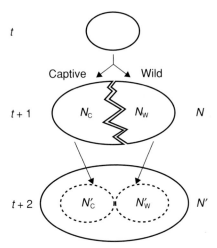

Figure 18.6 Schematic representation of supportive breeding. The total population of N individuals is divided into a captive and a wild group of size N_c and N_w that reproduce in captivity and in the wild, respectively. The N'_c and N'_w offspring are mixed before breeding in generation $t + 2$. From Ryman et al. (1995).

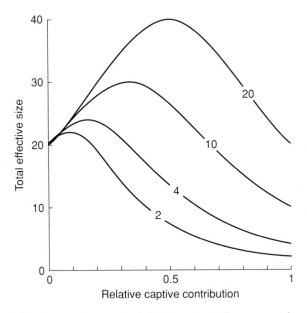

Figure 18.7 Total effective population size (wild + captive) when a natural population of 20 effective parents is supported by offspring from different numbers of captive parents, as indicated by the numbers on the different curves. The x-axis is the proportion of parents contributed by the captive parents. From Ryman and Laikre (1991).

captivity. The overall effective size may be substantially smaller than the effective number of parents reproducing in the wild when the contribution of the captive population is high. For example, consider the case where the wild population consists of 22 effective parents and that two of these parents are taken into captivity and then produce 50% of the total progeny. In this case, the total effective population size will be approximately six rather than the 22 that it would have been in the absence of a supportive breeding program.

Consideration of this problem has been extended to multiple generations (Wang and Ryman 2001; Duchesne and Bernatchez 2002). The effects of supportive breeding on N_e is complex. Moreover, the effects of supportive breeding on the inbreeding and variance effective population sizes (see Section 7.6) may differ. Nevertheless, supportive breeding, when carried out successfully over multiple generations, may increase not only the census but also the effective size of the supported population as a whole. If supportive breeding does not result in a substantial and continuous increase of the census size of the breeding population, however, it might be genetically harmful because of elevated rates of inbreeding and genetic drift.

18.7.2 Natural selection and supportive breeding

Supportive breeding can also have important genetic effects on supplemented populations because alleles that are harmful in the wild but advantageous in captivity may rise to high frequencies in captive populations (Lynch and O'Hely 2001). This genetic supplementation load will be especially severe when a captive population that is largely closed to import makes a contribution to the breeding pool of individuals in the wild. Moreover, theory indicates this load may become substantial in a wild supplemented population when the captive breeders are always derived from the wild.

Many recent papers have modeled possible harmful genetic effects of supportive breeding programs of natural populations (e.g., Duchesne and Bernatchez 2002; Ford 2002; Theodorou and Couvet 2004). These populations can be managed to increase rather than decrease effective population size. Nevertheless, the effects of supportive breeding on adaptation of wild populations are more difficult to predict. Selection in captivity can substantially reduce the fitness of a wild population during supportive breeding. The continual introduction of wild individuals into the captive population can reduce, but is not expected to eliminate, this effect. These programs can reduce the probability of local extirpations, but it is essential to carefully design the genetic aspects of these programs.

18.8 Reintroductions and translocations

To insure a successful **reintroduction, introduction,** or supplemental **translocation,** we should consider several issues: (1) where to release the individuals; (2) how many populations to establish; (3) how many individuals to release; (4) age and sex of individuals to release; (5) which and how many source population to use; and (6) how to monitor the population after the release of individuals. Genetics should play a role in all of these issues (Sarrazin and Barbault 1996).

Monitoring after the release of individuals is crucially important in insuring the success of reintroductions and translocations. Unfortunately, post-release studies of genetic contribution or of population status are seldom conducted. Molecular markers can help

---

Example 18.6 Rapid genetic decline in a translocated plant (Krauss et al. 2002)

The Corregin grevillea is one of the world's rarest plant species; only five plants were known in the wild in 2000. These plants occurred in degraded and isolated remnants of natural vegetation on road verges in Western Australia. In 1995, 10 plants were selected from the 47 plants known at the time to act as genetically representative founders for translocation into secure sites. Hundreds of ramets (tissue-cultured propagules of these 10 clones) were produced from these plants. By late 1998, 266 plants had been successfully translocated and were producing large numbers of seeds.

Krauss et al. (2002) used AFLPs to determine the genetic contribution of the 10 founders to this translocated population and their first-generation progeny. They found that only eight clones, not 10, were present in the translocated population. In addition, 54% of all plants were from a single clone. They also found that F_1 hybrids produced between founders were on average 22% more inbred and 20% less heterozygous than their founders, largely because 85% of all seeds were the product of only four clones. They estimated that the effective population size of the translocated population was approximately two. That is, the loss in heterozygosity from the founders to the next generation was what would be expected if two founders had been used.

These results demonstrate the importance of genetic monitoring of translocation programs.

---

monitor generic diversity, effective population size, and reproductive contribution of released individuals. For example, if few founders actually reproduce, due to extreme polygamy, paternity analysis could detect the problem by identifying only a few males as fathers. If paternity analysis is not feasible, then monitoring for loss of alleles, rapid genetic change, and small effective population size could help determine if few founders reproduce (Luikart et al. 1999) (Example 18.6).

18.8.1 Reintroductions

Where to release individuals depends on habitat suitability and availability. To maximize the chances of a successful reintroduction, the habitat should be similar to that to which the individuals to be released are adapted. Obviously, sufficient food, water, breeding habitat, and shelter or escape terrain should be available. Furthermore, the habitat should be free from exotic predators or competitive invasive species. For example, when threatened marsupials are reintroduced in Australia, exotic foxes and domestic cats should not be present because they are highly efficient at killing the marsupials and preventing reestablishment of the population. In the case of the African rhino, there is abundant habitat, but little habitat free of human predators, i.e., poachers (see Section 16.1).

How many populations? At least two, and preferable several populations should be established and maintained. Populations should be independent demographically and environmentally to avoid a catastrophic species-wide decline due to severe weather, floods, fire, or disease epizootics, for example. Two or more populations should be established within each different environment or for each divergent genetic lineage, whenever divergent environments or lineages exist within a species' range. The preservation of multiple populations across multiple diverse environments can help insure long-term persistence of a species (Hilborn et al. 2003; see also Example 15.2). Genetics can help determine if populations are independent demographically through genetic mark–recapture to identify migrants. Molecular genetic markers are widely used to assess gene flow, an indicator of the degree of population independence or isolation.

How many individuals to release depends, in part, on the breeding system, effective population size, and population growth rate after the reintroduction (Example 18.7). When feasible, at least 30–50 individuals should be reintroduced (see Guest Box 18). More individuals will be required if the breeding system is strongly sex biased (e.g., strong polygamy) or the effective population size is small compared to the census population size. Also, when population size does not increase above approximately 100–200 individuals within a few generations, more individuals should be released, when possible.

The sex and age of individuals can influence the success of the reintroductions and translocations. For example, it is often important to release more females than males to maximize population growth, which limits demographic stochasticity and subsequent genetic drift. Many reintroductions of large game animals in western North America have used about 60–80% females. For supplemental translocations in polygynous species, it is better in reintroduce females than males if we want only limited gene flow, because a single male can potentially breed with many females thereby swamping a population with introduced genes. Further, a male in a polygynous population might never breed, if he is not dominant, for example, making male-mediated gene flow highly variable and unpredictable. In territorial carnivores such as grizzly bears it is often best to translocation

---

Example 18.7 Genetic management of a reintroduction: Guam rails (Haig et al. 1990)

Over 50,000 Guam rails were estimated to be present on Guam in the 1960s. However, the introduction of the brown tree snake to Guam during World War II caused extinction or severe endangerment of all Guam's native forest birds. By 1986, Guam rails became extinct in the wild. However, 21 birds had been brought into captivity in 1983 and 1984 to initiate a captive breeding program (Haig et al. 1990).

The birds bred very successfully in captivity. By 1989, 113 birds were in the captive population and plans began to introduce Guam rails to the nearby island of Rota. Environmental conditions on Rota are similar to Guam except that the brown tree snake is not present. Initial plans were to introduce 90 birds to Rota. A number of factors were considered in designing an introduction program (e.g., behavior, demography, genetics, and the physical conditions of each animal) (Griffith et al. 1989).

Haig et al. (1990) compared six possible genetic mating schemes to produce 90 chicks planned for introduction: (1) select chicks at **random** from the captive population; (2) select on the basis of **fitness**; that is, use chicks from those birds that produced the greatest number of progeny in captivity; (3) select chicks that would

Table 18.1 Comparison of six breeding options (see text for explanation) for creating a group of 90 Guam rails for an introduction program (Haig et al. 1990).

Option	H_e	No. of alleles	Founder genome equivalents	Breeding pairs needed
Founders	1.00	42	21	–
Current population	0.98	31.5	10.5	–
Random (no selection)	0.95	24.1	9.4	23
Select for fitness	0.98	20.5	8.3	8
Maximize allozyme heterozygosity	1.00	18.9	7.1	13
Equalize founder contribution	0.98	27.2	13.4	8
Maximize allelic diversity	1.00	29.3	13.7	23
Founder genome equivalents	1.00	29.2	14.4	16

maximize the heterozygosity of the introduce population at 23 **allozyme** loci; (4) select chicks to **equalize founder contributions**; (5) select chicks to maximize the **allelic diversity** in the introduced population; or (6) select chicks to maximize the **founder genome equivalents** (see Section 18.6.1). Each option was evaluated in terms of how well it would maintain genetic diversity in the introduced population using a gene drop analysis (see Section 13.2).

The results indicated that some strategies would have done a poor job of maintaining genetic variation in the introduced population (Table 18.1). Selecting for reproductive fitness in captivity or heterozygosity at allozyme loci would have resulted in a substantial decline in allelic diversity in the introduced population. This shows the importance of minimizing differences in reproductive success among individuals in captivity.

The other three active options (equalize founder contributions, maximize allelic diversity, and maximize founder genome equivalents) all performed fairly equally (Table 18.1). The founder genome equivalent strategy seems best because it would retain nearly as much allelic diversity, maintain more founder genome equivalents, and require fewer breeding pairs, which would make it logistically preferable.

Some authors have suggested individuals should be chosen for breeding in captivity to increase genetic variation at certain loci, which can be examined with molecular techniques, that may have particular adaptive importance (Wayne et al. 1986; Hughes 1991). However, the above results show that selecting for increased variation at a few detectable loci can reduce the effective population size and reduce genetic variation throughout the genome (see Chapter 7).

As of June 2005, there were nearly 200 Guam rails in captivity in Guam and US facilities. Over 100 Guam rails have been introduced to Rota. Sixteen rails were reintroduced to Guam in 1998 to a 24 ha enclosure that is surrounded by a 2 m snake barrier. Five of these birds were still alive in October 2000.

females because males are more likely to fight for territory, sometimes to the death. Molecular genetic sexing can help determine sex before translocation in some species (birds and reptiles), where sex is cryptic.

Age can influence the likelihood that a translocated individual remains in the location of release and integrates socially into the new population. In large mammals, young juvenile or yearling individuals are often more likely than adults to integrate socially and/or not leave the release area. Currently there is no way to obtain age information for molecular genetic approaches, although the amount of telomere DNA on chromosomes is correlated with age and quantifying the amount might become feasible one day.

Which and how many source populations? For reintroductions and supplemental translocations, the source population generally should have high genetic diversity, genetic similarity, and environment similarity when compared to the new or recipient population. Environmental similarity helps limit chances of maladaptation of the translocated individuals in the site of release. However, if populations have recently become fragmented and differentiated, multiple differentiated source populations can help maximize genetic diversity in reintroductions or translocations, with little risk of outbreeding depression. For example, a source population with greater genetic divergence from the recipient population will result in a greater increase in heterozygosity in the recipient population.

If no individuals are available from a similar environment, then individuals from several source populations could be mixed upon release to maximize diversity for natural selection to act upon. Mixing of individuals from multiple sources is less desirable in supplemental translocations where some locally adapted individuals still persist because releasing many mixed individuals could swamp the local gene pool and lead to loss of locally adapted alleles.

18.8.2 Restoration of plant communities

These same genetic principles apply to developing sources to be used in restoration projects with plants (Fenster and Dudash 1994; Lesica and Allendorf 1999; Kephart 2004) (Example 18.8). Restoration is an important tool for the preservation of native plant communities (Hufford and Mazer 2003). Restoration ecology is a synthesis of ecology and population genetics.

In general, native local plants are the preferred source for restoration projects because of the potential importance of local adaptations (Linhart and Grant 1996). A variety of studies have found evidence that plants of relatively local origin are preferred as sources of reintroduction and restoration (Keller et al. 2000; Vergeer et al. 2004).

In some cases, local source populations may not be available. In addition, restoration projects may involve highly disturbed sites to which local genotypes are not adapted. In such cases, hybrids between populations, or mixtures of genotypes from different populations, may provide the best strategy (Guerrant 1996; Vergeer et al. 2004) (Figure 18.9). Mixtures of genotypes from ecologically distinct populations or hybrids of these genotypes will possess high levels of genetic variation. Introduced populations with enhanced variation are more likely to rapidly evolve genotypes adapted to the novel ecological challenges of severely disturbed sites.

Strains of plants that have been selected for captive conditions are a common source of plants for restoration (Keller et al. 2000). Such **cultivars** are often readily available, and are much less expensive than acquiring progeny from wild seed sources. However, the

Example 18.8 Genetic management of a reintroduction: Mauna Kea silversword (Robichaux et al. 1998)

The Mauna Kea silversword is a member of the silversword alliance, a group of Hawai'ian endemic plants that is one of the premier examples of adaptive radiation (Baldwin and Robichaux 1995). This plant is named for its mountain habitat and its striking rosette of dagger-shaped leaves covered with jewel-like silvery hairs (Robichaux et al. 1998) (Figure 18.8). The Mauna Kea silversword historically was common in exposed subalpine and alpine habitats high on the 4,205 m volcano on the Island of Hawai'i. The introduction of sheep and other ungulates devastated this plant, presumably because of heavy browsing. By the 1970s, only a small remnant population confined to cliffs and rocks persisted.

Three plants from this remnant population of an estimated less than 100 plants flowered in 1973. Most Mauna Kea silverswords live up to 50 years and are monocarpic (i.e., they flower only once before dying). Seeds from two of these plants were removed, and over 800 plants resulted from outplanting seedlings from these seeds on Mauna Kea. Today there are over 1,500 plants in the reintroduced population that are first- or second-generation offspring of the two maternal founders. This intervention and subsequent reintroduction dramatically increased the size of the silversword population on Mauna Kea.

This large, reintroduced population went through a severe genetic bottleneck because it is based on just two maternal plants. Analysis of seven variable microsatellite loci indicated substantial loss of genetic variation in the outplanted population in comparison to the native population (Friar et al. 2000). Three of the

Figure 18.8 The Mauna Kea silversword.

seven loci variable in the native plants, are fixed for a single allele. A total of eight of the total of 21 alleles over all loci were not detected in the outplanted population. The expected average heterozygosity in the outplanted population (0.074) was 70% less than that of the native population (0.250).

The greatest immediate genetic concern for recovery is the loss of allelic variation at the self-incompatibility locus in Mauna Kea silverswords (Robichaux et al. 1998; see Section 14.4.2). Loss of variation at this locus in the outplanted population may greatly reduce seed production and reduce the species' long-term chances for recovery.

Efforts are underway to increase genetic variation by hand transferring pollen from native plants that flower into the outplanting program. This is not an easy task; collecting pollen often involves perching precariously on steep cliffs because the remaining plants exist because they are out of the way of browsing ungulates. Two plants flowered in 1997 and a large number of seeds were produced by hand transfer of pollen. This doubled the number of founding maternal plants for the outplanting program.

More founders are expected to be added in the future. The program is currently concerned with balancing the genetic contributions of founders by equalizing founder contributions. This may be difficult because so many plants have already been outplanted and so few plants flower in any given year. In addition, the long generation interval will make this even more challenging! This program is a clear example of the importance of taking genetic concerns into consideration in the recovery of species that have reached small numbers.

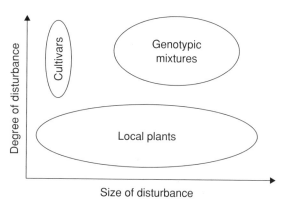

Figure 18.9 General relationship to degree and size of disturbance of three possible sources of plants to be used for restoration projects. In general, local plants should be preferred. However, cultivars may be appropriate in small but highly disturbed areas because they are more likely to quickly establish themselves. Hybrids between populations, or mixtures of genotypes from different populations, may provide the best strategy for highly disturbed sites to which local genotypes are not adapted. Introduced populations with enhanced variation are more likely to rapidly evolve genotypes adapted to the novel ecological challenges of severely disturbed sites. From Lesica and Allendorf (1999).

widespread use of cultivars is likely to lead to the introduction of genes into the adjacent resident population through cross-pollination, although the degree of genetic introgression will depend on the breeding system. Thus, widespread introductions of cultivars could alter the resident neutral gene pool. For these reasons, the use of cultivars should be restricted.

Guest Box 18 Effects of population bottlenecks on introduced species of birds
James V. Briskie

Colonial expansion in the 19th century not only brought a flood of European immigrants to the new world, it also lead to the establishment of many European birds in these foreign lands. The deliberate introduction of exotic birds was driven by both utilitarian needs (e.g., gamebirds) and sentimental reasons (e.g., settlers missed the song of familiar birds). Nowhere was this done with such sustained and organized effort as New Zealand. Acclimatization societies imported over a hundred species of birds, and although many failed to spread, today about 30 species of exotic birds range throughout New Zealand. Such species are now generally viewed as a nuisance (and certainly not worthy for conservation purposes), but they provide an exceptional opportunity to study how bottlenecks affect the fitness of populations.

The great difficulty and expense of transporting birds around the world meant that many species went through a severe bottleneck during their establishment. For example, the cirl bunting was established by a release of only 11 birds. Other species, such as the dunnock (250 founders), starling (653 founders), and blackbird (800 founders) were released in greater numbers. Variation in the size of bottlenecks experienced by each species provides a way to assess how the severity of a bottleneck affects the fitness of a post-bottleneck population. Such information is most valuable for the management of endangered native birds, though endangered native species seldom have large and non-bottlenecked populations to use as controls. In contrast, species introduced to New Zealand maintain large populations in their native range. The fitness effects of severe bottlenecks can therefore be quantified by comparing introduced birds in New Zealand ("post-bottleneck") with populations of the same species in their native range ("pre-bottleneck").

Hatching failure is one fitness trait known to be sensitive to the increased inbreeding expected in a severely bottlenecked population. Typically, 5–10% of eggs fail to hatch in a normal population, but highly inbred populations can have much higher rates of hatching failure as a result of infertility and embryo death. For example, the kakapo is a highly endangered parrot endemic to New Zealand that declined to a low of only 50 birds (see Example 18.1). Although intensive conservation efforts have paid off and the population is now increasing, recent breeding attempts have seen less than half their eggs hatch successfully. Levels of hatching failure in other endangered New Zealand species are also high, especially when populations drop below about 150 individuals. However, levels of hatching failure in these species before they passed through a bottleneck are unknown simply because they became endangered before any baseline information could be collected. Perhaps island birds like the kakapo may have naturally high levels of hatching failure? We simply don't know.

An examination of introduced species indicates that endangered New Zealand birds are not unique in this regard, and that high levels of hatching failure are indeed the product of severe population bottlenecks (Figure 18.10). All introduced species have levels of hatching failure below 10% (most below 5%) in their native range, yet some of the same species in New Zealand experience levels of failure up to 3–4 times higher.

These results also provide insight into the number of founders needed to avoid fitness decline due to bottleneck effects for introduced birds. As high levels of hatching failure in New Zealand populations are greatest in those species that passed through the most severe bottlenecks, it is possible to estimate the number of founders that would be required to start a new population and not induce higher levels of hatching failure in later generations. Based on hatching failure rates alone, at least 150 founders may be needed, a number much higher than that currently used (usually about 40–70 individuals) by most conservation biologists to found new populations of endangered birds. Whether other fitness traits are similarly affected remains to be determined, but it is clear that managers may be unwittingly contributing to the further endangerment of the species they are trying to save by not fully understanding the genetic effects of population bottlenecks.

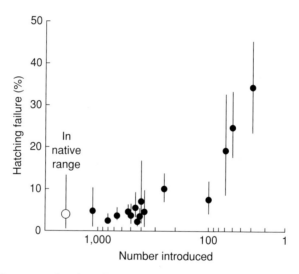

Figure 18.10 Increase in hatching failure of 15 introduced birds species with decreased numbers of individuals released by 19th century New Zealand acclimatization societies. Circles are means ±95% confidence intervals. The open circle shows mean hatching failure in the same species in their native range. Note that the x-axis is ordered from large to small bottleneck size on a log scale. From Briskie and Mackintosh (2004).

Problem 18.1

As we saw in Example 18.1, the kakapo conservation breeding program faces a variety of unusual challenges. All but one of the founding kakapo came from a population discovered on Stewart Island, which has very little genetic variation. Richard Henry is the only founding bird from a population on the South Island that contained more genetic variation. This population also was substantially genetically divergent from the Stewart Island population. Ideally, what proportion of the genetic composition of the kakapo breeding population should come from Richard Henry? The answer would be approximately 1/50 if we equalize the genetic contributions of founders. The answer would be 1/2 if we equalize the genetic contributions of founding populations. What would you recommend?

Problem 18.2

Under what circumstances should captive breeding be considered as a useful tool for conservation?

Problem 18.3

How could captive breeding be detrimental to the protection of habitat?

Problem 18.4

Why (and how) can the potential negative influence of natural selection and genetic drift be balanced?

Problem 18.5

Give at least three ways molecular markers can facilitate conservation breeding.

19

Invasive Species

St. John's Wort, Guest Box 19

Biological invaders are now widely recognized as one of our most pressing conservation threats.

Ingrid M. Parker et al. (2003)

The synergism arising from combining ecological, genetic, and evolutionary perspectives on invasive species may be essential for developing practical solutions to the economic and environmental losses resulting from these species.

Ann K. Sakai et al. (2001)

Invasion by **nonindigenous (alien) species** is recognized as second only to loss of habitat and landscape fragmentation as a major cause of loss of global biodiversity (Walker and Steffen 1997). The economic impact of these species is a major concern throughout the world. For example, an estimated 50,000 nonindigenous species established in the United

States cause major environmental damages and economic losses that total more than an estimated 125 billion dollars per year (Pimentel et al. 2000). Management and control of nonindigenous species is perhaps the biggest challenge that conservation biologists will face in the next few decades.

A chapter on **invasive species** may seem out of place in a book on conservation genetics. However, we have chosen to include such a consideration for several reasons. Molecular genetic analysis of introduced species can provide valuable information about the source and number of introduced populations. Also, understanding the ecological genetics of invasive species biology may provide helpful insights into developing methods of eradication or control. In addition, the study of species introductions offers exceptional opportunities to answer fundamental questions in population genetics that are important for the conservation of species. For example, how crucial is the amount of genetic variation present in introduced populations for their establishment and spread?

Molecular genetic analysis of introduced species (including diseases and parasites) can provide valuable information (Walker et al. 2003). Understanding the "epidemiology of invasions" (Mack et al. 2000) is crucial to control current invasions and prevent future invasions. Understanding the source of the introduced population, the frequency with which a species is introduced into an area, the size of each introduction, and the subsequent pattern of spread is important in order to develop effective mechanisms of control. However, observing such events is particularly difficult and assessment of the relative frequency of introductions or pattern of spread is extremely difficult. Molecular markers provide an important opportunity to answer these questions. In addition, many problematic diseases and parasites have been introduced and spread. Molecular markers are being used extensively to monitor and control these diseases (Criscione et al. 2005).

There is evidence that native species evolve and adapt to the presence of invasive species. For example, the highly toxic cane toad has had devastating ecological effects after their introduction to Australia in 1935. Phillips et al. (2003) concluded that cane toads threaten populations of approximately 30% of Australian terrestrial snakes because of their toxicity. Phillips and Shine (2004) predicted that eating cane toads would exert selection that would favor larger body size and a decrease in relative head size to reduce the relative prey mass of ingested cane toads in predatory snakes. A comparison of two high risk and two low risk snake species supported this prediction and provided strong evidence of adaptive changes in native predators resulting from the introduction of a toxic prey.

An understanding of genetics may also help predict which species are most likely to become invasive. There are two primary stages in the development of an invasive species (Figure 19.1). The first stage is the introduction, colonization, and establishment of a nonindigenous species in a new area. In other words, the introduced species must arrive, survive, and establish. The second stage is the spread and replacement of native species by the introduced species. The genetic principles that may help us predict whether or not a nonindigenous species will pass through these two stages to become invasive are the same principles that apply to the conservation of species and populations threatened with extinction: (1) genetic drift and the effects of small populations; (2) gene flow and hybridization; and (3) natural selection and adaptation.

In this chapter, we consider the possible importance of genetic change in the establishment and spread of invasive species. We also examine the significant role that hybridization may play in the development of invasive species. Finally, we consider ways in which

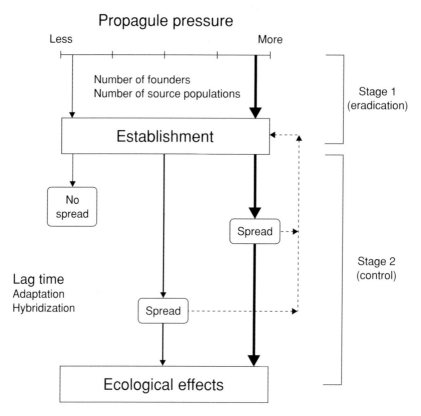

Figure 19.1 The two stages of invasion that generally coincide with different management responses. Propagule pressure is a continuum with greater pressure leading to an increased chance of establishment and spread with shorter lag times. If spread involves small groups of dispersing individuals, each group must be able to establish in a different area. Establishment or subsequent spread may be inhibited where groups reach the limits of particular environmental conditions. From Allendorf and Lundquist (2003).

genetic understanding may be applied to help predict which species are likely to be successful invaders and to help control invasive species.

19.1 Why are invasive species so successful?

Not all introduced species become invasive. A general observation is that only one out of every 10 introduced species becomes established, and only one out of every 10 newly established species becomes invasive. Therefore, roughly only one out of every 100 introduced species becomes a pest. The next few sections consider what factors may influence whether a species becomes established and becomes invasive.

Invasive species provide an exceptional opportunity for basic research in the population biology and short-term evolution of species. Many of the best examples of rapid evolutionary change come from the study of introduced populations (Lee 2002). For example, *Drosophila subobscura* evolved a north–south cline in wing length 20 years after introduction into North America that paralleled the pattern present in their native Europe of increased wing length with latitude (Huey et al. 2000). Similarly, two species of goldenrods evolved a cline in flowering time that resembled a cline in their native North America after being introduced into Europe (Weber and Schmid 1998).

Many unresolved central issues in the application of genetics to conservation – such as the inbreeding effects of small populations and the importance of local adaptation – can be much better experimentally addressed with introduced species. Two apparent paradoxes emerge from comparison of our previous conclusions of the effects of small population size and local adaptation with the successful invasions by introduced species.

19.1.1 If population bottlenecks are harmful, then why are invasive species that have gone through a founding bottleneck so successful?

Much of the concern in conservation genetics relates to the potential harmful effects of small population sizes. The loss of genetic variation through genetic drift and the inbreeding effect of small populations contribute to the increased extinction rate of small populations (e.g., Frankham and Ralls 1998). However, colonization of introduced species often involves a population bottleneck since the number of initial colonists is often small. Thus, a newly established population is likely to be much less genetically diverse than the population from which it is derived (Barrett and Kohn 1991).

The reduced genetic diversity can have two harmful consequences. First, inbreeding depression may limit population growth, and lower the probability that the population will persist. Second, reduced genetic diversity will limit the ability of introduced populations to evolve in their new environments. Thus we face a paradox: if population bottlenecks are harmful, then why are invasive species that have gone through a founding bottleneck so successful? One answer to this paradox is that introduced species often have greater genetic variation than native species because they are a mixture of several source populations (Example 19.1).

One solution to the first of the two genetic paradoxes lies in the strong observed effect of propagule pressure on the invasiveness of species. That is, the clear association between the greater number of introduced individuals and the number of release events and the probability of an introduced species becoming invasive suggests that many invasive species are not as genetically depauperate as expected. In addition, plant species can avoid the reduction in genetic variation associated with colonization by their means of reproduction (Barrett and Husband 1990).

Many invasive plant species reproduce asexually by apomixis or vegetative reproduction (Baker 1995; Calzada et al. 1996). In both cases, the effects of inbreeding depression are avoided because the progeny are genetically identical to the parental plants. In addition, many invasive plant species are polyploids and can reproduce by selfing. In this situation, genetic variation is maintained in the form of fixed heterozygosity because of genetic divergence between the genomes combined in the formation of the allopolyploid (Brown and Marshall 1981).

19.1.2 If local adaptation is important, then why are introduced species so successful at replacing native species?

The presence of local adaptations is often an important concern in the conservation of threatened species (McKay and Latta 2002). That is, adaptive differences between local

---

Example 19.1 Genetic variation increases during invasion of a lizard Kolbe et al. (2004)

The brown anole is a small lizard that is native to the Caribbean, but has been introduced widely throughout the world (Hawai'i, Taiwan, and mainland USA). Introduced populations often reach high population densities, show exponential range expansion, and are often competitively superior and predators of native lizards.

The brown anole first appeared in the Florida Keys in the late 19th century. Its range did not expand appreciably for 50 years, but widespread expansion throughout Florida began in the 1940s and increased in the 1970s.

Genetic analyses of mtDNA suggest that at least eight introductions have occurred in Florida from across this lizard's native range. This has resulted in an admixture from different geographic source populations and has produced populations that are substantially more genetically variable than native populations. Moreover, recently introduced brown anole populations around the world originate from Florida, and some have maintained these elevated levels of genetic variation.

Kolbe et al. (2004) suggest that one key to the invasive success of this species may be the occurrence of multiple introductions that transform among-population variation in native ranges to within-population variation in introduced areas. These genetically variable populations appear to be particularly potent sources for introductions elsewhere.

---

populations are expected to evolve in response to selective pressures associated with different environmental conditions. The presence of such local adaptations in geographically isolated populations often plays an important role in the management of threatened species (Crandall et al. 2000).

When a species invades a new locality it will almost certainly face a novel environment. However, many introduced species often outcompete and replace native species. For example, introduced brook trout are a serious problem in the western United States where they often outcompete and replace ecologically similar native trout species. However, the situation is reversed in the eastern USA where brook trout are native. They are in serious jeopardy because of competition and replacement from introduced rainbow trout that are native to the western USA. Thus we face a second paradox: if local adaptation is common and important, then why are introduced species so successful at replacing native species?

A variety of explanations have been proposed to explain why introduced species often outperform indigenous species. First, some species may be intrinsically better competitors because they evolved in a more competitive environment. Second, the possible absence of enemies (e.g., herbivores in the case of plants) allows nonindigenous species to have more resources available for growth and reproduction and thereby outcompete native species. Siemann and Rogers (2001) found that an invasive tree species, the Chinese tallow tree, had evolved increased competitive ability in their introduced range. Invasive genotypes were larger than native genotypes and produced more seeds; however, they had lower quality leaves and invested fewer resources on defending them. Thus, there are a number of reasons why introduced species may fare well even though native species may be locally adapted.

In addition, local adaptation of native populations might only be essential during periodic episodes of extreme environmental conditions (e.g., winter storms, drought, or fire). For example, Rieman and Clayton (1997) have suggested that the complex life histories of some fish species (mixed migratory behaviors, etc.) are adaptations to periodic disturbances such as fire and flooding. Thus, introduced species may be able to outperform native species in the short term (a few generations) because the performance of native species in the short term is constrained by long-term adaptations that may come into play every 50 or 100 years.

19.2 Genetic analysis of introduced species

Molecular genetic analysis of introduced species can provide valuable information about the origin of introduced taxa. In addition, study of the amount and distribution of genetic variation in introduced species can provide valuable insight into the mechanisms of establishment and spread. In some cases, even identifying the species of invasive organisms may be difficult without genetic analysis. In other cases, populations of a native species may become invasive when introduced into a new ecosystem. Such populations would technically not be considered to be alien since conspecific populations were already present. Nevertheless, such populations may become invasive when introduced outside of their natural area (Genner et al. 2004). Current regulations dealing with invasive organisms are based upon species classification. However, recognizing biological differences between populations within the same species is important for the control of invasive species.

19.2.1 Molecular identification of invasive species

In some cases, genetic identification may be necessary to identify the species of introduced species. For example, populations of Asian swamp or rice eels (genus *Monopterus*) have been found throughout the southeastern United States since 1994 (Collins et al. 2002). Swamp eels have a variety of characteristics that make them a potentially disruptive species. They are large predators (up to 1 m in length) that are capable of breathing out of water and dispersing over land. They are also extremely tolerant of drought because they produce large amounts of mucous that can prevent desiccation and burrow when water levels drop.

The morphological similarity of swamp eels makes identification difficult. Collins et al. (2002) sampled four locations in Georgia and florida to see if these eels were the result of a single introduction or multiple introductions. Examination of mtDNA revealed that introduced populations, even in close proximity (<40 km), were genetically distinct. These genetically distinct populations represent at least two and possibly three different species.

Genetics may also allow the detection of an invasive species that is conspecific with a native species. For example, Genner et al. (2004) used mtDNA to detect a non-native morph from Asia of the gastropod *Melanoides tuberulata* in Lake Malawi, Africa, which is sympatric with indigenous forms of the same species. This non-native morph was not present in historical collections and appears to be spreading rapidly and replacing the indigenous form.

19.2.2 Distribution of genetic variation in invasive species

Examination of published descriptions of the amount and patterns of genetic variation in introduced species reveals two contrasting patterns. In the first, introduction and establishment is often associated with a population bottleneck, or bottlenecks, so that introduced populations have less genetic variation than populations in the native range of the species. Under some circumstances, this reduced genetic variation may actually stimulate invasion (Example 19.2). In the second pattern, introduction and establishment is associated with admixture of more than one local population in the native range so that populations in the introduced range have greater genetic variation. The bottleneck and admixture situations result in very different patterns of genetic variation in introduced species.

Bottleneck model

In many cases, introduced species may only have a few founders so that genetic variation is reduced by the founder effect. The land snail that we examined in Example 6.2 is an excellent example of this pattern (Johnson 1988). This species was introduced from Europe to Perth in western Australia. The Perth population has reduced heterozygosity and allelic diversity compared to a population from France (see Figure 6.9; Problem 6.10). In addition, another population was founded on Rottnest Island by a limited number of founders from the Perth population. This second bottleneck further reduced heterozygosity and allelic diversity.

Example 19.2 Loss of genetic variation in an introduced ant species promotes a successful invasion (Tsutsui et al. 2000)

Ants are among the most successful, widespread, and harmful invasive taxa. Highly invasive ants are often unicolonial, and form supercolonies in which workers and queens mix freely among physically separate nests. By reducing costs associated with territoriality, unicolonial species can attain high worker densities, allowing them to achieve interspecific dominance.

Tsutsui et al. (2000) examined the behavior and population genetics of the invasive Argentine ant (*Linepithema humile*) in its native and introduced ranges. They demonstrated with microsatellites that population bottlenecks have reduced the genetic diversity of introduced populations. This loss is associated with reduced intraspecific aggression among spatially separate nests, and leads to the formation of interspecifically dominant supercolonies. In contrast, native populations are more genetically variable and exhibit pronounced intraspecific aggression.

These findings provide an example of how a genetic bottleneck associated with introduction can lead to widespread ecological success.

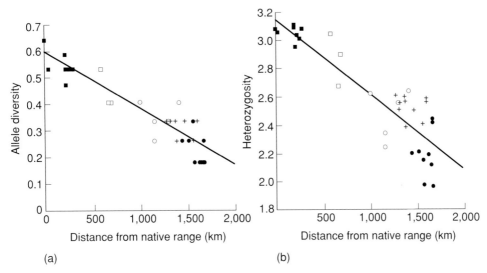

Figure 19.2 Relationship in an invasive gallwasp between distance from its native Hungary and (a) allelic diversity and (b) expected heterozygosity. Populations further from the native range show reduced genetic variation and patterns of allele frequency differentiation that suggests a stepping-stone invasion process rather than multiple introductions from its native range. From Stone and Sunnucks (1993).

Stone and Sunnucks (1993) have described the invasion of northern and western Europe of the gallwasp *Andricus quercusalicis* following human introduction of an obligate host plant, the Turkey oak from southeastern Europe. Populations further from the native range show reduced allelic diversity and heterozygosity (Figure 19.2). This suggests that this species has experienced a series of bottlenecks as it spread throughout Europe over the last 300–400 years. Patterns of allele frequency differentiation suggest that the invasion of this species followed a stepping-stone process rather than multiple introductions from its native range.

Admixture model

In contrast to the bottleneck model, many introduced species actually have greater variation in comparison with populations from the native range because their founders come from different local populations within the native range. Admixing individuals from genetically divergent populations will increase genetic variation by converting genetic differences between populations to genetic variation between individuals within populations. We saw in Chapter 9 that the total heterozygosity (H_T) within a species can be partitioned into genetic variation within and between subpopulations:

$$H_T = H_S + D_{ST} \tag{19.1}$$

where H_S is the average heterozygosity within subpopulations, and D_{ST} is the average gene diversity between subpopulations. D_{ST} is related to the more familiar F_{ST}:

$$D_{ST} = (F_{ST})(H_T) \tag{19.2}$$

and,

$$F_{ST} = \frac{D_{ST}}{H_T} \tag{19.3}$$

Thus, D_{ST} is the proportion of the total heterozygosity due to genetic divergence between subpopulations (Nei 1987, p. 189).

We saw in Example 9.2 that two separate demes of brown trout existed within a single Swedish lake and had substantial genetic differentiation at the *LDH-A2* locus (and many other loci as well) (Ryman et al. 1979):

$$H_T = H_S + D_{ST} = H_S + (F_{ST})(H_T)$$

$$= 0.128 + (0.728)(0.489) = 0.128 + 0.356 = 0.489$$

Thus, if we introduced an equal number of fish from each of these demes into a new lake and the fish mated at random, the expected heterozygosity in the newly founded population would be 0.489, nearly four times as great as in the original populations at this locus.

This effect can be seen in many introduced populations (see Example 19.1). Approximately 400 chaffinches were imported from England into New Zealand between 1862 and 1877. Overwintering birds from several populations on the European continent were included in the birds collected for introduction. Baker (1992) reported that chaffinches from eight populations in New Zealand have an average heterozygosity that is 38% greater (0.066 versus 0.048) than 10 native European populations at 42 allozyme loci. As expected, chaffinches in New Zealand have greatly reduced differentiation among subpopulations ($F_{ST} = 0.040$) compared to chaffinches in their native Europe ($F_{ST} = 0.222$).

19.2.3 Mechanisms of reproduction

Molecular genetic analysis can also be used to determine if an introduced plant species is reproducing sexually or asexually, the breeding system, and the ploidy level of introduced plants. Further examination could determine how many different clonal lineages are present if an invader is reproducing asexually. This information, along with an understanding of genetic population structure, is essential for the development of effective control measures for invasive weed species (Chapman et al. 2004).

For example, most strains of the marine green algae *Caulerpa taxifolia* are not invasive. However, a small colony of *C. taxifolia* was introduced into the Mediterranean in 1984 from a public aquarium and spread widely and seriously reduced biological diversity in the northwestern Mediterranean (Jousson et al. 2000). The invasive strain differs from native tropical strains because it reproduces asexually, grows more vigorously, and is resistant to lower temperatures. Colonies of *C. taxifolia* have recently been reported on the coast of California and this has raised concerns about the danger of an invasion similar to that in the Mediterranean. Genetic analysis of the California alga has shown that it is the same strain as the one responsible for the Mediterranean invasion (Jousson et al. 2000). Thus, the rapid eradication of this introduced alga should receive high priority in order to reduce the probability of a new invasion.

19.2.4 Quantitative genetic variation

Although much information can be gained from molecular markers, characterization of the genetic variation controlling those life history traits most directly related to establishment and spread is also crucial. These traits are likely to be under polygenic control with strong interactions between the genotype and the environment; they cannot be analyzed directly with molecular markers, although mapping quantitative trait loci (QTLs) affecting fitness, colonizing ability, or other traits affecting invasiveness may be possible (Barrett 2000). For example, variation in the number of rhizomes producing above-ground shoots, a major factor in the spread of the noxious weed johnsongrass, is associated with three QTLs (Paterson et al. 1995). This knowledge may provide opportunities for predicting the location of corresponding genes in other species and for growth regulation of major weeds.

Application of the methods of quantitative genetics could be useful for those species in which information can be obtained from breeding design or from parent–offspring comparisons. For example, one could compare the additive genetic variance/covariance structure of a set of life history traits of different populations to evaluate the role of genetic constraints on the evolution of invasiveness. Comparisons of the heritability of a trait could be made among different, newly established populations or between invasive populations and the putative source population. Consideration of both the genetic and ecological context of these traits is critical, given the potentially strong interaction of genetic and environmental effects (Barrett 2000).

19.3 Establishment and spread of invasive species

One common feature of invasions is a lag time between initial colonization and the onset of rapid population growth and range expansion (Sakai et al. 2001) (see Figure 19.1). This lag time is often interpreted as an ecological phenomenon (the lag phase in an exponential population growth curve). Lag times are also expected if evolutionary change is an important part of the colonization process. This process could include the evolution of adaptations to the new habitat, the evolution of invasive life history characteristics, or the purging of genetic load responsible for inbreeding depression (Figure 19.3). It appears likely that in many cases there are genetic constraints on the probability of a successful invasion, and the lag times of successful invasives could be a result of the time required for adaptive evolution to overcome these genetic constraints (Ellstrand and Schierenbeck 2000; Mack et al. 2000).

19.3.1 Propagule pressure

Propagule pressure has emerged as the most important factor for predicting whether or not a nonindigenous species will become established (Kolar and Lodge 2001). Propagule pressure includes both the number of individuals introduced and the number of release events. Propagule pressure is expected to be an important factor in the establishment of introduced species on the basis of demography alone. That is, it is unclear what role, if any, genetic factors may play in the effect of propagule pressure.

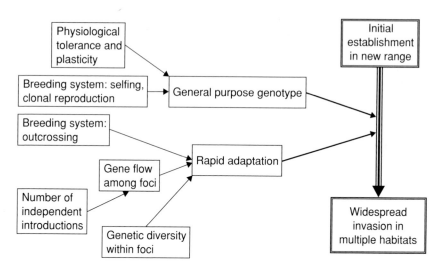

Figure 19.3 Factors that influence the process by which an introduced species moves from initial establishment in a new range to widespread invasion of multiple habitats. Two alternative, but not mutually exclusive, mechanisms are presented: rapid adaptation and the general purpose genotype. Characteristics of the invading species (e.g., breeding system) or of the invasion process (e.g., number of introductions) that influence these two mechanisms are outlined. From Parker et al. (2003).

There are two primary ways in which the genetics of an introduced species may be affected by propagule pressure. First, a greater number of founding individuals would be expected to reduce the effect of any population bottleneck so that the newly established population would have greater genetic variation. Second, and perhaps most importantly, different releases may have different source populations. Therefore, hybridization between individuals from genetically divergent native populations may result in introduced populations having more genetic variation than native populations of the same species (see Section 19.4).

19.3.2 Spread

Many recently established species often persist at low, and sometimes undetectable, numbers and then "explode" to become invasive years or decades later (Sakai et al. 2001). Adaptive evolutionary genetic changes may explain the commonly observed lag time that is seen in many species that become invasive (García-Ramos and Rodríguez 2002). Many of the best examples of rapid evolutionary change come from the study of recently introduced populations (Lee 2002).

19.4 Hybridization as a stimulus for invasiveness

Hybridization may play an important role in introduced species becoming invasive (Example 19.3). As we have seen in the previous sections, many species become invasive only after: (1) an unusually long lag time following initial arrival; and (2) multiple

> **Example 19.3** Hybrid mosquitoes may spread West Nile virus by acting as a bridge between humans and birds
>
> The rapid spread of West Nile virus in humans in North America provides an unusual example of the possible role of hybridization in the spread of invasive species (Couzin 2004). Mosquitoes in the *Culex pipens* complex are the primary vectors for the spread of West Nile virus to humans in North America and Europe (Fonseca et al. 2004). However, there have been few human outbreaks in Europe even though the virus is endemic there and the birds that harbor it and the mosquitoes that spread it are also present.
>
>   Microsatellite analysis of *Culex* mosquitoes in northern Europe indicated several reproductively isolated taxa of mosquitoes that differ in biting behavior and physiology (Fonseca et al. 2004). In contrast, hybrids between these distinct taxa are found throughout the United States. It appears that hybrid mosquitoes in North America bite both humans and birds and apparently serve as bridge vectors of the disease from birds to humans.

introductions. Ellstrand and Schierenbeck (2000) have proposed that hybridization between species, or genetically divergent source populations, may serve as a stimulus for the evolution of invasiveness on the basis of these observations. Ellstrand and Schierenbeck (2000) proposed four genetic mechanisms to explain how hybridization may stimulate invasiveness.

Evolutionary novelty

Hybridization may result in the production of novel genotypes and phenotypes that do not occur in either of the parental taxa. Evolutionary novelty can result either from the combination of different traits from both parents or from traits in the hybrids that transgress the phenotypes of both parents (transgressive segregation; see Section 17.1.4).

Genetic variation

An increase in the amount of genetic variation may in itself be responsible for the evolutionary success of hybrids. That is, the greater genetic variation (heterozygosity and allelic diversity) in hybrid populations may provide more opportunity for natural selection to bring about adaptative evolutionary change. This mechanism is a group selection.

Fixed heterosis

Many invasive plant species have genetic or reproductive mechanisms that stabilize first-generation hybridity and thus may fix genotypes at individual or multiple loci that demonstrate heterosis. These mechanisms include alloployploidy, permanent translocation heterozygosity, **agamospermy**, and clonal reproduction. The increased fitness resulting from fixed heterozygosity may contribute to the invasiveness of many plant species.

  Common cordgrass (*Spartina anglica*) has been identified as one of the world's worst invasive species by the World Conservation Union (IUCN 2001). Common cordgrass is a perennial salt marsh grass that has been planted widely to stablize tidal mud flats. Its invasion and spread leads to the exclusion of native plant species and the reduction of suitable feeding habitat for wildfowl and waders. This species originated by chromosome doubling of the sterile hybrid between the Old World *S. maritime* and the New World *S. alterniflora*. Genetic analysis has found almost a total lack of genetic differences among individuals. However, the allopolyploid origin of this species has resulted in fixed heterozygosity at all loci for which these two parental species differed.

Reduction of genetic load

As we have seen, small isolated populations will accumulate deleterious recessive mutations so that mildly deleterious alleles become fixed and lead to a slow erosion of average fitness (see Section 14.7). Hybridization between populations would lead to a reduction in this mutational genetic load (Whitlock et al. 2000; Morgan 2002). Ellstrand and Schierenbeck (2000) have suggested that the increase in fitness of this effect may under some circumstances be sufficient to account for invasiveness.

19.5 Eradication, management, and control

Understanding the biology of invasive species is not necessary, and in many circumstances will not even be helpful for their management and control (Simberloff 2001). Simberloff has described this as a policy of "Shoot first, ask questions later" (see also Ruesink et al. 1995). This recommendation is in agreement with basic population biology. The best way to reduce the probability that an introduced species becomes invasive is to eliminate it before it becomes abundant, widespread, and has had sufficient time to evolve any adaptations that may allow it to outcompete native species. Nevertheless, understanding population biology, genetics, and evolution may be helpful in the prediction of the potential for invasive species to evolve responses to management practices, and in the development of policy.

Genetics may play an important role in the potential of an established invader to evolve defenses against the effects of a control agent (e.g., the evolution of resistance to herbicides or biological control agents). The rate of change in response to natural selection is proportional to the amount of genetic variation present (Fisher 1930). Therefore, the amount of heterozygosity or allelic diversity at molecular markers that are likely to be neutral with respect to natural selection may provide an indication of the amount of genetic variation at loci that potentially could be involved in response to a control agent.

The amount of molecular genetic variation may not be a reliable general indicator of the amount of heritable variation for adaptive traits (Frankham 1999; McKay and Latta 2002). However, molecular genetic variation is likely to be a reliable indicator for invasive species of the potential for adaptive change because of the genetic effects of recent colonization. For example, greatly reduced molecular variation in an invasive population relative to native populations of the same species is a good indicator of a small effective population during the founding event; this is expected to reduce the amount of variation at adaptive loci. In addition, greater molecular variation in an invasive population relative to native populations of the same species is a good indicator of introductions from multiple source populations. This would indicate that the invasive species likely has substantial amounts of adaptive genetic variation to escape the effects of a control agent.

Genetics should play a more central role in developing policy to manage and control invasive species. Regulations generally have not taken into account that some genotypes may be more invasive than others of the same species. According to the standards set by the International Plant Protection Convention, import cannot be restricted for species that are already widespread and not the object of an "official" control program (Baskin 2002). For example, several well-known noxious range weeds (e.g., the yellow star thistle) are on the list of permitted imports in the State of Western Australia because they are widespread and the government is not officially trying to control them. However, they are subject to control attempts by landowners for whom they are a problem. Allowing the future import of additional strains that could be more invasive seems unwise in situations such as this.

19.5.1 Units of eradication

Eradication of introduced species is a potentially valuable tactic in restoration (Myers et al. 2000). The eradication of rats, mice, and other introduced mammals is becomingly increasingly common on oceanic islands and isolated portions of continental land masses. The New Zealand Department of Conservation applied 120 metric tons of poison bait onto Campbell Island, a large subantarctic island (11,300 ha) approximately 700 km south of the New Zealand mainland. A survey of the island in 2003 found no trace of brown rats on the island, and an incredible recovery of bird and insect life.

Successful eradication requires a low risk of recolonization. Isolated island populations or populations limited to isolated "habitat islands" have a low risk of recolonization. However, other islands or regions that display no distinct geographic structure or barriers are more problematic. The eradication of a portion of a population, or a sink population within an unidentified source–sink dynamic, would result in rapid recolonization and a waste of resources (see examples in Myers et al. 2000).

Genetics can be used to identify isolated reproductive units that are appropriate groups for eradication on the basis of patterns of genetic divergence (Calmet et al. 2001; Robertson and Gemmell 2004). Little genetic differentiation between spatially isolated populations is indicative of significant gene flow, while significant differentiation between adjacent populations indicates limited dispersal. Examination of the patterns of genetic variation can allow the identification of distinct population units with negligible immigration. With appropriate care, these population units could be eradicated with little chance of recolonization. The identification of "units of eradication" is an interesting analogue to the identification of units of conservation as seen in Chapter 16 (Robertson and Gemmell 2004). Genetic analysis would also allow distinction between an eradication failure (i.e., recovery by a few surviving individuals) and recolonization.

Robertson and Gemmell (2004) examined 18 microsatellite loci in two populations of brown rats separated by a glacier on South Georgia Island in the Antarctic. Rats were unintentionally introduced to South Georgia when commercial sealing started there in the late 1700s. The brown rats have devastated the island's avifauna. In addition, remaining rat-free areas on the island contain unusual assemblages of plants and animals that appear to be unable to sustain populations in the presence of rats.

The eradication of rats from the entire island is a daunting task because of its great size (400,000 ha). However, appropriate rat habitat is limited to coastal regions that are often separated by glaciers, permanent snow and ice, and icy waters. If such barriers preclude dispersal, then each discrete population could be considered as an eradication unit. Eradication of rats from South Georgia could then proceed sequentially with low risk of natural recolonization.

One population, Greene Peninsula, was earmarked for an eradication trial. Genetic diversity in 40 rats sampled from Greene Peninsula and a nearby population showed a pronounced genetic population differentiation that allowed individuals to be assigned to the correct population of origin (Robertson and Gemmell 2004). The results suggested limited or negligible gene flow between the populations and that glaciers, permanent ice, and icy waters restrict rat dispersal on South Georgia. Such barriers define eradication units that could be eradicated with low risk of recolonization, hence facilitating the removal of brown rats from South Georgia.

19.5.2 Genetics and biological control

Invasive species can undergo rapid adaptive evolution during the process of range expansion. Here, such evolutionary change during invasions has important implications for biological control programs (Wilson 1965). The degree to which such evolutionary processes might affect biological control efficacy remains largely unknown (Müller-Scharer et al. 2004).

The first application of genetics in the control of invasive species was the use of genetics in association with the sterile insect technique. One approach has been to introduce genotypes that could subsequently facilitate control or render the pest innocuous (Foster et al. 1972). Another approach was to release genotypes with chromosomal aberrations whose subsequent segregation would result in reduced fertility and damage the population (Foster et al. 1972). Recent efforts using the sterile insect technique have used transgenic insects homozygous for repressible female-specific lethal effects (Thomas et al. 2000). The release of males with this system may be an effective mechanism of control for some insect pests.

19.5.3 Pesticides and herbicides

Invasive species have the ability to evolve quickly in response to human control efforts (Example 19.4). Therefore, application of genetic principles is important for developing effective controls for invasive species. The evolution of resistance to insecticides and herbicides has increased rapidly in many species over the last 50 years (Denholm et al. 2002). Reducing the evolution of resistance to control measures in invasive species will require an understanding of the origin, selection, and spread of resistant genes. Comparison of the genomes of insect species such as *Drosophila* and malaria vector mosquitoes should aid the development of new classes of insecticides, and should also allow the lifespan of current pesticides to be increased (Hemingway et al. 2002).

Example 19.4 Evolution of herbicide resistance in the invasive plant hydrilla

In the early 1950s, a female form of a dioecious strain of hydrilla was released into the surface water of Florida in Tampa Bay and spread rapidly throughout the state (Michel et al. 2004). Today hydrilla is one of the most serious aquatic weed problems in the USA. This invasive plant can rapidly cover thousands of contiguous hectares, displacing native plant communities and causing significant damage to the ecosystems.

Hydrilla has been controlled by the sustained use of the pesticide fluridone for several weeks in lake water. Fluridone is an inhibitor of phytoene desaturase (PDS), a rate-limiting enzyme in carotenoid biosynthesis. PDS is a nuclear-encoded protein and has activity in the chloroplasts, the site of carotenoid synthesis. Under high light intensities, carotenoids stabilize the photosynthetic apparatus by quenching the excess excitation energy. Inhibition of PDS decreases colored carotenoid concentration and causes photobleaching of green tissues.

An apparent decrease in the effectiveness of fluridone to control hydrilla has been observed in a number of lakes. Evolution of herbicide resistance was considered unlikely to occur in the absence of sexual reproduction. Nevertheless, a major effort was undertaken in 2001 and 2002 to test for herbicide-resistant hydrilla in 200 lakes throughout florida. No within-site variation in fluridone resistance was detected, and approximately 90% of the lakes contained fluridone-sensitive hydrilla. However, three phenotypes of fluridone-resistant hydrilla populations were discovered in 20 water bodies of central florida. A hydrilla phenotype with low level resistance was found in eight lakes, a phenotype with intermediate resistance was found in seven lakes, and the most resistant phenotype was found in five lakes (Figure 19.4).

Sequencing of the phytoene desaturase (*pds*) locus indicated that the three fluridone-resistant types had different amino acid substitutions at codon 304 of the sensitive form of PDS (Figure 19.5). The three PDS variants had specific activities similar to the sensitive form of the enzyme but were 2–5 times less sensitive to fluridone. *In vitro* activity levels of the enzymes correlated with *in vivo* resistance of the corresponding populations.

It appears that fluridone resistance has arisen by somatic mutations that caused a single biotype to quickly become the dominant type within a lake. The establishment of herbicide-resistant biotypes as the dominant forms in these lakes was not anticipated. Asexually reproducing plants are under strong uniparental constraints that limit their ability to respond to environmental changes (Holsinger 2000).

The future expansion of resistant biotypes poses significant environmental challenges in the future. Weed management in large water bodies relies heavily on fluridone, the only Environmental Protection Agency (EPA)-approved synthetic herbicide available for systemic treatments of lakes in the USA. Current plans include regular monitoring to detect resistance and prevent the spread of these herbicide-resistant biotypes.

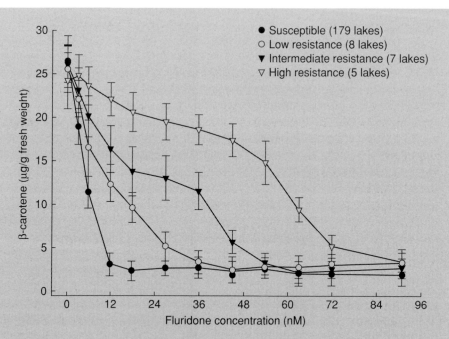

Figure 19.4 Differential response of hydrilla populations from lakes in Florida to the herbicide fluridone which causes photobleaching of green tissues by decreasing β-carotene content. Low β-carotene content indicates high susceptibility to the effects of flurodine. Plots show mean and standard deviations of β-carotene content of hydrilla shoot apices following a 14-day laboratory exposure to fluridone at different concentrations in four different phenotypes.

Rainbow River	cDNA/genomic DNA	ATTGCCTTAAACCGTTTCCTTCAGGAA
	Amino acid	- I - - A - - L - - N - - **R** - - F - - L - - Q - - E -
Lulu	cDNA/genomic DNA	ATTGCCTTAAACAGTTTCCTTCAGGAA
	Amino acid	- I - - A - - L - - N - - **S** - - F - - L - - Q - - E -
Pierce	cDNA/genomic DNA	ATTGCCTTAAACTGTTTCCTTCAGGAA
	Amino acid	- I - - A - - L - - N - - **C** - - F - - L - - Q - - E -
Okahumpka	cDNA/genomic DNA	ATTGCCTTAAACCATTTCCTTCAGGAA
	Amino acid	- I - - A - - L - - N - - **H** - - F - - L - - Q - - E -

Figure 19.5 DNA sequences and their corresponding deduced amino acid substitutions at codon 304 of the hydrilla *pds* gene that convert the susceptible biotype Rainbow River into resistant biotypes: Lulu (low), Pierce (intermediate), and Okahumpka (high).

Guest Box 19 Rapid adaptation of invasive populations of St John's Wort
John L. Maron

St John's Wort is a short-lived perennial plant native to Europe, North Africa, and Asia. This plant has been introduced to North and South America, Australia, New Zealand, and South Africa where it can grow at very high density, particularly in disturbed or overgrazed grasslands. Exotic species like St John's Wort are commonly introduced into communities where they confront different abiotic and biotic conditions from where they are native. A key question in invasion biology concerns how successful exotics cope with these novel environmental circumstances. Common-garden experiments in the native and introduced range coupled with genetic analyses of plants can provide insight into the role of contemporary evolution versus phenotypic plasticity and founder effects in shaping or constraining exotic plant success.

We compared phenotypes of St John's Wort collected from widely distributed populations across North America and Europe in common-gardens in Washington, California, Sweden, and Spain (Maron et al. 2004a). Both introduced and native populations exhibited latitudinally based clines in size and fecundity in common-gardens. Populations collected from northern latitudes generally outperformed those collected from southern latitudes when grown in northern latitude gardens of Washington and Sweden. Conversely, populations from southern latitudes outperformed plants from northern latitudes in southern gardens in Spain and California. In contrast to size and fecundity, introduced plants from western North American showed no relationship between leaf area and latitude of population origin in any common-garden. European populations formed clines in leaf area but these clines did not change between gardens; plants from northern populations had larger leaves than plants from southern populations in both a northern and southern common-garden.

We also analyzed AFLP variation to infer the invasion history into North America. Molecular genetic analyses revealed evidence of multiple introductions of St John's Wort into North America. Introduced plants possess as much molecular genetic variation as do natives, indicating that there has not been a large genetic bottleneck in the introduced range.

The presence of geographic clines when plants are grown in a constant environment provides classic evidence for adaptation to broad-scale climatic conditions across the native range. Clines in North American populations are notable because they suggest rapid postintroduction evolution. The alternative – climate matching – can be ruled out because introduced plants do not always occur at similar latitudes as their most closely related native progenitor (Maron et al. 2004b).

While introduced and native plants do not differ significantly in size or fecundity, they do have genetically based differences in some secondary defensive compounds that are thought to provide resistance to generalist pathogens and herbivores. North American St John's Wort produces consistently lower levels of hypericin and hypericide than do European conspecifics in common-gardens. Furthermore, based on a separate experiment, in Spain, where we experimentally manipulated native pathogen pressure (by treating plants with fungicides), we found that North American populations have higher probabilities of pathogen infection and, if infected, pathogen-induced mortality, compared to European populations (Maron et al. 2004a). These results suggest that North American St John's Wort has lost enemy resistance, perhaps in response to less enemy pressure in the introduced range.

Problem 19.1

We saw in Chapters 6 and 15 that the loss of genetic variation caused by population bottlenecks is harmful. Introduced species generally go through a founding population bottleneck. Why are introduced species often so successful if population bottlenecks are harmful?

Problem 19.2

We saw in Chapters 8 and 16 that local adaptation is an important concern for conservation. Invasive species that have not had the opportunity to develop local adaptations often replace native species that we would expect to be locally adapted. Why?

Problem 19.3

Assume there are two conspecific populations that each has an average heterozygosity (H_S) of 0.10. These two populations are moderately genetically diverged from each other ($F_{ST} = 0.20$). What would be the average heterozygosity in a new population that is founded by a large number of founders from each of these populations? Assume an equal number of founders are derived from each population.

Problem 19.4

A symposium in 1964 brought together a collection of the world's best population and evolutionary biologist to discuss the genetics of colonizing and invasive species (Baker and Stebbins 1965). The following exchange took place between R. C. Lewontin and Ernst Mayr (p. 481):

> **Lewontin:** I would like to be a spokesman for the geneticists and clear up the confusion that I think we've spread about the effect of small numbers in colonizations. If there is colonization by a single fertilized female, there will be a loss of genes and a radical change in gene frequencies at loci where alleles are at intermediate frequencies. But the one thing that will not happen is a profound change in the total amount of genetic variation available.

Mayr: But isn't that based on certain assumptions? Suppose you had a thousand loci each with 25 isoalleles, are you still telling us that you get 75% of that variation in that one single pregnant female?

How do you think Lewontin replied? Who is correct?

Problem 19.5

How much gene flow do you think is needed before two geographically isolated populations should be considered to be the same "unit of eradication"?

Problem 19.6

According to standards set by the International Plant Protection Convention, imports cannot be restricted for species that are already widespread and are not the object of an "official" control program. Do you think that this policy is biologically sound? Why?

20

Forensic and Management Applications of Genetic Identification

Red king crab, Section 20.4.2

Illegal wildlife "trade" represents the world's third largest illegal trafficking after drugs and weapons.

Jean Robert (2000)

This laboratory can track you down years later. We can detect a little bit of blood on your clothing invisible to the naked eye and match it back to that killed animal with absolute statistical certainty.

Ken Goddard (quoted in Pahl (2003))

Genetic identification is the use of molecular analyses to identify the species, individual, or even the population of origin of a sample (e.g., tissue or blood stain). Genetic identification greatly aids law enforcement and wildlife management. For example, molecular identification of an individual (from its multilocus genotype) or a species (from its mtDNA sequence) can provide information to help convict poachers of protected plants and animals. Genetic identification can also help monitor the presence of an endangered species in a nature reserve or national park.

Applications of genetic identification in conservation genetics include the following:

1   Identification of the **species** of origin of tissues or products (e.g., whale meat or tiger bones sold in open markets).
2   Identification of **individuals** or matching of tissue samples (e.g., matching blood stains in a national park to a trophy animal in a taxidermy shop).
3   Determination of the **parents** of individuals (e.g., paternity or maternity of animals claimed to be born in captivity, but possibly taken from the wild).
4   Determination of the **sex** of individuals (e.g., to monitor for illegal harvest of females when only male harvest is allowed).
5   Identification of the **population of origin of a group** of individuals (e.g., a boatload of fish or lobster), or of a **single individual** or tissue (e.g., an individual claimed to be from a legal hunting area, rather than from a protected nature reserve).
6   Estimation of **population composition** of a mixed population (e.g., estimating the per cent contribution of each of several breeding populations to a mixed population).

These applications are discussed in turn below. But first we provide brief background information on the need for, and usefulness of, genetic identification in wildlife forensics and management.

Poaching and trafficking are among the most serious threats to the persistence of many wildlife populations, as suggested by the quotes above. Poaching and the illegal trade of pets and wildlife products threaten taxa ranging from plants (e.g., orchids) to insects (exotic tropical beetles and butterflies), reptiles (snakes, turtles, and lizards), fish (sturgeon for making caviar), birds (parrots and canaries), and mammals (especially trophy-horned ungulates, large carnivores, primates, elephants, rhinos, and cetaceans).

The most important international treaty prohibiting the trade of endangered species is CITES (Convention on International Trade in Endangered Species), established in 1973 in association with the United Nations Environmental Program (UNEP). The main international program for monitoring wildlife trade is TRAFFIC – a network of dozens of staff and researchers across 20 countries jointly sponsored by the World Wide Fund for Nature (WWF) and the World Conservation Union (IUCN). Other organizations that work to control illegal wildlife trade include the international organization WildAid, headquartered in San Francisco, California, and PAW (Partnership for Action against Wildlife Crime) in the United Kingdom. Unfortunately, even with such programs, it is difficult to detect poaching and to enforce treaties and antipoaching laws.

Wildlife genetic identification shares much in common with human forensic genetics (Jobling and Gill 2004). However, wildlife forensics more often involves the identification of species than does human forensics. A wildlife forensics laboratory in Ashland, Oregon (US Fish and Wildlife Service, USFWS) is the only crime lab in the world dedicated entirely to wildlife. The Wildlife Forensics DNA Laboratory at Trent University in Ontario, Canada was the first lab to produce DNA evidence to be used in a North American court, in 1991. It has been involved in over 50 cases a year with convictions and fines ranging from 1,000 to $US50,000. There are more than a dozen labs in North America, Europe, Australia, and other countries, that conduct wildlife forensics testing to help solve wildlife crimes such as illegal trafficking.

In wildlife management, genetic identification (of species or individuals) is the first step in many applications of molecular markers. For example, individualization of samples is required in most noninvasive genetic studies, which use DNA from feces, shed hair, urine, saliva, sloughed skin, or feathers. Once individuals have been identified, we can estimate the abundance of individuals (see Section 14.1), monitor their movements, identify immigrants, and estimate sex ratios (see below).

Both molecular and statistical technologies are rapidly improving; together they provide enormous potential to facilitate wildlife forensics and conservation management. Nonetheless, this potential is not fully exploited. There is much room for further development and application of genetic approaches to help combat poaching and to improve wildlife management. We hope this chapter will encourage agency biologists, academic researchers, students, and funding organizations to further develop and apply genetic identification approaches wherever useful for biodiversity conservation.

20.1 Species identification

Many kinds of molecular markers and polymerase chain reaction (PCR)-based analysis systems are being used to identify species for both forensics (law enforcement) and wildlife management applications. Mitochondrial DNA analysis is the most widely used molecular approach for animal species identification because many species have distinctive mtDNA sequences, and because "universal" primers exist that work among taxa, e.g., among mammals or even all vertebrates. Furthermore, mtDNA is relatively easy to extract from most tissues, including hair, elephant tusks, and old skins, because of its high copy number per cell (see Section 4.1).

Chloroplast DNA markers are commonly used in plants for species identification. Universal primers for noncoding regions are widely used (Taberlet et al. 1991). These primers work in a range of taxa from algae to gymnosperms and angiosperms.

A mtDNA fragment is going to be sequenced for a large proportion of the world's species as part of an ambitious and controversial initiative called "DNA barcoding" (Moritz and Cicero 2004). The goal is to develop a huge data base of DNA sequences from a single gene (e.g., cytochrome c oxidase I, COI) for use in species identification and species discovery. This would facilitate biodiversity inventory, conservation, and the detection of illegal trafficking of wildlife.

Recently, nuclear DNA markers have been used to help identify species. Nuclear markers are especially useful when interspecific hybridization is possible, because mtDNA analysis cannot detect male-mediated gene flow or introgression. A disadvantage of single-copy nuclear DNA (e.g., intron sequences) is the low rate of evolution, compared to mtDNA, making it relatively difficult to find species-diagnostic nucleotide sites (Palumbi and Cipriano 1998). However, nuclear microsatellites do have high rates of evolution, and species-diagnostic alleles have been identified in several species (e.g., sharks (Shivji et al. 2002) and mountain ungulates (Maudet et al. 2004)).

Other nuclear markers include species-diagnostic "fingerprinting" approaches that involve PCR amplification of mammalian-wide interspersed repeat sequences. This offers a quick method for identifying known and unknown species from tiny tissue samples or processed meat samples, for example, using an automatic sequencing machine (Buntjer and Lenstra 1998). Similar "fingerprinting" approaches using SINEs (short interspersed elements) have been developed for salmonids (Spruell et al. 2001; see also Section 4.3). AFLPs have been used as nuclear DNA markers for species identification; for example, in detecting the trafficking of marijuana and endangered plants (Miller Coyle et al. 2003).

Recently, nuclear DNA single nucleotide polymorphisms (SNPs) and real-time PCR (quantitative PCR) have been used to identify species of crab harvested in commercial fisheries (Smith et al. 2005). Advantages of SNPs are that they are typically biallelic and relatively easy to code in a data base (0 or 1), and they can be transferred between laboratories with little error in genotype scoring (unlike some microsatellites, AFLPs, and other markers). Thus SNPs are especially useful for forensic applications where data base errors and genotyping errors would be highly problematic.

20.1.1 Forensic genetics

One of the most widely publicized forensic applications of wildlife DNA analysis was the identification of illegally traded whale meat sold in Japanese and Korean markets (Baker et al. 1996). PCR-based analysis of mtDNA control region sequences revealed that about 50% of the whale meat sampled from markets had originated from protected species and not from the southern minke whale species that Japan is allowed to harvest under their scientific whaling program. For this study, the researchers were not allowed to export the tissue samples from Japan because many species of whale are protected by CITES (which forbids transportation without a permit). Consequently, the researchers set up a portable PCR machine and amplified the mtDNA in a hotel in Japan. They subsequently transported the synthetic DNAs (not regulated by CITES) back to laboratories in the USA and New Zealand for sequence analysis.

A particularly laudable application of genetics in species identification is in the web-based *DNA Surveillance* software program (Ross et al. 2003). *DNA Surveillance* is a computer package that applies phylogenetic methods to the identification of species of whales, dolphins, and porpoises. One advantage of *DNA Surveillance* is it contains a data base of validated, prealigned sequences with wide taxonomic and geographic representation, developed specifically for taxonomic identification (unlike GenBank, with limited sampling and variable data quality). The user typically pastes a mtDNA sequence (e.g., 400–500 base pairs of control region) of unknown origin into a data input window, and then receives back an alignment, genetic distance estimates, and a tree (Example 20.1). This type of service is badly needed for other taxonomic groups.

Other examples of using DNA analysis for detecting illegal wildlife trade involves turtle meat (Example 20.2), and pinniped penises. Pinniped (seals, sea lions, fur seals, and walrus) penises are purchased in traditional Chinese medicine shops in Asia and North America. To investigate the trade of pinniped penises, researchers purchased 21 samples of unknown origin (labeled as pinnipeds) and sequenced 261 base pairs (bp) of the cytochrome *b* gene

Example 20.1 Use of the *DNA Surveillance* website for taxonomic classification of unknown cetacean samples

The web-based *DNA Surveillance* computer program is useful for identifying the cetacean species of origin of a tissue sample of unknown origin. To use the program, you must obtain a mtDNA sequence (control region or cytochrome *b* fragment) from your individual tissue sample, and then simply cut and paste the sequence into the "Data Entry" window of the program.

Figure 20.1a shows a control region sequence of unknown origin (">Unknown 1") pasted into the "Data Entry" window. The sequence is from a meat sample purchased in Japanese markets in 1999 (see *DNA Surveillance* website and Baker et al. 2002b). Assuming it is from a baleen whale (Mysticetes), we simply "click" the circle under "ctrl" and under "Mysticetes", and then click on the "submit" button in the *DNA surveillance* program window (Figure 20.1a). The program then aligns the user-submitted sequence against a set of validated sequences, and outputs (to the computer screen) a cluster dendrogram (Figure 20.1b). If we do not assume the sequence was from a baleen whale, we could click the circle "All Cetaceans", but we would obtain a much larger tree (e.g., including dolphins, etc.).

Data Entry

```
>Unknown1
GAAAATATATATTGTACAATAACCACAAGGCCACAGTATTA
ATGTAACTTGTGCATGTATGTACTCCCACATAACCCATAGTA
TATGTATAATTGTGCATTCAATTATCTTCACTACGGAAGTTAA
AATATTTATTAATAGTACAATAGTACATGTTCTTATGCATCCT
```

Select a database:

Database	ctrl	cyt b
All Cetaceans	○	○
Mysticetes	⊙	○
Odontocetes	○	○
Ziphiidae	○	○
Phocoenidae	○	○
Delphinidae (subgroups)		
Delphininae + Stenoninae	○	○
Globicephalinae + Orcininae	○	○
Lissodelphininae	○	○
Humpback Whale Populations	○	

Genomic regions:
ctrl = mtDNA control Region (= D-Loop);
(a)   cyt b = cytochrome b

Figure 20.1 (a) Data entry and data base window in the *DNA Surveillance* web-based computer program.

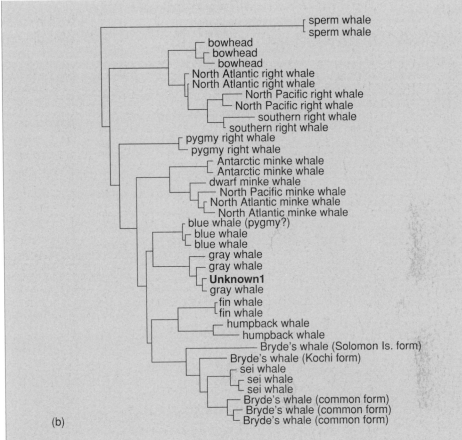

Figure 20.1 (*Cont'd*) (b) Distance phenogram of mtDNA control region sequences from Mysticetes whales.

It turns out that "unknown 1" is actually a gray whale product purchased in Japan. This is presumably a "Korean" western North Pacific gray whale, one of the most endangered populations of whales, unlike the "California" or eastern North Pacific population. This finding was published as a likely infraction of international agreements (Baker et al. 2002b; C. Baker, personal communication).

(Malik et al. 1997). One sample from Bangkok turned out to be from domestic cattle, and six could not be identified because of lack of published reference sequences, although two were most similar to the African wild dog. The remaining samples were from seals. This study suggests that the lucrative market for pinniped penises may be encouraging the unregulated hunting of seals and other unidentified mammalian species. It also illustrates the importance of a large reference data base. Information on the size of the international market suggests that the trade of penises, bacula, and testes is lucrative and growing. For example, Australia exports nearly 5,000 tons of domestic cattle penises to Chinese aphrodisiac markets each year (Malik et al. 1997).

Example 20.2 Identification of illegally traded marine turtle meat

An example of species identification involves mtDNA sequencing of turtle meat from markets in southeastern United States. Although all marine turtles are protected, most species of freshwater turtles in North America (e.g., the alligator snapping turtle) can be legally traded. Biologists have feared that the remaining legal trade in turtle products would act as a cover for illegally harvested species. Roman and Bowen (2000) sequenced segments of the mitochondrial control region and the cytochrome *b* gene to assess the composition of species in commerce. Of 36 purchased putative turtle meat products, eight were from the American alligator, 19 were from the common snapping turtle, three from the Florida softshell, one from the spiny softshell, and one from an alligator snapping turtle (Figure 20.2). This study provides another example of molecular methods showing that animal trade is not entirely legitimate, and mislabeling could potentially threaten rare species. Clearly, genetic monitoring would be helpful to prevent the illegal trade of rare and endangered turtle species.

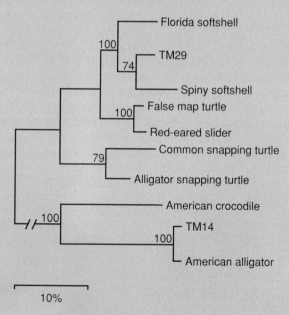

Figure 20.2 Neighbor-joining tree of reference mtDNA (cytochrome *b*) sequences and putative turtle-meat samples (TM14 and TM29). TM14 is clearly not from a turtle, but is from an alligator. TM29 is apparently from a spiny softshell turtle (although only 74% of bootstrap-resample trees group TM29 with the spiny softshell). The 10% bar (lower left) refers to percent sequence divergence.

In a final example, more than 20% of caviar samples purchased from New York markets had mislabeled species names, including some threatened species of sturgeon. PCR-based mtDNA analysis was used for species identification (Birstein et al. 1998). The largest of the sturgeon, the beluga from the Caspian Sea, was recently listed as "threatened" under the ESA. Endangered status could end the importation of beluga caviar into the USA. Caviar from beluga sturgeon has been sold in New York for more than $US75 per ounce.

20.1.2 Wildlife management

Species identification via DNA analysis is increasingly used in wildlife research and management. Sequencing of mtDNA is often used to monitor for the presence of endangered species in wildlife management areas. If an endangered species is detected in an area, that area might be granted protection from logging or development. For example, when the endangered long-footed potoroo (a small kangaroo) was first detected in forests of southeastern Australia, logging was halted in some areas. Porotoos, like many marsupials, are nocturnal, elusive, and their presence is difficult to detect. Thus biologists detected potoroos by using field signs (e.g., diggings), feces, and "hair traps" (consisting of baited plastic tubes with sticky tape around the tube entrance to recover hairs). DNA analysis of hair and feces is necessary for species identification, because related potoroo species occur in sympatry (B. Sherwin, personal communication).

DNA analysis of feces or shed hairs also helps monitor and map the spatial distribution of hybridization in endangered species. For example, microsatellite analysis of fecal (and tissue) DNA from lynx and bobcats has allowed identification of hybrids in Minnesota (Schwartz et al. 2004). Hybridization, previously unknown in the wild, could limit the recovery of endangered lynx populations. Three out of 20 individuals sampled were hybrids (for more details, see Example 17.2). A similar application of fecal DNA analysis for monitoring hybridization is being used to help prevent introgression of coyote genes into the endangered red wolf (Adams et al. 2003).

Species identification from the analysis of fecal DNA is increasingly used to identify bears or canids, for example, for management purposes. Confirming the presence of black versus brown bears (or wolf versus coyotes) from feces often involves mtDNA testing – e.g., sequencing of a species-diagnostic fragment of 100–300 bp. An interesting complication could arise if, for example, a brown bear consumes a black bear and the black bear's DNA (or another species' DNA) shows up in the fecal sample. Or, for example, if one wolf urinates on another wolf's feces, it could make individual identification difficult, because DNA sequences from both wolves might be amplified during PCR.

20.2 Individual identification and probability of identity

Individual identification (DNA "fingerprinting") is one of the most widely used applications of molecular markers in conservation genetics, forensics, and molecular ecology. For example, in the lynx–bobcat hybridization study above, researchers had to individualize fecal samples to know the number of different individuals sampled. As another example, wildlife officers might need to match a tissue sample (gut pile or blood stain) at the scene of a wildlife crime to a trophy animal being transported, e.g., through an airport or highway check point.

To match individual samples, we must first genotype them with highly polymorphic molecular markers (or with many moderately polymorphic markers). We then compute a match probability (or probability of identity, see below) by using allele frequencies estimated from the population of reference, e.g., the national park or the geographic location from which the sample at the "crime" scene originated. If allele frequencies from the reference population are not available, we can still estimate the match probability, but it requires additional markers to achieve a reasonably high power to resolve individuals with high certainty (Menotte-Raymond et al. 1997; Box 20.1).

Box 20.1 Computation of the match probability (*MP*) for an individual sample (genotype) "in hand"

Here we consider a scenario where we have one sample in hand (e.g., a blood stain at a wildlife crime scene) and we want to compute the probability of sampling a different individual that has an identical multilocus genotype (in the same population). This is often called the "match probability".

To compute the *MP*, consider two loci that each has two alleles at the following frequencies: $p_1 = 0.50$, $q_1 = 0.50$; and $p_2 = 0.90$, $q_2 = 0.10$, respectively. A blood stain from the scene of a wildlife crime (e.g., poaching in a national park) has a genotype that is heterozygous at both of these loci. What is the probability that an individual sampled at random from the same population has the same genotype (as the individual whose blood stain is "in hand")?

First we compute each single locus *MP*:

Locus one: $2p_1q_1 = 2\,(0.50 \times 0.50) = 0.50$
Locus two: $2p_2q_2 = 2\,(0.90 \times 0.10) = 0.18$

Then, the multilocus *MP* is the product of the two single locus probabilities: $0.50 \times 0.18 = 0.09$ (assuming independence between loci). We conclude there is a 9% chance of sampling a different individual with a double-heterozygous genotype identical to the one "in hand". Thus there is a 9% chance of matching the blood stain to the wrong individual. Clearly, many more loci (and perhaps more highly polymorphic loci) are needed to have a reasonably low chance (e.g., <1/10,000) of a match to the wrong individual. Recall that here we are assuming unrelated individuals, no substructure, and that the allele frequencies are for the population considered.

What if the wildlife crime occurs in a population with no reference data (i.e., allele frequencies are unknown)? How can we estimate the probability that an individual sampled at random has the same double-heterozygous genotype as the individual "in hand"? Here we could **assume** that the frequency of the observed heterozygous genotype at each locus is high (e.g., 0.50). This is the highest frequency possible (assuming a biallelic locus and Hardy–Weinberg proportions), and gives the least power for individualization. Assuming that the heterozygote genotype frequency is 0.50 is conservative and generally overestimates the true *MP* (Menotte-Raymond et al. 1997). This is especially true if a locus is multiallelic, because the two alleles in a heterozygote "in hand" could never have a population frequency as high as 0.50.

We then could compute the multilocus match probability as follows: $0.50 \times 0.50 = 0.25$. Here, the estimated 25% chance of sampling this multilocus genotype is much greater than the 9% chance estimated (above) by using the reference allele frequencies. This illustrates the power benefit of having reference allele frequencies for the population. Note that if two samples match and are homozygous, we cannot use the locus because we have no evidence the locus is polymorphic (and thus informative) within the population. Thus we need many more loci when we do not know population allele frequencies, in order to achieve a low *MP*.

Microsatellites are the most widely used markers in forensics and genetic management because: (1) their short length (<300 bp) makes them relatively easy to PCR-amplify from partially degraded DNA (unlike AFLP markers, for example, which are longer and more difficult to amplify; see Section 4.3); (2) they are generally highly polymorphic; and (3) alleles from the same locus can be easily identified (unlike some AFLPs and the multi-locus DNA fingerprinting probes first used in human DNA forensic applications; Jeffreys et al. 1985; see Section 4.3). For human forensic investigations in the USA and Britain, a standard set of 13 and 10 microsatellite loci are used, respectively (Watson 2000; Reilly 2001). These marker sets provide a chance of a match (between two random people) that is between about one in a million and one in a billion. For these marker sets, the genotyping of one individual sample costs approximately $US100 (Watson 2000). This is similar to the cost in some wildlife genetics laboratories.

An example of DNA-based individualization for wildlife management is the identification of problem animals. For example, when a wolf or bear attacks humans, kills livestock, or steals a picnic basket, the problem animal will often be removed from the population. Removing the wrong bear could waste resources and eventually result in the removal of several animals before removing the correct one. This could negatively impact on the population, especially if the individuals removed are reproductive females. Furthermore, knowing with certainty that the true problem individual was removed would satisfy some of the public. Thus it is critical to identify the correct individual before removing it. Here, matching DNA from the scene of the "crime" to an individual can help. For an example, see the description of DNA matching in the case of the problem family of grizzly bears from Glacier National Park, Montana (see Guest Box 20).

Another example use of DNA individualization is the identification of logs illegally removed from forest preserves. DNA typing is used to match stumps in forest preserves to logs being sold or transported illegally. DNA matching of stumps to illegally trafficked logs can help stop illegal deforestation. Illegal logging is estimated to cost British Columbia 10–20 million US dollars annually.

Other uses of individual identification in conservation management include **genetic tagging** for studying movements and estimating population census size from mark–recapture methods (see Section 14.1). Individualization also helps identify clonal plants (genets) and animals (corals, anemone, and fishes). The identification of clones is required for accurate estimation of patterns and rates of gene flow, geographic distributions of clones, and inbreeding versus outcrossing rates.

The statistical power of molecular markers to identify all individuals from their multilocus genotype is estimated as the average probability of identity (PI_{av}). PI_{av} is the probability of randomly sampling two individuals that have the same genotype (for the loci being studied). If we use highly polymorphic molecular markers, there is a low probability of two individuals sharing the same genotype at multiple loci. Thus, if we find two samples (e.g., blood stains or tissues) with matching genotypes, we can determine with high probability that they come from the same animal or plant (e.g., found at a crime scene).

PI_{av} is computed using the following expression:

$$PI_{av} = \sum_{i=1}^{n} p_i^4 + \sum_{i>j=1}^{n} (2p_i p_j)^2 \tag{20.1}$$

where p_i and p_j are the frequencies of ith and jth allele at the locus (Waits et al. 2001; Ayres and Overall 2004). Here, p_i^4 is simply the average probability of randomly sampling two homozygotes (e.g., aa), and $(2p_i p_j)^2$ is the average probability of randomly sampling two heterozygotes (e.g., Aa). This equation assumes Hardy–Weinberg proportions and that no substructure exists in the population. The multilocus PI_{av} is computed by using the product rule (i.e., multiplication rule, see Appendix Section A1), and multiplying together the single locus probabilities (see Box 20.1). A reasonably low multilocus PI_{av} for forensics applications (e.g., matching blood from a wildlife crime scene to blood on a suspect's clothes) is approximately $1/10,000$ to $1/100,000$. Achieving this low a PI_{av} would require approximately 5–20 markers, depending on their polymorphism level (Figure 20.3).

PI_{av} is also often used to quantify the power of molecular markers for studies involving genetic tagging. A reasonably low PI_{av} for genetic tagging is approximately $1/100$ (Waits et al. 2001). This is not as low as for forensics, because it is less problematic to misidentify individuals in genetic tagging than in a law enforcement case (where someone might be fined or imprisoned). To achieve a reasonably low PI_{av} for genetic tagging, approximately 4–8 highly polymorphic markers are often sufficient (Figure 20.3). For less polymorphic markers such as allozymes, SNPs, and AFLPs (with heterozygosity typically from 0.20 to

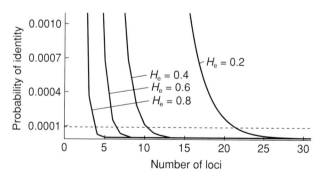

Figure 20.3 Relationship between probability of identity (PI_{av}) and the number of loci (for each of four heterozygosities). PI_{av} was computed using expression 20.1 and allele frequencies that result in a heterozygosity (H_e) of 0.20, 0.40, 0.60 or 0.80, in the following equation: $H_e = 1 - \sum p_i^2$ (where p_i is the frequency of the ith allele). For example, for $H_e = 0.2$, two alleles must have the frequency of 0.885 and 0.115. From Waits et al. (2001).

0.40), more markers would be required. Power is lower for dominant markers like AFLPs than for codominant markers like microsatellites and SNPs (see Chapter 4). Interestingly, the power of a set of markers is better predicted by heterozygosity than allelic richness; loci with the same heterozygosity but 10 versus three alleles will have nearly the same power to resolve individuals (Waits et al. 2001).

It is important to note that expression 20.1 used to estimate PI_{av} assumes that individuals are unrelated (e.g., no siblings), sampled randomly, and that no substructure or gametic disequilibrium exists. These assumptions are often violated in natural populations. The violation of assumptions could cause an underestimation of the true PI_{av}. For example, in data sets from wolves and bears, PI_{av} was underestimated by up to three orders of magnitude (e.g., 1/100,000, which underestimates the true value of 1/100; Waits et al. 2001). To avoid problems with underestimation, other PI_{av}-related statistics such as PI_{av}-sibs, should also be used to compute the probability of identity. Other PI_{av} statistics (e.g., accounting for potential substructure) can be computed with the user-friendly software API-CALC (Ayres and Overall 2004).

The match probability (MP) is a useful statistic related to PI_{av} (i.e., the average probability of identity, expression 20.1). While PI_{av} is the **average** probability of randomly sampling **two** individuals consecutively that have the same genotype, the MP is the **actual** probability of sampling **one** individual identical to the one already "in hand" (i.e., sampled previously). PI_{av} is for computing the average power of a set of markers (considering all genotypes, homozygotes, and heterozygotes in a given study), whereas MP gives a probability of sampling the individual genotype in question, that was sampled previously (see Box 20.1). MP requires the same assumptions (no substructure, no gametic disequilibrium, and no siblings) as does PI_{av}, although more sophisticated MP statistics exist that do not require all these assumptions. It is confusing that sometimes the "average probability of identity" (PI_{av}) is referred to as the "average MP" in the literature.

We have thus far considered only nuclear DNA markers for determining match probabilities and the average probability of identity. However, mtDNA can also be useful for individual identification. For example, the mtDNA control region in canids and felids has tandemly repeated sequences (Fridez et al. 1999; Savolainen et al. 2000). These repeats are highly polymorphic and heteroplasmic (i.e., multiple clones with different repeat lengths are found within an individual). Thus, mtDNA analysis will occasionally be useful for individual differentiation because different individuals often have different mtDNA repeat profiles. But since mtDNA represents only one "locus", it will provide much less certainty than multilocus nuclear DNA methods. An advantage of mtDNA is that it can be amplified from hair shafts, whereas nuclear DNA is only found in the hair root bulb (Watson 2000). Animal hairs are often found at poaching crime scenes and on people's clothing.

20.3 Parentage testing

Parentage analysis is the determination of the mother (maternity) or father (paternity) of an individual. Parentage testing can help wildlife conservation by verifying that an individual originates from captive parents, as might be claimed by some pet trade industry workers. An enormous problem for wildlife conservation is the illegal capture of individuals from the wild, which occasionally involves killing of the wild parents (e.g., gorillas and orangutans). DNA typing captive individuals to analyze potential parents could help detect the illegal capture of individuals from the wild, and thereby reduce threats to wild populations. One example of using parentage analysis in wildlife management is given in the example of the "problem bear family" (see Guest Box 20).

Another example comes from Australia where the owners of an adult female northern hairy-nosed wombat claimed that their juvenile wombat was the offspring of their adult female. The owners had a legal permit for the adult, but not for the young wombat. The owners claimed that their female must have been impregnated by a wild wombat somewhere near their backyard. Wildlife law enforcement officials questioned this story and conducted maternity testing in a genetics laboratory. The laboratory typed nine microsatellite loci on the mother and offspring and found no incompatibilities, i.e., the mother had an allele at each locus compatible with the offspring (A. Taylor, personal communication). Thus there was no evidence that the offspring was taken from the wild, and the owners were allowed to keep the young wombat. The statistical certainty of a maternity (or paternity) assignment can be computed, based on allele frequencies at the loci (e.g., Slate et al. 2000).

Other applications of parentage analysis include understanding a species' mating system, estimating variance in reproductive success, and detecting multiple paternities. Such information is helpful for population management. Variance in reproductive success influences the effective population size and thus the rate of loss of genetic variation, inbreeding, and efficiency of selection. Knowing that variance in reproductive success is high, for example, can help biologists predict that the N_e is much smaller than the census population size (see Section 7.10),

Parentage analysis (e.g., paternity exclusion or assignment) requires more molecular markers than does individual identification (often twice as many). Low polymorphism SNP markers or allozymes are often not highly useful (Morin et al. 2004). To quantify the statistical power of molecular markers for parentage analysis (e.g., paternity exclusion), researchers often compute the expected paternity exclusion probability (*PE*) or the probability of excluding a randomly chosen nonfather (e.g., Double et al. 1997). As an alternative to paternity exclusion, paternity assignment (using probabilities and likelihood computation) is widely used to estimate the probability that a given male is the father (Slate et al. 2000; and see the CERVUS and PARENTE software, listed on this book's website).

The power of a set of molecular markers for paternity exclusion (*PE*) is often quantified by simply plugging allele frequencies into the following expression (Jamieson and Taylor 1997):

$$PE = \sum_{i=1}^{n} p_i^2(1 - p_i)^2 + \sum_{i>j=1}^{n} 2p_i p_j(1 - p_i - p_j)^2 \tag{20.2}$$

where p_i and p_j are the frequency of the *i*th and *j*th allele, respectively. For one locus, this expression gives the average probability of excluding (as the father) a randomly sampled nonfather, when the mother and offspring genotypes are both known. To compute the multilocus *PE*, we multiply together the *PE* for each locus, assuming independence among loci. Often 10–15 highly variable markers (heterozygosity 0.50–0.60) are required to achieve a high probability of paternity exclusion (99.9–99.99%). Many more markers are required if the polymorphism is low ($H_e = 0.20$ to 0.40) or if the mother is not known

(Morin et al. 2004). Other expressions are available for estimating power for parent exclusion when neither parental genotype is known (Jamieson and Taylor 1997).

20.4 Sex identification

Assessment of an individual's sex can be difficult in species with little sexual dimorphism or with internal gonads, as in birds. Nonetheless, information on sex and sex ratios is often critical for behavioral studies, wildlife management, and captive breeding programs, and for estimating N_e (see Section 7.2). Fortunately, molecular sexing is possible and becoming widely used in forensics, conservation, and wildlife management. Sex identification and sex ratio estimation is even possible using noninvasive samples (e.g., feces, urine, shed hair, or feathers), which can greatly facilitate wildlife studies. Molecular sexing principles and techniques are discussed in Section 4.4. Here we give brief examples of the use of molecular sexing in wildlife forensics and management.

In wildlife forensics, DNA-based sex identification is useful to detect illegal harvesting of one sex. For example, in leopards from Tanzania, researchers used molecular sexing methods to detect violations of hunting regulations that prohibit harvesting females. The researchers tested for X-specific and Y-specific DNA sequences (ZFX and ZFY) in 77 skins from animals shot between 1995 and 1998. Despite tags indicating that all the skins were from males, 29% were actually from females (Spong et al. 2000).

In wildlife management, sexing is used to monitor individual movements and home range size, which often is different for different sexes. For example, Taberlet et al. (1997) used noninvasive sampling of feces and shed hairs to assess the movements of individual male and female brown bears in the Pyrenees Mountains of western France. Y chromosome markers (and microsatellites) were used to identify the sex (and individual) of origin of the feces and hairs collected over a wide geographic area. Individual males were found to move over a much larger geographic area than females, consistent with other bear research.

20.5 Population assignment

Genetic markers can help identify the population of origin (i.e., birth) of individuals or groups of individuals. Determining the population or geographic region of origin of wildlife products can help identify populations threatened by poaching, and the trade routes used by traffickers. Such information could help law enforcement officials target poaching. Recently developed assignment tests, based on multilocus genotypes, can determine the population of origin of individuals (Waser and Strobeck 1998; Manel et al. 2002). Assignment tests work by assigning an individual (or tissue) to the population in which its genotype has the highest expected frequency (see Section 9.8; and Manel et al. 2005).

20.5.1 Assignment of individuals

An excellent example of assigning individuals for wildlife forensics comes from a fishing competition in Scandinavia. A fisherman claimed to have caught a large salmon in Lake Saimaa from Finland. However, the organizers of the competition questioned the origin of the salmon because of its unusually large size (5.5 kg). A genetic analysis of seven microsatellite loci was conducted on the large fish and on 42 fish from the tournament lake, Saimaa. A statistical analysis was conducted using the exclusion–simulation assignment test (Primmer et al. 2000). The exclusion test suggested that the probability of finding the large fish's genotype in Lake Saimaa was less than one in 10,000. Thus the competition organizers excluded Lake Saimaa as the origin of the salmon. Subsequently, the fisherman confessed to having purchased the fish in a bait shop.

Population assignment, in the fishing tournament example, is based on the exclusion principle and computer simulations to assess statistical confidence. In the exclusion–simulation approach, we "assign" an individual to one population only if all other populations can be excluded with high certainty (e.g., $P < 0.001$). We exclude a population if the genotype (in question) is unlikely to occur in the population ($P < 0.001$), i.e., if the genotype is observed in less that one per 1,000 randomly simulated genotypes (assuming Hardy–Weinberg proportions and gametic equilibrium; see Section 9.8).

An advantage of the exclusion–simulation approach is that it is feasible when only one population (e.g., the suspected source population) has been sampled. Further, it does not require the assumption that the true population of origin has been sampled. Other approaches, e.g., Bayesian (Pritchard et al. 2000) and likelihood ratio tests (Banks and Eichert 2000), generally require samples from at least two populations, and assume that the true population has been sampled. If the true population of origin has not been sampled, the assignment probabilities from Bayesian and likelihood ratio approaches could be misleading. It seems prudent to apply both the exclusion-based and Bayesian assignment approaches (Manel et al. 2002).

An advantage of the Bayesian approach is that it is generally more powerful than exclusion-based methods and other methods (Manel et al. 2002; Maudet et al. 2002). User-friendly software exists for all three (and other) approaches (Rannala and Mountain 1997; Banks and Eichert 2000; Pritchard et al. 2000; Piry et al. 2004).

20.5.2 Assignment of groups

Assignment of groups of individuals to a population or region of origin is also increasingly feasible. For example, the Alaska Department of Fish and Game confiscated a boatload of red king crab that they suspected was caught in an area closed to harvest near Bristol Bay in the Bering Sea (Seeb et al. 1989). The captain of the skipper claimed that the crabs were caught near Adak Island in the Aleutian Islands, over 1,500 km away from the closed area.

Thirteen populations of king crab from Alaskan waters had been previously examined at 42 allozyme loci (14 polymorphic loci) to describe the genetic population structure of red king crab (Seeb et al. 1989). Genetic data at eight loci indicated that the confiscated crabs could not have been caught near Adak Island (in the Aleutian Islands), the only area open to harvest. Allele frequencies at Adak Island significantly differed from the allele frequencies among the confiscated crab, as inferred using a chi-square test (e.g., $X^2 = 21.6$, $P < 0.001$, and $X^2 = 88.5$, $P < 0.001$, for the *Pghd* and *Alp* loci, respectively). Discriminate function analysis was a second statistical approach used to conclude that the confiscated crabs did not originate from the Adak Island area (see Seeb et al. 1989 for details). The allele frequencies in the confiscated sample matched the samples from farther north in the Bering Sea (Figure 20.4). Based upon these results, the vessel owner and captain agreed to pay the State of Alaska a $US565,000 penalty for fishing violations.

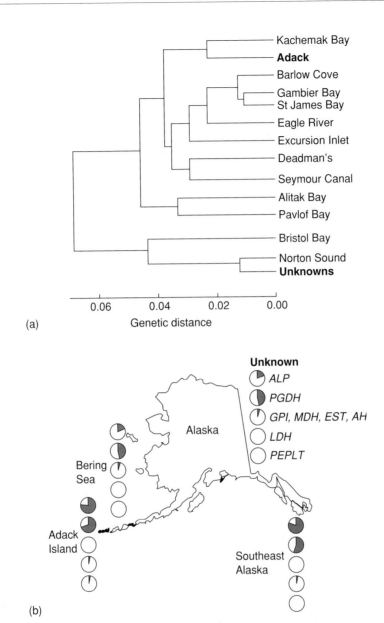

Figure 20.4 (a) Tree estimating relationships among king crab populations using allele frequency similarity (modified from Seeb et al. 1989). Note the "unknown" confiscated crabs do not cluster with the Adack population where the crab harvest was permitted (in the Aleutian Islands, see text). Allele frequency similarity is based on the Cavalli-Sforza and Edwards chord distance (Cavalli-Sforza and Edwards 1967). Future methods should include some measure of statistical confidence in clustering (e.g., bootstrap percentages). (b) Map of locations of populations (Adack and Bering Sea) in Alaska, and five allele frequency pie charts for five allozyme loci.

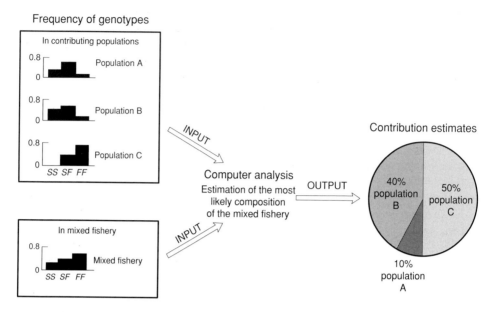

Figure 20.5 Outline of the procedure for estimating the population (stock) composition on the basis of genetic data at a single locus with two alleles (*S* and *F*). The mixture is composed of three populations. Actual application of this procedure requires multiple loci. From Milner et al. (1985).

Assigning a group of individuals can be easier (i.e., yield a higher statistical certainty) than assigning a single individual, because more information is available in a group of genotypes than in a single individual's genotype. A user-friendly software program for assigning groups (as well as individuals) to a population of origin is available in *GeneClass 2.0* (Piry et al. 2004).

20.6 Population composition analysis

Many species are harvested in mixed populations, such as mixed stock fisheries (salmon, marine mammals, and others) and waterfowl. Other species migrate in mixed groups (neotropical song birds, butterflies, and others). Effective management of mixed stock fisheries and mixed populations requires that the populations or stocks that compose the mixture be identified and the extent of their contribution determined (Larkin 1981; Pella and Milner 1987) (Figure 20.5). Stocks are generally analogous to management units (e.g., demographically independent populations) that were discussed in Section 16.5.

The fundamental unit of replacement or recruitment for anadromous salmon is the local breeding population because of homing (Rich 1939; Ricker 1972). That is, an adequate number of individuals for each local reproductive population are needed to insure persistence of the many reproductive units that make up a fished stock of salmon. The homing of salmon to their natal streams produces a branching system of local reproductive populations that are demographically and genetically isolated. The demographic dynamics of a fish population are determined by the balance between reproductive potential (i.e., biological and physical limits to production) and losses due to natural death and fishing. "Population persistence requires replacement in numbers by the recruitment process" (Sissenwine 1984), so fishery scientists have focused on setting fishing intensity so that adequate numbers of individuals "escape" fishing to provide sufficient recruitment to replace losses.

The distinction between a local breeding population and a fished (harvested) stock is critical (Beverton et al. 1984). A local breeding population has a specific meaning: a local population in which mating occurs. A stock is essentially arbitrary and can refer to any recognizable group of population units that are fished (Larkin 1972). The literature has often been unclear on this distinction. In practice, it is extremely difficult to regulate losses to fishing on the basis of individual local breeding populations. Thousands of local breeding populations make up the US west coast salmon fishery, and many of these are likely to be intermingled in any particular catch. Nevertheless, the result of regulating fishing on a stock basis and ignoring the reproductive units that together constitute a stock is the disappearance or extirpation of some of the local breeding populations (Clark 1984).

The loss of local populations could lead to the crash or eventual extirpation of the entire metapopulation or species, with negative consequences for the larger ecosystem and regional economy (commercial fisheries, sports fisheries or hunting, ecotourism). The importance of maintaining numerous diverse local populations for insuring long-term metapopulation viability is well illustrated in the study of salmon by Hilborn et al. (2003) (see Example 15.2).

An important application of mixed population analysis in conservation management is illustrated by a study of sockeye salmon in Alaska. Seeb et al. (2000) genotyped 27 allozyme loci in all major spawning populations from the upper Cook Inlet and found substantial among-population differentiation (e.g., $F_{ST} = 0.075$ among nursery lakes). The salmon from these major populations are harvested in a mixed stock aggregation that forms in the upper Cook Inlet (Figure 20.6). A mixed stock analysis (based on maximum likelihood, see Manel et al. 2005) allowed estimation of the proportion of genes (and thus individuals) from each population in the pool of harvested fish. Impressively, the genotyping and statistical analysis can be conducted within 48 hours after harvest! This is critical because it allows real-time monitoring of harvest from each major population. It allows biologists to close the harvest if too many fish are harvested from any one major breeding population. This is critical to help prevent overfishing, longer term closures of fishing, and the extinction of a major source population.

Population composition analysis differs from individual-based assignment tests in that composition analysis estimates the percentage of the gene pool (or alleles) that originates from each local breeding population. Whereas individual-based assignment methods estimate the actual number and identity of individuals originating from each breeding population (Manel et al. 2005). The Bayesian method programmed in the STRUCTURE computer program (Pritchard et al. 2000) even computes the percentage of an individual's genome that originates from different breeding populations. The relative performance of the different assignment and composition analysis approaches depends on the question. More studies are needed to evaluate the relative performance of these analysis methods under different scenarios relevant to management and conservation.

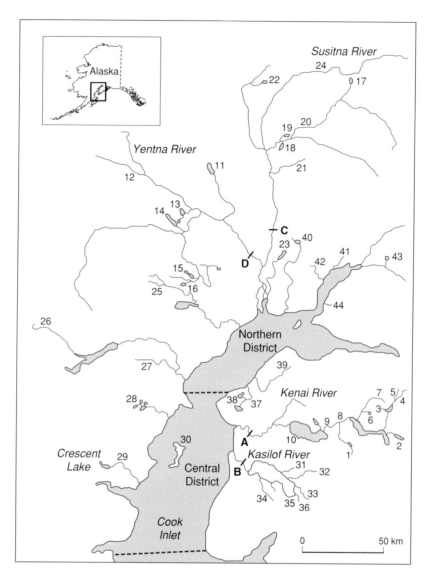

Figure 20.6 Locations (numbers) of major populations of sockeye salmon sampled for genetic monitoring of harvest. The mixed stock fish harvest goes on in the upper Cook Inlet, and the Central and Northern Districts. From Seeb et al. (2000).

Guest Box 20 Microsatellite DNA genotyping identifies problem bear and cubs
Lisette P. Waits

In 1998, a hiker was killed and partially consumed by a bear in Glacier National Park, Montana. Following park policy, managers needed to find and remove this bear. Park biologists noted that the killing occurred within the home range of a radio-collared grizzly bear female (235; Table 20.1) and hypothesized that she might have killed the hiker. They also were aware that this female had two 2-year-old cubs that may have been with her at the time of attack. Since grizzly bears are protected under the ESA, park biologists did not want to kill this grizzly and her cubs unless they could obtain conclusive evidence that she had killed and consumed the hiker.

To address these questions, park biologists turned to DNA analysis of hair and fecal samples. Hair samples were taken from female 235 and one (238) of the cubs and sent to our repository at the Laboratory for Ecological and Conservation Genetics at the University of Idaho. Park biologists then collected two bear hair and 11 fecal samples from the kill site. My laboratory was asked to evaluate: (1) did the suspect bear's genotype match that from the bear hair or feces found at the attack site; and (2) were there two cubs present at the kill site?

After extracting DNA from the hair and fecal samples in a room dedicated to low quantity/quality DNA samples, we performed PCR of a section of the mtDNA control region that is ~148 bp in grizzly bears and 160 bp in black bears (Murphy et al. 2000) and verified that all samples from the kill site originated from a grizzly bear. For individual identification, we attempted to generate data from five highly variable microsatellite loci (*G1A*, *G10B*, *G10C*, *G1D*, and *G10L*; Paetkau et al. 1995) using the multiple tubes approach of Taberlet et al. (1996). Success rates for individual identity genotyping were low overall (31%), but we were able to obtain reliable partial genotypes for one hair (H-14) and three fecal samples (S-37, S-34b,

Table 20.1 Genotypes at five microsatellite loci for the female (suspected mother), and one cub caught with the mother and another cub that was the suspected missing cub. Bold alleles are those that do not match either allele in the female.

	Locus				
Sample	**G1A**	**G10B**	**G10C**	**G1D**	**G10L**
Female 235	189/189	155/155	102/110	171/175	153/155
Cub 238	189/**193**	155/**159**	**104**/110	175/**180**	153/155
Unknown cub	189/**193**	155/**159**	**104**/110	171/**180**	155/155
H-14	NS	155/**159**	**104**/110	175/**180**	153/155
S-37	189/189	155/155	102/110	NS	NS
S-34b	189/189	155/155	102/110	NS	153/155
S-3	NS	155/155	NS	171/175	153/155

NS, not scored.

and S-3; Table 20.1). Sample H-14 matched the genotype of cub 238 and samples S-37, S-34b, S-3 matched the female 235. The probability of identity for the 3–5 loci that successfully amplified in different samples ranged from approximately one in 2,000 to one in 40,000 (see expression 20.1).

While it would be better to collect data at more loci, the probability of identity values are convincing considering the fact that there are only 400–700 bears in this ecosystem. Thus, the park biologists attempted to capture all three bears. Female 235 and cub 238 were captured quickly and removed from the park. But, the second cub was not captured. A few weeks later a 2-year-old grizzly charged a group of people and was killed by park biologists in the general home range area of the female and cubs. Park biologists believed this was the missing cub, so we analyzed this bear to evaluate this hypothesis. His genotype ("unknown cub") is shown in Table 20.1. This cub shares one allele with female 235 at all loci consistent with the hypothesis that he is her offspring. Also, relatedness statistics (Queller and Goodnight 1989) reveal a pairwise relatedness of 0.70 between the mother and cub providing strong support that this bear was the missing cub.

Problem 20.1

Consider a wildlife forensics case where a suspected poacher possesses a trophy mountain sheep he claims to have legally harvested from a hunting area. Law enforcement officials suspect that the horns match those from a missing mountain sheep often seen by tourists in the national park several hundred miles away. How could molecular (or other) methods help resolve this wildlife forensics case? What molecular markers and statistical methods could be useful? What samples are required? List four factors that will influence the power of molecular marker-based approaches to help resolve such a forensics case?

Problem 20.2

The sequence below is from mtDNA (control region, "Unknown5") of whale meat purchased in a Japanese market in 2002 (from the Surveillance website).

(a) Identify the species of origin using the "simple" search option on the web-based tool *Surveillance*, by cutting and pasting "Unknown5" into the search window. (The *Surveillance* website is available via a Google search or from this book's home page.) From what species does this sequence originate?

(b) Conduct an "Advanced" search with 100 bootstrap replicates. Remove the first two lines of sequence (106 bp) and reconduct the "Advanced" search with 100 bootstrap replicates. What is the bootstrap value with and without the 106 bp? What is the effect of using shorter sequences for species identification?

(c) Can you identify the species of origin of this sequence by conducting a "BLAST" search on GenBank?

```
>Unknown5
ATATTGTACAATAACCACAAGACTACAGTACTATGTCCGTATTAAAAATAATT
TATCTTATTACATACTATTATGTACTCGTGCATGTATGCATGTCCACATAACC
AATAATAACGATGTCCTCCTGTAAATATGTATATGTATACATACTATGTATAA
TTGTGCATTCAATTATCTTCACCACGAGCAGTTGAAGCCCGTATTAAAATTCA
TTAATTTTACATATTACATAATATTCATTGATAGTACATTAGCGCATGTTCTTA
TGCATCCACAAGTTAATTTAGTAAAACTAATTCTTATGGCCGCTCCATTAGAT
CACGAGCTTAGTCAGCATGCCGCGTGAAACCAGCAACCCGCTCGGCAGGGAT
CCCTCTTCTCGCACCGGGCCCATCAATCGTGGGGGTAGCTATTTAATGATCTT
TATAAGACATCTGGTTCTTACTTCAGGACCATATTAACTTAAAATC
GCCCACTCGTTCCCCTTAAAT
```

Problem 20.3

Imagine that customs officials confiscate an illegal shipment of suspected tiger bones at one of the London airports (England is a signatory of CITES). Microsatellite DNA analysis revealed the following two-locus genotype for one confiscated bone: *AaBB*. Compute the expected two-locus genotype frequency for this individual in the candidate source population in India versus China. Allele frequencies in India are as follows: frequency of $A = 0.40$, frequency of $B = 0.90$; frequencies in China are: $A = 0.60$, frequency of $B = 0.40$. (See Box 20.1.)

Problem 20.4

What might be some limitations for using assignment tests in natural populations for forensics investigations? (Hint: assumptions.) How might you quantify the possible consequences of violating the assumption on the reliability of the tests (see Appendix Section A3.2)?

Problem 20.5

What are some practical considerations for sampling populations and the storage of samples if DNA evidence is to be used in a court of law? How might researchers insure that their genotyping data (or DNA sequence data) is reliable?

Problem 20.6

In the Yellowstone National Park population of grizzly bears there are approximately 300–500 individuals. Given this number, what would be a reasonably small probability of identity (*PI*) for use in forensic cases in this population? Would a *PI* of 1/300 or 1/500 be sufficiently low? Why or why not?

Problem 20.7

Chimpanzee populations in Africa are rapidly declining due to deforestation, mining, and hunting for food or trade on the international pet market. It has been estimated that in international trade, 10 infant chimpanzees die for every one that survives to its final destination (Goldberg 1997). Chimpanzees (and other primates) have often been transported through tortuous routes only to be confiscated by customs officials and subsequently kept at overcrowded zoos or captive facilities. How might DNA typing for individual identification and population assignment help to: (a) reduce the illegal trade of chimpanzees; and (b) reintroduce confiscated chimpanzees back into the wild?

Problem 20.8

Imagine that a waterfowl species is hunted only around one group of lakes in the middle of its migratory route (central United States) but that migrants come from numerous breeding grounds in Canada. What are the special risks of harvesting from such a mixed population compared to nonmixed populations? How and what molecular genetic analysis methods might help to monitor harvest (numbers of individuals from different breeding populations) in such mixed populations? Can you list some potential advantages of molecular genetic versus traditional demographic approaches (banding of birds) for managing the harvest of mixed populations?

Problem 20.9

Consider a case where we need to determine paternity of offspring, when the mother is known. Compute the paternity exclusion power of four loci. Use expression 20.2 and the following biallelic allele frequencies for each of the four loci ($P1 = 0.50$, $P2 = 0.45$, $P3 = 0.35$, $P4 = 0.30$). How could we increase the power for paternity exclusion?

Problem 20.10

It is important to know if the wolf population in Yellowstone National Park, like any island population, receives migrants. How might noninvasive sampling of feces help detect immigrants? What statistical methods might be helpful? How might you detect "effective" gene flow (reproduction) from an outside population?

Problem 20.11

It would be useful to detect immigration of wolves into Yellowstone from Montana so that biologists would know if the Yellowstone population is isolated. Small isolated populations have increased risks of extinction due to demographic and genetic factors. Multilocus genotypes and assignment tests can be used to iden-tify immigrants into Yellowstone. This requires genetic data (allele frequencies) from Yellowstone and candidate source populations of immigrants.

Go to this book's web page and download the microsatellite data set for wolves from Yellowstone National Park and Northwestern Montana. Use *GeneClass 2.0* software (see this book's web site) to determine the percentage of individuals that can be assigned back to their population of origin. Is this percentage sufficient to have >90% probability of detecting an immigrant into Yellowstone from northwest-ern Montana? How might power (the percentage of individuals correctly assigned) be increased?

Glossary

ABC   *See* **approximate Bayesian computation**.

acrocentric   Chromosomes and chromatids with a centromere near one end.

addition rule   *See* **sum rule**.

additive genetic variation   The portion of total genetic variation that is the average effect of substituting one allele, responsible for a phenotypic trait, for another. The proportion of genetic variation that responds to natural selection.

admixture   The formation of novel genetic combinations through hybridization of genetically distinct groups.

AFLP   *See* **amplified fragment length polymorphism**.

agamospermy   The asexual formation of seeds without fertilization in which mitotic division is sometimes stimulated by male gametes.

alien species   A non-native or nonindigenous species.

allele   Alternative form of a gene.

allelic diversity   A measure of genetic diversity based on the average number of alleles per locus present in a population.

allelic richness   A measure of the number of alleles per locus; allows comparison between samples of different sizes by using various statistical techniques (e.g., rarefaction).

allopatric   Species or populations that occur in geographically separate areas.

allopolyploid   A polyploid originating through the addition of unlike chromosome sets, often in conjunction with hybridization between two species.

allozygous   An individual whose alleles at a locus are descended from different ancestral alleles in the base population. Allozygotes may be either homozygous or heterozygous in state at this locus.

allozyme   An allelic enzyme detected through protein electrophoresis used in many genetic applications such as hybrid identification and estimation of genetic variation.

AMOVA   *See* **analysis of molecular variation**.

amplified fragment length polymorphism (AFLP)   A technique that uses PCR to amplify genomic DNA, cleaved by restriction enzymes, in order to generate DNA fingerprints; it is a combination of RFLP and arbitrary primer PCR. It does not require prior sequence knowledge.

amplify   To use PCR to make many copies of a segment of DNA.

anagenesis Evolutionary changes that occur within a single lineage through time. *See* **cladogenesis**.

analysis of molecular variation (AMOVA) A statistical approach to partition the total genetic variation in a species into components within and among populations or groups at different levels of hierarchical subdivision. Analogous to ANOVA in statistics.

aneuploid A chromosomal condition resulting from either an excess or deficit of a chromosome or chromosomes so that the chromosome number is not an exact multiple of the typical haploid set in the species.

anneal The joining of single strands of DNA because of the pairing of complementary bases. In PCR, primers anneal to complementary target DNA sequences during cooling of the DNA (after DNA is made single stranded by heating).

ANOVA Analysis of variance.

apomixis Seed development without fertilization and meiosis. An apomict or apomictic plant produces seeds that are genetically identical to the parent plant.

approximate Bayesian computation (ABC) A statistical framework using simulation modeling to approximate the Bayesian posterior distribution of parameters of interest (e.g., N_e, Nm) often by using multiple summary statistics (H_e, number of alleles, F_{ST}). It is far faster computationally than fully Bayesian approaches but generally slightly less accurate and precise.

artificial selection Anthropogenic selection of phenotypes, with a heritable genetic basis, to elicit a desired phenotypic change in succeeding generations.

ascertainment bias Selection of loci for marker development (e.g., SNPs or microsatellites) from an unrepresentative sample of individuals, or using a particular method, which yields loci that are not representative of the spectrum of allele frequencies in a population. For example, the choice of loci with high heterozygosity may bias assessments of allele frequency distributions in future studies using the loci such that alleles at low frequency (rare alleles) are underrepresented.

assignment tests A statistical method using multilocus genotypes to assign individuals to the population from which they most likely originated (i.e., in which their expected multilocus genotype frequency is highest).

associative overdominance An increase in fitness of heterozygotes at a neutral locus because it is in gametic disequilibrium at a locus that is under selection. Also known as **pseudo-overdominance**. Compare with **hitchhiking**.

assortative mating Preferential mating between individuals with a similar (or a different) phenotype is referred to positive (or negative) assortative mating. *See also* **disassortative mating**.

autogamy Self-fertilization in a hermaphroditic species where the two gametes fused in fertilization come from the same individual.

autosomal A locus that is located on an autosome (i.e., not on a sex chromosome).

autosomes Chromosomes that do not differ between sexes.

autozygosity A measure of the expected homozygosity where alleles are identical by descent.

autozygous Individuals whose alleles at a locus are identical by descent from the same ancestral allele.

B chromosome *See* **Supernumerary chromosome**.

balancing selection Diversifying selection that maintains polymorphism resulting from such mechanisms as frequency-dependent selection, spatially heterogeneous selection, or heterozygous advantage.

Barr bodies Inactivated X chromosomes in female mammals that condense to form a darkly colored structure in the nuclei of somatic cells.

Bayesian inference A procedure of statistical inference in which observed data are interpreted not as frequencies or proportions, but rather are used to compute the probability that a hypothesis is true, given what was observed. Bayesian inference also allows for the incorporation of prior data or information. Bayes' theorem is named after the Reverend Thomas Bayes.

binomial proportion   A population will be in binomial proportions when it conforms to the binomial distribution so that the occurrence of a given event X, r_i times with a probability (p_i) of success, in a population of n total events, is not significantly different than that which would be expected based on random chance alone.

biogeography   The study of the geographic distribution of species and the principles and factors influencing these distributions.

biological species concept (BSC)   Groups of naturally occurring interbreeding populations that are reproductively isolated from other such groups or species.

BLAST   *Basic Local Alignment Search Tool.* Software program to search a DNA sequence data base for sequence similar to the one in hand.

Bonferroni correction   A correction used when several statistical tests are being performed simultaneously (since while a given α-value may be appropriate for each individual comparison, it is not for the set of all comparisons). In order to avoid a lot of spurious positives, the α-value needs to be adjusted to account for the number of comparisons being performed. Suppose we are testing for Hardy–Weinberg proportions at 20 loci. Instead of using the traditional 0.05 α-level, we would test at α of $0.05/20 = 0.0025$ level. This insures that the overall chance of making a Type I error is still less than 0.05.

bootstrap analysis   A nonparametric statistical analysis for computing confidence intervals for a phylogeny or a point estimate (e.g., of F_{ST}). Re-sampling with replacement to estimate the proportion of times an event (such as the positioning of a node on a phylogenetic tree) appears during multiple re-sampling of a data set.

bottleneck   A special case of strong genetic drift where a population experiences a loss of genetic variation by temporarily going through a marked reduction in effective population size. In demography, a severe transient reduction in population size.

branch length   Length of branches on a phylogenetic tree. Often proportional to the amount of genetic divergence between species or groups.

broad sense heritability (H_B)   The proportion of phenotypic variation within a population that is due to genetic differences among individuals.

BSC   *See* **biological species concept**.

cDNA   Complementary DNA.

census population size   The number of individuals in a population.

centromere   An constricted region of a chromosome containing spindle microtubules responsible for chromosomal movement during mitosis and meiosis.

chi-square test   A test of statistical significance based on the chi-squared statistic, which determines how closely experimental observed values fit theoretical expected values.

chloroplast DNA (cpDNA)   A circular DNA molecule located in chloroplasts. Forty to 80 copies occur per organelle and replication occurs throughout the cell cycle.

chromosome   A molecule of DNA in association with proteins (histones and non-histones) constituting a linear array of genes. In prokaryotes, the circular DNA molecule contains the set of instructions necessary for the cell.

CITES   Convention on International Trade in Endangered Species of Wild Fauna and Flora.

clade   A species, or group of species that has originated and includes all the descendents from a common ancestor. A monophyletic group.

cladistics   The classification of organisms based on phylogeny.

cladogenesis   The splitting of a single evolutionary lineage into multiple lineages.

cladogram   A diagram illustrating the relationship between taxa that is built using synapomorphies. Also called a **phylogeny**.

cline   A gradual directional change in a character across a geographic or environmental gradient.

coalescent   The point at which the ancestry of two alleles converge at a common ancestral sequence.

codon   A three nucleotide sequence on a strand of mRNA that gets translated into a specific amino acid forming a protein.

conservation breeding   Efforts to manage plant and animal species breeding that do not necessarily involve captivity.

conservation collections   Living collections of rare or endangered organisms established for the purpose of contributing to the survival and recovery of a species.

conspecific   A member of the same species.

continuous characters   Phenotypic traits that are distributed continuously throughout the population (e.g., height or weight).

continuous distribution model of migration   Individuals are continuously distributed across the landscape; neighborhoods of individuals exist that are areas within which panmixia occurs, and across which genetic differentiation occurs due to isolation by distance.

Convention on International Trade in Endangered Species of Wild Fauna and Flora (CITES)   An agreement among 145 countries that bans commercial international trade in an agreed-upon list of endangered species, and that regulates and monitors trade in others that might become endangered.

converged   The point where a MCMC simulation has become independent of starting parameter biases, or has been "burnt in". Typically, thousands of simulation steps are required (and discarded) before the MCMC simulation is used to estimate a parameter (e.g., N_e, Nm, etc.)

countergradient variation   Occurs when genetic effects on a trait oppose or compensate for environmental effects so that phenotypic differences across an environmental gradient among populations are minimized.

cpDNA   *See* **chloroplast DNA**.

CU   Conservation unit.

cultivars   A human-cultivated plant that was derived through athropogenenic selection.

cytogenetics   A discipline of science combining cytology, the study of cells (their structure, function, and life history), and genetics.

cytoplasmic genes   Genes located in cellular organelles such as mitochondria and chloroplasts.

degrees of freedom   The total number of items in a data set that are free to vary independently of each other. In testing for Hardy–Weinberg proportions this is the number of possible genotypes minus the number of alleles because the frequency of homozygous genotypes are determined by the frequency of heterozygous genotypes.

deme   A local conspecific group of individuals that mate at random.

demographic   Topics relating to the structure and dynamics of populations, such as birth, death, and migration rates.

demographic stochasticity   Differences in the dynamics of a population that are the effects of random events on individuals in the population.

dendrogram   A tree diagram that serves as a visual representation of the relationships between populations within a species.

derived   A derived character is one found only in a particular lineage within a larger group. For example, feathers are derived characters that distinguish birds from their reptile ancestors.

deterministic   Events that have no random or probabilistic aspects but rather occur in a completely predictable fashion.

diagnostic locus   A locus that is fixed, or nearly fixed for different alleles allowing differentiation between parental species, populations, or their hybrids.

dioecious   Varieties or species of plants that have separate male and female reproductive organs on unisexual individuals.

diploid   The condition in which a cell or individual has two copies of every chromosome.

directional selection   The selective increase in the frequency of an advantageous allele, gene, or phenotypic trait in a population.

disassortative mating   Preferential mating of individuals with different phenotypes.

discrete generations   Generations that can be defined by whole integers and in which all individuals will breed only with individuals in their generation (e.g., pink salmon or annual flowers without a seed bank).

dispersal   In ecological literature dispersal is the movement of individuals from one genetic population (or birth place) into another. Dispersal is also known as **migration** in genetics literature.

distinct population segment (DPS)   A level of classification under the **ESA** that allows for legal protection of populations that are distinct, relatively reproductively isolated, and represent a significant evolutionary lineage to the species.

DNA   Deoxyribonucleic acid.

DNA barcoding   The use of a short gene sequence from a standardized region of the genome that can be used to help discover, characterize, and distinguish species, and to assign unidentified individuals to species.

DNA fingerprinting   Individual identification through the use of multilocus genotyping.

Dobzhansky–Muller incompatibilities   Genic interactions between alleles at multiple loci in which alleles that enhance fitness within their parental genetic backgrounds may reduce fitness in the novel genetic background produced by hybridization.

dominance genetic variation   The proportion of total genetic variation that can be attributed to the interactions of alleles at a locus in heterozygotes.

dominant   An allele (A) whose phenotypic effect is expressed in both homozygotes (AA) and heterozygotes (Aa).

DPS   *See* **distinct population segment**.

ecosystem   A community of organisms and its environment.

ecosystem services   The products and services humans receive from functioning ecosystems.

ecotone   The region that encompasses the shift between two biological communities.

effective number of alleles   The number of equally frequent alleles that would create the same heterozygosity as observed in the population.

effective population size (N_e)   The size of the ideal, panmictic population that would experience the same loss of genetic variation, through genetic drift, as the observed population.

electrophoresis   The movement of molecules through a medium across an electric field. Electrophoresis is used to separate allelic enzymes (allozymes) and DNA molecules of differing charge, size, or shape.

EM   *See* **expectation maximization algorithm**.

Endangered Species Act of the United States ESA.

endonuclease   An enzyme that cleaves either a single, or both, strands of a DNA molecule. Bacterial endonucleases are used to split genomic DNA at specific sites for analysis. *See* **restriction enzyme**.

Environmental Protection Agency of the United States EPA.

environmental stochasticity   Random variation in environmental factors that influence population parameters affecting all individuals in that population.

EPA   Environmental Protection Agency of the United States.

epidemiology   The study of the spread and control of a disease in a population.

epistatic genetic variation   The proportion of total genetic variation that can be attributed to the interaction between loci producing a combined effect different from the sum of the effects of the individual loci.

ESA   Endangered Species Act of the United States.

ESPA   Endangered Species Protection Act of Australia.

ESU   *See* **evolutionary significant unit**.

evolutionary significant unit (ESU)   A classification of populations that have substantial reproductive isolation which has led to adaptive differences so that the population represents a significant evolutionary component of the species. Evolutionary significant units have also been classified as populations that exhibit reciprocal monophyly and no recent gene flow. The original term used was "evolutionarily" rather than "evolutionary" (Ryder 1986). However, both terms are currently used in the literature.

***ex situ* conservation**   The conservation of important evolutionary lineages of species outside the species natural habitat.

exact tests   An approach to compute the exact *P*-value for an observed result rather than use an approximation, such as the chi-square distribution.

exon   A coding portion of a gene that produces a functional gene product (e.g., a peptide).

expectation maximization algorithm (EM)   A computational tool in statistics for finding maximum likelihood estimates of parameters in probabilistic models, where the model depends on unobserved variables. It can provide an estimate of the most likely allele frequencies assuming the sample is in Hardy–Weinberg proportions. Bayesians also use the EM algorithm to optimize the *a posteriori* distribution to compute the maximum *a posteriori* (MAP) estimate of an unknown parameter.

extant   Currently living; not extinct.

extinction   The disappearance of a species or other taxon so that it no longer exists anywhere.

extirpation   The loss of a species or subspecies from a particular area, but not from its entire range.

FCA   Frequency correspondence analysis.

fecundity   The potential reproductive capacity of an individual or population (e.g., the number of eggs or young produced by an individual per unit time).

fertility   The ability to conceive and have offspring. Sometimes used for **fecundity**.

Fisher–Wright model   *See* **Wright–Fisher model**.

fitness   The ability of an individual, or genotype to survive and produce viable offspring. Quantified as the number of offspring contributed to the next generation, or as proportion of the individual's genes in all the genes contributed to the next generation.

fitness rebound   Following an episode of inbreeding depression, successive generations of breeding may result in a rebound in fitness due to the selective decrease in frequency of deleterious alleles (purging). If inbreeding depression is due to deleterious recessive alleles (with negative fitness effects in a homozygous state) then successive generations of inbreeding may result in a rebound in fitness due to the selective decrease in frequency of deleterious alleles.

fixation index   The proportional increase of homozygosity through population subdivision. F_{ST} is sometimes referred to as the fixation index.

fluctuating asymmetry (FA)   Asymmetry in which deviations from symmetry are randomly distributed about a mean of zero. FA provides a simple measure of developmental precision or stability.

forensics   The use of scientific methods and techniques, such as genetic fingerprinting, to solve crimes.

founder effect   A loss of genetic variation in a population that was established by a small number of individuals that carry only a fraction of the original genetic diversity from a larger population. A special case of genetic drift.

frequency-dependent selection   Natural selection in which fitness varies as a function of the frequency of a phenotype.

gametic disequilibrium   Nonrandom association of alleles at different loci within a population. Also known as **linkage disequilibrium**.

gametic equilibrium   Random association of alleles at different loci within a population. Also known as **linkage equilibrium**.

gametogenesis   The creation of gametes through meiosis.

gene A segment of DNA whose nucleotide sequence codes for protein or RNA, or regulates other genes.

gene drop Simulation of the transmission of alleles in a pedigree. Each founder is assigned two unique alleles, and the alleles are then passed on from parent to offspring, with each offspring receiving one allele chosen at random from each parent (modeling Mendelian segregation), until all individuals in the pedigree have an assigned genotype.

gene flow Exchange of genetic information between demes through migration.

gene genealogies The tracing of the inherited history of the genes in an individual. Gene genealogies are most easily constructed using nonrecombining DNA such as mtDNA or the mammalian Y chromosome.

genet A genetically unique individual.

genetic assimilation A process in which phenotypically plastic characters that were originally "acquired" become converted into inherited characters by natural selection. This term also has been applied to the situation in which hybrids are fertile and displace one or both parental taxa through the production of hybrid swarms (i.e., genomic extinction).

genetic distance matrix A pairwise matrix composed of differentiation between population (or individual) pairs that is calculated using a measure of genetic divergence such as F_{ST}.

genetic divergence The evolutionary change in allele frequencies between reproductively isolation populations.

genetic draft A stochastic process in which selective substitutions at one locus will reduce genetic diversity at neutral linked loci through hitchhiking.

genetic drift Random changes in allele frequencies in populations between generations due to binomial sampling of genes during meiosis. Genetic drift is more pronounced in small populations.

genetic engineering A processes in which an organism's genes are selectively modified, often through splicing DNA fragments from different chromosomes or species, to achieve a desired result.

genetic exchange *See* **gene flow**.

genetic load The decrease in the average fitness of individuals in a population due to deleterious genes or heterozygous advantage.

genetic rescue The recovery in the average fitness of individuals through increased gene flow into small populations, typically following a fitness reduction due to inbreeding depression.

genetic stochasticity Random changes in the genetic characteristics of populations through genetic drift and binomial sampling of alleles during Mendelian segregation.

genetic swamping The loss of locally adapted alleles or genotypes caused by constant immigration and gene flow.

genetics The study of how genes are transmitted from one generation to the next and how those genes affect the phenotypes of the progeny.

genomic extinction The situation in which hybrids are fertile and displace one or both parental taxa through the production of hybrid swarms so that the parental genomes no longer exist even though the parental alleles are still present.

genomic ratchet A process where hybridization producing fertile offspring will result in a hybrid swarm over time, even in the presence of outbreeding depression and with relatively few hybrids per generation.

genomics The study of the structure or function of large numbers of genes in a genome.

genotype An organism's genetic composition.

gynodioecy The occurrence of female and hermaphroditic individuals in a population of plants.

haploid The condition in which a cell or individual has one copy of every chromosome.

haplotype The combination of alleles at loci that are found on a single chromosome or DNA molecule.

Hardy–Weinberg principle   The principle that allele and genotype frequencies will reach equilibrium, defined by the binomial distribution, in one generation and remain constant in large random mating populations that experience no migration, selection, mutation, or nonrandom mating.

Hardy–Weinberg proportion   A state in which a population's genotypic proportions equal those expected with the binomial distribution.

hemizygous   A term used to denote the presence of only one copy of an allele due a locus being in a haploid genome, on a sex chromosome, or only one copy of the locus being present in an aneuploid organism.

heritability   The proportion of total phenotypic variation within a population that is due to individual genetic variation (H_B; broad sense heritability). Heritability is more commonly referred to as the proportion of phenotypic variation within a population that is due to additive genetic variation (H_N; narrow sense heritability).

hermaphrodite   An individual that produces both female and male gametes.

heterochromatin   Highly folded chromosomal regions that contain few functional genes. When these traits are characteristic of an entire chromosome, it is a heterochromosome or supernumerary chromosome.

heterogametic   The sex that is determined with different sex chromosomes (e.g., the male in mammals (XY) and female in birds (ZW)).

heteroplasmy   The presence of more than one mitochondrial DNA haplotype in a cell.

heterosis   A case when hybrid progeny have higher fitness than either of the parental organisms. Also called **hybrid vigor**.

heterozygosity   A measure of genetic variation that accounts for either the observed, or expected proportion of individuals in a population that are heterozygotes.

heterozygote   An organism that has different alleles at a locus (e.g., *Aa*).

heterozygous advantage   A situation where heterozygous genotypes are more fit than homozygous genotypes. This fitness advantage can create a stable polymorphism. Also called **overdominance**.

heterozygous disadvantage   A situation where heterozygous genotypes are less fit than homozygous genotypes. Also called **underdominance**.

HFC   Heterozygosity–fitness correlation.

Hill–Robertson effect   An effect where selection at one locus will reduce the effective population size of linked loci; increasing the chance of genetic drift forming negative genetic associations that reduce the ability of associating loci to respond to selection. *See also* **genetic draft**.

hitchhiking   The increase in frequency of a selectively neutral allele through gametic disequilibrium with a beneficial allele that selection increases in frequency in a population.

homogametic   The sex that possesses the same sex chromosomes (e.g., the female in mammals (XX) and male in birds (ZZ)).

homoplasmy   The presence of a single mitochondrial DNA haplotype within a cell.

homoplasy   Independent evolution or origin of similar traits, or gene sequences. At a locus, homoplasy can result from back mutation or mutation to an existing allelic state.

homozygosity   A measure of the proportion of individuals in a population that are homozygous; it is the reciprocal of heterozygosity.

homozygote   An organism that has two or more copies of the same alleles at a locus (e.g., *AA*).

HW   Hardy–Weinberg.

hybrid sink   The situation where immigration of locally unfit genotypes produces hybrids with low fitness that reduces local density and thereby increases the immigration rate.

hybrid swarm   A population of individuals that are all hybrids by varying numbers of generations of backcrossing with parental types and matings among hybrids.

hybrid vigor   *See* **heterosis**.

hybrid zone   An area of sympatry between two genetically distinct populations where hybridization occurs without forming a hybrid swarm in either parental population beyond the area of co-occurrence.

hybridization   Mating between individuals of two genetically distinct populations.

identical by descent   Alleles that are identical copies of the same allele from a common ancestor.

inbreeding   The mating between related individuals that results in an increase of homozygosity in the progeny because they possess alleles that are identical by descent.

inbreeding coefficient   A measure of the level of inbreeding in a population that determines the probability that an individual possesses two alleles at a locus that are identical by decent. It can also be used to describe the proportion of loci in an individual that are homozygous.

inbreeding depression   The reduction in fitness of progeny from matings between related individuals compared to progeny from unrelated individuals.

inbreeding effect   Inbreeding eventually will occur in panmictic small populations due to the individuals becoming increasingly related through time.

inbreeding effective number (N_{ei})   The size of the ideal panmictic population that loses heterozygosity at the same rate as the observed population.

introduction   The placement, or escape, of a species or individual into a novel habitat. Often introductions are used in conservation to aid genetic rescue of isolated populations.

introgression   The incorporation of genes from one population to another through hybridization that results in fertile offspring that further hybridize and backcross to parental populations.

intron   A portion of a gene that produces a nonfunctional RNA strand that is cleaved prior to translation; a noncoding region between the exons.

invasive species   An introduced alien species that is likely to cause harm to the natural ecosystem, the economy, or human health.

island model of migration   A model of migration in which a population is subdivided into a series of demes, of size N, that randomly exchange migrants at a given rate, m.

isolation by distance   The case where genetic differentiation is greater the further individuals (or populations) are from each other because gene flow decreases as geographic distance increases. Originally individuals used in the case where individuals are distributed continuously across large landscapes (e.g., coniferous tree species across boreal forests) and are not subdivided by sharp barriers to gene flow.

ISSR   Intersimple sequence repeat markers that use similar PCR methods as PINE fragments, but have primers based on simple sequence repeats of microsatellites.

IUCN   World Conservation Union (formerly International Union for Conservation of Nature).

karyotype   The composition of the chromosomal complement of a cell, individual, or species.

landscape genetics   The study of the interaction between landscape or environmental features and population genetics, such as gene flow.

LE   *See* **lethal equivalent**.

lek   A specific area where the males of a population that exhibits female sexual selection will congregate and display for females.

lethal equivalent   The number of deleterious alleles in an individual whose cumulative effect is the same as that of a single lethal allele. For example, four alleles each of which would be lethal 25% of the time (or to 25% of their bearers), are equivalent to one lethal allele.

library   Collection of DNA fragments from a given organism "stored" in a virus or bacteria.

likelihood statistics   An approach for parameter estimation and hypothesis testing that involves building a model (i.e., a likelihood function) and the use of the raw data (not a summary statistic), which often provides more precision and accuracy than frequentist statistic approaches (method of moments). The parameter of interest is estimated as the member of the parameter space that maximizes the probability of obtaining your observed data. Likelihood approaches facilitate comparisons between different models (e.g., via likelihood ratio tests) and thus the testing of alternate hypotheses (e.g., stable versus declining population size).

lineage sorting A process where different gene lineages within an ancestral taxon are lost by drift or replaced by unique lineages evolving in different derived taxa.

linkage disequilibrium The nonrandom association of alleles between linked loci. Also called **gametic disequilibrium**.

linkage equilibrium Random association of alleles between liked loci. Also called **gametic equilibrium**.

local adaptation Greater fitness of individuals in their local habitats due to natural selection.

local scale The spatial scale at which individuals routinely interact with their environment.

locus The position on a chromosome of a gene or other marker.

LOD *See* **log of odds ratio**.

log of odds ratio (LOD) The odds ratio is the odds of an event occurring in one group to the odds of it occurring in another group. For example, if 80% of the individuals in a population are *Aa* and 20% are *AA*, then the odds of *Aa* over *AA* is four; there are four (4.0) times as many *Aa* as *AA* genotypes. The natural log of this ratio is often computed because it is convenient to work with statistically.

management unit A local population that is managed as a unit due to its demographic independence.

marginal overdominance Greater fitness of heterozygous genotypes, which are not the most fit in any single environment, due to an organism's interactions with multiple environments that each favor different alleles.

match probability (MP) The probability of sampling an individual with an identical multilocus genotype to the one already sampled ("in hand").

maternal effects The influence of the genotype or phenotype of the mother on the phenotype of the offspring. Because it has no genetic basis, maternal effects are not heritable.

maximum likelihood A statistical method of determining which of two or more competing alternative hypotheses (such as alternative phylogenetic trees) yields the best fit to the data.

maximum likelihood estimate (MLE) A method of parameter estimation that obtains the parameter value that maximizes the likelihood of the observed data.

MCMC Markov chain Monte Carlo. A tool or algorithm for sampling from probability distributions based on constructing a Markov chain. The state of the chain after many steps is then used as a sample from the desired distribution. Sometimes called a random walk Monte Carlo method.

MDS Multidimensional scaling. A statistical graphing technique used to represent genetic distances between samples in two or three dimensions, and thereby visualizing similarities and differences between different groups or samples.

Mendelian segregation The random separation of paired alleles (or chromosomes) into different gametes.

meristic character A trait of an organism that can be counted using integers (e.g., fin rays or ribs).

metacentric A chromosome in which the centromere is centrally located.

metapopulation A collection of spatially divided subpopulations that experience a certain degree of gene flow among them.

metapopulation scale The spatial scale at which individuals migrate between local subpopulations, often across habitat that is unsuitable for colonization.

microchromosomes Small chromosomes found in many bird species which, unlike heterochromosomes, carry functional genes.

microsatellite Tandemly repeated DNA consisting of short sequences of one to six nucleotides repeated between approximately five and 100 times. Also known as **VNTRs**, **SSRs**, or **STRs**.

migration   The movement of individuals from one generically distinct population to another resulting in gene flow.

minimum viable population (MVP)   The minimum population size at which a population is likely to persist over some defined period of time.

minisatellite   A tandemly repeated sequence of approximately 10–100 nucleotides that are 500 to 30,000 base pairs in length.

mitochondrial DNA (mtDNA)   A small, circular, haploid DNA molecule found in the mitochondria cellular organelle of eukaryotes.

ML   *See* **maximum likelihood**.

MLE   *See* **maximum likelihood estimation**.

molecular clock   The observation that mutations sometimes accumulate at relatively constant rates, thereby allowing researchers to estimate the time since two species diverged (TMRCA).

molecular genetics   The branch of genetics that studies the molecular structure and function of genes, or that (more generally) uses molecular markers to test hypotheses.

molecular mutations   Changes to the genetic material of a cell, including single nucleotide changes, deletions, and insertions of nucleotides as well as recombinations and inversions of DNA sequences.

monoecious   A plant in which male and female organs are found on the same plant but in different flowers (for example maize).

monomorphic   The presence of only one allele at a locus, or the presence of common allele at a high frequency (>95% or 99%) in a population.

monophyletic   A group of taxa that include all species, ancestral and derived, from a common ancestor.

monophyly   The presence of a monophyletic group.

monotypic   A taxonomic group that encompasses only one taxonomic representative. The reptile family that contains tuatara (Sphenodontia) is currently monotypic.

morphology   The study of the physical structures of an organism, including the evolution and development of these structures.

MP   *See* **match probability**.

MRCA   Most recent common ancestor. In cladistics, the organism at the base of a clade, from which that clade arose.

mRNA   Messenger ribonucleic acid.

MSD   Multiple factor sex determination.

mtDNA   *See* **mitochondrial DNA**.

MU   *See* **management unit**.

mutagenesis   The natural or intentional formation of mutations in a genome.

mutation   An error in the replication, or transmission, of DNA that cause a structural change in a gene. *See also* **molecular mutations**.

mutational meltdown   The process by which a small population accumulates deleterious mutations, which leads to loss of fitness and decline of the population size, which leads to further accumulation of deleterious mutations. A population experiencing mutational meltdown is trapped in a downward spiral and will eventually go extinct.

MVP   *See* **minimum viable population**.

narrow sense heritability (H_N)   The amount of individual phenotypic variation that is due to additive genetic variation.

native species   A species that was not introduced and historically, or currently, occurs in a given ecosystem.

natural catastrophes   Natural events causing great damage to populations and that increase their probability of extinction.

natural selection   Differential contribution of genotypes to the next generation due to differences in survival and reproduction.

NCA   Nested clade analysis. A statistical approach to describe how genetic variation is distributed spatially within a species' geographic range. This method uses a haplotype tree to define a nested series of branches (clades), thereby allowing a nested analysis of the spatial distribution of genetic variation, often with the goal of resolving between past fragmentation, colonization, or range expansion events.

nearly exact test   A method of using a nearly exact P-value to test if the observed test statistic deviates from the expected value under the null hypothesis, For example, a test of whether populations are in Hardy–Weinberg proportions by comparing the observed chi-squared value to the chi-squared values of random computer permutations of genotypes from the population's allele frequencies.

neighborhood   The area in a continuously distributed population that call be considered panmictic.

neutral allele   An allele that is not under selection because it does not affect fitness.

NMFS   National Marine Fisheries Service.

NOAA   National Oceanic Atmospheric Administration.

node   A branching point or end point on a phylogenic tree that represents either an ancestral taxon (internal node) or an extant taxon (external node).

nonindigenous species   Species present in a given ecosystem that were introduced and did not historically occur in that ecosystem.

nuclear DNA (nDNA)   DNA that forms chromosomes in the cell nucleus of eukaryotes.

nuclear gene   A gene located on a chromosome in the nucleus of a eukaryotic cell.

nucleotides   The building blocks of DNA and RNA made up of a nitrogen-containing purine or pyrimidine base linked to a sugar (ribose or deoxyribose) and a phosphate group.

null allele   An allele that is not detectable either due to a failure to produce a functional product or a mutation in a primer site that precludes amplification during PCR analysis.

Ocham's razor   The principle that the least complicated explanation (most parsimonious hypothesis) generally should be accepted to explain the data at hand.

offsite conservation   *See ex situ* **conservation**.

outbreeding depression   The relative reduction in the fitness of hybrids compared to parental types.

outlier loci   Loci that may be under selection (or linked to loci under selection) that are detected because they fall outside the range of expected variation for a given summary statistic (e.g., extremely high or low F_{ST} compared to most "neutral" loci in a sample).

overdominance   *See* **heterozygous advantage**.

overlapping generations   A breeding system where sexual maturity does not occur at a specific age, or where individuals breed more than once, causing individuals from different brood years to interbreed in a given year.

panmictic   A population that is randomly mating.

paracentric inversion   A chromosomal inversion that does not include the centromere because both breaks were on the same chromosomal arm.

paraphyletic   A clade that does not include all of the descendants from the most recent common ancestor taxon. For examples, reptiles are paraphyletic because they do not include birds.

parentage analysis   The assessment of the maternity and/or paternity of a given individual.

parsimony   The principle that the preferred phylogeny of an organism is the one that requires the fewest evolutionary changes; the simplest explanation.

PAW   Partnership for Action against Wildlife Crime.

PCA   Principal component analysis.

PCoA   Principle coordinates analysis.

PCR   *See* **polymerase chain reaction**.

pdf   Probability density function.

PE Paternity exclusion (probability of).

pericentric inversion A chromosomal inversion that includes the centromere because the breaks were on opposite chromosomal arms.

phenetics Taxonomic classification solely based on overall similarity (usually of phenotypic traits), regardless of genealogy.

phenogram A branching diagram or tree that is based on estimates of overall similarity between taxa derived from a suite of characters.

phenotype The observable characteristics of an organism that are the product of the organism's genotype and environment.

phenotypic Relating to an aspect of an individual's phenotype.

phenotypic plasticity Variation in the phenotype of individuals with similar genotypes due to differences in environmental factors during development. For example, cod in areas with red algae develop a reddish color.

philopatry A characteristic of reproduction of organisms where individuals faithfully home to natal sites. Individuals exhibiting philopatry are philopatric.

phylogenetic Evolutionary relationships between taxa or gene lineages. These relationships are often expressed visually in phylogenetic trees with nodes representing taxa or lineages (ancestral or derived), and branch lengths often corresponding to the amount of divergence between groups.

phylogenetic species concept (PSC) States that a species is a discrete lineage or recognizable monophyletic group.

phylogeny *See* **cladogram**.

phylogeography The assessment of the geographic distributions of the taxa of a phylogeny to understand the evolutionary history (e.g., origin and spread) of a given taxon.

PI *See* **probability of identity**.

PINEs Paired interspersed nuclear elements. Use of PCR primes that bind one end of a transposable element (along with a few adjacent single-copy nucleotides), to generate DNA markers for studies in population genetics (e.g., hybridization or admixture).

pleiotropy The case where one gene affects more than one phenotypic trait.

Poisson distribution A probability distribution, with identical mean and variance, that characterizes discrete events occurring independently of one another in time, when the mean probability of that event on any one trial is very small. Earthquake hazards, radioactive decay, and mutation events follow a Poisson distribution. The Poisson is a good approximation to the binomial distribution when the probability is small and the number of trials is large.

polygenic Affected by more than one gene.

polymerase A molecule that catalyzes the synthesis of DNA or RNA from a single-stranded template and free deoxynucleotides (e.g., during PCR).

polymerase chain reaction (PCR) A technique to replicate a desired segment of DNA. PCR starts with primers that flank the desired target fragment of DNA. The DNA strands are first separated with heat, and then cooled allowing the primers bind to their target sites. Polymerase then makes each single strand into a double strand, starting from the primer. This cycle is repeated multiple times creating a 10^6 increase in the gene product after 20 cycles and a 10^9 increase over 30 cycles.

polymorphic The presence of more than one allele at a locus. Generally defined as having the most common allele at a frequency less than 95% or 99%.

polymorphism The presence of more than one allele at a locus. Polymorphism is also used as a measure of the proportion of loci in a population that are genetically variable or polymorphic (P).

polyphyletic A group of taxa classified together that have descended from different ancestor taxa (i.e., taxa that do not all share the same recent common ancestor).

polyploid Individuals whose genome consists of more than two sets of chromosomes (e.g., tetraploids).

population viability   The probability that enough individuals in a population will survive to reproductive age to prevent extirpation of the population.

population viability analysis (PVA)   The general term for the application of models that account for multiple threats facing the persistence of a population to access the likelihood of the population's persistence over a given period of time. PVA helps identify the threats faced by a species, plan research and data collection, prioritize management options, and predict the likely response of species to management actions (e.g., reintroduction, captive breeding, or prescribed burning).

primer   A small oligonucleotide (typically 18–22 base pairs long) that anneals to a specific single-stranded DNA sequence to serve as a starting point for DNA replication (e.g., extension by polymerase during PCR).

private allele   An allele present in only one of many populations sampled.

probability   The certainty of an event occurring. The observed probability of an event, r, will approach the true probability as the number of trials, n, approaches infinity.

probability of identity (PI)   The probability that two unrelated (randomly sampled) individuals would have an identical genotype. This probability becomes very small if many highly polymorphic loci are considered.

product rule   A statistical rule that states that the probability of n_i independent events occurring is equal to the product of the probability of each n independent event.

propagule   A dispersal vector. Any disseminative unit or part of an organism capable of independent growth (e.g., a seed, spore, mycelial fragment, sclerotium bud, tuber, root, or shoot).

propagule pressure   A measure of the introduction of nonindigenous individuals that includes the number of individuals (or propagules) introduced and the number of introductions.

proportion of admixture   The proportion of alleles in a hybrid swarm that come from each of the parental taxa.

protein   A polypeptide molecule.

PSC   *See* **phylogenetic species concept**.

pseudo-overdominance   *See* **associative overdominance**.

purging   The removal of deleterious recessive alleles from a population through inbreeding which increases homozygosity which in turn increases the ability of selection to act on recessive alleles.

PVA   *See* **population viability analysis**.

QTLs   *See* **quantitative trait loci**.

quantitative trait loci (QTLs)   Genetic loci that affect phenotypic variation (and potentially fitness), which are identified by a statistically significant association between genetic markers and measurable phenotypes. Quantitative traits are often influenced by multiple loci as well as environmental factors.

RAPD   Randomly amplified polymorphic DNA. A method of analysis where PCR amplification using two copies of an arbitrary oligonucleotide primer is used to create a multilocus fingerprint (i.e., band profile).

reciprocal monophyly   A genetic lineage is reciprocally monophyletic when all members of the lineage share a more recent common ancestor with each other than with any other lineage on a phylogenetic tree.

recombination   The process that generates a haploid product of meiosis with a genotype differing from both the haploid genotypes that originally combined to form the diploid zygote.

reintroduction   The introduction of a species or population into a historical habitat from which it had previously been extirpated.

relative fitness   A measure of fitness that is the ratio of a given genotype's absolute fitness to the genotype with the greatest absolute fitness. Relative fitness is used to model genetic change by natural selection.

rescue effect   When immigration into an isolated deme (either genetically or demographically) reduces the probability of the extinction of that deme.

restriction enzyme   An enzyme (*see* **endonuclease**), isolated from bacteria, that cleaves DNA at a specific four or six nucleotide sequence. Over 400 such enzymes exist that recognize and cut over 100 different DNA sequences; used in RFLP, AFLP, and RAPD analysis and to construct recombinant DNA (in genetic engineering).

restriction fragment length polymorphism (RFLP)   A method of genetic analysis that examines polymorphisms based on differences in the number of fragments produced by the digestion of DNA with specific endonucleases. The variation in the number of fragments is created by mutations within restriction sites for a given endonuclease.

reverse mutation rate   Back mutation rate. The rate at which a gene's ability to produce a functional product is restored. This rate is much lower than the forward mutation rate because there are many more ways to remove the function of a gene than restore it. Also used to describe mutation at microsatellite loci where (under the stepwise mutation model, for example) a back mutation yields an allele of length that already exists (i.e., homoplasy) in the population.

RFLP   *See* **restriction fragment length polymorphism**.

ribonucleic acid (RNA)   A polynucleotide similar to DNA that contains ribose in place of deoxyribose and uracil in place of thymine. RNA is involved in the transfer of information from DNA, programming protein synthesis, and maintaining ribosome structure.

Robertsonian fission   An event where a metacentric chromosome breaks near the centromere to form two acrocentric chromosomes.

Robertsonian fusion   An event where two acrocentric chromosomes fuse to form one metacentric chromosome.

Robertsonian translocation   A special type of translocation where the break occurs near the centromere or telomere and involves the whole chromosomal arm so balanced gametes are usually produced.

SARA   *See* **Species at Risk Act of Canada**.

selection coefficient   The reduction in relative fitness, and therefore genetic contribution to future generations, of one genotype compared to another.

selection differential   The difference the mean value of a quantitative trait found in a population as a whole compared to the mean value of the trait in the breeding population.

selective sweep   The rapid increase in frequency by natural selection of an initially rare allele that also fixes (or nearly fixes) alleles at closely linked loci thus reduces the genetic variation in a region of a chromosome.

sensitivity testing   A method used in population viability analyses where the effects of parameters on the persistence of populations are determined by testing a range of possible values for each parameter.

sequential Bonferroni correction   A method, similar to the Bonferroni correction, that is used to reduce the probability of a Type I statistical error when conducting multiple simultaneous tests.

sex chromosomes   Chromosomes that pair during meiosis but differ in the hererogametic sex.

sex-linked locus   A locus that is located on a sex chromosome.

sexual selection   Selection due to differential mating success either through competition for mates or mate choice.

shadow effect   A case usually caused by low marker polymorphism in mark–recapture studies in which a novel capture is labeled as a recapture due to identical genotypes at the loci studied.

SINEs   Short interspersed nuclear elements.

single nucleotide polymorphism (SNP)   A nucleotide site (base pair) in a DNA sequence that is polymorphic in a population either due to transitions or transversions and can be used as a marker to assess genetic variation within and among populations. Usually only two alleles exist for a SNP in a population.

SMM   *See* **stepwise mutation model**.

SNP   *See* **single nucleotide polymorphism**.

species   A group of organisms with a high degree of physical and genetic similarity, that naturally interbreed among themselves and can be differentiated from members of related groups of organisms.

Species at Risk Act of Canada (SARA)   Legislation (passed in 2002) to prevent wildlife species from becoming extinct and secure the necessary actions for their recovery. It provides for the legal protection of wildlife species and the conservation of critical habitat.

species concepts   The ideas of what constitutes a species, such as reproductive isolation (BSC), or monophyly of a lineage (PSC).

species scale   The spatial scale encompassing an entire species' distribution.

SSRs   Simple sequence repeats. *See* **microsatellite**.

stable polymorphism   A polymorphism that is maintained at a locus through natural selection.

stabilizing selection   Selection for a phenotype with a more intermediate state.

stepping stone model of migration   A model of migration in which the probability of migration between nearby or adjacent populations is higher than the probability of migration between distant populations.

stepwise mutation model (SMM)   A model of mutation in which the microsatellite allele length has an equal probability of either increasing or decreasing (usually by a single repeat unit, as in the strict one-step SMM).

stochastic   The presence of a random variable in determining the outcome of an event.

stock   A term generally used in fisheries management that refers to a population that is demographically independent and often represents a subunit (e.g., MU) of an ESU.

STR   Short tandem repeat. *See* **microsatellite**.

subpopulations   Groups within a population delineated by reduced levels of gene flow with other groups.

subspecies   A taxonomically defined subdivision within a species that is physically or genetically distinct, and often geographically separated.

sum rule   A statistical rule that states that the probability of n_i mutually exclusive, independent events occurring is equal to the sum of the probabilities of each n event.

supergene   Allelic combinations found at closely linked loci that affect related traits and are inherited together. An example of a supergene is the major histocompatibility complex (MHC), which in humans contains more than 200 genes adjacently located over several megabases of sequence on chromosome 6.

supernumerary chromosome   A chromosome, often present in varying numbers, that is not needed for normal development, lacks functional genes, and does not segregate during meiosis. These small chromosomes, which are also called B chromosomes, are present in addition to the normal complement of functional chromosomes in an organism.

supportive breeding   The practice of removing a subset of individuals from a wild population for captive breeding and releasing the captive-born offspring back into their native habitat to intermix with wild-born individuals and increase population size or persistence.

sympatric   Populations or species that occupy the same geographic area.

synapomorphy   A shared derived trait between evolutionary lineages. A homology that evolved in an ancestor common to all species on one branch of a phylogeny, but not common to species on other branches.

Taq   The bacterium *Thermus aquaticus* from which a heat stable DNA polymerase used in PCR was isolated.

telomere   A tandemly repeated segment of a short DNA sequences, one strand of which is G-rich and the other strand is C-rich, that form the ends of linear eukaryotic chromosomes.

threshold   The point at which environmental (or genetic) changes produce large phenotypic changes in an organism (or population). For example, there could be a threshold effect of inbreeding on fitness such that after a certain level of inbreeding is reached, individual fitness declines increasingly rapidly.

threshold character　A phenotypic character that contains a few discrete states that are controlled by many genes underlying continuous variation, which affects a character phenotypically only when a certain physiological threshold is exceeded.

Time since the most recent common ancestor　TMRCA.

TMRCA　Time since the most recent common ancestor.

TRAFFIC　A wildlife trade monitoring network sponsored by the WWF and IUCN.

transgressive segregation　Hybridization events that produce progeny that express phenotypic values outside the range of either parental phenotypic value. These differences are usually due to the disruption of polygenic traits.

transition　The more common single nucleotide mutation (or polymorphism) that results from a point mutation in which a purine is substituted with a purine (G↔A) or a pyrimidine is substituted with a pyrimidine (C↔T).

translocation　(1) The movement of individuals from one population (or location) to another that is usually intended to achieve either genetic or demographic rescue of an isolated population. (2) A rearrangement occurring when a piece of one chromosome is broken off and joined to another chromosome.

transposable element　Any genetic unit that can insert into a chromosome, exit, and relocate; includes insertion sequences, transposons, some bacteriophages, and controlling element. A region of the genome, flanked by inverted repeats, a copy of which can be inserted at another place; also called a **transposon** or a jumping gene.

transposon　A mobile element of DNA that jumps to new genomic locations through a DNA intermediate and which usually carries genes other than those that encode for transposase proteins used to catalyze movement.

transversion　The replacement of a purine with a pyrimidine (A or G to C or T) or vice versa (C or T to A or G). Less common than a transition.

Type I statistical error　The probability of rejecting a true null hypothesis. Usually chosen, by convention, to be 0.05 or 0.01.

Type II statistical error　The probability of accepting a false null hypothesis.

underdominance　*See* **heterozygous disadvantage**.

UNEP　United Nations Environmental Program.

United Nations Environmental Program　UNEP.

UPGMA　Unweighted pair group method with arithmetic averages.

USFWS　United States Fish and Wildlife Service.

variance effective number (N_{ev})　The size of the ideal population that experiences changes in allele frequency at the same rate as the observed population.

viability　The probability of the survival of a given genotype to reproductive maturity (or of a population to persist through a certain time interval).

VNTRs　Variable number of tandem repeats. *See* **microsatellite**.

Wahlund principle　The deficit of heterozygotes in subdivided populations, compared to expected Hardy–Weinberg proportions, due to subdivision into small panmictic (random mating) demes within the large population.

Wright–Fisher model　A random mating population model with complete random union of gametes (including the possibility of selfing).

WWF　World Wide Fund For Nature (formerly known as the World Wildlife Fund).

Appendix

Probability and Statistics

He has a lot of extremely abstruse, in fact almost esoteric mathematics. Mathematics, incidentally, of a kind which I certainly do not claim to understand. I am not a mathematician at all. My way of reading Sewall Wright's papers, which I think is perfectly defensible, is to examine the biological assumptions the man is making, and to read the conclusions he arrives at, and hope to goodness that what comes in between is correct.

Theodosius Dobzhansky (1962)

Current research in population genetics employs advanced mathematical methods that are beyond the reach of most biology students.

James F. Crow (1986)

The gulf between mathematical population genetics and the understanding of most biologists has greatly increased over the last 20 years because of the introduction of a variety of new theoretical and computational approaches (see Guest Box A). This can make it difficult for conservation geneticists to analyze new data sets with recent computational approaches, and to publish their results in peer-reviewed conservation journals (Example A1).

The purpose of this appendix is to provide biologists with a basic understanding of the mathematical and statistical approaches used in this book. We have modeled this section after the appendices that appear in Crow and Kimura (1970) and Crow (1986), with substantial use of Dytham (2003). We have not tried to provide mathematical rigor, but rather intend to make clear the general nature and limitations of the mathematical and statistical approaches used in this book. We aim to provide a conceptual understanding of different statistical approaches and show how to interpret results, rather than to teach details about how to actually conduct a certain statistical test or likelihood estimation.

There are three main approaches or paradigms to statistical inference: frequentist, likelihood based, and Bayesian approaches. Likelihood methods are sometimes classified within the frequentist approach (see Section A4). Here, we first give a brief historical perspective and explain the major differences between the three approaches. Then, we present concepts of probability and basic statistics including hypothesis testing. Finally, we return to discuss in more detail likelihood and Bayesian approaches, along with the coalescent and MCMC (Markov chain Monte Carlo), and their importance in conservation genetics.

Example A1 Problems understanding sophisticated computational approaches

The senior author of this book was an Associate Editor in the initial days of the journal *Conservation Biology*. In the early 1990s, he handled a manuscript that applied some fairly sophisticated mathematical population genetics theory to a problem in conservation. He received the following review comments from a well-known population geneticist: "According to the Instructions to Reviewers for this journal, manuscripts should be understandable to conservation managers and government officials. It is not reasonable to expect either of these groups to understand stochastic theory of population genetics." This problem is much worse now than in the early 1990s because of the increasing sophistication of computational approaches as presented in this appendix. It is becoming increasingly important to analyze empirical data with complex statistical approaches. However, it is also becoming increasingly difficult to evaluate the reliability of these analyses.

The Bayesian philosophy and statistical approach to data analysis was developed in the 18th century by the Reverend Thomas Bayes. The classic frequentist approach was formalized later, during the early 1900s, by K. Pearson, R. A. Fisher (from England), and J. Neyman (from Poland); it quickly became dominant in science. Modern likelihood analysis was developed almost single-handedly by R. A. Fischer between 1912 and 1922. A revival of the Bayesian approach has occurred during the last 5–10 years, thanks to advances in computer speed and simulation-based algorithms such as MCMC (see Section A5) that allow the analysis of complex probabilistic models containing multiple interdependent parameters such as genotypes, population allele frequencies, population size, migration rates, and variable mutation rates across loci.

The **frequentist approach** to statistical inference often involves four steps: stating a hypothesis, collecting data, computing a summary statistic (e.g., $F_{ST} = 0.01$), and then inferring how frequently we would observe our statistic (0.01) by chance alone if our null hypothesis (H_0) is true (e.g., H_0: $F_{ST} = 0.00$). If our statistic is so large, e.g., $F_{ST} = 0.10$, that we expect to observe it very infrequently by chance alone (e.g., only once per 100 independent experiments), we would reject the null hypothesis. The frequentist approach determines the expected long-term frequency of an observation or a summary statistic, if we were to repeat the experiment or observation many times. Frequentist approaches typically use the moments of the distribution (of a summary statistic) and thus are called "methods of moments". The moments are the mean and variance, as well as skewness and kurtosis. These concepts are discussed in detail below.

Likelihood approaches typically involve four steps: collecting data, developing a mathematical model with parameters (e.g., F_{ST}), plugging into the model the raw data (not a summary statistic), and computing the likelihood of the data for each of all possible parameter values, for example $F_{ST} = 0.00, 0.01, 0.02$, up to 1.00. This requires many computations or iterations. We then identify the parameter value that maximizes the likelihood of obtaining our actual data under the model. The main advantage of likelihood over frequentist approaches is that likelihood uses the raw data (e.g., allele counts at **each locus** separately) and not a summary of it, e.g., F_{ST} **averaged across loci** (see Section A4). Thus more information is used from the data (e.g., interlocus variation in F_{ST}), and therefore the estimates of parameters (and inference in general) should be more accurate and precise.

The **Bayesian approach** is distinct in that: (1) it can incorporate prior information (e.g., data from previous studies) to compute a probability estimate (i.e., a "posterior probability"); and (2) it **directly** yields the probability that the hypothesis of interest is true, e.g., H_A: $F_{ST} > 0.00$. Thus it more directly tests a hypothesis than frequentist methods that assess how frequently we expect to observe a summary statistic (e.g., $F_{ST} = 0.10$) if the null hypothesis is true (recall that the null hypothesis is not the direct hypothesis of interest in the frequentist approach, but rather is the hypothesis we try to reject, see Section A3).

The Bayesian approach is model based, like likelihood. In fact it combines likelihood computation with prior information to obtain a modified likelihood estimate called the posterior probability (see Section A5). Further, Bayesian approaches compute the probability (posterior probability) of the parameter given the data, whereas likelihood computes the probability of the data for a given parameter value (to find the maximum value). For example, when estimating N_e, the Bayesian approach outputs the (posterior) probability for different N_e values (e.g., for $N_e = 0$ to 500) given the data (see Section A5), whereas likelihood finds the parameter values that maximize the probability of the data (see Section A4).

We will return to Bayesian and likelihood methods again, after considering the important concepts of probability, statistical distributions, and hypothesis testing. Such concepts will help in the understanding of the different methods of statistical inference and modeling.

A1 Probability

Probability was defined in 1812 by a French mathematician, Pierre Simon Laplace, as a number between 0 and 1 that measures our certainty of some event. A probability of 1.0 means the event is 100% certain to occur. An example use of probability in conservation genetics is in estimating the probability of loss (by genetic drift) of an allele at frequency 0.95. It is simply the frequency of the allele (0.95) in the population, assuming no selection. For another example see Table 6.1. Probability concepts (including probability distributions, such as posterior distributions) are important in statistics for using samples from a population to make inferences about the population, based on the sample characteristics (see next section below).

Two important probability rules that we often use in genetics are the addition and product rule. The **addition rule** is illustrated in Box 5.1. The addition rule (also known as the "**either/or**" probability rule) states that the probability of any of several mutually exclusive events occurring equals the **sum** of separate probabilities of each event. In conservation genetics, we often study the probability of mutually exclusive events, such as being male or female, or of originating from population X versus population Y or Z (see also Box 5.1). The sum of mutually exclusive events adds to one (1.0). For example, using Bayesian assignment tests (e.g., Section 9.8), the estimated probability of an individual (multilocus genotype) originating from population X, versus Y or Z, might be 0.00, 0.01, and 0.99, respectively, all of which sum to a total probability of 1.0.

The **product rule** says that the probability of two independent events is equal to the product of the probabilities of the two events. The product rule (also called the "**both/and**" rule) is illustrated as follows: the probability a heterozygous parent will transmit both the A allele at a locus (Aa) **and** the B allele at another locus (Bb) is $(0.50 \times 0.50) = 0.25$, assuming independent loci. For an example application, consider a wildlife forensics case where the four-locus genotype from a blood stain is $Aa/Bb/CC/dd$. What is the probability of randomly sampling a second individual with an identical genotype from this population, if the genotype frequencies are as follows: $Aa = 0.25$, $Bb = 0.50$, $CC = 0.10$, and $dd = 0.10$? Using the product rule (and assuming four independent loci), P(Aa Bb CC dd) = $(0.25)(0.50)(0.10)(0.10) = 0.00125$ (see also Example 20.1).

The probability of an event can be estimated from a large number of observations – e.g., flipping a coin hundreds of times and computing the long-term frequency of heads versus tails. This is called an empirical probability because it is obtained through empirical observations. This conceptual framework involving repeated events and their "long-run" frequency is known as the "frequentist approach" to probability and statistics.

The above concepts of probability are "objective probabilities". That is, there is no subjectivity, best guess, or intuition involved in computing the probability. For example, we know from Mendel's laws that each allele at a locus generally has an equal probability of being transmitted. Furthermore, if we did not know the probability (50 : 50), we could empirically estimate the probability via repeated observations (e.g., repeated transmissions of alleles through genealogies or pedigrees).

A disadvantage of this frequentist approach is that it cannot give probability estimates for rare or infrequent events. Further, frequentist probability estimates cannot incorporate common sense or prior knowledge because the estimates are based only on a sample. For example, if you flip a coin 10 times and obtain only three heads your probability estimate will be 0.30. However, prior knowledge that unfair coins are rare would lead us to suspect that the estimate of 0.30 is too low (and should be close to 0.50). In this case a more subjective approach to estimating probability could be used to incorporate of all available information (prior information that unfair coins are rare), and thereby obtain an estimate closer to 0.50.

"Subjective probability" is an important concept because it facilitates an alternative approach for describing probabilities. It can take into account previous knowledge or "best guesses". For example, when computing the probability of extinction for a certain population, we can use input parameters in a population viability model (e.g., VORTEX, Chapter 14), which include "best guesses" or intuitive predictions. When modeling population viability and the cost of inbreeding on population growth, we might use the average cost measured across mammals in captivity, if no data exist for our particular mammal species. The average cost of inbreeding is approximately a 30% reduction in juvenile survival of progeny produced by mating full-sibs ($F = 0.25$) for mammals in captivity (see Section 13.5). This best guess of the cost is a somewhat "subjective probability" if we do not measure the cost in the actual species and population being studied.

Another example of a "subjective probability" (and nonfrequentist approach) is when estimating the probability of a $1°$ temperature increase due to global warming. This type of computation often is conducted using a somewhat subjective model and parameter values (e.g., including uncertainties inherent in the feedback processes that must be included in climate models).

Subjective probabilities are used in the Bayesian statistical approach (described below) that uses Bayes' theorem to incorporate prior information. The Bayesian approach uses a modifiable (or relativist) view of probability by using prior probability estimates (from prior knowledge) and then updating them with new data (from new observations) to give an "improved" posterior probability estimate.

A1.1 Joint and conditional probabilities

We often must compute the probability of two events (E) occurring at the same time. This leads us to consider joint and conditional probabilities. For an example, in order for inbreeding to increase the risk of population extinction, it is necessary that inbreeding reduces individual fitness ($E1$ = inbreeding depression) and that the reduced individual fitness also leads to reduced population fitness ($E2$ = reduced population growth rate). Here, $P(E1$ and $E2)$ is the joint probability of $E1$ and $E2$. Joint probabilities are important in the modeling of complex processes (e.g., Bayesian inference of processes) that have multiple sources of variation; for example, allele frequency changes are influenced by multiple sources of variation such as drift, selection, and migration (Beaumont and Rannala 2004).

A conditional probability is the probability of an event given that another event has happened. Conditional probabilities are used whenever considering events that are not independent. For example, if the effect of inbreeding on fitness increases with environmental stress, then we could compute the probability of inbreeding depression conditional upon a certain stress such as temperature change (resulting from global warming or an unusually hot summer). A conditional probability, the probability of *E2* given *E1*, i.e., conditioned on *E2*, is defined as follows:

$$P(E2 \mid E1) = \frac{P(E1 \; and \; E2)}{P(E1)} \qquad (A1.1)$$

Note that conditioning on an independent event does not change the probability of the event, i.e., $P(E1 \mid E2) = P(E1)$.

Bayes' theorem is used to obtain a posterior probability conditioned on the data available from a sample. The posterior probability $P(E1 \mid E2)$ uses the prior probability $P(E1)$ conditioned on the event *E2* (the sample of data). Thus, the Bayesian approach computes revised (updated) estimates of the probability of event *E1* by conditioning on new data (*E2*), as data become available. A prior probability can be "flat" and thus uninformative; For example, we could consider that microsatellite mutation rates range from 10^{-2} to 10^{-6}, with all values having an equal probability (a flat probability distribution). Alternatively, we could use a bell-shaped prior probability distribution with a higher probability for mutation rates between 10^{-3} and 10^{-4}, which is consistent with published observations suggesting that mutation rates are most often near 10^{-3} or 10^{-4}.

A1.2 Odds ratios and LOD scores

Another probability concept important in conservation genetics is that of "odds". The probability of an event can be expressed as the odds of an event. The odds ratio for an event *E* is computed as the probability that *E* will happen divided by the probability that *E* will not happen. Thus, for example, the probability 0.01 has the odds of 1 to 99 (or 1/99). Odds ratios (also called likelihood odds ratios) are used, for example, in paternity analysis to decide if one candidate father is more likely than another candidate to be the true father (Marshael et al. 1998).

Odds ratios also are used in assignment tests to decide if population X is more likely than population Z to be the origin of an individual (Banks and Eichert 2000). For example, we can compute the probability (expected genotype frequency, e.g., 2pq, for a heterozygote) of a multilocus genotype originating (occurring) in Pop X versus Pop Z. If the logarithm of the ratio of the probabilities is very large (e.g., $Log_{10}\{P[Pop \; X]/P[Pop \; Z]\}$), we can conclude that Pop X is the origin of the individual. For example, we might decide to assign individuals to Pop X if the log of the odds (LOD) ratio is at least 2.0. In this case, with LOD = 2.0, we expect only 1/100 erroneous assignments where an individual assigned to Pop X actually originates from Pop Z. If the LOD score is 3.0, we expect only 1 in 1,000 erroneous assignments (e.g., Banks and Eichert 2000).

A2 Statistical measures and distributions

A statistic is any descriptor of some characteristic of a "population" of observations. Statistics are computed from samples because the entire population of observations usually can not be collected. We can divide statistics into five categories based on the questions they address: descriptive, tests for differences, tests for relationship, multivariate exploratory methods, and estimators of population parameters (Dytham 2003).

A2.1 Kinds of statistics

Descriptive statistics are computed to describe and summarize sample data during the initial stages of data analysis, without fitting the data to a probability distribution or model (e.g., the normal distribution or model). Since no probability models are involved, descriptive statistics are not used to test hypotheses or to make testable predictions about the whole population. Nevertheless, computing descriptive statistics is an important part of data analysis that can reveal interesting features in the sample data. Examples of descriptive statistics are the mean and variance, which are described below (see Section A2.2).

Tests of difference address questions like "is genetic variation (heterozygosity) different in population A and B?". Here, the null hypothesis becomes "A and B are not different". Tests for differences can also be used to compare distributions. For example we might ask if the shape of the distribution of allele frequencies is different in population A and B (e.g., if the proportion of low frequency alleles is the same in A and B). There are many statistical tests for differences, including parametric and nonparametric tests described below.

Tests for relationship ask questions like "is fitness related to inbreeding level or heterozygosity?". A null hypothesis might be: "heterozygosity is not associated with juvenile survival." Two classes of tests for relationships are correlation and regression. Correlation assesses the degree of association without implying a cause and effect. Regression fits a relationship (e.g., linear or curvilinear) between two variables so that one can be predicted from the other, implying a cause and effect relationship. The effect of inbreeding on fitness traits can be predicted via regression (lethal equivalents, see Section 13.5.1 and Figure 13.12). We could imagine a scenario where inbreeding is associated with reduced fitness, but inbreeding is not the direct cause. For example, if individuals from population A are more inbred, but also have poorer nutrition than individuals from population B, a correlation (between populations) for individual growth rate versus inbreeding could be caused by the environment, not genetics. A factor complicating the assessment of relationships is interactions (e.g., genetic by environment interactions). There are many ways to test for correlations, compute regressions, and account for interactions.

Multivariate exploratory techniques ask questions such as "are there major patterns in the data?", or "can we assign individuals to groups (based on multilocus genotypes)?", or "which variable (e.g., locus) is most useful (i.e., explains most the variance) when assigning individuals to groups?". Multivariate exploratory techniques can help identify hypotheses to test. In large data sets with multiple variables (e.g., many loci, morphological, or environmental measurements) we might not initially test a specific hypothesis because so many potential hypotheses exist. Exploratory techniques are more appropriate for generating hypotheses than for formally testing them (i.e., they do not yield P-values, likelihoods, or probability values). A wide range of statistical approaches exist such as principal component analysis (PCA), frequency correspondence analysis (FCA), multidimensional scaling (MDS), cluster analysis, analysis of variance (ANOVA), or analysis of molecular variance (AMOVA) (e.g., Section 9.7).

Statistical **estimators** infer a population parameter using data that are related to that parameter. For example, we could infer the effective population size (N_e) from data on the temporal change in allele frequencies between two generations. Change in allele frequencies is influenced by N_e, but might also be influenced by sample size, population structure, demographic status (expanding/declining), and selection or mutation rates. There are different approaches to statistical estimation (method of moments, maximum likelihood,

Bayesian, and approximate Bayesian methods based on summary statistics; see below, e.g., Sections A3, A4, and A5).

Statistical tests (e.g., for differences or relationships) can be divided into two classes: parametric and nonparametric. **Parametric** statistics assume that the data follow a known distribution (a probability distribution) – usually the normal distribution. Parametric distributions can be defined completely using very few parameters (e.g., only the mean and variance, in a function or formula). Parametric statistical tests are generally more powerful than nonparametric tests, and thus are preferred (see below). An example parametric test is the t-test, which assumes a normal distribution; it can be used to compare mean heterozygosity from two population samples, if the distribution of heterozygosity among the loci is similar to the normal distribution (Archie 1985).

Nonparametric statistics require no knowledge (no assumptions) about the distribution of the data or test statistic. Therefore nonparametric statistics are called "distribution-free" tests. They are also called "ranking tests" because they often involve ranking observations to generate an empirical cumulative distribution. These tests are generally less powerful, but safer than parametric tests if the data might not follow a parametric distribution. An important example is the Wilcoxon's signed-ranks test (a nonparametric version of the t-test), which often is used to compare mean heterozygosity from each of two population samples.

A2.2 Measures of location and dispersion

In statistics, the "population" is defined as the totality of the observations of some characteristic we are studying. The sample is a subset of observations. We compute sample statistics to infer the population parametric value of a parameter (e.g., the mean). For any trait X, the general formula for the sample mean and population mean are as follows:

$$\bar{x} = \frac{\Sigma x_i}{n} \tag{A1.2}$$

$$\mu = \frac{\Sigma x_i}{N} \tag{A1.3}$$

where i is the individual number, the bar over the x is the mean, and N and n are the population size and sample size, respectively.

The mean is a statistical measure of "central value" or the central location (of a distribution). The arithmetic mean is given by expression A1.2. Another kind of mean important in population genetics is the harmonic mean (see expression 7.8), which gives more weight to observations with small values. The harmonic mean is used for computing the effective population size from successive N_c estimates. An interesting controversy in conservation genetics results from, in part, confusing the arithmetic and harmonic mean when computing the ratio of $N_e : N_c$. The N_e, averaged across generations, is always computed as a harmonic mean, whereas N_c averaged across generations is often computed as an arithmetic mean. The harmonic mean is strongly influenced by low values causing the (harmonic) mean N_e estimates to be lower than (arithmetic) mean N_c. Thus, the estimates of $N_e : N_c$ ratios (averaged across generations) can be biased low due to the statistical artifact of using the harmonic mean of N_e but the arithmetic mean of N_c (see Chapters 7 and 15).

Other familiar measures of central location are the median and mode. An advantage of the median is that it is less influenced than is the mean by the skewness of the distribution of the statistic (i.e., the median is resistant to extreme high or low outlier values). Thus, the median is said to be a relatively robust (or resistant) measure of central location.

A statistical measure of variability (or "dispersion") of points around the mean is the variance. If all points have the same value there is no dispersion and the variance is zero. If points have only very high and very low values, the variance would be quite high. The variance is the average of the squared deviations from the mean – i.e., the mean is subtracted from each observation point, this difference is squared, and the average of the squares in computed. The population variance (σ_x^2) and sample variance (s_x^2) are computed as follows:

$$\sigma_x^2 = \frac{\Sigma(x - \mu)^2}{N} \tag{A1.4}$$

$$s_x^2 = \frac{\Sigma(x - \bar{x})^2}{n - 1} \tag{A1.5}$$

where n (and N) is the number of sample (and population) observations, as above.

The standard deviation is another important measure of dispersion. It is computed as the square root of the variance ($s_x = \sqrt{[V(x)]}$, where s_x is the standard deviation). We take the square root of the variance to avoid having to think in terms of squared measures, which are less interpretable (for example, it is easier to interpret the "height" of individuals than the "height squared"). Furthermore, recall that one standard deviation under the normal (bell-shaped) distribution encompasses 68% of the central area, while two standard distributions encompasses 95%, and three standard deviations contain 99% (99% fall between $\mu \pm 3\sigma$; see Section A2.3). Probability distributions such as the normal distribution, and their use for describing dispersion, are discussed more in the next section.

The standard error is a measure of the dispersion of a sample statistic (e.g., the sample mean, $\bar{x}$). The standard error of the mean should not be confused with the standard deviation of a variable, which describes the probability distribution of the underlying raw data or parameter (x). For example, the standard error describes the distribution of the sample mean heterozygosity, whereas the standard deviation describes the sampling distribution of the raw parameter heterozygosity (see Section A2.4 and Example A2 below). Probability distributions are discussed in the next section. Unfortunately, in publications, standard error and standard deviation are often confused or not well differentiated.

A2.3 Probability distributions

Probability distributions are important to understand because statistical tests and estimators require the use of a probability distribution. Different types of variables (mean, variance, F_{ST}) have different probability distributions (Figure A1).

A probability distribution for a discrete variable x (e.g., number of subpopulations represented in a sample of individuals) gives the probability of all the possible values of s

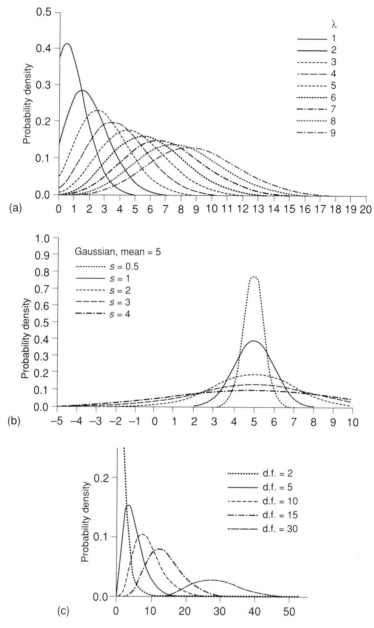

Figure A1 Probability distributions important in conservation genetics. (a) The Poisson distribution with a mean from 1 up to 9 (nine curves). (b) Normal (Gaussian) distributions with variance (*s*) from 0.5 to 4, and mean 5. (c) The chi-square distribution (d.f. refers to the degrees of freedom). Modified from P. Bourke (personal communication).

($x = 1, 2, 3 \ldots$). Probability distributions are generally illustrated graphically as a curve (or frequency histogram). The total area under a probability curve is 1.0. The probability of a rare or unusual observation is represented as a small area (e.g., 0.05) in the tail(s) of the distribution. We can obtain an empirical estimate of a probability distribution by plotting the relative frequency (histogram) of occurrence of each observation (for example, height of each individual) in a sample.

Binomial

An important probability distribution in genetics is the binomial distribution. The binomial is one of several theoretical probability distributions used for modeling (approximating) the distribution of observed data that occur in discrete classes (e.g., genotypes at a locus), as opposed to a continuous distribution of observations (e.g., height). The binomial is useful for modeling the proportion of binary events (male versus female births; transmission of allele A versus a; or survival versus death) that occur in a population sample of size n. Note that when more than two events are possible, we can use the multinomial distribution – a simple extension of the binomial.

The binomial distribution contains information on the number of times, x, an event with probability π occurs in a fixed number of observations n. The binomial distribution can be described as:

$$P(X = m) = \frac{n!}{(n - m)!m!} \pi^m (1 - \pi)^{n-m} \qquad (A1.6)$$

The factorials in the fraction (left side) give the number of ways m positive outcomes (transmission of A) can occur out of n events (offspring). The binomial has a variance of:

$$V(X) = n\pi (1 - \pi) \qquad (A1.7)$$

For example, if the probability of transmitting the A allele is $\pi = 0.50$, then out of 100 transmissions (offspring), we expect a mean of $100 \times 0.50 = 50$ transmissions of the A allele, with a variance of $100 \times 0.50 \times 0.50 = 25$ (standard deviation $= 5.0$). When the number of observations (n) becomes large, the binomial approaches the normal distribution.

Poisson

The Poisson distribution is another discrete distribution that is widely used in conservation genetics and ecology (Figure A1). The Poisson assumes an event is rare (relative to the maximum number of possible events), and that events are independent. Thus, the Poisson is used to model rare and independent events that occur in a spatial or temporal sample. For example, in genetics, the Poisson is used to model the probability of mutations through time (e.g., under the coalescent, see Section A9), because mutations are rare events that arise randomly (among individuals or lineages). The Poisson also is used to model variance in family size (reproductive success), as in expression 7.5 (see Figure 7.4). Ecologists use the Poisson to test if the distribution of organisms over space is uniform versus random. For example, if the observed variance in distance between individuals is less than the mean distance, then the spacing is more uniform than random.

An important property of the Poisson is that the mean equals the variance. For example, when using the Poisson to model a stable-sized (stationary) population, the mean family size (number of offspring per mating pair) equals two, as does the variance. This widely used model is called the "Wright–Fisher model" (see Section 6.1). In such an ideal model, the effective population size (N_e; see Section 7.1) equals the census size (N_c). Although the Poisson is useful here, we know that in natural populations N_e is generally less than N_c because, for example, the variance in family size is often high (>2.0) (see Figure 7.5). Thus the Poisson is not always the most appropriate distribution for modeling N_e or variance in reproductive success in natural populations.

Under the Poisson, the probability of any number x of occurrences is:

$$P(x) = \frac{e^{-\mu}\mu^x}{x!}$$

(A1.8)

where the mean number of occurrences μ equals N (the population size).

Normal

The normal (Gaussian) distribution is the most widely used continuous distribution – it is the famous symmetric "bell-shaped" curve (Gauss 1809; see Figure A1b). The binomial distribution approaches the normal as sample sizes increase. Thus, for example, the shape of the distribution of the observed heterozygosity (H_0) at a locus approaches a smooth bell shape, when sample size approaches 50–100 individuals.

The normal distribution is useful for modeling many observed variables because of the central limit theorem, which states that the distribution of the sample mean will approach the normal distribution as the sample size of observations increases (even if the observed variable itself is not normally distributed!). A normally distributed random variable is described by the following function:

$$P(x) = \frac{1}{\sigma\sqrt{2\pi^2}} e^{\frac{-(x-\mu)^2}{2\sigma^2}}$$

(A1.9)

For any continuously distributed variable, the probability distribution is defined as the probability of a random variable being less than or equal to a particular value $P(X \le x) = P(x)$. Here, $P(x)$ is called the probability distribution function. The derivative of the probability distribution is called the probability density function (pdf). The area under any segment of a pdf curve is the probability of X being in a certain interval. Note that a pdf is the output of Bayesian analyses (posterior distribution) and also of maximum likelihood estimation (likelihood curve) where we are estimating the probability of some parameter (e.g., N_e, F_{IS}, or mutation rate; see below).

The population probability distribution can be estimated empirically by computing the cumulative frequencies of observations in a sample (e.g., by plotting a histogram of cumulative frequencies of observations having values less than x). The accuracy of the empirical distribution (as an estimate of the population probability distribution) increases with large sample sizes.

Chi-square

The chi-square distribution is another continuous distribution widely used in statistics and in conservation genetics. It is asymmetric, unlike the normal, and ranges from zero to infinity. The chi-square distribution is used to model and conduct tests comparing variance measures; thus the chi-square probability distribution is used when studying, for example, the variance in allele frequencies (F_{ST}). The chi-square can be used to compute confidence intervals around F_{ST} or around N_e estimates that are based on the temporal variance in allele frequencies. Chi-square tests (using the chi-square probability distribution) are discussed extensively in Section 5.3 and Example 5.1.

We remind readers that chi-square tests use numbers (not proportions), and that if the "expected number" in any class (e.g., genotype class) is less than approximately one (<1.0), we should consider using an exact multinomial test (based on the multinomial probability distribution). Exact tests are explained in Example 5.3. Exact tests are performed by determining the exact probabilities of all possible sample outcomes, and then summing the probabilities of all equal and less probable sample outcomes, to obtain the exact probability of the observed outcome.

A sampling distribution is a probability distribution of a sample statistic (e.g., the mean). Readers should not confuse probability distributions of sample statistics (e.g., mean heterozygosity of a sample) with probability distributions of the underlying parameter (e.g., heterozygosity at loci). Two important characteristics of sampling distributions are: (1) they have lower variance than parameter distributions, simply because each sample includes multiple observations; and (2) they approach the normal distribution for large sample sizes, no matter what the parameter – a surprising principle of the central limit theorem. This interesting phenomenon (and the central limit theorem) explains why we so often see the normal distribution used in statistical tests and for computing confidence intervals.

A2.4 Interval estimates: confidence intervals and support limits

Interval estimates are usually more useful than point estimates. In fact, without an interval estimate, a point estimate (e.g., mean H_e, F_{ST}, or N_e) is generally of little value. Two kinds of interval estimates that often are used in conservation genetics are confidence intervals (for frequentist approaches) and support limits (e.g., in likelihood-based and Bayesian approaches).

Confidence intervals give the range of values within which the true population parameter (e.g., population mean) is likely to occur, with some chosen probability (usually 95 or 99%). Thus, confidence intervals (CIs) are a measure of spread. Publications often report 95% CIs, which should span all but 5% of outcomes from repeated, independent sampling events. Note that error bars (e.g., on histograms) often report ±1 standard errors (±1 SE, or standard deviations of the mean), which represent 68% CIs for normal/Gaussian distributed statistics (Example A2). Note also that 95% CIs are nearly twice as wide as 68% CIs, i.e., a 95% CI represent approximately ±2 SE (Figure A3).

To compute a 95% CI, we choose an alpha value of 0.05. Alpha (α) is the critical threshold P-value used for rejecting the null hypothesis (e.g., if $P < 0.05$). For a sample statistic $t(x)$, we can compute a $[(1 - \alpha)100\%]$ confidence interval as $[t_{\alpha/2}, t_{1-(\alpha/2)}]$, with lower and upper confidence limits of $t_{\alpha/2}$ and $t_{1-(\alpha/2)}$, respectively (where t_n is the nth quantile of the sampling distribution of the population parameter T).

Support limits are used in likelihood and Bayesian approaches instead of CIs. Support limits can be computed, like confidence limits, such that the estimated sampling distribution (likelihood or posterior distribution) has cut-off points placing 2.5% of the probability density area in each tail. For a graphic illustration, see Figure A3b. Support limits are

Example A2 Comparison of different types of error bars

Consider a hypothetical study where you discover a brain protein (language destroying enzyme, LDE) that causes people to utter strange words (Streiner 1996). You think LDE is in higher concentrations in administrators than in other people. You sample 25 administrators and 25 other people (as a control group) and compute the mean and standard deviation (Table A1). You present the data in a bar graph to make it more visually interpretable (Figure A2, from Streiner 1996).

But how do you compute the error bars to extend above and below each histogram bar? In all studies it is important to report the standard deviation because this shows the dispersion of the actual raw data points. However, the reader generally also wants to know the sample-to-sample variation. For example, if we repeat this study 100 times, how much variation between the means of each study would we expect? Stated another way, how much confidence do we have in the

Table A1 Levels of LDE in the cerebrospinal fluid of administrators and controls (Streiner 1996).

Group	Number	Mean	SD
Administrators	25	25.83	5.72
Controls	25	17.25	4.36

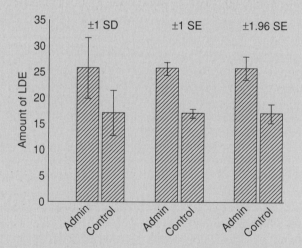

Figure A2 Computing error bars using standard deviations, standard errors (i.e., standard deviations of the mean), and 95% confidence intervals (assuming a normal distribution). Note that 1.96 SE represents 95% confidence intervals. Because the error bars do not overlap for the ±1.96 SE, we can conclude with 95% confidence that the administrators and controls are significantly different. From Streiner (1996).

estimation of the population mean from our sample mean? For this we must compute a standard error (i.e., a standard deviation of the mean).

Should we report one or two standard errors? We are generally interested in a range of values in which we are 95% certain. Thus we could report 2 SE, which should contain approximately 95% of the study means (Figure A2). Furthermore, 2 SE are used to compute exact 95% confidence intervals (assuming a normal distribution) when testing for statistically significant differences between populations means.

For example, using our table of the normal distribution, we find that 95% of the area falls between −1.96 and +1.96 SE (standard deviations of the means, for this example). We compute 95% CIs as follows:

$$95\% \text{ CI} = M \pm (1.96 \times \text{SE})$$

where M is the mean.

Of course, ±1.96 SD of the mean nearly equals ±2 SD of the mean. Confidence intervals show the range in which statistically significant differences exist between means. Showing 95% confidence intervals (or ±2 SE) supports statistical testing (see Section A3) and allows for an "eyeball test" of significance. Note that this eyeball approach does not work accurately when more than two groups are compared because of issues of multiple tests.

How do we interpret the error bar results? If the top of the lower bar (controls) and the bottom of the upper bar (administrators) do not overlap, then the difference between the groups is significant at the 5% level (see Section A2.4). We could then conclude that administrators have higher concentrations of LDE.

generally reported with (or plotted on) a probability curve (likelihood or posterior distribution) allowing easy visualization of the probability of different outcomes just by "eye-balling" the curve (Figure A3b). This makes interpretation of probability estimates (from probability curves) more straightforward than frequentist CIs.

A3 Frequentist hypothesis testing, statistical errors, and power

Hypothesis testing is widely used across scientific disciplines. It requires a formal statement called the null hypothesis (H_0), followed by a statistical test of the null, which determines the probability of null being true, by computing a P-value or a likelihood (probability) distribution. The null hypothesis is a negative statement that mirrors the alternative hypothesis. For example, a null hypothesis might be: "population X is stable or growing". The alternative hypothesis is: "population X is declining".

Errors in rejecting the null hypothesis can arise because we usually have only a small sample from an entire population and because statistical tests give only a probability that the hypothesis is true. Two kinds of errors, Type I and Type II, are possible when conducting

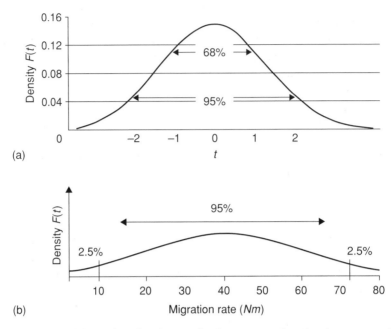

Figure A3 (a) A normal sampling distribution for the statistic t showing the upper and lower 68% and 95% confidence limits $[t_{\alpha/2}, t_{1-(\alpha/2)}]$, where α (critical/threshold P-value) equals 0.32 and 0.05, respectively (see text). (b) Hypothetical probability (likelihood) distribution output from a likelihood estimation (e.g., of Nm), and 95% support limits identified by placing 2.5% of the area in each tail of the distribution.

a statistical test; they are, respectively: (I) rejecting the null when it is true; and (II) failing to reject the null when it if false (Table A2).

The lower the P-value, the more confident you are that the H_0 is false. For example, if $P < 0.001$, you expect that in only one in 1,000 independent experiments would you observe an outcome (statistic) as unusual as the one observed. A P-value of 0.05 is often used in hypothesis testing as the threshold (α value) for rejecting the null hypothesis. When $P = 0.05$, we have five chances in 100 of rejecting the null when it is true (Type I error). The use of 0.05 is arbitrary and other α values can be used (0.10, 0.01 or 0.001) depending on the importance of avoiding a Type I error.

A decrease in the Type I error rate (choosing a low critical α value) will increase the Type II error rate. Therefore choosing the appropriate α depends on the relative importance of avoiding a Type I versus Type II error. For example, consider the following null

Table A2 Type I and Type II errors that can result when testing a null hypothesis.

	Accept H_0	**Reject H_0**
H_0 True	Correct	**Type I error**
H_0 False	**Type II error**	Correct

hypothesis: "Population X is **stable or growing**". An important question is, "Would it more risky to erroneously reject the H_0 (wrongly accept that "population X is declining") or to erroneously fail to reject the H_0 (wrongly conclude that "population X is stable or growing")? If we wrongly conclude the population is stable (a Type II error), and it is actually declining, it could lead to extinction of the population or species.

In conservation biology it often is more risky to make a Type II error than to make a Type I error. Type I errors can be the more risky kind of error in other sciences, such as human medicine, where we must not reject the null when it is true. For example, we would not want to reject the following null: "H_0: medication 'X' has no side effects", unless we are highly certain ($P < 0.001$) the null is false and there are no side effects.

A3.1 One-versus two-tailed tests

Two-tailed and one-tailed hypothesis tests exist. In a **one-tailed test**, the alternative hypothesis is a deviation in only **one** direction (Figure A4b). For example, "H_A: population X is **declining**". However in a **two-tailed test**, the alternative hypothesis would be: "H_A: population X is **declining or growing**" (i.e., changing in size) (Figure A.4a). Thus a two-tailed test tests for deviations in either of two directions. A one-tailed test is appropriate when: (1) biological evidence suggests a deviation in one direction (e.g., a population has declined so we conduct a one-tailed test for reduced allelic diversity); or (2) we only care about a deviation in one direction. For example, we might use a one-tailed test for reduced heterozygosity in a population that recently became isolated, if we care only about detecting a reduction of heterozygosity.

One-tailed tests generally have more power than two-tailed tests. Thus it is important to understand the difference between one- and two-tailed tests, and to use one-tailed tests when possible and appropriate. A one-tailed test (e.g., t-test) is more powerful, because more of the "rejection region" (all 5%, not just 2.5%, in Figure A4b) is located in the one tail that we are interested in, making it easier to reject the null hypothesis.

A3.2 Statistical power

An important consideration when choosing a statistical approach or test is its statistical power (see also Section A8). Power is the probability of detecting an effect when the effect or phenomenon occurs. For example, the power of a statistical test for detecting a population decline (given that a decline occurs) is obviously important in conservation genetics.

Power is related to the Type II error rate as follows: Power $= 1 - \beta$. Thus, the power of a test depends on the choice of beta (and alpha), such that choosing a small β leads to more power, but requires a larger α. Other factors that influence power, besides α and β, are the effect size (strength of the effect, e.g., severity of population decline) and the sample size (e.g., number of individuals or loci sampled).

Power is also influenced by the chosen statistical test itself. For example, we mentioned that parametric tests (e.g., t-test for loss of heterozygosity) are expected to be more powerful that nonparametric tests. A relevant example for conservation genetics is that the most powerful test for detecting a decline in heterozygosity is not the standard t-test, but rather a paired t-test. The paired test is more powerful because it treats each locus individually and thereby reduces the influence of interlocus variation that often is high. For example, different loci in a sample might have H_e ranging from 0.2 to 0.8, but the between-sample

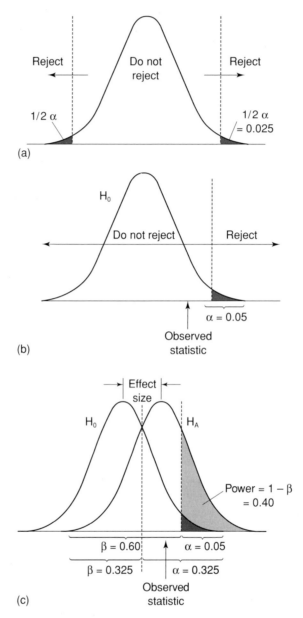

Figure A4 Illustration of a two-tailed test (a) in contrast to a one-tailed test (b). A two-tailed test is appropriate when we do not know the direction of deviation expected (e.g., we do not expect H_e to be lower (or higher) in a certain population). Panel (b) shows the conventional $P < 0.05$ (alpha $= \alpha = 0.05$) as a threshold to reject the null hypothesis, whereas panel (c) shows a more balanced approach of choosing an α value leading to similar risk of Type I versus Type II errors. Note that the risk of a Type II error (beta, β) is 0.60 when α is 0.05. However, if we choose an α of 0.325, β will also be 0.325. Further, note that the "observed statistic" does not fall in the tail (right side of vertical dotted line for $\alpha = 0.5$) in (b) ($P > 0.05$), so we would not reject the null hypothesis. However in (c), we would reject the null because the statistic is smaller than the threshold of rejection, α ($P < 0.325$). Modified from Taylor and Dizon (1999).

difference in mean H_e that we are testing might be only 0.6 versus 0.5 (e.g., in a large versus small population). For a more thorough explanation, see a statistics textbook. Interestingly, Wilcoxon's nonparametric test often is not less powerful than the parametric t-test when monitoring for loss of heterozygosity using two temporally spaced samples (Luikart et al. 1998).

A statistical power of 0.80 is often considered by statisticians as a "reasonably high" power for detecting the event of interest (e.g., population decline, migration, fragmentation, etc.), thus making it worth conducting the study of interest. A problem in science, and particularly in conservation biology, is the failure of researchers to compute the power of statistical tests. Fortunately, power analyses are becoming easier to conduct, thanks to the increasing availability of computer simulation programs that allow simulation of various population scenarios (e.g., population declines) and marker numbers and types (dominant, codominant). See the simulation programs listed on this book's website.

A3.3 Problems with *P*-values

A problem with P-values and hypothesis testing via the frequentist approach is that P-values can be difficult to interpret (compared to Bayesian posterior probabilities, see below). A P-value should be interpreted as the chance, assuming the null is true, that you will get a similar or more extreme result if you repeat an experiment thousands of times. A value of $P < 0.05$ is sometimes misinterpreted to mean that there is 95% probability that the alternative hypothesis is true. This is different from the actual definition, given in the previous sentence. Furthermore, P-values tend to overstate the strength of evidence, compared to Bayesian approaches (Malakoff 1999).

Another problem of P-values often arises when the P-value is low, but not "significant". If $P = 0.06$, researchers might not "reject the null" and subsequently conclude there is no effect, e.g., no evidence the population is declining. However, as mentioned above the choice of $\alpha = 0.05$ is generally arbitrary with no theoretical basis, and in fact, $P = 0.06$ is suggestive of an effect (especially if the power of the test is low). Recall that if the effect size is small, we are unlikely to obtain a significant P-value (e.g., $P < 0.05$), unless sample sizes are very large (see Section A3.2).

Another problem with P-values is that "negative results" ($P > 0.05$) are sometimes difficult to publish, and can lead to a bias in the scientific literature, and an underrepresentation of studies that find no "significant" effect. For example, there might be more studies published that find a correlation between heterozygosity and fitness than do not, thereby leading to a biased proportion of (published) studies finding a correlation. This potential lack of publication of "negative results" has been called the "file drawer effect", because negative results might often end up in a file drawer, unpublished.

A4 Maximum likelihood

Likelihood is the probability of observing the data given some parameter value (e.g., $Nm = 50$), under a certain statistical model (e.g., island model of migration). Maximum likelihood (ML) methods estimate the parameter value that maximizes the probability of obtaining the observed data under a given model. For example, we might compute the likelihood of each of many migration rates ($Nm = 10, 11, 12 \ldots$, up to 500), and then choose the best (point) estimate of Nm as the value that has the highest (maximum) likelihood (e.g., approximately $Nm = 40$ in Figure A3).

An advantage of likelihood analysis is that it is model based and thus allows easy comparison of different models (even complex models), thereby improving inference about complex processes (e.g., different dispersal patterns, mutation models, stable versus declining population size) that might explain the data. Likelihood analysis is often used to test the fit of two different models by using the ratio of the MLE (maximum likelihood estimate) for one versus the other model. For example, if one model is far more likely [e.g., $\log_{10}(\text{MLE1}/\text{MLE2}) > 3$], we might reject the second model for MLE2 (see Section A1.2). The two models might be, for example, a stable versus declining population, or alternatively the existence of two versus three subpopulations. Note that when "$\log_{10}(\text{MLE1}/\text{MLE2}) > 3$", the probability of MLE1 is generally 1,000 times more likely that MLE2 (e.g., $P < 0.001$); when >2 the probability of MLE1 is considered to be 100 times more likely that MLE2 ($P < 0.01$).

Likelihood methods are sometimes classified as "frequentist". For example, when we compute the expected long-run frequency of a likelihood ratio (or a likelihood value), e.g., as part of a statistical test, this is a frequentist approach.

The main advantage of maximum likelihood approaches is they use "all the data", in their raw form, and not some summary statistic (e.g., H_e or F_{IS}). Because likelihood methods use a maximum of information from the data, they should, in theory, be more accurate and precise than moments-based methods. For example, likelihood-based methods use the raw data (number and genealogical divergence of each allele) to estimate N_e (or Nm), and not a single summary statistic, e.g., H_e (or F_{ST}), as in classic moments-based estimators of N_e (or Nm) (see Guest Box 7, or expression 9.12).

Different (raw) data sets can give the same summary statistic, e.g., F_{ST}, whereas different raw data sets are less likely to yield the same ML estimates. For example, two independent sets of temporally spaced samples can have the same F_{ST} (temporal F_{ST}) even though they have different numbers of alleles. When using the summary statistic F_{ST} to estimate N_e (as in the classic temporal variance method; Waples 1989), we would not be using the information about the proportion of rare alleles, and thus might not achieve the most accurate or precise estimate of N_e. In another example, two independent metapopulations could have the same F_{ST}, but have different proportions of rare alleles. Information about the proportions of rare alleles can help infer if a metapopulation is stable, fragmenting, or growing in size (e.g., Ciofi et al. 1999).

In actual practice, ML methods often are more accurate and precise than moment-based methods. For example, estimators of N_e based on likelihood provide tighter confidence intervals and less biased point estimates (Williamson and Slatkin 1999; Berthier et al. 2002). However, likelihood-based estimators generally require large sample sizes and can be biased and less precise than simpler summary statistics (moments-based methods) if sample sizes are small, e.g., less than 40 or 50 individuals (see, for example, Lynch and Ritland 1999).

A5 Bayesian approaches and MCMC (Markov chain Monte Carlo)

There are two main ways Bayesian inference differs from classic frequentist statistics. First, probabilities are defined and interpreted differently. In frequentist statistics, P-values (probability values) are interpreted as the long-term average outcome of a repeated experiment. *P*-values are interpreted as the probability of the test statistic being that extreme (or more extreme) if the null hypothesis is true. A frequentist test might yield $P = 0.05$, meaning there is 5% chance of observing the test statistic simply by chance alone.

Bayesians computations yield a more straightforward and informative probability answer that is easier to interpret than a *P*-value. For example, a Bayesian posterior distribution might yield a probability of $P = 0.95$, meaning there is 95% probability that N_e is less than 100. Recall that in the more complicated (less direct) frequentist approach, we would construct a null hypothesis (e.g., H_0: N_e is ≥ 100), and then reject the null if the *P*-value is low (e.g., $P < 0.05$); thereby finding support for the alternative hypothesis of interest "N_e is less than 100".

Furthermore, Bayesian posterior probability distributions (and support limits) are easier to interpret than confidence intervals because probability distributions show visually the probability as the area under a curve (e.g., in the tails of a probability distribution). We immediately get a feel for the width and degree of skewness of the probability distribution by observing the posterior distribution, which we cannot get from reading confidence limits. Thus, a probability distribution (posterior probability distribution) carries more information than a classic confidence interval and it gives a better feel for the relative probability of different parameter values (e.g., small versus large N_e, or Nm, or F_{IS}) (Ayres and Balding 1998).

Second, perhaps the main advantage of the Bayesian approach is the ability to factor in prior data or information when estimating the posterior probability that a hypothesis is correct. Bayes' theorem was developed to allow easy "updating" of an existing estimation when presented with new data such as observations from a new experiment. Classic frequentist statistics generally require each experiment to be totally independent and without reference to previous experiments. Prior information (previous data or even a hunch) can be incorporated into the computation of a probability (posterior probability) by multiplying the likelihood function by the prior information (Figure A5).

An example use of the Bayesian approach to incorporate prior information is estimating N_e when the population census size is known (e.g., $N_c = 250$). Here, we can use the prior knowledge of N_c, and knowledge that N_e cannot be more than twice the census size ($N_e \leq 2N_c$; see Chapter 7). Thus the prior probability of N_e being greater than 500 equals zero ($P[N_e > 500] = 0.0$; as in Berthier et al. 2002). Further, we know that N_e is often less than $^1\!/_2 N_c$ (Frankham 1995; see Section 7.10). This information can be used to give more "weight" to N_e estimates near or below $^1\!/_2 N_c$ (e.g., using a prior probability distribution, see below).

Another example use of prior information is in models that incorporate mutation dynamics. Published data suggest that most microsatellites have mutation rates between 10^{-2} and 10^{-5}. So, we might use a flat prior ranging between 10^{-2} and 10^{-5} when modeling humans or other mammals. We also know that the average mutation rate is near 5×10^{-4}. Thus we might use a more informative prior – e.g., a bell-shaped prior (not flat) with a high probability peak near 5×10^{-4}. For an actual example, Beaumont (1999) used a prior mutation rate greater than zero for monomorphic loci, thereby allowing the use of monomorphic loci when testing for population bottlenecks. Other bottleneck inference tests do not use monomorphic loci (Luikart and Cornuet 1998). See Lewis (2001) for a simple example of Bayesian computation.

The Bayesian approach to incorporating prior information can be especially useful in conservation biology because it facilitates decision making when data are few and we

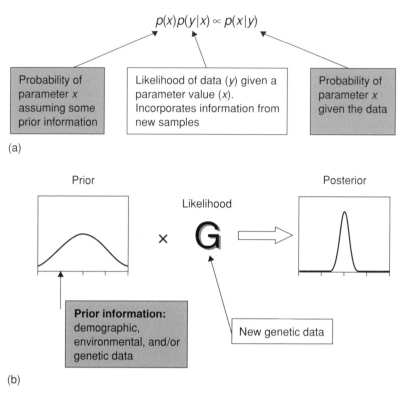

$$p(x)p(y|x) \propto p(x|y)$$

| Probability of parameter x assuming some prior information | Likelihood of data (y) given a parameter value (x). Incorporates information from new samples | Probability of parameter x given the data |

(a)

Prior

Likelihood

Posterior

× **G** ⟹

Prior information: demographic, environmental, and/or genetic data

New genetic data

(b)

Figure A5 (a) Simplified Bayesian mathematical expression showing how the Bayesian approach allows us to combine the information from the data with prior information about the parameters of the model in order to obtain their **posterior** distribution (to estimate a parameter). (b) Illustration of how prior information (in the prior probability distribution) is modified by the multiplication of it by the likelihood function (from the standard likelihood-based approach) to obtain a posterior probability distribution. Modified from O. Gaggiotti (personal communication).

want to integrate all available knowledge. In conservation biology, we often must make decisions based on limited data. For example, wildlife managers often must decide if a population's size is large enough to allow harvest, or alternatively if the population needs protection, monitoring, or supplementation. Interestingly, the United States National Academy of Sciences panel recommended that fisheries scientists consider Bayesian methods to help estimate fish population status and guide management policies (Malakoff 1999). Harvest quotas could be more appropriate and flexible if the risk of population decline were calculated directly via Bayesian statistics (incorporating prior information such as the probability that harvest actions might endanger a stock).

The main criticism of Bayesian approaches is that they can be strongly influenced by prior information, and thus be less objective than classic approaches. For example, two different people could use different prior information and obtain different results. A counterargument is that we can quantify the effects of different priors (e.g., via sensitivity analysis using different priors); thus we can (and should) consider the magnitude of influence of the prior when making management decisions. Often prior information has

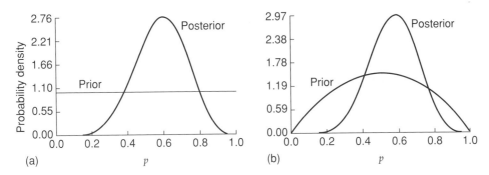

Figure A6 Probability of heads (p) in a coin-flipping experiment illustrating (a) a flat and (b) an informative prior distribution. Here, the prior has little effect, as is the case when extensive data exist and the likelihood function (alone) is relatively informative. From Lewis (2001).

little influence on the posterior, especially if data are extensive (Figure A6). Unfortunately, such sensitivity analysis is not always conducted. It seems reasonable to use both Bayesian and classic frequentist approaches in many applications (e.g., estimation of F_{ST}, N_e, or Nm, especially when one or both have been poorly validated).

An important general contribution of Bayesian approaches is they allow for computations using complex models that that could not be achieved using the other statistical approaches (Beaumont and Rannala 2004). Baysian computation using complex models has been greatly facilitated by Markov chain Monte Carlo (MCMC) computational methods.

A5.1 Markov chain Monte Carlo (MCMC)

MCMC is a simulation-based methodology to generate probability distributions that are difficult or impossible to obtain from analytical equations (including likelihood equations). Analytical equations often cannot be developed to describe complex processes with many variables (e.g., population size, allele frequencies, and mutation rates). MCMC allows simulation of a special kind of stochastic process known as a Markov chain. A Markov chain generates a series of random variables whose future state depends only on the current state at any point in the chain (Beaumont and Rannala 2004).

MCMC allows us to obtain random samples from "sample space", even when the sample space is enormous (e.g., billions of phylogenies or genealogies). MCMC combines: (1) a Markov chain model – i.e., a model involving a random walk (chain of random steps) in which the next step is determined by the characteristics of the current or previous step; and (2) the "Monte Carlo" process of drawing a random number that is necessary at each step of the random walk (Monte Carlo is a city famous for gambling, which also uses random events like the rolling of die).

MCMC is well illustrated by an analogy of a robot taking a random walk in a square field (Figure A7). Each step of the robot varies in length and direction, randomly. Eventually, the robot visits every space within the field. However the robot spends more time in spaces that are on hill tops at higher elevation (i.e., having higher probability). This is achieved by using a model with the following two main rules: (1) if a step takes the robot uphill, the robot will automatically take it; and (2) if a step would take the robot downhill,

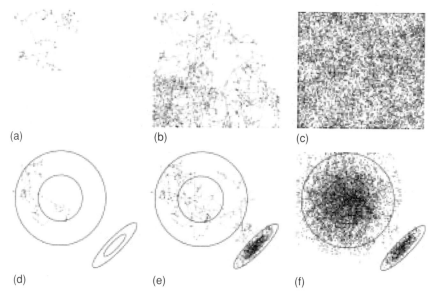

(a) (b) (c)

(d) (e) (f)

Figure A7 Illustration of the principles behind the Markov chain Monte Carlo (MCMC) methods using a simple analogy of a "random walk" in a square field by a robot (a–c). The robot begins its walk in the upper left corner and continues for 100 steps (a), 1,000 steps (b), and 10,000 steps (c) until nearly every portion of the field has been covered. Now supposing that two hills are present, represented by concentric circles and smaller concentric ovals (d–f). The robot will take steps to points in proportion to their elevation, and thus higher points will be visited more often than lower ones. The proportion of time spent in any place approximates the probability of that location. From Lewis (2001).

the robot only takes the step with a probability depending on the elevation reduction (this probability can be computed several ways, e.g., via "Metropolis" or "Metropolis–Hastings" methods).

The first few steps (usually thousands of steps) are called the "burn in", and are discarded to reduce the influence of the starting point (bias). Once burn in is achieved, the MCMC simulation has **converged** (i.e., become independent of the starting point). The remaining steps (after convergence) give a good approximation of the landscape (probability space). This simulation of a random walk allows for estimation of the parts of the sample space with the highest probability (e.g., maximizing the probability of the data, given the model, as in maximum likelihood estimation; see Section A4). Under the Bayesian approach (see below), MCMC simulation is often used to sample from the posterior distribution of a parameter in order to generate the posterior probability estimate of the parameter.

The main problem with MCMC approaches is we sometimes are not sure we have conducted a long enough burn in to achieve convergence and thus avoid bias. Also, MCMC simulation programs are generally difficult to write in computer code, and thus errors (bugs) are relatively likely to occur and can be difficult to detect.

MCMC is primarily used within Bayesian approaches, but can also be used in maximum likelihood estimation. For example, some available software programs can use flat priors (or no priors) and give as output a likelihood (probability) curve or a posterior distribution if prior information is used.

A6 Approximate Bayesian computation (ABC)

Approximate Bayesian computation (ABC) employs a Bayesian framework (e.g., incorporating prior information) to output an "approximate" posterior probability distribution. This posterior distribution is only an estimation of the full posterior, because all the raw data is not used to compute the posterior. Instead, the posterior is approximated by summarizing the data using multiple different summary statistics. For example, a full (exact) Bayesian approach would conduct MCMC simulations to obtain the exact posterior probability of the raw sample (allele number and frequency distribution), using each of thousands of simulated data sets (e.g., genealogies) for the population model under consideration (e.g., a stable, isolated population). Here, for example, we might consider population models with $N_e = 10, 20, 30$, etc., if we were estimating N_e for our observed data.

Conversely, ABC would: (1) replace (summarize) the raw observed data with multiple summary statistics of the data (e.g., F_{ST}, H_e, and number of alleles); then (2) compute the same summary statistics for each of the thousands of simulated population data sets (under the population model under consideration); and finally (3) match the observed data summary statistics to those from simulated populations in order to chose the population parameter estimate that best fits our data. This ABC approach is also called "summary statistic matching" because we match our observed summary stats to those from simulated population data sets to find the population parameter estimate (e.g., N_e, or Nm) most similar to that computed from our data.

ABC methods are becoming increasingly popular because they use nearly "all the information" from the data (Beaumont et al. 2002), yet they are far less computationally demanding than fully Bayesian (MCMC) approaches. Thus their performance can be evaluated thoroughly (see Section A8), and they can be used with large data sets with many loci or when conducting complex analyses with numerous parameters (e.g., population size, dispersal, and sex ratio). Finally, an experienced modeler can construct an ABC model in hours or days, whereas it can take weeks to construct a fully Bayesian MCMC model (M. Beaumont, personal communication).

A7 Parameter estimation, accuracy, and precision

Here we consider statistical frameworks (moments, likelihood, Bayesian) for inferring population parameters. To estimate a population parameter (e.g., the mean, μ), we usually compute a sample statistic ($\bar{x}$) from a sample of individuals. We can estimate a population parameter using different sample statistics (arithmetic mean, harmonic mean, median, or mode). To further complicate things, to compute an estimator, such as the mode, we can use different approaches, including moment methods, maximum likelihood estimation, or Bayesian estimation.

The sample moments, e.g., $\bar{x}$, $\bar{x}^2$, and $\bar{x}^3$, are used to obtain estimates of location, variance (scale), and shape of the population distribution, respectively. Moment-based estimators are widely used (e.g., in classic frequentist statistics), but can yield biased estimates when the underlying population distribution is non-normal, especially when the "higher" moments ($\bar{x}^2$, $\bar{x}^3$) are not considered. An example of such bias is the classic F_{ST}-based estimator of N_e, which is often biased because: (1) the underlying probability distribution of

F_{ST} is often skewed with a long tail (unlike the normal); and (2) the moment estimator (F_{ST}) incorporates information only from the first two moments, which do not contain information on skewness of the sampling distribution.

Maximum likelihood estimation (MLE) infers a parameter by finding the parameter value that maximizes the likelihood of obtaining the sample data (assuming some model such as Mendelian inheritance or a Wright–Fisher equilibrium population). MLE is increasingly used in population genetics because:

1 It yields probability distributions that are easy to interpret (see Section A4 and Figure A3), rather than just a point estimate and confidence interval, as in moments methods.
2 MLE can help evaluate and chose the best estimators (including moment-based estimators of the mean or variance, when data are normally distributed).
3 Faster computers and computer programs increasingly allow the computation of MLE estimates, e.g., see LEMARK, MIGRATE, MSVAR, and other computer programs on this book's web page.

Which estimator and approach performs best? This is a critical question in conservation genetics that often is ignored or underappreciated. It is especially important in light of the many new methods and computer programs published in recent years. The performance of an estimator (accuracy, precision, and robustness, see Section A8) depends on the question, sample size, sample characteristics, and the parameter being estimated. For example, MLE approaches are generally most efficient (see Section A4) with large samples, but can be less efficient than moment methods when using small sample sizes (e.g., less than 40 individuals). Efficiency refers to ability to extract information from the data and to achieve high accuracy and precision in estimating the true population parameter.

Identifying the best estimator generally requires a performance evaluation comparing estimators. For examples of performance evaluations, see Section A8 and publications such as Tallmon et al. (2004) and Wang (2002).

Accuracy (bias) and precision are critical concepts related to estimators of central tendency and dispersion, respectively. Accuracy of an estimator is its tendency to yield estimates near the true population parametric value. For example, if we estimate the mean heterozygosity (H_e) for each of four independent samples, the accuracy is good if 50% of estimates are high and 50% low. Otherwise the estimator is biased. If an estimator has poor precision, the four estimates will be scattered widely – often far from the true value. A precise statistical estimator will have relatively narrow confidence intervals, and the point estimates from independent estimations will cluster tightly together (see below). An estimator can have low precision but high accuracy, or vice versa (Figure A8).

Several different estimators should often used whenever assessing a given question. For example, it is useful to estimate both the mean and median because if they are different we can infer that the distribution might be skewed. It is also useful to compute both moment-based and likelihood-based estimators, as we sometimes do not know which is most reliable or accurate. In general, when estimating parameters, it is prudent to use multiple methods and software programs, to avoid errors and to increase confidence in results (e.g., if the same result is obtained from different methods).

Random and representative sampling is critical, and often assumed without testing (or discussing) the assumption. If sampling is not random or not representative, the

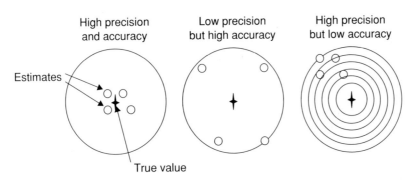

High precision
and accuracy

Low precision
but high accuracy

High precision
but low accuracy

Estimates

True value

Figure A8 Cartoon illustration of the difference between accuracy and precision. Imagine these are archery targets with the bullseye in the middle (i.e., the true value).

statistical estimate may be biased. For an extreme example, imagine that we sample only 10 individuals (F_1 offspring) from within only one family from a population containing hundreds of family groups. The sample is clearly not random or representative of the population. The allelic richness statistic we compute will often be low compared to the true population value, simply because the individuals we sampled are closely related compared to individuals from a true population-wide sample (with random representation of all family groups).

A8 Performance testing

Performance testing is the quantification of the accuracy (bias), precision, power, and robustness of a statistical estimator or test. This includes quantifying the bias caused by violating assumptions (random sampling, no selection, etc.); such violations often occur in real data sets from natural populations.

Performance testing involves four main steps: (1) generate a test data set (simulated or real) with a known parameter value for the parameter of interest (N_e, Nm, etc.); (2) estimate the parameter (e.g., with a confidence interval); (3) repeat 1,000 times both steps one and two; and (4) compute the proportion of the 1,000 estimates that give the true parameter (most accurately and precisely) (Figure A9).

Performance testing is critically important to allow conservation biologists to use statistical methods on real populations with minimal risk of making erroneous management decisions. Unfortunately, performance testing is rarely conducted thoroughly. Fortunately, the growing availability of computer simulation programs (e.g., EASYPOP, METASIM; see this book's website) makes performance testing increasingly feasible, even for undergraduate students or as part of a PhD degree program, for example.

Without performance evaluations, statistical methods are often used, and later are found to be biased. For example, some assignment tests and N_e estimators were shown to produce misleading or erroneous results (e.g., underestimated N_e, erroneous Type I error rates for assignment tests), long after they were being used in natural populations (see Paetkau et al. 2004; Waples, in press).

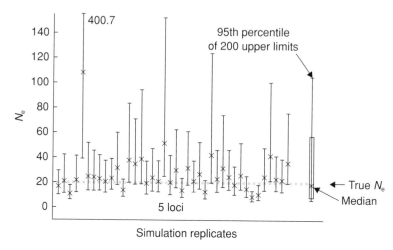

Figure A9 Example of power analysis where hundreds of independent populations with $N_e = 20$ were simulated, and then N_e (CI 90%) was estimated for each simulation replicate. Point estimates (crosses) of N_e, with confidence intervals (vertical lines), for each of 25 independent simulated populations are shown. The box plot graph on the far right summarizes the accuracy of point estimates by comparing the median of the many point estimates with the true N_e. The median is biased low. The box plot upper limit is the upper 95th percentile of the upper confidence interval limits (over 200 simulations). Modified from Berthier et al. (2002).

A9 The coalescent and genealogical information

The coalescent is a powerful modeling approach for analyzing population genetic data. It involves a different way of thinking about population genetics compared to classic approaches. Classic approaches for modeling populations typically trace the inheritance of genes in a "forward direction". For example, individual parents are randomly mated to produce offspring; the offspring are eventually mated to produce the next generation (as in individual-based simulation modeling). On the contrary, the coalescent approach looks backwards in time and traces gene copies (alleles) back from offspring to parents, to grandparents, and eventually to a single most recent common ancestor.

To "coalesce" means to fuse, unite, or come together. This refers to the process of tracing backward through time the joining of (coalescence of) homologous gene copies from different individuals into the same parent or ancestor (Figure A10). The word coalescent is used in several ways in the genetics literature. The "coalescent theory" was developed (mainly by Kingman 1982) to model a genealogy of gene copies so that allele frequency patterns and genealogical patterns (e.g., shapes of genealogies; see below) could be used to infer population parameters and demographic history (e.g., gene flow, population expansion, and selection).

The most important contribution of the coalescent to population genetics is that it allows for extraction of genealogical information from DNA data (i.e., information on the genealogical relationships among alleles at a locus). Many classic (noncoalescent-based)

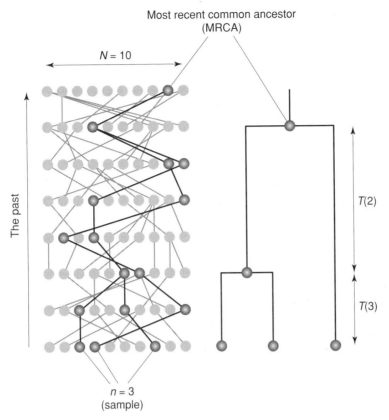

Figure A10 The coalescent approach for modeling the genealogy of individuals in a population. The complete genealogy of 10 haploid individuals (clones) is on the left. The dark lines trace back through time (from bottom to top) the ancestries of the three sampled lineages (gene copies). On the right is the "subgenealogy" (sample, $n = 3$) showing the coalescence pattern and times (e.g., the two genes on the left coalesce first at time $T(3)$). The coalescence time durations are proportional to the branch lengths. The average (and distribution) of branch lengths provides information about the tree shape, which is used to make inferences about demographic history (see Figure 20.7). Modified from Rosenberg and Nordborg (2002).

statistical estimators (e.g., F_{ST}) do not use genealogical information; they use only allele frequency information. With the advent of DNA sequencing (and restriction enzyme analysis, see Chapter 4) most data sets contain information on relationships (i.e., divergence) among alleles. Even microsatellite data contain genealogical information in the number of repeat unit differences between two alleles (assuming the stepwise mutation model discussed in Sections 4.2.1 and 12.1.2).

The "coalescent approach" is the modeling of gene transmission between generations in populations, by modeling a coalescent process (Example A3). The coalescent can be used in frequentist, maximum likelihood, and Bayesian statistical approaches, for example to generate the expected distribution of allele frequencies to test hypotheses and estimate parameters, e.g., N_e, Nm, etc. (Example A4).

Example A3 Coalescent modeling

Coalescent modeling involves two main steps: first we generate a random genealogy of individuals backward through time. Here it helps to envision clonal individuals (or haploid chromosomes such as mtDNA). We start with a sample of clones and randomly connect them to parents, grandparents, great grandparents, etc., until all clones coalesce into a single ancestor (the most recent common ancestor, MRCA; Figure A10). Going back in time, two lineages will coalesce whenever two clones are produced by the same parent. Going forward in time, lineages branch whenever a parent has two or more offspring, and branches end when no offspring are produced (i.e., lineage sorting, see Section 7.8).

Second, we randomly place mutations on branches (e.g., using Monte Carlo simulations and a random number generator while considering the mutation rate). We start by assigning some allelic state to the MRCA and then "drop" mutations along branches randomly moving forward. If a mutation is placed on a branch, then the allelic state (e.g., length for a microsatellite) must be determined by following rules of a model. For example, under a stepwise mutation model, a mutation will cause the allele length to increase or decrease (50 : 50 chances) by a single repeat unit (see Section 12.1.2).

Coalescent modeling is computationally efficient because we only simulate the sampled lineages ("subgenealogy" in Figure A10), and not the entire population as is done for individual-based forward models. Simulating only the subgenealogy requires less "record keeping" and saves computer time compared with the forward (individual-based) simulation modeling approach that requires record keeping for all individuals including those not sampled.

In coalescent modeling, we often want to separate the two stochastic genealogical processes: (1) random neutral mutation; and (2) random reproduction and population demography (which cause genetic drift). These two processes determine the genetic make up of the population of lineages. Separation of the two is important because we often are interested in the biological phenomena of demography and reproduction, but not mutation processes (Rosenberg and Nordborg 2002). For example, we are often interested in testing for population expansion or population subdivision, not mutation dynamics.

Example A4 The coalescent used in frequentist, likelihood, and Bayesian approaches

The coalescent can be used for modeling or conducting statistical tests under different statistical frameworks including frequentist, likelihood, or Bayesian. For example, a frequentist coalescent approach might be used to test if N_e is significantly smaller than 100. For this, we might: (1) use the coalescent to simulate 1,000 independent data sets for a population with $N_e = 100$; (2) compute N_e

for each simulated population (to obtain a distribution of possible N_e estimates consistent with a true N_e of 100); and (3) calculate how frequently (out of 1,000 data sets) we obtain a simulation estimate of N_e as small as our N_e estimate from our study population. If our population's estimated N_e is so small that it occurs only once in 1,000 simulated data sets, then we would conclude that that our population's N_e is significantly ($P < 0.001$) less than 100. This kind of approach was used in Funk et al. (1999) to test for small N_e in a salamander population.

In a maximum likelihood approach to test if N_e is significantly smaller than 100, the coalescent could be used to help compute the likelihood of $N_e = 1, 2, 3, \ldots,$ up to $N_e = 200$, given our raw data. Here, the coalescent could be used to simulate thousands of data sets for each N_e, and then compute the likelihood of each N_e ($N_e = 1, 2, 3$, etc.) given our real data set. This would yield a probability (likelihood) distribution of N_e values (with $N_e = 1, 2, 3, \ldots,$ up to $N_e = 200$ on the x-axis). If all the area under the likelihood (probability) curve was less than 100 (i.e., did not include $N_e = 100$), we could conclude that our population's effective size is less than 100. The resulting likelihood curve for inferring N_e is similar to that for inferring *Nm* in Figure A3b.

In a Bayesian approach, we would conduct the same computations as in the maximum likelihood approach just described, using the coalescent. However, we then would modify the resulting likelihood distribution by multiplying it times a prior distribution to obtain a posterior distribution, as illustrated in Figure A5.

This example illustrates how the coalescent can be used within different statistical frameworks to conduct statistical tests or estimate a population parameter.

Genealogical methods, such as the coalescent, do not estimate evolutionary trees (as when "inferring a phylogeny"), but rather they estimate parameters of the random evolutionary processes that give rise to trees, such as gene flow rates, population size, or population growth rates. For example, different population demographic histories yield different-shaped genealogies (Figure A11). Consequently, genealogical shape can be used to infer a population's demographic history.

Population growth yields star-like genealogies with many long branches (Figure A11b). Many long (similar-length) branches are expected to arise during a long-term population expansion because new alleles (mutations) tend to persist for a long time because drift is negligible in growing populations. Thus, in a real study, if we detect a star-like gene tree for each of many independent genes, we can infer that the population has been growing.

Random genealogical processes lead to many possible random genealogies for different genes under a given demographic history (Figure A12). Therefore, we must study many genes to obtain accurate and precise estimates of demographic history. We can simulate thousands of random genealogies for each population history (e.g., a stable versus growing population) to test if one history best fits our empirical data set. If one history best fits our observed field data, then other histories might be rejected (e.g., using likelihood ratio tests).

Selection can also cause distinctive-shaped genealogies. For example, a selective sweep will first remove many alleles (like a bottleneck signature) and subsequent mutation can lead to a star phylogeny (like an expansion signature). If the genealogy of one locus differs

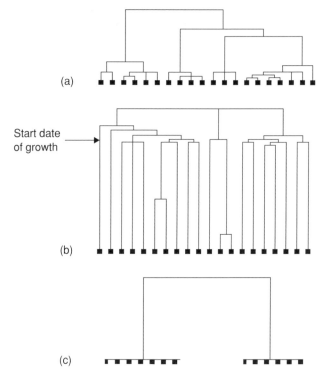

Figure A11 Example gene tree for a population that is (a) at constant size (many variable branch lengths), (b) growing (mostly long branches originating at the time of population growth), and (c) declining (fewer alleles and mostly at even frequencies). Each genealogy is only a single representative of thousands of possible genealogies from each population model (constant, growing, or declining). The vertical axis is the same for all trees, but the bottom one has many shallow branches that are invisible. Modified from Harpending et al. (1998).

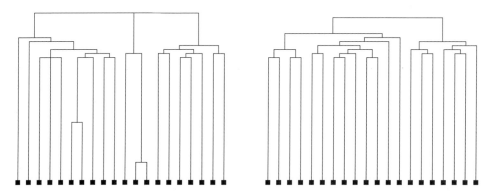

Figure A12 Two genealogies for the same demographic history (population growth). Note that hundreds of similar genealogies are possible for a single demographic history. Thus, many independent genes (genealogies) must be studied to infer population history. Therefore, **distributions** of genealogies are used to infer (or exclude) different demographic histories. From Harpending et al. (1998).

significantly from other loci, we might infer that selection has influenced the locus (see Section 9.6.3). Selection and "outlier genealogies" make it important to study many independent genome segments when inferring demographic history.

Guest Box A Is mathematics necessary?
James F. Crow

Much of our understanding of the application of genetics to problems in conservation depends upon the field of population genetics. Population genetics used to consist of two quite different disciplines. One utilized observations of populations in nature or laboratory studies. These were often descriptive and involved no mathematics. This area is epitomized by the early work of Theodosius Dobzhansky, Ernst Mayr, and G. Ledyard Stebbins. At the same time a mathematical theory was being developed by J. B. S. Haldane, R. A. Fisher, and Sewall Wright. One of the earliest bridges was built in 1941 when Dobzhansky and Wright collaborated in a joint experimental paper with lots of theory.

Since that time, most work in population genetics has had some mathematical involvement. Almost every experiment or field observation now utilizes quantitative measurements, and that means statistics. The day is past when one can simply report results with no test of their statistical reliability. Increasingly, experiments are performed or observations are made based on some underlying theory. The person doing the experiments may develop the theory or make use of existing mathematical theory. Finally, there is the development of ever deeper, more general, and more sophisticated theory. Much of this is being done by people with professional mathematics training.

We cannot all be mathematicians. But we can learn a minimum amount. Every population geneticist must know some mathematics and some statistics. I have done both experimental (usually driven by theory) and theoretical work. But my mathematics is limited and some of the research that I most enjoyed was done in collaboration with better mathematicians, notably Motoo Kimura.

There are two recent changes in the field. Computers have altered everything, and it is hardly necessary for me to mention that you need to know how to use them. It used to be that theoretical work was regularly stymied by insoluble problems. The computer has greatly broadened the range of problems that can be solved, not in the mathematical sense but numerically (e.g., MCMC and other simulation-based approaches), which is often what is wanted. At the same time, the mathematical theory itself is advancing as mathematicians enter the field.

The second change is the advent of molecular methods. Population genetics used to have a theory that was too rich for the data. That is no longer true. DNA analysis can yield mountains of data that call for improved, computerized analyses. Even in nonmodel species, data sets are becoming large enough that some sophisticated statistical methods can take days to conduct computations, and some analyses might not be feasible because, for example, computer programs take too long or do not converge.

If you are going to be an experimenter or analyze data with modern statistical tools, you need to know some mathematics and statistics, and be adept at computers. If you are going to develop theory (even if for application to natural populations and conservation), you usually need to be a real mathematician or collaborate with one.

Most readers of this book are primarily interested in understanding, but not contributing to, the primary literature in population and conservation genetics. Much of the current literature in population genetics employs advanced mathematical methods that are beyond the reach of most biology students. Dobzhansky's method of reading and understanding the papers of Sewall Wright is one possible approach (see quote at the beginning of this appendix). Examining the biological assumptions being made is crucial, but not sufficient. However, a healthy amount of skepticism is probably a good thing. There was only one Sewall Wright!

References

Adams, J. M., G. Piovesan, S. Strauss, and S. Brown. 2002. The case for genetic engineering of native and landscape trees against introduced pests and diseases. Conservation Biology 16:874–879.

Adams, J. R., B. T. Kelly, and L. P. Waits. 2003. Using faecal DNA sampling and GIS to monitor hybridization between red wolves (*Canis rufus*) and coyotes (*Canis latrans*). Molecular Ecology 12:2175–2186.

Agapow, P. M., and A. Burt. 2001. Indices of multilocus linkage disequilibrium. Molecular Ecology Notes 1:101–102.

Akey, J. M., G. Zhang, K. Zhang, L. Jin, and M. D. Shriver. 2002. Interrogating a high-density SNP map for signatures of natural selection. Genome Research 12:1805–1814.

Alatalo, R. V., and A. Lundberg. 1986. Heritability and selection on tarsus length in the pied flycatcher (*Ficedula hypoleuca*). Evolution 40:574–583.

Allen, P. J., W. Amos, P. P. Pomeroy, and S. D. Twiss. 1995. Microsatellite variation in grey seals (*Halichoerus grypus*) shows evidence of genetic differentiation between two British breeding colonies. Molecular Ecology 4:653–662.

Allendorf, F. W. 1983. Isolation, gene flow, and genetic differentiation among populations. Pp. 51–65 *in* Schonewald-Cox, C., S. Chambers, B. MacBryde, and L. Thomas, eds. Genetics and Conservation. Benjamin/Cummings, Menlo Park, CA.

Allendorf, F. W. 1986. Genetic drift and the loss of alleles versus heterozygosity. Zoo Biology 5:181–190.

Allendorf, F. W. 1993. Delay of adaptation to captive breeding by equalizing family size. Conservation Biology 7:416–419.

Allendorf, F. W., W. A. Gellman, and G. H. Thorgaard. 1994. Sex-linkage of two enzyme loci in *Oncorhynchus mykiss* (rainbow trout). Heredity 72:498–507.

Allendorf, F. W., K. L. Knudsen, and G. M. Blake. 1982. Frequencies of null alleles at enzyme loci in natural populations of ponderosa and red pine. Genetics 100:497–504.

Allendorf, F. W., and R. F. Leary. 1988. Conservation and distribution of genetic variation in a polytypic species: the cutthroat trout. Conservation Biology 2:170–184.

Allendorf, F. W., R. F. Leary, P. Spruell, and J. K. Wenburg. 2001. The problems with hybrids: setting conservation guidelines. Trends in Ecology and Evolution 16:613–622.

Allendorf, F. W., and L. L. Lundquist. 2003. Introduction: population biology, evolution, and control of invasive species. Conservation Biology 17:24–30.

Allendorf, F. W., and S. R. Phelps. 1981. Use of allelic frequencies to describe population structure. Canadian Journal of Fisheries and Aquatic Sciences 38:1507–1514.

Allendorf, F. W., and N. Ryman. 2002. The role of genetics in population viability analysis. Pp. 50–85 in Beissinger, S. R., and D. R. McCullough, eds. Population Viability Analysis. University of Chicago Press, Chicago.

Allendorf, F. W., and L. W. Seeb. 2000. Concordance of genetic divergence among sockeye salmon populations at allozyme, nuclear DNA, and mitochondrial DNA markers. Evolution 54:640–651.

Allendorf, F. W., and C. Servheen. 1986. Genetics and conservation of grizzly bears. Trends in Ecology and Evolution 1:88–89.

Allendorf, F. W., G. Ståhl, and N. Ryman. 1984. Silencing of duplicate genes: a null allele polymorphism for lactate dehydrogenase in brown trout (*Salmo trutta*). Molecular Biology and Evolution 1:238–248.

Allendorf, F. W., and R. S. Waples. 1996. Conservation and genetics of salmonid fishes. Pp. 238–280 in Avise, J. C., and J. L. Hamrick, eds. Conservation Genetics: Case Histories from Nature. Chapman Hall, New York.

Alstad, D. N. 2001. Basic *Populus* Models of Ecology. Prentice-Hall, Upper Saddle River, NJ.

Antilla, C. K., C. C. Daehler, N. E. Rank, and D. R. Strong. 1998. Greater male fitness of a rare invader (*Spartina alterniflora*, Poaceae) threatens a common native (*Spartina foliosa*) with hybridization. American Journal of Botany 85:1597–1601.

Antolin, M. F. 1999. A genetic perspective on mating systems and sex ratios of parasitoid wasps. Reseaches on Population Ecology 41:29–37.

Archie, J. W. 1985. Statistical analysis of heterozygosity data: independent sample comparisons. Evolution 39:623–637.

Armbruster, P., R. A. Hutchinson, and T. Linvell. 2000. Equivalent inbreeding depression under laboratory and field conditions in a tree-hole-breeding mosquito. Proceedings of the Royal Society of London B 267:1939–1945.

Armbruster, P, and D H Reed. 2005. Inbreeding depression in benign and stressful environments. Heredity 95:235–242.

Arnold, M. L. 1997. Natural hybridization and evolution. Oxford University Press, New York.

Arnold, S. J. 1981. Behavioral variation in natural populations. 2. The inheritance of a feeding response in crosses between geographic races of the garter snake *Thamnophis elegans*. Evolution 35:510–515.

Ashley, M. V., M. F. Willson, O. R. W. Pergams, D. J. O'Dowd, S. M. Gende, and J. S. Brown. 2003. Evolutionarily enlightened management. Biological Conservation 111:115–123.

Aspinwall, N. 1974. Genetic analysis of North American populations of the pink salmon (*Oncorhynchus gorbuscha*), possible evidence for the neutral mutation-random drift hypothesis. Evolution 28:295–305.

Avise, J. C. 1986. Mitochondrial DNA and the evolutionary genetics of higher animals. Philosophical Transactions of the Royal Society of London B 312:325–342.

Avise, J. C. 1990. Mitochondrial gene trees and the evolutionary relationship of mallard and black ducks. Evolution 44:1109–1119.

Avise, J. C. 1992. Molecular population structure and the biogeographic history of a regional fauna – a case history with lessons for conservation biology. Oikos 63:62–76.

Avise, J. C. 1994. Molecular markers, natural history, and evolution. Chapman & Hall, New York.

Avise, J. C. 2000a. Cladists in wonderland. Evolution 54:1828–1832.

Avise, J. C. 2000b. Phylogeography: the History and Formation of Species. Harvard University Press, Cambridge, MA.

Avise, J. C. 2004. Molecular Markers, Natural History, and Evolution, 2nd edn. Chapman & Hall, New York.

Avise, J. C., and 7 others. 1987. Intraspecific phylogeography: the mitochondrial DNA bridge between population genetics and systematics. Annual Review of Ecology and Systematics 18:489–522.

Avise, J. C., C. Giblin-Davidson, J. Laerm, J. C. Patton, and R. A. Lansman. 1979a. Mitochondrial DNA clones and matriarchal phylogeny within and among geographic populations of the pocket gopher, *Geomys oinetis*. Proceeding of the National Academy of Sciences, USA 76:6694–6698.

Avise, J. C., R. A. Lansman, and R. O. Shade. 1979b. The use of restriction endonucleases to measure mitochondrial DNA sequence relatedness in natural populations. I. Population structure and evolution in the genus *Peromyscus*. Genetics 92:279–295.

Awadalla, P., A. Eyrewalker, and J. M. Smith. 1999. Linkage disequilibrium and recombination in hominid mitochondrial DNA. Science 286:2524–2525.

Ayala, F. J., and J. R. Powell. 1972. Allozymes as diagnostic characters of sibling species of *Drosophila*. Proceedings of the National Academy Sciences, USA 69:1094–1096.

Ayre, D. J., R. J. Whelan, and A. Reid. 1994. Unexpectedly high levels of selfing in the Australian shrub *Grevillea barklyana* (Proteaceae). Heredity 72:168–174.

Ayres, K. L., and D. J. Balding. 1998. Measuring departures from Hardy–Weinberg: a Markov chain Monte Carlo method for estimating the inbreeding coefficient. Heredity 80:769–777.

Ayres, K. L., and A. D. J. Overall. 2004. API-CALC 1.0: a computer program for calculating the average probability of identity allowing for substructure, inbreeding and the presence of close relatives. Molecular Ecology Notes 4:315–318.

Baer, C. F., and J. Travis. 2000. Direct and correlated responses to artificial selection on acute thermal stress tolerance in a livebearing fish. Evolution 54:238–244.

Bahlo, M., and R. C. Griffiths. 2000. Inference from gene trees in a subdivided population. Theoretical Population Biology 57:79–95.

Baker, A. J. 1992. Genetic and morphometric divergence in ancestral European and descent New Zealand populations of chaffinches (*Fringilla coelebs*). Evolution 46:1784–1800.

Baker, A. N., A. N. H. Smith, and F. B. Pichler. 2002a. Geographical variation in Hector's dolphin: recognition of new subspecies of *Cephalorhynchus hectori*. Journal of the Royal Society of New Zealand 32:713–727.

Baker, C. S., F. Cipriano, and S. R. Palumbi. 1996. Molecular genetic identification of whale and dolphin products from commercial markets in Korea and Japan. Molecular Ecology 5:671–687.

Baker, C. S., M. L. Dalebout, G. M. Lento, and N. Funahashi. 2002b. Gray whale products sold in commercial markets along the Pacific Coast of Japan. Marine Mammal Science 18:295–300.

Baker, C. S., L. florezgonzalez, B. Abernethy, H. C. Rosenbaum, R. X. Slade, J. Capella, and J. L. Bannister. 1998. Mitochondrial DNA variation and maternal gene flow among humpback whales of the southern hemisphere. Marine Mammal Science 14:721–737.

Baker, C. S., G. M. Lento, F. Cipriano, and S. R. Palumbi. 2000. Predicted decline of protected whales based on molecular genetic monitoring of Japanese and Korean markets. Proceedings of the Royal Society of London B 267:1191–1199.

Baker, H. G. 1995. Aspects of the genecology of weeds. Pp. 189–224 *in* Kruckeberg, A. R., R. B. Walker, and A. E. Leviton, eds. Genecology and Ecogeographic Races. Pacific Division of the American Association for the Advancement of Science, San Francisco.

Baker, H. G., and Stebbins, G. L., eds. 1965. The Genetics of Colonizing Species. Academic Press, New York.

Baldwin, B. G., and R. H. Robichaux. 1995. Historical biogeography and ecology of the Hawaiian silversword alliance (Asteraceae). Pp. 259–287 *in* Wagner, W. L., and V. A. Funk, eds. Hawaiian Biogeography: Evolution on a Hot Spot Archipelago. Smithsonian Institution Press, Washington, DC.

Ballard, J. W. O., and M. C. Whitlock. 2004. The incomplete natural history of mitochondria. Molecular Ecology 13:729–744.

Ballou, J. 1983. Calculating inbreeding coefficients from pedigrees. Pp. 509–520 in Schonewald-Cox, C., S. Chambers, B. MacBryde, and L. Thomas, eds. Genetics and Conservation. Benjamin/Cummings, Menlo Park, CA.

Ballou, J. D. 1997. Ancestral inbreeding only minimally affects inbreeding depression in mammalian populations. Journal of Heredity 88:169–178.

Ballou, J. D., and T. J. Foose. 1996. Demographic and genetic management of captive populations. Pp. 263–283 in Kleiman, D. G., M. E. Allen, K. V. Thompson, and S. Lumpkin, eds. Wild Mammals in Captivity. University of Chicago Press, Chicago.

Ballou, J. D., M. Gilpin, and T. J. Foose, eds. 1994. Population Management for Survival and Recovery. Columbia University Press, New York.

Ballou, J. D., and R. C. Lacy. 1995. Identifying genetically important individuals for management of genetic variation in pedigreed populations. Pp. 76–111 in Ballou, J. D., M. Gilpin, and T. J. Foose, eds. Population Management for Survival and Recovery. Columbia University Press, New York.

Ballou, J., and J. D. Seidensticker. 1987. The genetic and demographic characteristics of the 1983 captive population of Sumatran tigers (*Panthera tigris sumatrae*). Pp. 329–347 in Tilson, R. L., and U. S. Seal, eds. Tigers of the World: the Biology, Biopolitics, Management and Conservation of an endangered species. Noyes Publications, Park Ridge, NJ.

Balloux, F. 2001. EASYPOP (Version 1.7): a computer program for population genetics simulations. Journal of Heredity 92:301–302.

Balloux, F., N. LugonMoulin, and J. Hausser. 2000. Estimating gene flow across hybrid zones: how reliable are microsatellites? Acta Theriologica 45:93–101.

Balmford, A., N. Leader-Williams, and M. J. B. Green. 1995. Parks or arks: where to conserve threatened mammals? Biodiversity and Conservation 4:595–607.

Balmford, A., G. M. Mace, and N. Leader-Williams. 1996. Designing the ark: setting priorities for captive breeding. Conservation Biology 10:719–727.

Bangert, R. K., R. J. Turek, G. D. Martinsen, G. M. Wimp, J. K. Bailey, and T. G. Whitham. 2005. Benefits of conservation of plant genetic diversity to arthropod diversity. Conservation Biology 19:379–390.

Banks, M. A., and W. Eichert. 2000. WHICHRUN (Version 3.2): a computer program for population assignment of individuals based on multilocus genotype data. Journal of Heredity 91:87–89.

Banks, S. C., S. D. Hoyle, A. Horsup, P. Sunnucks, and A. C. Taylor. 2003. Demographic monitoring of an entire species (the northern hairy-nosed wombat, *Lasiorhinus krefftii*) by genetic analysis of non-invasively collected material. Animal Conservation 6:101–107.

Barash, D. P. 1973. The ecologist as Zen master. American Midland Naturalist 89:214–217.

Barker, J. S. F. 1994. Animal breeding and conservation genetics. Pp. 381–395 in Loeschcke, V., J. Tomiuk, and S. K. Jain, eds. Conservation Genetics. Birkhauser Verlag, Basel, Switzerland.

Barrett, S. C. H. 2000. Microevolutionary influences of global changes on plant invasions. Pp. 115–139 in Mooney, H. A., and R. J. Hobbs, eds. Invasive Species in a Changing World. Island Press, Washington, DC.

Barrett, S. C. H., and B. C. Husband. 1990. The genetics of plant migration and colonization. Pp. 254–278 in Brown, A. H. D., M. T. Clegg, A. L. Kahler, and B. S. Weir, eds. Plant Population Genetics, Breeding, and Genetic Resources. Sinauer, Sunderland, MA.

Barrett, S. C. H., and J. R. Kohn. 1989. Quoted in "How to get plants into the conservationists' ark" by R. Lewin. Science 244:32–33.

Barrett, S. C. H., and J. R. Kohn. 1991. Genetic and evolutionary consequences of small population size in plants – implications for conservation. Pp. 3–30 in Falk, D. A., and K. E. Holsinger, eds. Genetics and Conservation of Rare Plants. Oxford University Press, New York.

Barton, N. H. 2000. Estimating multilocus linkage disequilibria. Heredity 84:373–389.

Barton, N. H., and K. S. Gale. 1993. Genetic analysis of hybrid zones. Pp. 13–45 in Harrison, R. G., ed. Hybrid Zones and the Evolutionary Process. Oxford University Press, Oxford.

Barton, N. H., and G. M. Hewitt. 1985. Analysis of hybrid zones. Annual Review of Ecology and Systematics 16:113–148.

Barton, N. H., and P. D. Keightley. 2002. Understanding quantitative genetic variation. Nature Reviews Genetics 3:11–21.

Baskin, Y. 2002. A Plague of Rats and Rubbervines: the Growing Threat of Species Invasions. Island Press, Washington, DC.

Bass, R. A. 1979. Chromosomal polymorphism in cardinals, Cardinalis cardinalis. Canadian Journal of Genetics and Cytology 21:549–553.

Bassett, S. M., M. A. Potter, R. A. Fordham, and E. V. Johnston. 1999. Genetically identical avian twins. Journal of Zoology, London 247:475–478.

Bataillon, T. 2003. Shaking the "deleterious mutations" dogma? Trends in Ecology and Evolution 18:315–317.

Battaglia, E. 1964. Cytogenetics of the B-chromosomes. Caryologia 8:205–213.

Bawa, K. S., S. Menon, and L. R. Gorman. 1997. Cloning and conservation of biological diversity: paradox, panacea, or Pandora's box? Conservation Biology 11:829–830.

Beaumont, M. A. 1999. Detecting population expansion and decline using microsatellites. Genetics 153:2013–2029.

Beaumont, M., E. M. Barratt, D. Gottelli, A. C. Kitchener, M. J. Daniels, J. K. Pritchard, and M. W. Bruford. 2001. Genetic diversity and introgression in the Scottish wildcat. Molecular Ecology 10:319–336.

Beaumont, M. A., and B. Rannala. 2004. The Bayesian revolution in genetics. Nature Reviews Genetics 5:251–261.

Beaumont, M. A., W. Zhang, and D. J. Balding. 2002. Approximate Bayesian computation in population genetics. Genetics 162:2025–2035.

Beerli, P., and J. Felsenstein. 2001. Maximum likelihood estimation of a migration matrix and effective population sizes in n subpopulations by using a coalescent approach. Proceedings of the National Academy of Sciences, USA 98:4563–4568.

Begun, D. J., and C. F. Aquadro. 1992. Levels of naturally occuring DNA polymorphism correlate with recombination rates in D. melanogaster. Nature 356:519–520.

Beissinger, S. R., and M. I. Westphal. 1998. On the use of demographic models of population viability in endangered species management – invited paper. Journal of Wildlife Management 62:821–841.

Bellemain, E., J. E. Swenson, D. Tallmon, S. Brunberg, and P. Taberlet. 2005. Estimating population size of elusive animals with DNA from hunter-collected feces: four methods for brown bears. Conservation Biology 19:150–161.

Bellen, H. J., and 6 others. 1992. The Drosophila couch potato gene: an essential gene required for normal adult behavior. Genetics 131:365–375.

Belovsky, G. E. 1987. Extinction models and mammalian persistence. Pp. 225–242 in Soulé, M. E., ed. Viable Populations for Conservation. Sinauer, Sunderland, MA.

Benirschke, K., and A. T. Kumamoto. 1991. Mammalian cytogenetics and conservation of species. Journal of Heredity 82:187–191.

Bennett, D. C., and M. L. Lamoreux. 2003. The color loci of mice – a genetic century. Pigment Cell Research 16:333–344.

Bennuah, S. Y., T. Wang, and S. N. Aitken. 2004. Genetic analysis of the Picea sitchensis × glauca introgression zone in British Columbia. Forest Ecology and Management 197:65–77.

Berlocher, S. H. 1984. Genetic changes coinciding with the colonization of California by the walnut husk fly, Rhagoletis completa. Evolution 38:906–918.

Bernatchez, L., H. Glemet, C. C. Wilson, and R. G. Danzmann. 1995. Introgression and Fixation of Arctic char (Salvelinus alpinus) mitochondrial genome in an allopatric population of brook trout (Salvelinus fontinalis). Canadian Journal of Fisheries and Aquatic Sciences 52:179–185.

Berry, O., M. D. Tocher, D. M. Gleeson, and S. D. Sarre. 2005. Effect of vegetation matrix on animal dispersal: genetic evidence from a study of endangered skinks. Conservation Biology 19:855–864.

Berthier, P., M. A. Beaumont, J.-M. Cornuet, and G. Luikart. 2002. Likelihood-based estimation of the effective population size using temporal changes in allele frequencies: a genealogical approach. Genetics 160:741–751.

Berthold, P. 1991. Genetic control of migratory behaviour in birds. Trends in Ecology and Evolution 6:254–257.

Berthold, P., and A. J. Helbig. 1992. The genetics of bird migration – stimulus, timing, and direction. Ibis 134:35–40.

Berven, K. A., D. E. Gill, and S. J. Smith-Gill. 1979. Counter gradient selection in the green frog *Rana clamitans*. Evolution 33:609–623.

Beverton, R. J. H., and 11 others. 1984. Dynamics of single species: group report. Pp. 13–58 in May, R. M., ed. Dahlem Konferenzen. Springer Verlag, Berlin.

Bijlsma, R., J. Bundgaard, A. C. Boerema, and W. F. van Putten. 1997. Genetic and environmental stress, and the persistence of populations. Pp. 193–207 in Bijlsma, R., and V. Loeschcke, eds. Environmental Stress, Adaptation and Evolution. Birkhauser Verlag, Basel, Switzerland.

Bijlsma, R., J. Bundgaard, and W. F. van Putten. 1999. Environmental dependence of inbreeding depression and purging in *Drosophila melanogaster*. Journal of Evolutionary Biology 12:1125–1137.

Birky, C. W. Jr., T. Maruyama, and P. Fuerst. 1983. An approach to population and evolutionary genetic theory for genes in mitochondria and chloroplasts, and some results. Genetics 103:513–527.

Birstein, V. J., P. Doukakis, B. Sorkin, and R. DeSalle. 1998. Population aggregation analysis of three caviar-producing species of sturgeons and implications for the species identification of black caviar. Conservation Biology 12:766–775.

Bittner, T. D., and R. B. King. 2003. Gene flow and melanism in garter snakes revisited: a comparison of molecular markers and island vs. coalescent models. Biological Journal of the Linnean Society 79:389–399.

Black, F. L., and P. W. Hedrick. 1997. Strong balancing selection at HLA loci: evidence from segregation in South Amerindian families. Proceedings of the National Academy of Sciences, USA 94:12452–12456.

Black, R., and M. S. Johnson. 1979. Asexual viviparity and population genetics of *Actinia tenebrose*. Marine Biology 53:27–31.

Boecklen, W. J., and D. J. Howard. 1997. Genetic analysis of hybrid zones: numbers of markers and power of resolution. Ecology 78:2611–2616.

Bogart, J. P. 1980. Evolutionary implications of polyploidy in amphibians and reptiles. Pp. 341–378 in Lewis, W., ed. Polyploidy: biological relevance. Plenum, New York.

Bowen, B. W. 1999. Preserving genes, species, or ecosystems? Healing the fractured foundations of conservation policy. Molecular Ecology 8:S5–S10.

Bowen, B. W., and S. A. Karl. 1996. Population structure, phylogeography, and molecular evolution. Pp. 29–50 in Lutz, P. L., and J. A. Musick, eds. The Biology of Sea Turtles. CRC Press, Boca Raton, FL.

Bowen, B. W., A. B. Meylan, J. P. Ross, C. J. Limpus, G. H. Balazs, and J. C. Avise. 1992. Global population structure and natural history of the green turtle (*Chelonia mydas*) in terms of matriarchal phylogeny. Evolution 46:865–881.

Bowman, J. C., and D. S. Falconer. 1960. Inbreeding depression and heterosis of litter size in mice. Genetical Research 1:262–274.

Brannon, E. L., and 10 others. 2004. The controversy about salmon hatcheries. Fisheries 29(9):12–31.

Brewer, B. A., R. C. Lacy, M. L. Foster, and G. Alaks. 1990. Inbreeding depression in insular and central populations of *Peromyscus* mice. Journal of Heredity 81:257–266.

Briscoe, D. A., and 6 others. 1992. Rapid loss of genetic variation in large captive populations of *Drosophila* flies: implications for the genetic management of captive populations. Conservation Biology 6:416–425.

Briskie, J. V., and M. Mackintosh. 2004. Hatching failure increases with severity of population bottlenecks in birds. Proceedings of the National Academy of Sciences, USA 101:558–561.

Britten, H. B. 1996. Meta-analyses of the association between multilocus heterozygosity and fitness. Evolution 50:2158–2164.

Brodie, E. D. 1992. Correlational selection for color pattern and antipredator behavior in the garter snake *Thamnophis ordinoides*. Evolution 46:1284–1298.

Brook, B. W., J. J. Ogrady, A. P. Chapman, M. A. Burgman, H. R. Akcakaya, and R. Frankham. 2000. Predictive accuracy of population viability analysis in conservation biology. Nature 404:385–387.

Brookfield, J. F. Y. 1996. A simple new method for estimating null allele frequency from heterozygote deficiency. Molecular Ecology 5:453–455.

Brown, A. H. D. 1979. Enzyme polymorphisms in plant populations. Theoretical Population Bioliolgy 15:1–42.

Brown, A. H. D., and J. D. Briggs. 1991. Sampling strategies for genetic variation in *ex situ* collections of endangered plant species. Pp. 99–119 *in* Falk, D. A., and K. E. Holsinger, eds. Genetics and Conservation of Rare Plants. Oxford University Press, New York.

Brown, A. H. D., and D. R. Marshall. 1981. Evolutionary changes accompanying colonization in plants. Pp. 351–363 *in* Scudder, G. G. E., and J. L. Reveal, eds. Evolution Today, Proceedings of Second International Congress of Systematic and Evolutionary Biology. Hunt Institute for Botanical Documentation, Carnegie-Mellon University, Pittsburgh, PA.

Brown, A. H. D., and A. G. Young. 2000. Genetic diversity in tetraploid populations of the endangered daisy *Rutidosis leptorrhynchoides* and implications for its conservation. Heredity 85:122–129.

Brown, C. R., and M. B. Brown. 1998. Intense natural selection on body size and wing and tail asymmetry in cliff swallows during severe weather. Evolution 52:1461–1475.

Brown, J. H., and A. Kodric-Brown. 1977. Turnover rates in insular biogeography: effect of immigration on extinction. Ecology 58:445–449.

Brown, W. M., M. George, and A. C. Wilson. 1979. Rapid evolution of animal mitochondrial DNA. Proceedings of the National Academy of Sciences, USA 76:1967–1971.

Brown, W. M., and J. W. Wright. 1979. Mitochondrial DNA analysis and the origin and relative age of parthenogenetic lizards (Genus *Cnemidophorus*). Science 203:1247–1249.

Bruford, M. W., O. Hanotte, J. F. Y. Brookfield, and T. Burke. 1998. Multi-locus and single-locus DNA fingerprinting. Pp. 287–336 *in* Hoelzel, A. R., ed. Molecular Genetic Snalysis of Populations. Oxford University Press, New York.

Brumfield, R. T., P. Beerli, D. A. Nickerson, and S. V. Edwards. 2003. The utility of single nucleotide polymorphisms in inferences of population history. Trends in Ecology and Evolution 18:249–256.

Brünner, H., and J. Hausser. 1996. Genetic and karyotypic structure of a hybrid zone between the chromosomal races Cordon and Valais in the common shrew, *Sorex araneus*. Hereditas 125:147–158.

Bryant, E. H., S. A. McCommas, and L. M. Combs. 1986. The effect of an experimental bottleneck upon quantitative genetic variation in the housefly. Genetics 114:1191–1211.

Bryant, E. H., and D. H. Reed. 1999. Fitness decline under relaxed selection in captive populations. Conservation Biology 13:665–669.

Bryant, E. H., V. L. Vackus, M. E. Clark, and D. H. Reed. 1999. Experimental tests of captive breeding for endangered species. Conservation Biology 13:1487–1496.

Buckley, P. A. 1987. Mendelian genes. Pp. 1–44 *in* Cooke, F., and P. A. Buckley, eds. Avian Genetics: a Population and Ecological Approach. Academic Press, London.

Buerkle, C. A., D. E. Wolf, and L. H. Rieseberg. 2003. The origin and extinction of species through hybridization. Pp. 117–141 *in* Brigham, C. A., and M. W. Schwartz, eds. Population Viability in Plants: Conservation, Management, and Modeling of Rare Plants. Springer Verlag, Berlin.

Buktenica, M. W. 1997. Bull trout restoration and brook trout eradication at Crater Lake National Park, Oregon. Pp. 127–136 *in* Mackay, W. C., M. K. Brewin, and M. Monita, eds. Friends of the

Bull Trout Conference Proceedings. Bull Trout Task Force (Alberta), Trout Unlimited Canada, Calgary.

Bull, J. J. 1983. Evolution of Sex Determining Mechanisms. Benjamin/Cummings, Menlo Park, CA.

Buntjer, J. A., and J. A. Lenstra. 1998. Mammalian species identification by interspersed repeat PCR fingerprinting. Journal of Industrial Microbiology and Biotechnology 21: 121–127.

Bürger, R., and M. Lynch. 1995. Evolution and extinction in a changing environment: a quantitative-genetic analysis. Evolution 49:151–163.

Busby, P. J., O. W. Johnson, T. C. Wainwright, F. W. Waknitz, and R. S. Waples. 1993. Status review for Oregon's Illinois River winter steelhead. NOAA Technical Memoir NMFS-NWFSC-10. US Department of Commerce.

Busby, P. J., T. C. Wainwright, and R. S. Waples. 1994. Status review for Klamath Mountains Province steelhead. NOAA Technical Memoir NMFS-NWFSC-10. US Department of Commerce.

Bush, G. L. 1994. Sympatric speciation in animals: new wine in old bottles. Trends in Ecology and Evolution 9:285–288.

Bush, M., B. B. Beck, J. Dietz, et al. 1996. Radiographic evaluation of diaphragmatic defects in gloden lion tamarins (*Leontopithecus rosalia rosalia*): implications for reintroduction. Journal of Zoo and Wildlife Medicine 27:346–357.

Byers, D. L., and D. M. Waller. 1999. Do plant populations purge their genetic load? Effects of population size and mating history on inbreeding depression. Annual Review of Ecology and Systematics 30:479–513.

Byers, J. A. 1997. American Pronghorn. University of Chicago Press, Chicago.

Cade, T. J. 1983. Hybridization and gene exchange among birds in relation to conservation. Pp. 288–310 *in* Schonewald-Cox, C., S. Chambers, B. MacBryde, and L. Thomas, eds. Genetics and Conservation. Benjamin/Cummings, Menlo Park, CA.

Calmet, C., M. Pascal, and S. Samadi. 2001. Is it worth eradicating the invasive pest *Rattus norvegicus* from Molene archipelago? Genetic structure as a decision-making tool. Biodiversity and Conservation 10:911–928.

Calzada, J. P. V., C. F. Crane, and D. M. Stelly. 1996. Botany – Apomixis: the asexual revolution. Science 274:1322–1323.

Campton, D. E. 1987. Natural hybridization and introgression in fishes: methods of detection and genetic interpretations. Pp. 161–192 *in* Ryman, N. and F. Utter, eds. Population Genetics and Fishery Management. University of Washington Press, Seattle, WA.

Capy, P., G. Gasperi, C. Biemont, and C. Bazin. 2000. Stress and transposable elements: co-evolution or useful parasites? Heredity 85:101–106.

Carney, S. E., K. A. Gardner, and L. H. Rieseberg. 2000. Evolutionary changes over the fifty-year history of a hybrid population of sunflowers (*Helianthus*). Evolution 54:462–474.

Caro, T. M., and M. K. Laurenson. 1994. Ecological and genetic factors in conservation – a cautionary tale. Science 263:485–486.

Carr, D. E., and M. R. Dudash. 2003. Recent approaches into the genetic basis of inbreeding depression in plants. Philosophical Transaction of the Royal Society of London B 358:1071–1084.

Carr, G. D., and A. C. Medeiros. 1998. A remnant greensword population from pu'u 'alaea, maui, with characteristics of *Argyroxiphium virescens* (Asteraceae). Pacific Science 52:61–68.

Carrington, M., and 9 others. 1999. *HLA* and HIV-1: heterozygote advantage and *B*35-Cw*04* disadvantage. Science 283:1748–1752.

Carroll, S. P., H. Dingle, T. R. Famula, and C. W. Fox. 2001. Genetic architecture of adaptive differentiation in evolving host races of the soapberry bug, *Jadera haematoloma*. Genetics 112:257–272.

Castle, W. E. 1903. The laws of hereidty of Galton and Mendel and some laws governing race improvment by selection. Proceedings of the American Academy of Arts and Science 39:223–242.

Castric, V., L. Bernatchez, K. Belkhir, and F. Bonhomme. 2002. Heterozygote deficiencies in small lacustrine populations of brook charr *Salvelinus fontinalis* Mitchill (Pisces, Salmonidae): a test of alternative hypotheses. Heredity 89:27–35.

Castric, V., F. Bonney, and L. Bernatchez. 2001. Landscape structure and hierarchical genetic diversity in the brook charr, *Salvelinus fontinalis*. Evolution 55:1016–1028.

Castric, V., and X. Vekemans. 2004. Plant self-incompatibility in natural populations: a critical assessment of recent theoretical and empirical advances. Molecular Ecology 13:2873–2889.

Caughley, G. 1994. Directions in conservation biology. Journal of Animal Ecology 63:215–244.

Cavalli-Sforza, L. L., and A. W. F. Edwards. 1967. Phylogenetic analysis: models and estimation procedures. American Journal of Human Genetics 19:233–257.

Ceballos, G., and P. R. Ehrlich. 2002. Mammal population losses and the extinction crisis. Science 296:904–907.

Center for Plant Conservation. 1991. Genetic sampling guidelines for conservation collections of endangered plants. Pp. 225–238 *in* Falk, D. A., and K. E. Holsinger, eds. Genetics and Conservation of Rare Plants. Oxford University Press, New York.

Chapman, H., B. Robson, and M. L. Pearson. 2004. Population genetic structure of a colonising, triploid weed, *Hieracium lepidulum*. Heredity 92:182–188.

Charlesworth, B. 1991. The evolution of sex chromosomes. Science 251:1030–1033.

Charlesworth, B. 1996. Background selection and patterns of genetic diversity in *Drosophila melanogaster*. Genetical Research 68:131–149.

Charlesworth, D. 2002. Plant sex determination and sex chromosomes. Heredity 88:94–101.

Charlesworth, D., and B. Charlesworth. 1987. Inbreeding depression and its evolutionary consequences. Annual Review of Ecology and Systematics 18:237–268.

Chilcote, M. W. 2003. Relationship between natural productivity and the frequency of wild fish in mixed spawning populations of wild and hatchery steelhead (*Oncorhynchus mykiss*). Canadian Journal of Fisheries and Aquatic Sciences 60:1057–1067.

Choler, P., B. Erschbamer, A. Tribsch, L. Gielly, and P. Taberlet. 2004. Genetic introgression as a potential to widen a species' niche: insights from alpine *Carex curvula*. Proceedings of the National Academy of Sciences, USA 101:171–176.

Ciofi, C., M. A. Beaumont, I. R. Swingland, and M. W. Bruford. 1999. Genetic divergence and units for conservation in the Komodo dragon *Varanus komodoensis*. Proceedings of the Royal Society of London B 266:2269–2274.

Clark, A. G. 1988. The evolution of the Y chromosome with X-Y recombination. Genetics 119:711–20.

Clark, C. W. 1984. Strategies for multispecies management: objectives and constraints. Pp. 303–312 *in* May, R. M., ed. Dahlem Konferenzen. Springer Verlag, Berlin.

Clarke, B. 1979. The evolution of genetic diversity. Proceedings of the Royal Society of London B 205:19–40.

Clarke, B., M. S. Johnson, and J. Murray. 1998. How "molecular leakage" can mislead us about island speciation. Pp. 181–195 *in* Grant, P. R., ed. Evolution on Islands. Oxford University Press, Oxford.

Clarke, B. C., and L. Partridge. 1988. Frequency dependent selection: a discussion. Philosophical Transactions of the Royal Socerty of London B 319:457–640.

Clarke, G. M. 1993. Fluctuating asymmetry of invertebrate populations as a biological indicator of environmental quality. Environmental Pollution 82:207–211.

Clausen, J. 1951. Stages in the Evolution of Plant Species. Cornell University Press, Ithaca, NY.

Clausen, J., D. D. Keck, and W. M. Hiesey. 1948. Experimental Studies on the Nature of Species: III. Environmental Responses of Climatic Races of *Achillea*. Carnegie Institute Publication No. 581, Carnegie Institute, Washington, DC.

Clegg, M. T. 1990. Molecular diversaity in plant populations. Pp. 98–115 *in* Brown, A. H. D., M. T. Clegg, A. L. Kahler, and B. S. Weir, eds. Plant Population Genetics, Breeding, and Genetic Resources. Sinauer, Sunderland, MA.

Clegg, M. T., and M. L. Durbin. 2000. flower color variation: a model for the experimental study of evolution. Proceedings of the National Academy of Sciences, USA 97:7016–7023.

Clout, M. N., G. P. Elliott, and B. C. Robertson. 2002. Effects of supplementary feeding on the offspring sex ratio of kakapo: a dilemma for the conservation of a polygynous parrot. Biological Conservation 107:13–18.

Cochran, R. G. 1954. Some methods for strengthening the common X^2 tests. Biometrics 10:417–451.

Cockerham, C. C., and B. S. Weir. 1977. Digenic descent measures for finite populations. Genetic Resource 30:121–147.

Cohan, F. M., and A. A. Hoffmann. 1986. Genetic divergence under uniform selection. II. Different responses to selection for knockdown resistance to ethanol among *Drosophila melanogaster* populations and their replicate lines. Genetics 114:145–163.

Collins, T. M., J. C. Trexler, L. G. Nico, and T. A. Rawlings. 2002. Genetic diversity in a morphologically conservative invasive taxon: multiple introductions of swamp eels to the southeastern United States. Conservation Biology 16:1024–1035.

Collyer, M. L. 2003. Ecological morphology of the White Sands pupfish. Ph.D. dissertation, North Dakota State University, Fargo, ND.

Collyer, M. L., J. M. Novak, and C.A. Stockwell. 2005. Morphological divergence in recently established populations of White Sands pupfish (*Cyprinodon tularosa*). Copeia 2005:1–11.

Coltman, D. W. 2005. Testing marker-based estimates of heritability in the wild. Molecular Ecology 14:2593–2599.

Coltman, D. W., M. Festa-Bianchet, J. T. Jorgenson, and C. Strobeck. 2002. Age-dependent sexual selection in bighorn rams. Proceedings of the Royal Society of London Series B 269:165–172.

Coltman, D. W., P. O'Donoghue, J. T. Jorgenson, J. T. Hogg, C. Strobeck, and M. Festa-Bianchet. 2003. Undesirable evolutionary consequences of trophy hunting. Nature 425:655–658.

Coltman, D. W., J. G. Pilkington, J. A. Smith, and J. M. Pemberton. 1999. Parasite-mediated selection against inbred Soay sheep in a free-living, island population. Evolution 53:1259–1267.

Comings, D. 1978. Mechanisms of chromosome banding and implications for chromosome structure. Annual Review of Genetics 12:25–46.

Comstock, K. E., and 6 others. 2002. Patterns of molecular genetic variation among African elephant populations. Molecular Ecology 11:2489–2498.

Conner, M. M., and G. C. White. 1999. Effects of individual heterogeneity in estimating the persistence of small populations. Natural Resources Modeling 12:109–127.

Conover, D. O., and S. B. Munch. 2002. Sustaining fisheries yields over evolutionary time scales. Science 297:94–96.

Conover, D. O., and E. T. Schultz. 1995. Phenotypic similarity and the evolutionary significance of countergradient variation. Trends in Ecology and Evolution 10:248–252.

Cooke, F. 1987. Lesser snow geese: a long-term population study. Pp. 407–432 *in* Cooke, F. and P. A. Buckley, eds. Avian Genetics: a Population and Ecological Approach. Academic Press, London.

Cooper, D. W. 1968. The significance level in multiple tests made simultaneously. Heredity 23:614–617.

Cornuet J.-M., S. Piry, G. Luikart, A. Estoup, and M. Solignac. 1999. New methods employing multilocus genotypes for selecting or excluding populations as origins of individuals. Genetics 153:1989–2000.

Cotgreave, P. 1993. The relationship between body size and population abundance in animals. Trends in Ecology and Evolution 8:244–248.

Couzin, J. 2004. Hybrid mosquitoes suspected in West Nile virus spread. Science 303:1451.

Cracraft, J. 1989. Speciation and its ontology: the empirical consequences of alternative species concepts for understanding patterns and processes of differentiation. Pp. 29–59 *in* Otte, D. and J. A. Endler, eds. Speciation and its Consequences. Sinauer, Sunderland, MA.

Craig, J. K., and C. J. Foote. 2001. Countergradient variation and secondary sexual color: Phenotypic convergence promotes genetic divergence in carotenoid use between sympatric anadromous and nonanadromous morphs of sockeye salmon (*Oncorhynchus nerka*). Evolution 55:380–391.

Craighead, J. J., Summner, J. S., and Mitchell, J. A. 1995. The Grizzly Bears of Yellowstone: Their Ecology in the Yellowstone Ecosystem, 1959–1992. Island Press, Washington, DC.

Craighead, J. J., Varney, J. R., and Craighead Jr., F. C. 1973. A computer analysis of the Yellowstone grizzly bear population. Montana Cooperative Wildlife Unit, Missoula, MT.

Crampe, H. 1883. Zuchtversuche mit zahmen Wanderratten. Landwirtschaftliches Jahrbuch 12:389–458.

Crandall, K. A., O. R. P. Binindaemonds, G. M. Mace, and R. K. Wayne. 2000. Considering evolutionary processes in conservation biology. Trends in Ecology and Evolution 15:290–295.

Cree, A., M. B. Thompson, and C. H. Daugherty. 1995. Tuatara sex determination. Nature 375:543.

Cresswell, M. (compiler). 1996. Kakapo Recovery Plan 1996–2005. Threatened Species Recovery Plan No. 21. Department of Conservation, Wellington, New Zealand.

Criscione, C. D., R. Poulin, and M. S. Blouin. 2005. Molecular ecology of parasites: elucidating ecological and microevolutionary processes. Molecular Ecology 14:2247–2257.

Crnokrak, P., and D. A. Roff. 1999. Inbreeding depression in the wild. Heredity 83:260–270.

Crow, J. F. 1948. Alternate hypotheses of hybrid vigor. Genetics 33:477–487.

Crow, J. F. 1954. Breeding structure of populations. II. Effective population number. Pp. 543–556 *in* Kempthorne, O., T. A. Bancroft, J. W. Gowen, and J. L. Lush, eds. Statistics and Mathematics in Biology. Iowa State College Press, Ames, IA.

Crow, J. F. 1957. Genetics of insect resistance to chemicals. Annual Review of Entomology 2:227–246.

Crow, J. F. 1986. Basic Concepts in Population Genetics. Freeman, New York.

Crow, J. F. 1993. Mutation, mean fitness, and genetic load. Oxford Surveys in Evolutionary Biology 9:3–42.

Crow, J. F. 2001. The beanbag lives on. Nature 409:771.

Crow, J. F., and K. Aoki. 1984. Group selection for a polygenic trait: estimating the degree of population subdivision. Proceedings of the National Academy of Sciences, USA 81:6073–6077.

Crow, J. F., and C. Denniston. 1988. Inbreeding and variance effective population numbers. Evolution 42:482–495.

Crow, J. F., and M. Kimura. 1970. An Introduction to Population Genetics Theory. Burgess Publishing Company, Minneapolis, MN.

Crozier, R. H. 1971. Heterozygosity and sex determination in haplo-diploidy. American Naturalist 105:399–412.

Crozier, R. H., and R. M. Kusmierski. 1994. Genetic distances and the setting of conservation priorities. Pp. 227–237 *in* Loeschcke, V., J. Tomiuk, and S. K. Jain, eds. Conservation Genetics. Birkhauser Verlag, Basel, Switzerland.

Currens, K. P., A. R. Hemmingsen, R. A. French, D. V. Buchanan, C. B. Schreck, and H. W. Li. 1997. Introgression and susceptibility to disease in a wild population of rainbow trout. North American Journal of Fisheries Management 17:1065–1078.

Darlington, C. D. 1969. The Evolution of Man and Society. Simon & Schuster, New York.

Darwin, C. 1859. The Origin of Species by Means of Natural Selection or the Preservation of Favored Races in the Struggle for Life. John Murray, London.

Darwin, C. 1896. The Variation of Animals and Plants under Domestication, Vol. II. D. Appleton & Co., New York.

Daugherty, C. H., A. Cree, J. M. Hay, and M. B. Thompson. 1990. Neglected taxonomy and continuing extinctions of tuatara (*Sphenodon*). Nature 347:177–179.

David, P. 1998. Heterozygosity-fitness correlations: new perspectives on old problems – short review. Heredity 80:531–537.

Dawson, S., K. Russell, F. B. Pichler, L. Slooten, and C. S. Baker. 2001. The North Island Hector's dolphin is vulnerable to extinction. Marine Mammal Science 17:366–371.

Degnan, S. M. 1993. The perils of single gene trees – mitochondrial versus single-copy nuclear DNA variation in white-eyes (Aves: Zosteropidae). Molecular Ecology 2:219–225.

De Boer, L. E. M. 1982. Karyological problems in breeding owl monkeys, *Aotus trivirgatus*. International Zoo Yearbook 22:119–124.

deJong, H. 2003. Visualizing DNA domains and sequences by microscopy: a fifty-year history of molecular cytogenetics. Genome 46:943–946.

Delpuech, J. M., Y. Carton, and R. T. Roush. 1993. Conserving genetic variability of a wild insect population under laboratory conditions. Entomologia Experimentalis et Applicata 67:233–239.

Demauro, M. M. 1993. Relationship of breeding system to rarity in the lakeside daisy (*Hymenoxys acaulis* var *glabra*). Conservation Biology 7:542–550.

Dempster, A. P., N. M. Laird, and D. B. Rubin. 1977. Maximum likelihood estimation from incomplete data via the *EM* algorithm. Journal of the Royal Statistics Society B 39:1–38.

Denholm, I., G. J. Devine, and M. S. Williamson. 2002. Insecticide resistance on the move. Science 297:2222–2223.

Devall, B., and G. Sessions. 1984. The development of natural resources and the integrity of nature. Environmental Ethics 6:293–322.

Devlin, R. H., and Y. Nagahama. 2002. Sex determination and sex differentiation in fish: an overview of genetic, physiological, and environmental influences. Aquaculture 208:191–364.

Diamond, J. 1997. Guns, germs, and steel. W. W. Norton, New York.

Dinerstein, E., and G. F. McCracken. 1990. Endangered greater one-horned rhinoceros carry high levels of genetic variation. Conservation Biology 4:417–422.

Dizon, A. E., C. Lockyer, W. F. Perrin, D. P. Demaster, and J. Sisson. 1992. Rethinking the stock concept – a phylogeographic approach. Conservation Biology 6:24–36.

Dobson, A. 2003. Metalife! Science 301:1488–1490.

Dobzhansky, Th. 1948. Genetics of natural populations, XVIII. Experiments on chromosomes of *Drosophila pseudoobscura* from different geographical regions. Genetics 33:588–602.

Dobzhansky, Th. 1951. Genetics and the Origin of Species, 3rd edn. Columbia University Press, New York.

Dobzhansky, Th. 1962. Oral History Memoir. Columbia University Press, New York (cited in Provine 1986).

Dobzhansky, Th. 1970. Genetics of the Evolutionary Process. Columbia University Press, New York.

Dobzhansky, Th., Ayala, F. J., Stebbins, G. L., and Valentine, J. W. 1977. Evolution. W. H. Freeman, San Francisco.

Dobzhansky, Th., and S. Wright. 1941. Genetics of natural populations. V. Relations between mutation rate and accumulation of lethals in populations of *Drosophila pseudoobscura*. Genetics 26:23–51.

Doremus, H. 1999. Restoring endangered species: the importance of being wild. Harvard Environmental Law Review 23:3–92.

Double, M. C., A. Cockburn, S. C. Barry, and P. E. Smouse. 1997. Exclusion probabilities for single-locus paternity analysis when related males compete for matings. Molecular Ecology 6:1155–1166.

Dournon, C., A. Collenot, and M. Lauthier. 1988. Sex-linked peptidase-1 patterns in *Pleurodeles waltlii* Michah. (Urodele amphibian): genetic evidence for a new codominant allele on the W sex chromosomes and identification of ZZ, ZW, and WW sexual genotypes. Reproduction, Nutrition, Development 28:979–987.

Dowling, T. E., and C. L. Secor. 1997. The role of hybridization and introgression in the diversification of animals. Annual Review of Ecology and Systematics 28:593–619.

Dozier, H. L. 1948. Color mutations in the muskrat (*Ondatra z. macrodon*) and their inheritance. Journal of Mammalogy 29:393–405.

Dozier, H. L., and R. W. Allen. 1942. Color, sex ratio, and weights of Maryland muskrats. J. Wildlife Management 6:294–300.

Drake, J. W., B. Charlesworth, D. Charlesworth, and J. F. Crow. 1998. Rates of spontaneous mutation. Genetics 148:1667–1686.

Duchesne, P., and L. Bernatchez. 2002. An analytical investigation of the dynamics of inbreeding in multi-generation supportive breeding. Conservation Genetics 3:47–60.

Dudash, M. R. 1990. Relative fitness of selfed and outcrossed progeny in a self-compatible, protandrous species, *Sabatia angularis* L. (Gentianaceae): a comparison in three environments. Evolution 44:1129–1140.

Dudash, M. R., and C. B. Fenster. 2000. Inbreeding and outbreeding depression in fragmented populations. Pp. 35–53 *in* Young, A. G., and G. M. Clarke, eds. Genetics, Demography and Viability of Fragmented Populations. Cambridge University Press, Cambridge, UK.

Dytham, C. 2003. Choosing and Using Statistics: a Biologists Guide, 2nd edn. Blackwell Publishing, Oxford.

Eanes, W. F. 1987. Allozymes and fitness: evolution of a problem. Trends in Ecology and Evolution 2:44–48.

Ebenhard, T. 1995. Conservation breeding as a tool for saving animal species from extinction. Trends in Ecology and Evolution 10:438–443.

Eberhardt, L. L., and R. R. Knight. 1996. How many grizzlies in Yellowstone? Journal of Wildlife Management 60:416–421.

Eberhart, S. A., E. E. Roos, and L. E. Towill. 1991. Strategies for long-term management of germplasm collections. P. 135 *in* Falk, D. A., and K. E. Holsinger, eds. Genetics and Conservation of Rare Plants. Oxford University Press, New York.

Echt, C. S., L. L. Deverno, M. Anzidei, and G. G. Vendramin. 1998. Chloroplast microsatellites reveal population genetic diversity in red pine, *Pinus resinosa* Ait. Molecular Ecology 7:307–316.

Edmands, S. 2002. Does parental divergence predict reproductive compatibility? Trends in Ecology and Evolution 17:520–527.

Edwards, S. V., and P. W. Hedrick. 1998. Evolution and ecology of MHC molecules: from genomics to sexual selection. Trends in Ecology and Evolution 13:305–311.

Edwards, S. V., and W. K. Potts. 1996. Polymorphism of genes in the major histocompatibility complex (MHC): implications for conservation genetics of vertebrates. Pp. 214–237 *in* Smith, T. B. and R. K. Wayne, eds. Molecular Genetic Approaches in Conservation. Oxford University Press, New York.

Eggert, L. S., J. A. Eggert, and D. S. Woodruff. 2003. Estimating population sizes for elusive animals: the forest elephants of Kalum National Park, Guana. Molecular Ecology 12:1389–1402.

Ehrlich, P. R., and P. H. Raven. 1969. Differentiation of populations. Science 165:1228–1232.

Eisen, J. A. 1999. Mechanistic basis for microsatellite instability. Pp. 34–48 *in* Goldsetin, D. B., and C. Schlotterer, eds. Microsatellites: Evolution and Applications. Oxford University Press, Oxford.

Eisner, T., J. Lubchenco, E. O. Wilson, D. S. Wilcove, and M. J. Bean. 1995. Building a scientifically sound policy for protecting endangered species. Science 268:1231–1233.

El Mousadik, A., and R. J. Petit. 1996. High level of genetic differentiation for alleleic richness among populations of the argan tree [*Argania spinosa* (L.) Skeels] endemic of Morocco. Theoretical and Applied Genetics 92:832–839.

Elena, S. F., V. S. Cooper, and R. E. Lenski. 1996. Punctuated evolution caused by selection of rare beneficial mutations. Science 272:1802–1804.

Ellegren, H. 1999. Inbreeding and relatedness in Scandinavian grey wolves *Canis lupus*. Hereditas 130:239–244.

Ellegren, H. 2000a. Microsatellite mutations in the germline: implications for evolutionary inference. Trends in Genetics 16:551–558.

Ellegren, H. 2000b. Evolution of the avian sex chromosomes and their role in sex determination. Trends in Ecology and Evolution 15:188–192.

Elliott, G. P., D. V. Merton, and P. W. Jansen. 2001. Intensive management of a critically endangered species: the kakapo. Biological Conservation 99:121–133.

Ellstrand, N. C. 2003. Dangerous Liaisons? When Cultivated Plants Mate with their Wild Relatives. John Hopkins University Press, Baltimore, MD.

Ellstrand, N. C., and D. R. Elam. 1993. Population genetic consequences of small population size – implications for plant conservation. Annual Review of Ecology and Systematics 24:217–242.

Ellstrand, N. C., and K. A. Schierenbeck. 2000. Hybridization as a stimulus for the evolution of invasiveness in plants? Proceedings of the National Academy of Sciences, USA 97:7043–7050.

Ellstrand, N. C., R. Whitkus, and L. H. Rieseberg. 1996. Distribution of spontaneous plant hybrids. Proceedings of the National Academy of Sciences, USA 93:5090–5093.

Emerson, S. 1939. A preliminary survey of the *Oenothera organensis* population. Genetics 24:524–537.

Emerson, S. 1940. Growth of incompatible pollen tubes in *Oenothera organensis*. Botanical Gazette 101:890–911.

England, P. R., G. H. R. Osler, L. M. Woodworth, M. E. Montgomery, D. A. Briscoe, and R. Frankham. 2003. Effects of intense versus diffuse population bottlenecks on microsatellite genetic diversity and evolutionary potential. Conservation Genetics 4:595–604.

Epifanio, J., and D. Philipp. 2001. Simulating the extinction of parental lineages from introgressive hybridization: the effects of fitness, initial proportions of parental taxa, and mate choice. Reviews in Fish Biology and Fisheries 10:339–354.

Epperson, B. K. 2005. Mutation at high rates reduces spatial structure within populations. Molecular Ecology 14:703–710.

Erwin, T. L. 1991. An evolutionary basis for conservation strategies. Science 253:750–752.

Escudero, A., J. M. Iriondo, and M. E. Torres. 2003. Spatial analysis of genetic diversity as a tool for plant conservation. Biological Conservation 113:351–365.

Estoup, A., and B. Angers. 1998. Microsatellites and minisatellites for molecular ecology: theoretical and empirical considerations. Pp. 55–86 *in* Carvalho, G. R., ed. Advances in Molecular Ecology. IOS Press, Amsterdam.

Etterson, J. R., and R. G. Shaw. 2001. Constraint to adaptive evolution in response to global warming. Science 294:151–154.

Evanno, G., S. Regnaut, and J. Goudet. 2005. Detecting the number of clusters of individuals using the software STRUCTURE: a simulation study. Molecular Ecology 14:2611–2620.

Ewens, W. J. 1982. On the concept of effective population size. Theoretical Population Biology 21:373–378.

Excoffier, L., and M. Slatkin. 1995. Testing for linkage disequilibrium in genotypic data using the expectation-maximization algorithm. Heredity 76:377–383.

Excoffier, L., P. E. Smouse, and J. M. Quattro. 1992. Analysis of molecular variance inferred from metric distances among DNA haplotypes – application to human mitochondrial DNA restriction data. Genetics 131:479–491.

Faith, D. P. 2002. Quantifying biodiversity: a phylogenetic perspective. Conservation Biology 16:248–252.

Falconer, D. S., and Mackay, T. F. C. 1996. Introduction to Quantitative Genetics, 4th edn. Longman Science and Technology, Harlow, UK.

Felsenstein, J. 2004. Inferring Phylogenies. Sinauer, Sunderland, MA.

Fenster, C. B., and M. R. Dudash. 1994. Genetic considerations for plant population restoration and conservation. Pp. 34–62 *in* Bowles, M. L., and C. J. Whelan, eds. Restoration of Endangered Species. Cambridge University Press, Cambridge, UK.

Fernando, P., and D. J. Melnick. 2001. Molecular sexing eutherian mammals. Molecular Ecology Notes 1:350–353.

Fisher, R. A. 1918. The correlation between relatives on the supposition of Mendelian inheritance. Transactions of the Royal Society of Edinburgh 52:399–433.

Fisher, R. A. 1930. The Genetical Theory of Natural Selection. Clarendon Press, Oxford.

Fisher, R. A. 1935. The logic of inductive inference. Journal of the Royal Statistics Society 98:39–54.

Fitzpatrick, J. W., and 16 others. 2005. Ivory-billed woodpecker (*Campephilus principalis*) persists in continental North America. Science 308:1460–1462.

FitzSimmons, N. N., A. R. Goldizen, J. A. Norman, C. Moritz, J. D. Miller, and C. J. Limpus. 1997a. Philopatry of male marine turtles inferred from mitochondrial markers. Proceedings of the National Academy of Sciences, USA 94:8912–8917.

FitzSimmons, N. N., C. Moritz, C. J. Limpus, L. Pope, and R. Prince. 1997b. Geographic structure of mitochondrial and nuclear gene polymorphisms in Australian green turtle populations and male-biased gene flow. Genetics 147:1843–1854.

Foltz, D. W. 1986. Null alleles as a possible cause of heterozygote deficiencies in the oyster *Crassostrea virginica* and other bivalves. Evolution 40:869–870.

Foltz, D. W., B. M. Schaitkin, and R. K. Selander. 1982. Gametic disequilibrium in the self-fertilizing slug *Derocera laeve*. Evolution 36:80–85.

Fonseca, D. M., and 7 others. 2004. Emerging vectors in the *Culex pipiens* complex. Science 303:1535–1538.

Foote, C. J., G. S. Brown, and C. W. Hawryshyn. 2004. Female colour and male choice in sockeye salmon: implications for the phenotypic convergence of anadromous and nonanadromous morphs. Animal Behaviour 67:69–83.

Forbes, S. H. 1990. Mitochondrial and nuclear genotypes in trout hybrid swarms: tests for gametic equilibrium and effects on phenotypes. PhD Dissertation, University of Montana, Missoula, MT.

Forbes, S. H., and F. W. Allendorf. 1991. Associations between mitochondrial and nuclear genotypes in cutthroat trout hybrid swarms. Evolution 45:1332–1349.

Forbes, S. H., and J. T. Hogg. 1999. Assessing population structure at high levels of differentiation: microsatellite comparisons of bighorn sheep and large carnivores. Animal Conservation 2:223–233.

Forbes, S. H., K. L. Knudsen, T. W. North, and F. W. Allendorf. 1994. One of two growth hormone genes in coho salmon is sex-linked. Proceedings of the National Academy of Sciences, USA 91:1628–1631.

Ford, E. B. 1971. Ecological Genetics. Chapman & Hall, London.

Ford, M. J. 2002. Selection in captivity during supportive breeding may reduce fitness in the wild. Conservation Biology 16:815–825.

Foster, G. G., M. J. Whitten, T. Prout, and R. Gill. 1972. Chromosome rearrangement for the control of insect pests. Science 176:875–880.

Fowler, D. P., and D. T. Lester. 1970. Genetics of red pine. USDA Forest Service Research Paper WO-8. USDA, Washington, DC.

Fowler, D. P., and R. W. Morris. 1977. Genetic diversity in red pine: evidence for low genic heterozygosity. Canadian Journal of Forest Research 7:343–347.

Frankel, O. H. 1970. Genetic conservation in perspective. Pp. 469–489 in Frankel, O. H., and E. Bennett, eds. Genetic Resources in Plants – their Exploration and Conservation. Blackwell Scientific Publications, Oxford.

Frankel, O. H. 1974. Genetic conservation: our evolutionary responsibility. Genetics 78:53–65.

Frankel, O. H., and M. E. Soulé. 1981. Conservation and Evolution. Cambridge University Press, Cambridge, UK.

Frankham, R. 1995. Effective population size/adult population size ratios in wildlife: a review. Genetical Research 66:95–107.

Frankham, R. 1999. Quantitative genetics in conservation biology. Genetical Research 74:237–244.

Frankham, R. 2003. Genetics and conservation biology. Comptes Rendus Biologies 326:S22–S29.

Frankham, R., H. Hemmer, O. A. Ryder, E. G. Cothran, M. E. Soulé, N. D. Murray, and M. Snyder. 1986. Selection in captive populations. Zoo Biology 5:127–138.

Frankham, R., M. Lees, M. E. Montgomery, P. R. England, E. H. Lowe, and D. A. Briscoe. 1999. Do population size bottlenecks reduce evolutionary potential? Animal Conservation 2:255–260.

Frankham, R., and D. A. Loebel. 1992. Modeling problems in conservation genetics using captive Drosophila populations – rapid genetic adaptation to captivity. Zoo Biology 11:333–342.

Frankham, R., H. Manning, S. H. Margan, and D. A. Briscoe. 2000. Does equalization of family sizes reduce genetic adaptation to captivity? Animal Conservation 3:357–363.

Frankham, R., and K. Ralls. 1998. Conservation biology – inbreeding leads to extinction. Nature 392:441–442.

Franklin, D. C., and P. L. Dostine. 2000. A note on the frequency and genetics of head colour morphs in the Gouldian Finch. Emu 100:236–239.

Franklin, I. R. 1980. Evolutionary changes in small populations. Pp. 135–149 in Soulé, M. E., and B. A. Wilcox, eds. Conservation Biology: an Evolutionary-Ecological Perspective. Sinauer, Sunderland, MA.

Franklin, I. R., and R. Frankham. 1998. How large must populations be to retain evolutionary potential? Animal Conservation 1:69–70.

Fraser, D. J., and L. Bernatchez. 2001. Adaptive evolutionary conservation: towards a unified concept for defining conservation units. Molecular Ecology 10:2741–2752.

Freeman, S., and Herron, J. C. 1998. Evolutionary Analysis, 1st edn. Prentice Hall, Upper Saddle River, NJ.

Friar, E. A., T. Ladoux, E. H. Roalson, and R. H. Robichaux. 2000. Microsatellite analysis of a population crash and bottleneck in the Mauna Kea silversword, Argyroxiphium sandwicense ssp sandwicense (Asteraceae), and its implications for reintroduction. Molecular Ecology 9:2027–2034.

Fridez, F., S. Rochat, and R. Coquoz. 1999. Individual identification of cats and dogs using mitochondrial DNA tandem repeats? Science and Justice 39:167–171.

Fu, Y.-X., and W.-H. Li. 1993. Maximum likelihood estimation of population parameters. Genetics 134:1261–1270.

Fu, Y.-X., and W.-H. Li. 1999. Coalescing into the 21st century: an overview and prospects of coalescent theory. Theoretical Population Biology 56:1–10.

Funk, W. C., D. A. Tallmon, and F. W. Allendorf. 1999. Small effective population size in the long-toed salamander. Molecular Ecology 8:1633–1640.

Funk, W. C., J. A. Tyburczy, K. L. Knudsen, K. R. Lindner, and F. W. Allendorf. 2005. Genetic basis of variation in morphological and life history traits of pink salmon (Oncorhynchus gorbuscha). Journal of Heredity 96:24–31.

Gabriel, W., and R. Bürger. 1994. Extinction risk by mutational meltdown: synergistic effects between population regulation and genetic drift. Pp. 69–84 in Loeschcke, V., J. Tomiuk, and S. K. Jain, eds. Conservation Genetics. Birkhauser Verlag, Basel, Switzerland.

Galbusera, P., L. Luc, T. Schenck, E. Waiyaki, and E. Matthysen. 2000. Genetic variability and gene flow in the globally, critically-endangered Taita thrush. Conservation Genetics 1:45–55.

Gams, W. 2002. Ex situ conservation of microbial diversity. Pp. 269–283 in Sivasithamparam, K., K. W. Dixon, and R. L. Barrett, eds. Microorganisms in Plant Conservation and Biodiversity. Kluwer, Dordrecht, Netherlands.

García-Ramos, G., and M. Kirkpatrick. 1997. Genetic models of adaptation and gene flow in peripheral populations. Evolution 51:21–28.

García-Ramos, G., and D. Rodríguez. 2002. Evolutionary speed of species invasions. Evolution 56:661–668.

Gaskin, J. F., and B. A. Schaal. 2002. Hybrid Tamarix widespread in U.S. invasion and undetected in native Asian range. Proceedings of the National Academy of Sciences, USA 99: 11256.

Gauss, C. F. 1809. Theoria motus corporum coelestium in sectionis conicis solem ambientum. Hamburg, Germany.

Gemmell, N. J., and F. W. Allendorf. 2001. Mitochondrial mutations may decrease population viability. Trends in Ecology and Evolution 16:115–117.

Genner, M. J., E. Michel, D. Erpenbeck, N. DeVoogd, F. Witte, and J. P. Pointier. 2004. Camouflaged invasion of Lake Malawi by an Oriental gastropod. Molecular Ecology 13:2135–2141.

Gerhardt, H. C., M. B. Ptacek, L. Barnett, and K. G. Torke. 1994. Hybridization in the diploid–tetraploid treefrogs *Hyla chrysoscelis* and *Hyla versicolor*. Copeia 1994:51–59.

Ghiselin, M. T. 1969. The Triumph of the Darwinian Method. University of Chicago Press, Chicago.

Gibbs, H. L., P. T. Boag, B. N. White, P. J. Weatherhead, and L. M. Tabak. 1991. Detection of a hypervariable locus in birds by hybridization with a mouse MHC probe. Molecular Biology and Evolution 8:433–446.

Gigord, L. D. B., M. R. Macnair, and A. Smithson. 2001. Negative frequency-dependent selection maintains a dramatic flower color polymorphism in the rewardless orchid *Dactylorhiza sambucina* (L.) Soň. Proceedings of the National Academy of Sciences, USA 98:6253–6255.

Gillespie, J. H. 1992. The Causes of Molecular Evolution. Oxford University Press, Oxford.

Gillespie, J. H. 2001. Is the population size of a species relevant to its evolution? Evolution 55:2161–2169.

Gilligan, D. M., D. A. Briscoe, and R. Frankham. 2005. Comparative losses of quantitative and molecular genetic variation in finite populations of *Drosophila melanogaster*. Genetical Research 85:47–55.

Gilligan, D. M., and R. Frankham. 2003. Dynamics of genetic adaptation to captivity. Conservation Genetics 4:189–197.

Gilpin, M. 1991. The genetic effective size of a metapopulation. Biological Journal of the Linnean Society 42:165–175.

Gilpin, M. E., and M. E. Soulé. 1986. Minimum viable populations: processes of species extinction. Pp. 19–34 *in* Soulé, M. E., ed. Conservation Biology: the Science of Scarcity and Diversity. Sinauer, Sunderland, MA.

Gilpin, M., and C. Wills. 1991. MHC and captive breeding – a rebuttal. Conservation Biology 5:554–555.

Godt, M. J. W., J. Walker, and J. L. Hamrick. 1997. Genetic diversity in the endangered lily *Harperocallis flava* and a close relative, *Tofieldia racemosa*. Conservation Biology 11:361–366.

Goldberg, T. L. 1997. Inferring the geographic origins of "refugee" chimpanzees in Uganda from mitochondrial DNA sequences. Conservation Biology 11:1441–1446.

Goldberg, T. L., E. C. Grant, K. R. Inendino, T. W. Kassler, J. E. Claussen, and D. P. Philipp. 2005. Increased infectious disease susceptibility resulting from outbreeding depression. Conservation Biology 19:455–462.

Goldman, E. 2003. Evolution: puzzling over the origin of species in the depths of the oldest lakes. Science 299:654–655.

Gomulkiewicz, R., and R. D. Holt. 1995. When does evolution by natural selection prevent extinction? Evolution 49:201–207.

Goodman, D. 1987. How do any species persist? Lessons for conservation biology. Conservation Biology 1:59–62.

Goodnight, K. F., and D. C. Queller. 1999. Computer software for performing likelihood tests of pedigree relationship using genetic markers. Molecular Ecology 8:1231–1234.

Gorman, G. C., and J. Renzi. 1979. Genetic distance and heterozygosity estimates in electrophoretic studies – effects of sample size. Copeia 1979:242–249.

Grandbastien, M. A. 1998. Activation of plant retrotransposons under stress conditions. Trends in Plant Science 3:181–187.

Grant, B. R., and P. R. Grant. 1998. Hybridization and speciation in Darwin's finches – the role of sexual imprinting on a culturally transmitted trait. Pp. 404–422 *in* Howard, D. J., and S. H. Berlocher, eds. Endless Forms. Oxford University Press, New York.

Grant, P. R. 1986. Ecology and Evolution of Darwin's Finches. Princeton University Press, Princeton, NJ.

Grant, P. R., and B. R. Grant. 1992a. Demography and the genetically effective sizes of two populations of Darwin finches. Ecology 73:766–784.

Grant, P. R., and B. R. Grant. 1992b. Hybridization of bird species. Science 256:193–197.

Grant, V. 1963. The Origin of Adaptations. Columbia University Press, New York.

Grant, W. S., and R. W. Leslie. 1993. Effect of metapopulation structure on nuclear and organellar DNA variability in semi-arid environments of southern Africa. South African Journal of Science 89:287–293.

Grauer, D., and Li, W.-H. 2000. Fundamentals of Molecular Evolution, 2nd edn. Sinauer, Sunderland, MA.

Graves, J. A. M., and S. Shetty. 2001. Sex from W to Z: evolution of vertebrate sex chromosomes and sex determining genes. Journal of Experimental Zoology 290:449–462.

Green, D. M. 1991. Supernumerary chromosomes in amphibians. Pp. 333–357 in Green, D. M., and S. K. Sessions, eds. Amphibian Cytogenetics and Evolution. Academic Press, San Diego, CA.

Griffith, B., J. W. Scott, J. W. Carpenter, and C. Reed. 1989. Translocation as a species conservation tool: status and strategy. Science 245:477–480.

Griffiths, R., and K. Orr. 1999. The use of amplified fragment length polymorphism (AFLP) in the isolation of sex-specific markers. Molecular Ecology 8:671–674.

Groombridge, J. J., C. G. Jones, M. W. Bruford, and R. A. Nichols. 2000. Conservation biology – "ghost" alleles of the Mauritius kestrel. Nature 403:616.

Groot, C., and Margolis, L. 1991. Pacific Salmon Life Histories. UBC Press, Vancouver.

Guerrant, E. O. 1996. Designing populations: demographic, genetic, and horticultural dimensions. Pp. 171–207 in Falk, D. A., C. I. Millar, and M. Olwell, eds. Restoring Diversity. Island Press, Washington, DC.

Gullion, G. W., and W. H. Marshall. 1968. Survival of ruffed grouse in a boreal forest. Living Bird 7:117–167.

Guo, S. W., and E. A. Thompson. 1992. Performing the exact test of Hardy–Weinberg proportion for multiple alleles. Biometrics 48:361–372.

Gutierrez-Espeleta, G. A., P. W. Hedrick, S. T. Kalinowski, D. Garrigan, and W. M. Boyce. 2001. Is the decline of desert bighorn sheep from infectious disease the result of low MHC variation? Heredity 86:439–450.

Gutschick, V. P., and H. BassiriRad. 2003. Extreme events as shaping physiology, ecology, and evolution of plants: toward a unified definition and evaluation of their consequences. New Phytologist 160:21–42.

Gyllensten, U., D. Wharton, A. Josefsson, and A. C. Wilson. 1991. Paternal inheritance of mitochondrial DNA in mice. Nature 352:255–257.

Haig, S. M., and F. W. Allendorf. In press. Listing and protection of hybrids under the Endangered Species Act in Scott, J. M., D. D. Goble, and F. W. Davis, eds. The Endangered Species Act at Thirty: Conserving Biodiversity on Human Dominated Landscapes. Island Press, Washington, DC.

Haig, S. M., J. D. Ballou, and S. R. Derrickson. 1990. Management options for preserving genetic diversity: reintroduction of Guam rails to the wild. Conservation Biology 4:290–300, 464.

Haig, S. M., T. D. Mullins, and E. D. Forsman. 2004. Subspecific relationships and genetic structure in the spotted owl. Conservation Genetics 5:683–705.

Hall, B. G. 2004. Phylogenetic Trees made Easy: a How-to Manual, 2nd edn. Sinauer, Sunderland, MA.

Hall, P., L. C. Orrell, and K. S. Bawa. 1994. Genetic diversity and mating system in a tropical tree, Carapa guianensis (Meliaceae). American Journal of Botany 81:1104–1111.

Halliburton, R. 2004. Introduction to Population Genetics. Pearson Prentice Hall, Upper Saddle River, NJ.

Hamrick, J. L., and M. J. Godt. 1990. Allozyme diversity in plant species. Pp. 43–63 *in* Brown, A. H. D., M. T. Clegg, A. L. Kahler, and B. S. Weir, eds. Plant Population Genetics, Breeding, and Genetic Resources. Sinauer, Sunderland, MA.

Hamrick, J. L., and M. J. Godt. 1996. Conservation genetics of endemic plants. Pp. 281–304 *in* Avise, J. C., and J. L. Hamrick, eds. Conservation Genetics: Case Histories from Nature. Chapman & Hall, New York.

Hankin, D. G., J. W. Nicholas, and T. W. Downey. 1993. Evidence for inheritance of age of maturity in chinook salmon (*Oncorhynchus tshawytscha*). Canadian Journal of Fisheries and Aquatic Sciences 50:347–358.

Hansen, M. M., E. E. Nielsen, D. E. Ruzzante, C. Bouza, and K. L. D. Mensberg. 2000. Genetic monitoring of supportive breeding in brown trout (*Salmo trutta* L.), using microsatellite DNA markers. Canadian Journal of Fisheries and Aquatic Sciences 57:2130–2139.

Hanski, I., and M. Gilpin. 1991. Metapopulation dynamics – brief history and conceptual domain. Biological Journal of the Linnean Society 42:3–16.

Hanski, I., and M. E. Gilpin, eds. 1997. Metapopulation Biology. Academic Press, San Diego, CA.

Hansson, B., S. Bensch, and D. Hasselquist. 2004. Lifetime fitness of short- and long-distance dispersing great reed warblers. Evolution 58:2546–2557.

Hansson, B., and L. Westerberg. 2002. On the correlation between heterozygosity and fitness in natural populations. Molecular Ecology 11:2467–2474.

Hardy, G. H. 1908. Mendelian proportions in a mixed population. Science 28:49–50.

Hardy, G. H. 1967. A Mathematician's Apology. Cambridge University Press, Cambridge, UK.

Harlin, A. D., B. Würsig, C. S. Baker, and T. Markowitz. 1999. Skin swabbing for genetic analysis: application to dusky dolphins (*Lagenorhynchus obscurus*). Marine Mammal Science 14:409–425.

Harpending, H. C., M. A. Batzer, M. Gurven, L. B. Jorde, A. R. Rogers, and S. T. Sherry. 1998. Genetic traces of ancient demography. Proceedings of the National Academy of Sciences, USA 95:1961–1967.

Harris, H. 1966. Enzyme polymorphism in man. Proceedings of the Royal Society of London B 164:298–310.

Harris, R. B., and F. W. Allendorf. 1989. Genetically effective population size of large mammals: an assessment of estimators. Conservation Biology 3:181–191.

Harris, S. A., and R. Ingram. 1991. Chloroplast DNA and biosystematics – the effects of intraspecific diversity and plastid transmission. Taxon 40:393–412.

Harrison, R. G. 1990. Hybrid zones: windows on evolutionary processes. Oxford Surveys in Evolutionary Biology 7:69–128.

Harrison, R. G. 1993. Hybrids and hybrid zones: historical perspective. Pp. 3–12 in Harrison, R. G., ed. Hybrid Zones and the Evolutionary Process. Oxford University Press, Oxford.

Harrison, R. G., and S. M. Bogdanowicz. 1997. Patterns of variation and linkage disequilibrium in a field cricket hybrid zone. Evolution 51:493–505.

Hartl, D. L., and Clark, A. G. 1997. Principles of Population Genetics, 3rd edn. Sinauer, Sunderland, MA.

Hartl, G. B., J. Markowski, A. Swiatecki, T. Janiszewski, and R. Willing. 1992. Studies on the European hare. 43. Genetic diversity in the Polish brown hare *Lepus europaeus pallas*, 1778 – implications for conservation and management. Acta Theriologica 37:15–25.

Hastings, A. 1981. Marginal underdominance at a stable equilibrium. Proceedings of the National Academy of Sciences, USA 78:6558–6559.

Hauffe, H. C., and J. B. Searle. 1998. Chromosomal heterozygosity and fertility in house mice (*Mus musculus domesticus*) from northern Italy. Genetics 150:1143–1154.

Hauser, L., G. J. Adcock, P. J. Smith, J. H. Bernal Ramirez, and G. R. Carvalho. 2002. Loss of microsatellite diversity and low effective population size in an overexploited population of New Zealand snapper (*Pagrus auratus*). Proceedings of the National Academy of Sciences, USA 99:11742–11747.

Hay, J. M., C. H. Daugherty, A. Cree, and L. R. Maxson. 2003. Low genetic divergence obscures phylogeny among populations of *Sphenodon*, remnant of an ancient reptile lineage. Molecular Phylogenetics and Evolution 29:1–19.

Heard, W. R. 1991. Life history of pink salmon (*Oncorhynchus gorbuscha*). Pp. 119–230 *in* Groot, C., and L. Margolis, eds. Pacific Salmon Life Histories. UBC Press, Vancouver.

Heath, D. D., J. W. Heath, C. A. Bryden, R. M. Johnson, and C. W. Fox. 2003. Rapid evolution of egg size in captive salmon. Science 299:1738–1740.

Hebert, P. D. N., E. H. Penton, J. M. Burns, D. H. Janzen, and W. Hallwachs. 2004. Ten species in one: DNA barcoding reveals cryptic species in the neotropical skipper butterfly *Astraptes fulgerator*. Proceedings of the National Academy of Sciences, USA 101:14812–14817.

Hedgecock, D. 1994. Does variance in reproductive success limit effective population sizes of marine organisms? Pp. 122–134 *in* Beaumont, A. R., ed. Genetics and Evolution of Aquatic Organisms. Chapman & Hall, London.

Hedrick, P. W. 1987. Gametic disequilibrium measures: proceed with caution. Genetics 117:331–341.

Hedrick, P. W. 1994. Purging inbreeding depression and the probability of extinction: full-sib mating. Heredity 73:363–372.

Hedrick, P. W. 1999. Perspective: highly variable loci and their interpretation in evolution and conservation. Evolution 53:313–318.

Hedrick, P. W. 2001. Conservation genetics: where are we now? Trends in Ecology and Evolution 16:629–636.

Hedrick, P. W. 2002. Pathogen resistance and genetic variation at MHC loci. Evolution 56:1902–1908.

Hedrick, P. W. 2005. Genetics of Populations, 3rd edn. Jones & Bartlett Publishers, Sudbury, MA.

Hedrick, P. W., R. Fredrickson, and H. Ellegren. 2001. Evaluation of d^2, a microsatellite measure of inbreeding and outbreeding, in wolves with a known pedigree. Evolution 55:1256–1260.

Hedrick, P. W., and M. E. Gilpin. 1997. Genetic effective size of a metapopulation. Pp. 165–181 *in* Hanski, I., and M. E. Gilpin, eds. Metapopulation Biology. Academic Press, San Diego, CA.

Hedrick, P. W., and S. T. Kalinowski. 2000. Inbreeding depression in conservation biology. Annual Review of Ecology and Systematics 31:139–162.

Hedrick, P. W., and P. S. Miller. 1992. Conservation genetics – techniques and fundamentals. Ecological Applications 2:30–46.

Hemingway, J., L. Field, and J. Vontas. 2002. An overview of insecticide resistance. Science 298:96–97.

Hernandez, J. L., and B. S. Weir. 1989. A disequilibrium coefficient approach to Hardy–Weinberg testing. Biometrics 45:53–70.

Herzog, S., A. Herzog, H. Hohn, B. Matern, and W. Hecht. 1992. Chromosome polymorphism in *Ateles geoffroyi* (Cebidae; Primates; Mammalia). Theoretical and Applied Genetics 84:986–989.

Hey, J. 2000. Anticipating scientific revolutions in evolutionary genetics. Evolutionary Biology 32:97–111.

Hey, J., R. S. Waples, M. L. Arnold, R. K. Butlin, and R. G. Harrison. 2003. Understanding and confronting species uncertainty in biology and conservation. Trends in Ecology and Evolution 18:597–603.

Higgins, K., and M. Lynch. 2001. Metapopulation extinction caused by mutation accumulation. Proceedings of the National Academy of Science, USA 98:2928–2933.

Hilborn, R., T. P. Quinn, D. E. Schindler, and D. E. Rogers. 2003. Biocomplexity and fisheries sustainability. Proceedings of the National Academy of Sciences, USA 100:6564–6568.

Hill, W. G. 1974. Estimation of linkage disequilibrium in random mating populations. Heredity 33:229–339.

Hill, W. G. 1979. A note on the effective population size with overlapping generations. Genetics 92:317–322.

Hill, W. G., and A. Robertson. 1966. The effect of linkage on limits to artificial selection. Genetical Research 8:269–294.

Hill, W. G., and A. Robertson. 1968. Linkage disequilibrium in finite populations. Theoretical and Applied Genetics 38:226–231.

Hobbs, R. J., and H. A. Mooney. 1998. Broadening the extinction debate: population deletions and additions in California and Western Australia. Conservation Biology 12:271–283.

Hoekstra, H. E., and M. W. Nachman. 2003. Different genes underlie adaptive melanism in different populations of rock pocket mice. Molecular Ecology 12:1185–1194.

Hoekstra, H. E., and T. Price. 2004. Parallel evolution in the genes. Science 303:1779–1781.

Hoelzel, A. R., ed. 1998. Molecular Genetic Analysis of Populations: a Practical Approach, 2nd edn. Oxford University Press, Oxford.

Hoffmann, A. A., R. J. Hallas, J. A. Dean, and M. Schiffer. 2003. Low potential for climatic stress adaptation in a rainforest Drosophila species. Science 301:100–102.

Hoffmann, A. A., and P. A. Parsons. 1997. Extreme Environmental Change and Evolution. Cambridge University Press, Cambridge, UK.

Hoffman, E. A., and M. S. Blouin. 2000. A review of colour and pattern polymorphisms in anurans. Biological Journal of the Linnean Society 70:633–665.

Hogan, Z. S., P. B. Moyle, B. May, M. J. Vander Zanden, and I. G. Baird. 2004. The imperiled giants of the Mekong: ecologists struggle to understand and protect Southeast Asia's large migratory catfish. American Scientist 92:238–237.

Hogbin, P. M., R. Peakall, and M. A. Sydes. 2000. Achieving practical outcomes from genetic studies of rare Australian plants. Australian Journal of Botany 48:375–382.

Hogg, J. T., S. H. Forbes, B. M. Steele, and G. Luikart. 2006. Genetic rescue of an insular population of large mammals. Proceedings of the Royal Society of London B. In press.

Holsinger, K. E. 2000. Reproductive systems and evolution in vascular plants. Proceedings of the National Academy of Sciences, USA 97:7037–7042.

Holsinger, K. E., R. J. Masongamer, and J. Whitton. 1999. Genes, demes, and plant conservation. Pp. 23–46 in Landweber, L. F., and A. P. Dobson, eds. Genetics and the Extinction of Species. Princeton University Press, Princeton, NJ.

Holt, W. V., A. R. Pickard, and R. S. Prather. 2004. Wildlife conservation and reproductive cloning. Reproduction 127:317–324.

Hosack, D. A., P. S. Miller, J. J. Hervert, and R. C. Lacy. 2002. A population viability analysis for the endangered Sonoran pronghorn, *Antilocapra americana sonoriensis*. Mammalia 66:207–229.

Houck, M. L., A. T. Kumamoto, D. S. Gallagher, and K. Benirschke. 2001. Comparative cytogenetics of the African elephant (*Loxodonta africana*) and Asiatic elephant (*Elephas maximus*). Cytogenetics and Cell Genetics 93:249–252.

Houlden, B. A., B. H. Costello, D. Sharkey, et al. 1999. Phylogeographic differentiation in the mitochondrial control region in the koala, *Phascolarctos cinereus* (Goldfuss 1817). Molecular Ecology 8:999–1011.

Huey, R. B., G. W. Gilchrist, M. L. Carlson, D. Berrigan, and L. Serra. 2000. Rapid evolution of a geographic cline in size in an introduced fly. Science 287:308–309.

Hufford, K. M., and S. J. Mazer. 2003. Plant ecotypes: genetic differentiation in the age of ecological restoration. Trends in Ecology and Evolution 18:147–155.

Hughes, A. L. 1991. MHC polymorphism and the design of captive breeding programs. Conservation Biology 5:249–251.

Hughes, J. B., G. C. Daily, and P. R. Ehrlich. 1997. Population diversity: its extent and extinction. Science 278:689–692.

Husband, B. C., and D. W. Schemske. 1996. Evolution of the magnitude and timing of inbreeding depression in plants. Evolution 50:54–70.

Hutt, F. B. 1979. Genetics for Dog Breeders. W.H. Freeman & Co., San Francisco, CA.

Huynen, L., C. D. Millar, and D. M. Lambert. 2002. A DNA test to sex ratite birds. Molecular Ecology 11:851–856.

Ingman, M., H. Kaessmann, S. Paabo, and U. Gyllensten. 2000. Mitochondrial genome variation and the origin of modern humans. Nature 408:708–713.

Ingvarsson, P. K. 2001. Restoration of genetic variation lost – the genetic rescue hypothesis. Trends in Ecology and Evolution 16:62–63.

Ingvarsson, P. K., and M. C. Whitlock. 2000. Heterosis increases the effective migration rate. Proceedings of the Royal Society of London B 267:1321–1326.

IUCN. 2001. IUCN Red List Categories and Criteria: Version 3.1. IUCN Species Survival Commission, Gland, Switzerland, and Cambridge, UK.

IUCN. 2002. IUCN technical guidelines on the management of ex situ populations for conservation. Approved at the 14th Meeting of the Programme Committee of Council, Gland, Switzerland, 19 December 2002.

IUDZG/CBSG (IUCN/SSC). 1993. The World Zoo Conservation Strategy: the Roles of Zoos and Aquaria of the World in Global Conservation. Chicago Zoological Society, Brookfield, IL.

Ives, A. R., and M. C. Whitlock. 2002. Inbreeding and metapopulations. Science 295:454–455.

Jaccoud, D., K. Peng, D. Feinstein, and A. Kilian. 2001. Diversity arrays: a solid state technology for sequence information independent genotyping. Nucleic Acids Research 29:e25.

Jackson, J. F., W. Ingram III, and H. W. Campbell. 1976. The dorsal pigmentation pattern of snakes as an antipredator strategy: a multivariate approach. American Naturalist 110:1029–1053.

Jacquard, A. 1975. Inbreeding: one word, several meanings. Theoretical Population Biology 7:338–363.

James, F. C. 1983. Environmental component of morphological differentiation in birds. Science 221:184–186.

James, F. C. 1991. Complementary descriptive and experimental studies of clinal variation in birds. American Zoologist 31:694–706.

James, J. W. 1971. The founder effect and response to artificial selection. Genetical Research 12:249–266.

Jamieson, A., and S. S. Taylor. 1997. Comparisons of three probability formulae for parentage exclusion. Animal Genetics 28:397–400.

Jeffreys, A. J., V. Wilson, and S. L. Thein. 1985. Hypervariable 'minisatellite' regions in human DNA. Nature 314:67–73.

Jiménez, J. A., K. A. Hughes, G. Alaks, L. Graham, and R. G. Lacy. 1994. An experimental study of inbreeding depression in a natural habitat. Science 266:271–273.

Jobling, M. A., and P. Gill. 2004. Encoded evidence: DNA in forensic analysis. Nature Reviews Genetics 5:739–751.

Johnson, M. S. 1982. Polymorphism for direction of coil in *Partula suturalis*: behavioral isolation and positive frequency dependent selection. Heredity 49:145–151.

Johnson, M. S. 1988. Founder effects and geographic variation in the land snail *Theba pisana*. Heredity 61:133–42.

Johnson, M. S., B. Clarke, and J. Murray. 1990. The coil polymorphism in *Partula suturalis* does not favor sympatric speciation. Evolution 44:459–464.

Johnson, N. A. 2000. Speciation: Dobzhansky–Muller incompatibilities, dominance and gene interactions. Trends in Ecology and Evolution 15:480–482.

Jones, J. C., M. R. Myerscough, S. Graham, and B. P. Oldroyd. 2004. Honey bee nest thermoregulation: diversity promotes stability. Science 305:402–404.

Jones, K. L., and 6 others. 2002. Refining the whooping crane studbook by incorporating microsatellite DNA and leg-banding analyses. Conservation Biology 16:789–799.

Jones, R. N. 1991. B-chromosome drive. American Naturalist 137:430–442.

Jones, W. G., K. D. Hill, and J. M. Allen. 1995. *Wollemia nobilis*, a new living Australian genus and species in the Araucariaceae. Telopea 6:173–176.

Joron, M., and P. M. Brakefield. 2003. Captivity masks inbreeding effects on male mating success in butterflies. Nature 424:191–194.

Joshi, J., and 16 others. 2001. Local adaptation enhances performance of common plant species. Ecology Letters 4:536–544.

Jousson, O. J., and 8 others. 2000. Invasive alga reaches California. Nature 408:157–158.

Kahler, A. L., R. W. Allard, and R. D. Miller. 1984. Mutation rates for enzyme and morphological loci in barley (*Hordeam vulgare* L.). Genetics 106:729–734.

Kalinowski, S. T., and P. W. Hedrick. 1999. Detecting inbreeding depression is difficult in captive endangered species. Animal Conservation 2:131–136.

Kalinowski, S. T., and P. W. Hedrick. 2001. Estimation of linkage disequilibrium for loci with multiple alleles: basic approach and an application using data from bighorn sheep. Heredity 87:698–708.

Kalinowski, S. T., P. W. Hedrick, and P. S. Miller. 2000. Inbreeding depression in the Speke's gazelle captive breeding program. Conservation Biology 14:1375–1384.

Kanda, N., R. F. Leary, and F. W. Allendorf. 2002. Evidence of introgressive hybridization between bull trout and brook trout. Transactions of the American Fisheries Society 131:772–782.

Kärkkäinen, K., H. Kuittinen, R. Vantreuren, C. Vogl, S. Oikarinen, and O. Savolainen. 1999. Genetic basis of inbreeding depression in *Arabis petraea*. Evolution 53:1354–1365.

Karl, S. A., and B. W. Bowen. 1999. Evolutionary significant units versus geopolitical taxonomy: molecular systematics of an endangered sea turtle (genus *Chelonia*). Conservation Biology 13:990–999.

Karl, S. A., B. W. Bowen, and J. C. Avise. 1992. Global population genetic structure and male-mediated gene flow in the green turtle (*Chelonia mydas*) – RFLP analyses of anonymous nuclear loci. Genetics 131:163–173.

Karron, J. D. 1991. Patterns of genetic variation and breeding systems in rare plant species. Pp. 87–98 *in* Falk, D. A., and K. E. Holsinger, eds. Genetics and Conservation of Rare Plants. Oxford University Press, New York.

Kaya, C. M. 1991. Rheotactic differentiation between fluvial and lacustrine populations of Arctic grayling (*Thymallus arcticus*), and implications for the only remaining indigenous population of fluvial "Montana grayling". Canadian Journal of Fisheries and Aquatic Sciences 48:53–59.

Keall, S. N., N. J. Nelson, P. Phillpot, S. Pledger, and C. H. Daugherty. 2001. Conservation in small places: reptiles on North Brother Island. New Zealand Journal of Zoology 28:367.

Kearsey, M. J. 1998. The principles of QTL analysis (a minimal mathematics approach). Journal of Experimental Botany 49:1619–1623.

Keightley, P. D., and M. Lynch. 2003. Toward a realistic model of mutations affecting fitness. Evolution 57:683–685.

Keller, L. F. 1998. Inbreeding and its fitness effects in an insular population of song sparrows (*Melospiza melodia*). Evolution 52:240–250.

Keller, L. F., and D. M. Waller. 2002. Inbreeding effects in wild populations. Trends in Ecology & Evolution 17:230–241.

Keller, M., J. Kollmann, and P. J. Edwards. 2000. Genetic introgression from distant provenances reduces fitness in local weed populations. Journal of Applied Ecology 37:647–659.

Kephart, S. R. 2004. Inbreeding and reintroduction: progeny success in rare *Silene* populations of varied density. Conservation Genetics 5:49–61.

Kidwell, M. G. 2002. Genome evolution – lateral DNA transfer mechanism and consequences, by F. Bushman. Science 295:2219–2220.

Kidwell, M. G., and D. R. Lisch. 1998. Hybrid genetics – transposons unbound. Nature 393:22–23.

Kimura, M. 1983. The Neutral Theory of Molecular Evolution. Cambridge University Press, Cambridge, UK.

Kimura, M., and J. F. Crow. 1963. The measurement of effective population number. Evolution 17:279–288.

Kimura, M., and J. F. Crow. 1964. The number of alleles that can be maintained in a finite population. Genetics 49:725–738.

Kimura, M., and T. Ohta. 1969. The average number of generation until fixation of a mutant gene in a finite population. Genetics 61:763–771.

Kimura, M., and G. H. Weiss. 1964. The stepping stone model of population structure and the decrease of genetic correlation with distance. Genetics 49:561–576.

King, D. R., A. J. Oliver, and R. J. Mead. 1978. The adaptation of some western Australian mammals to food plants containing fluoroacetate. Australian Journal of Zoology 26:699–712.

King, M. 1993. Species Evolution: the Role of Chromosome Change. Cambridge University Press, Cambridge, UK.

Kingman, J. F. C. 1982. The coalescent. Stochastic Processes and their Applications 13:235–248.

Kirby, G. C. 1975. Heterozygote frequencies in small populations. Theoretical Population Biology 8:31–48.

Kirpichnikov, V. S. 1981. Genetic Bases of Fish Selection. Springer-Verlag, Berlin.

Knowles, L. L., and W. P. Maddison. 2002. Statistical phylogeography. Molecular Ecology 11:2623–2635.

Kocher, T. D., and 6 others. 1989. Dynamics of mitochondrial DNA evolution in animals: amplification and sequencing with conserved primers. Proceedings of the National Academy of Sciences, USA 86:6196–6200.

Kohn, B., P. S. Henthorn, Y. Rajpurohit, M. P. Reilly, T. Asakura, and U. Giger. 1999. Feline adult beta-globin polymorphism reflected in restriction fragment length patterns. Journal of Heredity 90:177–181.

Kojima, K.-I. 1971. Is there a constant fitness value for a given genotype? No! Evolution 25:281–285.

Kolar, C. S., and D. M. Lodge. 2001. Progress in invasion biology: predicting invaders. Trends in Ecology and Evolution 16:199–204.

Kolbe, J. J., R. E. Glor, L. Rodriguez Schettino, A. Chamizo Lara, A. Larson, and J. B. Losos. 2004. Genetic variation increases during biological invasion by a Cuban lizard. Nature 431:177–181.

Korpelainen, H. 2002. A genetic method to resolve gender complements investigations on sex ratios in *Rumex acetosa*. Molecular Ecology 11:2151–256.

Kraak, S. B. M., B. Mundwiler, and P. J. B. Hart. 2001. Increased number of hybrids between benthic and limnetic three-spined sticklebacks in Enos Lake, Canada; the collapse of a species pair? Journal of Fish Biology 58:1458–1464.

Krauss, S. L., B. Dixon, and K. W. Dixon. 2002. Rapid genetic decline in a translocated population of the endangered plant *Grevillea scapigera*. Conservation Biology 16:986–994.

Kreitman, M. E. 1983. Nucleotide polymorphism at the alcohol dehydrogenase locus of *Drosophila melanogaster*. Nature 304:412–417.

Kruuk, L. E. B., T. H. Clutton-Bruck, J. Slate, J. M. Pemberton, S. Brotherstone, and F. E. Guiness. 2000. Heritability of fitness in a wild mammal population. Proceedings of the National Academy of Sciences, USA 97:698–703.

Kucera, T. E. 1991. Genetic variability in tule elk. California Fish and Game 77:70–78.

Kuhner, M. K., P. Beerle, J. Yamato, and J. Felsenstein. 2000. Usefulness of single nucleotide polymorphism data for estimating population parameters. Genetics 156:439–447.

Kurian, V., and A. J. Richards. 1997. A new recombinant in the heteromorphy "S" supergene in *Primula*. Heredity 78:383–390.

Lack, D. 1947. Darwin's Finches. Cambridge University Press, Cambridge, UK.

Lacy, R. C. 1988. A report on population genetics in conservation. Conservation Biology 2:245–247.

Lacy, R. C. 1989. Analysis of founder representation in pedigrees: founder equivalents and founder genome equivalents. Zoo Biology 8:111–123.

Lacy, R. C. 1997. Importance of genetic variation to the viability of mammalian populations. Journal of Mammalogy 78:320–335.

Lacy, R. C. 2000a. Should we select genetic alleles in our conservation breeding programs? Zoo Biology 19:279–282.

Lacy, R. C. 2000b. Structure of the VORTEX simulation model for population viability analysis. Ecological Bulletins 48:191–203.

Lacy, R. C., G. Alaks, and A. Walsh. 1996. Hierarchical analysis of inbreeding depression in *Peromyscus polionotus*. Evolution 50:2187–2200.

Lacy, R. C., and J. D. Ballou. 1998. Effectiveness of selection in reducing the genetic load in populations of *Peromyscus polionotus* during generations of inbreeding. Evolution 52:900–909.

Lacy, R. C., M. Borbat, and J. P. Pollak, 2003. VORTEX: a Stochastic Simulation of the Extinction Process. Version 9. Chicago Zoological Society, Brookfield, IL.

Lacy, R. C., A. Petric, and M. Warneke, 1993. Inbreeding and outbreeding in captive populations of wild animal species. Pp. 352–374 *in* Thorhill, N. W., ed. The Natural History of Inbreeding and Outbreeding. University of Chicago Press, Chicago.

Laikre, L. 1999. Conservation genetics of Nordic carnivores: lessons from zoos. Hereditas 130:203–216.

Laikre, L., N. Ryman, and E. A. Thompson. 1993. Hereditary blindness in a captive wolf (*Canis lupus*) population – frequency reduction of a deleterious allele in relation to gene conservation. Conservation Biology 7:592–601.

Lamborot, M. 1991. Karyotypic variation among populations of *Liolaemus monticola* (Tropiduridae) separated by riverine barriers in the Andean Range. Copeia 1991:1044–1059.

Lamont, B. B., T. He, N. J. Enright, S. L. Krauss, and B. P. Miller. 2003. Anthropogenic disturbance promotes hybridization between *Banksia* species by altering their biology. Journal of Evolutionary Biology 16:551–557.

Lande, R. 1979. Effective deme sizes during long-term evolution estimated from rates of chromosomal rearrangement. Evolution 33:234–251.

Lande, R. 1988. Genetics and demography in biological conservation. Science 241:1455–1460.

Lande, R. 1994. Risk of population extinction from fixation of new deleterious mutations. Evolution 48:1460–1469.

Lande, R. 1995. Mutation and conservation. Conservation Biology 9:782–791.

Lande, R. 1996. The meaning of quantitative genetic variation in evolution and conservation. Pp. 27–40 *in* Szaro, R. C. and D. W. Johnston, eds. Biodiversity in Managed Landscapes: Theory and Practice. Oxford University Press, New York.

Lande, R. 1999. Extinction risks from anthropogenic, ecological, and genetic factors. Pp. 1–22 *in* Landweber, L. F., and A. P. Dobson, eds. Genetics and the Extinction of Species. Princeton University Press, Princeton, NJ.

Lande, R., and G. F. Barrowclough. 1987. Effective population size, genetic variation, and their use in population management. Pp. 87–124 *in* Soulé, M. E., ed. Viable Populations for Conservation. Cambridge University Press, Cambridge, UK.

Lande, R., and D. W. Schemske. 1985. The evolution of self-fertilization and inbreeding depression in plants. 1. Genetic models. Evolution 39:24–40.

Landweber, L. F., and Dobson, A. P. 1999. Genetics and the Extinction of Species. Princeton University Press, Princeton, NJ.

Larkin, P. A. 1972. The stock concept and management of Pacific salmon. Pp. 11–15 in Simon, R. C., and P. A. Larkin, eds. The Stock Concept in Pacific Salmon. University of British Columbia, Vancouver.

Larkin, P. A. 1981. A perspective on population genetics and salmon management. Canadian Journal of Fisheries and Aquatic Sciences 38:1469–1475.

Leary, R. F., and F. W. Allendorf. 1989. Fluctuating asymmetry as an indicator of stress: implications for conservation biology. Trends in Ecology and Evolution 4:214–217.

Leary, R. F., F. W. Allendorf, and S. H. Forbes. 1993a. Conservation genetics of bull trout in the Columbia and Klamath River drainages. Conservation Biology 7:856–865.

Leary, R. F., F. W. Allendorf, and K. L. Knudsen. 1984. Major morphological effects of a regulatory gene: *Pgm1-t* in rainbow trout. Molecular Biology and Evolution 1:183–194.

Leary, R. F., R. W. Allendorf, and K. L. Knudsen. 1985. Inheritance of meristic variation and the evolution of developmental stability in rainbow trout. Evolution 39:308–314.

Leary, R. F., F. W. Allendorf, and K. L. Knudsen. 1993b. Null allele heterozygotes at two lactate dehydrogenase loci in rainbow trout are associated with decreased developmental stability. Genetica 89:3–13.

Leary, R. F., W. R. Gould, and G. K. Sage. 1996. Success of basibranchial teeth in indicating pure populations of rainbow trout and failure to indicate pure populations of westslope cutthroat trout. North American Journal of Fisheries Management 16:210–213.

Leberg, P. 1998. Influence of complex sex determination on demographic stochasticity and population viability. Conservation Biology 12:456–459.

Leberg, P. L. 1991. Effects of genetic variation on the growth of fish populations: conservation implications. Journal of Fish Biology 37(A):193–195.

Leberg, P. L. 1993. Strategies for population reintroduction: effects of genetic variability on population growth and size. Conservation Biology 7:194–199.

Lee, C. E. 2002. Evolutionary genetics of invasive species. Trends in Ecology and Evolution 17:386–391.

Leider, S. A., P. L. Hulett, J. J. Loch, and M. W. Chilcote. 1990. Electrophoretic comparison of the reproductive success of naturally spawning transplanted and wild steelhead trout through the returning adult stage. Aquaculture 88:239–252.

Lenormand, T. 2002. Gene flow and the limits to natural selection. Trends in Ecology and Evolution 17:183–189.

Lenski, R. E., and P. D. Sniegowski. 1995. "Adaptive mutation": the debate goes on. Science 269:285–286.

Leopold, A. S. 1944. The nature of heritable wildness in turkeys. Condor 46:133–197.

Les, D. H., J. A. Reinartz, and E. J. Esselman. 1991. Genetic consequences of rarity in *Aster furcatus* (Asteraceae), a threatened, self-incompatible plant. Evolution 45:1641–1650.

Lesica, P., and F. W. Allendorf. 1999. Ecological genetics and the restoration of plant communities: mix or match? Restoration Ecology 7:42–50.

Lessios, H. A. 1992. Testing electrophoretic data for agreement with Hardy–Weinberg expectations. Marine Biology 112:517–523.

Levene, H. 1949. On a matching problem arising in genetics. Annals of Mathematical Statistics 20:91–94.

Levin, D. A., J. Franciscoortega, and R. K. Jansen. 1996. Hybridization and the extinction of rare plant species. Conservation Biology 10:10–16.

Levin, D. A., K. Ritter, and N. C. Ellstrand. 1979. Protein polymorphism in the narrow endemic *Oenothera organensis*. Evolution 33:534–542.

Levins, R. 1970. Extinction. Pp. 77–107 in Gerstenhaber, M., ed. Some Mathematical Problems in Biology. American Mathematical Society, Providence, RI.

Lewis, P. O. 2001. Phylogenetic systematics turns over a new leaf. Trends in Ecology and Evolution 16:30–37.

Lewontin, R. C. 1964. The interaction of selection and linkage. I. General considerations; heterotic models. Genetics 49:49–67.

Lewontin, R. C. 1974. Genetic Basis of Evolutionary Change. Columbia University Press, New York.

Lewontin, R. C. 1988. On measures of gametic disequilibrium. Genetics 120:849–852.

Lewontin, R. C. 1991. 25 Years ago in genetics – electrophoresis in the development of evolutionary genetics – milestone or millstone. Genetics 128:657–662.

Lewontin, R. C. 1999. Evolutionary Genetics from Molecules to Morphology.

Lewontin, R. C. 2000. The triple helix: gene, organism, and environment. Harvard University Press, Cambridge, MA.

Lewontin, R. C, and C. C. Cockerham. 1959. The goodness-of-fit test for detecting natural selection in random mating populations. Evolution 13:561–564.

Lewontin, R. C., and J. Felsenstein. 1965. The robustness of homogeneity tests in $2 \times N$ tables. Biometrics 21:19–33.

Lewontin, R. C., L. R. Ginzburg, and S. D. Tuljapurkar. 1978. Heterosis as an explanation for large amounts of genetic polymorphism. Genetics 88:149–169.

Lewontin, R. C., and K. Kojima. 1960. The evolutionary dynamics of complex polymorphisms. Evolution 14:450–472.

Li, C. C. 1967. Castle's early work on selection and equilibrium. American Journal of Human Genetics 19:70–74.

Li, W.-H. 1978. Maintenance of genetic variability under the joint effect of mutation, selection and random drift. Genetics 90:349–382.

Li, Y. C., A. B. Korol, T. Fahima, A. Beiles, and E. Nevo. 2002. Microsatellites: genomic distribution, putative functions and mutational mechanisms: a review. Molecular Ecology 11:2453–2465.

Lichatowich, J. A. 1999. Salmon Without Rivers. Island Press, Washington, DC.

Lincoln, F. C. 1930. Calculating waterfowl abundance on the basis of banding returns. United States Department of Agriculture Circular 118:1–4.

Linder, C. R., I. Taha, G. J. Seiler, A. A. Snow, and L. H. Rieseberg. 1998. Long-term introgression of crop genes into wild sunflower populations. Theoretical and Applied Genetics 96:339–347.

Linhart, Y. B., and M. C. Grant. 1996. Evolutionary significance of local genetic differentiation in plants. Annual Review of Ecology and Systematics 27:237–277.

Lively, C. M., C. Craddock, and R. C. Vrijenhoek. 1990. The Red Queen hypothesis supported by parasitism in sexual and clonal fish. Nature 344:864–866.

Loi, P., G. Ptak, B. Barboni, J. Fulka Jr., P. Cappai, and M. Clinton. 2001. Genetic rescue of an endangered mammal by cross-species nuclear transfer using post-mortem somatic cells. Nature Biotechnology 19:962–964.

Louis, E. J., and E. R. Dempster. 1987. An exact test for Hardy–Weinberg and multiple alleles. Biometrics 43:805–811.

Luck, G. W., G. C. Daily, and P. R. Ehrlich. 2003. Population diversity and ecosystem services. Trends in Ecology and Evolution 18:331–336.

Luikart, G., and J. M. Cornuet. 1998. Empirical evaluation of a test for identifying recently bottlenecked populations from allele frequency data. Conservation Biology 12:228–237.

Luikart, G., J. M. Cornuet, and F. W. Allendorf. 1999. Temporal changes in allele frequencies provide estimates of population bottleneck size. Conservation Biology 13:523–530.

Luikart, G., P. R. England, D. Tallmon, S. Jordan, and P. Taberlet. 2003. The power and promise of population genomics: from genotyping to genome typing. Nature Reviews Genetics 4:900–910.

Luikart, G., W. B. Sherwin, B. M. Steele, and F. W. Allendorf. 1998. Usefulness of molecular markers for detecting population bottlenecks via monitoring genetic change. Molecular Ecology 7:963–974.

Luo, S.-J., and 21 others. 2004. Phylogeography and genetic ancestry of tigers (*Panthera tigris*). PLoS Biology 2:2275–2293.

Lush, J. L. 1937. Animal Breeding Plans. Iowa State University Press, Ames, IA.

Lynch, C. B. 1977. Inbreeding effects upon animals derived from a wild population of *Mus musculus*. Evolution 31:526–537.

Lynch, M. 1996. A quantitative-genetic perspective on conservation issues. Pp. 471–501 *in* Avise, J. C., and J. L. Hamrick, eds. Conservation Genetics: Case Histories from Nature. Chapman & Hall, New York.

Lynch, M. 2001. The molecular natural history of the human genome. Trends in Ecology and Evolution 16:420–422.

Lynch, M., J. Blanchard, D. Houle, T. Kibota, S. Schultz, L. Vassilieva, and J. Willis. 1999. Perspective: spontaneous deleterious mutation. Evolution 53:645–663.

Lynch, M., R. Burger, D. Butcher, and W. Gabriel. 1993. The mutational meltdown in asexual populations. Journal of Heredity 84:339–344.

Lynch, M., and J. S. Conery. 2000. The evolutionary fate and consequences of duplicate genes. Science 290:1151–1155.

Lynch, M., J. Conery, and R. Burger. 1995. Mutation accumulation and the extinction of small populations. American Naturalist 146:489–518.

Lynch, M., and W. Gabriel. 1990. Mutation load and the survival of small populations. Evolution 44:1725–1737.

Lynch, M., and R. Lande. 1998. The critical effective size for a genetically secure population. Animal Conservation 1:70–72.

Lynch, M., and M. O'Hely. 2001. Captive breeding and the genetic fitness of natural populations. Conservation Genetics 2:363–378.

Lynch, M., and K. Ritland. 1999. Estimation of pairwise relatedness with molecular markers. Genetics 152:1753–1766.

Lynch, M., and Walsh, B. 1998. Genetics and Analysis of Quantitative Traits. Sinauer, Sunderland, MA.

Lyons, L. A., T. F. Laughlin, N. G. Copeland, N. A. Jenkins, J. E. Womack, and S. J. O'Brien. 1997. Comparative anchor tagged sequences (CATS) for integrative mapping of mammalian genomes. Nature Genetics 15:47–56.

MacCluer, J. W., B. VandeBerg, B. Read, and O. A. Ryder. 1986. Pedigree analysis by computer simulation. Zoo Biology 5:147–160.

Mace, G. M., J. L. Gittleman, and A. Purvis. 2003. Preserving the tree of life. Science 300:1707–1709.

Mace, G. M., P. Pearcekelly, and D. Clarke. 1998. An integrated conservation programme for the tree snails (Partulidae) of Polynesia: a review of captive and wild elements. Journal of Conchology 2(Suppl.):89–96.

Mack, R. N., D. Simberloff, W. M. Lonsdale, H. Evans, M. Clout, and F. A. Bazzaz. 2000. Biotic invasions: causes, epidemiology, global consequences, and control. Ecological Applications 10:689–710.

Mackay, T. F. C. 2001. Quantitative trait loci in *Drosophila*. Nature Reviews Genetics 2:11–20.

Madsen, J. L., and G. M. Blake. 1977. Ecological genetics of ponderosa pine in the northern Rocky Mountains. Silvae Genetica 26:1–8.

Madsen, T., R. Shine, M. Olsson, and H. Wittzell. 1999. Conservation biology – restoration of an inbred adder population. Nature 402:34–35.

Maguire, L. A., and R. C. Lacy. 1990. Allocating scarce resources for conservation of endangered subspecies: partitioning zoo space for tigers. Conservation Biology 4:157–166.

Majerus, M. E. N., and N. I. Mundy. 2003. Mammalian melanism: natural selection in black and white. Trends Genetics 19:585–588.

Makela, M. E., and R. H. Richardson. 1977. The detection of sympatric sibling species using genetic correlation analysis. I. Two loci, two gamodemes. Genetics 86:665–678.

Malakoff, D. 1999. Bayes offers a "new" way to make sense of numbers. Science 286:1460–1464.

Malik, S., P. J. Wilson, R. J. Smith, D. M. Lavigne, and B. N. White. 1997. Pinniped penises in trade: a molecular-genetic investigation. Conservation Biology 11:1365–1374.

Manel, S., P. Berthier, and G. Luikart. 2002. Detecting wildlife poaching: identifying the origin of individuals using Bayesian assignment tests and multi-locus genotypes. Conservation Biology 16:650–657.

Manel, S., O. Gaggiotti, and R. Waples. 2005. Assignment methods: matching biological questions with appropriate techniques. Trends in Ecology and Evolution 20:136–142.

Manel, S., M. K. Schwartz, G. Luikart, and P. Taberlet. 2003. Landscape genetics: combining landscape ecology and population genetics. Trends in Ecology and Evolution 18:189–197.

Mann, C. C., and M. L. Plummer. 1999. A species' fate, by the numbers. Science 284:36–37.

Maron, J. L., M. Vilà, and J. Arnason. 2004a. Loss of enemy resistance among introduced populations of St. John's Wort, *Hypericum perforatum*. Ecology 85:3243–3253.

Maron, J. L. M. Vilà, R. Bommarco, S. Elmendorf, and P. Beardsley. 2004b. Rapid evolution of an invasive plant. Ecological Monographs 74:261–280.

Marshall, H. D., and K. Ritland. 2002. Genetic diversity and differentiation of Kermode bear populations. Molecular Ecology 11:685–697.

Marshall, T. C., J. Slate, L. E. B. Kruuk, and J. M. Pemberton. 1998. Statistical confidence for likelihood-based paternity inference in natural populations. Molecular Ecology 7:639–655.

Martinsen, G. D., T. G. Whitham, R. J. Turek, and P. Keim. 2001. Hybrid populations selectively filter gene introgression between species. Evolution 55:1325–1335.

Matayoshi, T., and 8 others. 1987. Heterochromatic variation in *Cebus apella* (Cebidae, Platyrrhini) of different geographical regions. Cytogenetics and Cell Genetics 44:158–162.

Mather, K. 1953. Genetical control of stability in development. Heredity 7:297–336.

Mattson, D. J., K. C. Kendall, and D. P. Reinhart. 2001. Whitebark pine, grizzly bears, and red squirrels. Pp. 121–136 *in* Tomback, D. F., S. F. Arno, and R. E. Keane, eds. Whitebark Pine Communities: Ecology and Restoration. Island Press, Washington, DC.

Mattson, D. J., and T. Merrill. 2002. Extirpations of grizzly bears in the contiguous United States, 1850–2000. Conservation Biology 16:1123–1136.

Maudet, C., G. Luikart, and P. Taberlet. 2001. Development of microsatellite multiplexes for wild goats using primers designed from domestic Bovidae. Genetics Selection Evolution 33:S193–S203.

Maudet, C., C. Miller, B. Bassano, et al. 2002. Microsatellite DNA and recent statistical methods in wildlife conservation management: applications in Alpine ibex [*Capra ibex* (Ibex)]. Molecular Ecology 11:421–436.

Maudet, C., and 10 others. 2004. A standard set of polymorphic microsatellites for threatened mountain ungulates (Caprini, Artiodactyla). Molecular Ecology Notes 4:49–55.

Maxted, N. 2003. Conserving the genetic resources of crop wild relatives in European Protected Areas. Biological Conservation 113:411–417.

May, B. 1998. Starch gel electrophoresis of allozymes. Pp. 1–28 *in* Hoelzel, A. R., ed. Molecular Genetic Analysis of Populations. Oxford University Press, Oxford.

May, R. M. 1990. Taxonomy as destiny? Nature 347:129–130.

May, R. M. 2004. Uses and abuses of mathematics in biology. Science 303:790–791.

Mayr, E. 1942. Systematics and the Origin of Species. Columbia University Press, New York.

Mayr, E. 1963. Animal Species and Evolution. Belknap Press, Cambridge, MA.

Mayr, E. 1981. Biological classification: toward of synthesis of classification. Science 214:510–516.

Mayr, E. 1982. The Growth of Biological Thought: Diversity, Evolution, and Inheritance. Belknap Press, Cambridge, MA.

McCauley, D. E. 1991. Genetic consequences of local population extinction and recolonization. Trends in Ecology and Evolution 6:5–8.

McCauley, D. E. 1994. Contrasting the distribution of chloroplast DNA and allozyme polymorphisms among local populations of *Silene alba*: implications for studies of gene flow in plants. Proceedings of the National Academy of Sciences, USA 91:8127–8131.

McCauley, D. E., M. F. Bailey, N. A. Sherman, and M. Z. Darnell. 2005. Evidence for paternal transmission and heteroplasmy in the mitochondrial genome of *Silene vulgaris*, a gynodioecious plant. Heredity 95:50–58.

McClintock, B. 1984. The significance of responses of the genome to challenge. Science 226:792–801.

McCullough, D. R. 1978. Population dynamics of the Yellowstone grizzly bear. Pp. 173–196 *in* Fowler, C. W., and T. D. Smith, eds. Dynamics of large mammal populations. John Wiley & Sons, New York.

McCullough, D. R., J. K. Fischer, and J. D. Ballou. 1996. From bottleneck to metapopulation: recovery of the tule elk in California. Pp. 375–403 *in* McCullough, D. R., ed. Metapopulations and Wildlife Conservation. Island Press, Covelo, CA.

McDowall, R. M. 1990. New Zealand Freshwater Fishes: a Natural History and Guide (revised). Heinemann Reed, Auckland, New Zealand.

McGinnity, P., and 9 others. 2003. Fitness reduction and potential extinction of wild populations of Atlantic salmon, *Salmo salar*, as a result of interactions with escaped farmed salmon. Proceedings of the Royal Society of London B 270:2443–2450.

McKay, J. K., and R. G. Latta. 2002. Adaptive population divergence: markers, QTL and traits. Trends in Ecology and Evolution 17:285–291.

McNeely, J. A., K. R. Miller, W. V. Reid, R. A. Mittermeier, and T. B. Werner. 1990. Conserving the world's biological diversity. IUCN, World Resources Institute, Conservation International, WWF-US, and the World Bank, Washington, DC.

Meagher, S., D. J. Penn, and W. K. Potts. 2000. Male–male competition magnifies inbreeding depression in wild house mice. Proceedings of the National Academy of Sciences, USA 97:3324–3329.

Meffe, G. K. 1992. Techno-arrogance and halfway technologies – salmon hatcheries on the Pacific coast of North America. Conservation Biology 6:350–354.

Mehrhoff, L. A. 1996. Reintroducing endangered Hawaiian plants. Pp. 101–120 *in* Falk, D. A., C. I. Millar, and M. Olwell, eds. Restoring Diversity. Island Press, Washington, DC.

Meine, C. D. 1998. Moving mountains: Aldo Leopold and a Sand County Almanac. Wildlife Society Bulletin 26:697–706.

Melampy, M. N., and H. F. Howe. 1977. Sex ratio in the tropical tree *Triplaris americana* Polygonaceae. Evolution 31:867–872.

Mendel, G. 1865. Versuche uber Pflanzen-hybriden. Verhandlungen des Naturforschenden Vereines, Abhandlungen, Brünn 4:3–37.

Menotte-Raymond, M., V. A. David, J. C. Stephens, and S. J. O'Brien. 1997. Genetic individualization of domestic cats using feline STR loci for forensic analysis. Journal of Forensics Science 42:1039–1051.

Merilä, J., and P. Crnokrak. 2001. Comparison of genetic differentiation at marker loci and quantitative traits. Journal of Evolutionary Biology 14:892–903.

Merton, D. V., R. D. Morris, and I. A. E. Atkinson. 1984. Lek behaviour in a parrot: the kakapo *Strigops habroptilus* of New Zealand. Ibis 126:277–283.

Meselson, M., and R. Yuan. 1968. DNA restriction enzyme from *E. coli*. Nature 217:1110–1114.

Meylan, A. B., B. W. Bowen, and J. C. Avise. 1990. A genetic test for the natal homing versus social facilitation models for green turtle migration. Science 248:724–729.

Michel, A., R. S. Arias, B. E. Scheffler, S. O. Duke, M. Netherland, and F. E. Dayan. 2004. Somatic mutation-mediated evolution of herbicide resistance in the nonindigenous invasive plant hydrilla (*Hydrilla verticillata*). Molecular Ecology 13:3229–3237.

Midgley, M. 1987. Keeping species on ice. Pp. 55–65 *in* McKenna, V., W. Travers, and J. Wray, eds. Beyond the Bars. The Zoo Dilemma. Thorsons, Wellingborough, UK.

Miller, C. R., P. Joyce, and L. P. Waits. 2005. A new method for estimating the size of small populations from genetic mark-recapture data. Molecular Ecology 14:1991–2005.

Miller, C. R., and L. P. Waits. 2003. The history of effective population size and genetic diversity in the Yellowstone grizzly (*Ursus arctos*): implications for conservation. Proceedings of the National Academy of Sciences, USA 100:4334–4339.

Miller, D. J., and M. J. H. van Oppen. 2003. A "fair go" for coral hybridization. Molecular Ecology 12:805–807.

Miller, H. C., D. M. Lambert, C. D. Millar, B. C. Robertson, and E. O. Minot. 2003. Minisatellite DNA profiling detects lineages and parentage in the endangered kakapo (*Strigops habroptilus*) despite low microsatellite DNA variation. Conservation Genetics 4:265–274.

Miller, P. S., and P. W. Hedrick. 1991. MHC polymorphism and the design of captive breeding programs – simple solutions are not the answer. Conservation Biology 5:556–558.

Miller, P. S., and Lacy, R. C. 2003. VORTEX: a Stochastic Simulation of the Extinction Process.

Version 9 User's Manual. Conservation Breeding Specialist Group (SSC/IUCN), Apple Valley, MN.

Miller Coyle, H., and 7 others. 2003. A simple DNA extraction method for marijuana samples used in amplified fragment length polymorphism (AFLP) analysis. Journal of Forensic Science 48:343–347.

Mills, L. S., and F. W. Allendorf. 1996. The one-migrant-per-generation rule in conservation and management. Conservation Biology 10:1509–1518.

Mills, L. S., J. J. Citta, K. P. Lair, M. K. Schwarz, and D. A. Tallmon. 2000. Estimating animal abundance using noninvasive DNA sampling: promise and pitfalls. Ecological Applications 10:283–294.

Mills, L. S., S. G. Hayes, C. Baldwin, M. J. Wisdom, J. Citta, D. J. Mattson, and K. Murphy. 1996. Factors leading to different viability predictions for a grizzly bear data set. Conservation Biology 10:863–873.

Mills, L. S., K. Pilgrim, M. K. Schwartz, and K. McKelvey. 2001. Identifying lynx and other North American felids based on mtDNA analysis. Conservation Genetics 1:285–289.

Mills, L. S., and P. E. Smouse. 1994. Demographic consequences of inbreeding in remnant populations. American Naturalist 144:412–431.

Milner, G. B., D. J. Teel, F. M. Utter, and G. A. Winans. 1985. A genetic method of stock indentification in mixed populations of Pacific salmon, *Oncorhynchus* spp. Marine Fisheries Review 47:1–8.

Mitton, J. B. 1989. Physiological and demographic variation associated with allozyme variation. Pp. 127–145 *in* Soltis, D., and P. Soltis, eds. Isozymes in Plant Biology. Dioscorides Press, Portland, OR.

Mitton, J. B. 1997. Selection in Natural Populations. Oxford University Press, New York.

Modiano, D., and 14 others. 2001. Haemoglobin C protects against clinical *Plasmodium falciparum* malaria. Nature 414:305–308.

Moore, A. J., and P. F. Kukuk. 2002. Quantitative genetic analysis of natural populations. Nature Reviews Genetics 3:971–978.

Morell, V. 1993. Evidence found for a possible "aggression gene". Science 260:1722–1723.

Morgan, M. T. 2002. Genome-wide deleterious mutation favors dispersal and species integrity. Heredity 89:253–257.

Morgan, T. J., M. A. Evans, T. Garland Jr., J. G. Swallow, and P. A. Carter. 2005. Molecular and quantitative genetic divergence among populations of house mice with known evolutionary histories. Heredity 94:518–525.

Morin, P., G. Luikart, R. K. Wayne, and SNP workshop group. 2004. SNPs in ecology, evolution and conservation. Trends in Ecology and Evolution 19:208–216.

Moritz, C. 1994. Defining "evolutionarily significant units" for conservation. Trends in Ecology and Evolution 9:373–375.

Moritz, C. 2002. Strategies to protect biological diversity and the evolutionary processes that sustain it. Systematic Biology 51:238–254.

Moritz, C., and C. Cicero. 2004. DNA barcoding: promise and pitfalls. PLoS Biology 2:e354.

Moritz, C., and D. P. Faith. 1998. Comparative phylogeography and the identification of genetically divergent areas for conservation. Molecular Ecology 7:419–429.

Moritz, C., S. Lavery, and R. Slade. 1995. Using allele frequency and phylogeny to define units for conservation and management. American Fisheries Society Symposium 17:249–262.

Morizot, D. C., J. C. Bednarz, and R. E. Ferrell. 1987. Sex linkage of muscle creatine kinase in Harris' hawks. Cytogenetics and Cell Genetics 44:89–91.

Morton, N. E., J. F. Crow, and H. J. Muller. 1956. An estimate of the mutational damage in man from data on consanguineous marriages. Proceedings of the National Academy of Sciences, USA 42:855–863.

Mosher, J. A., and C. J. Henny. 1976. Thermal adaptiveness of plumage color in screech owls. Auk 93:614–619.

Mosseler, A., K. N. Egger, and G. A. Hughes. 1992. Low levels of genetic diversity in red pine confirmed by random amplified polymorphic DNA markers. Canadian Journal of Forest Research 22:1332–1337.

Mosseler, A., D. J. Innes, and B. A. Roberts. 1991. Lack of allozymic variation in disjunct newfoundland populations of red pine (*Pinus resinosa*). Canadian Journal of Forest Research 21:525–528.

Mousseau, T. A., and D. A. Roff. 1987. Natural selection and the heritability of fitness components. Heredity 59:181–198.

Muir, C. C., B. M. F. Galdikas, and A. T. Beckenbach. 1998. Is there sufficient evidence to elevate the orangutan of Borneo and Sumatra to separate species? Journal of Molecular Evolution 46:378–379.

Muir, C. C., B. M. F. Galdikas, and A. T. Beckenbach. 2000. mtDNA sequence diversity of orangutans from the islands of Borneo and Sumatra. Journal of Molecular Evolution 51:471–480.

Müller-Scharer, H., U. Schaffner, and T. Steinger. 2004. Evolution in invasive plants: implications for biological control. Trends in Ecology and Evolution 19:417–422.

Mundy, N. I., N. S. Badcock, T. Hart, K. Scribner, K. Janssen, and N. J. Nadeau. 2004. Conserved genetic basis of a quantitative plumage trait involved in mate choice. Science 303:1870–1873.

Müntzing, A. 1966. Accessory chromosomes. Bulletin of the Botanical Society of Bengal 20:1–15.

Murphy, M. A., L. P. Waits, and K. C. Kendall. 2000. Quantitative evaluation of fecal drying methods for brown bear DNA analysis. Wildlife Society Bulletin 28:951–957.

Murray, B. G., and A. G. Young. 2001. Widespread chromosome variation in the endangered grassland forb *Rutidosis leptorrhynchoides* F. Muell. (Asteraceae: Gnaphalieae). Annals of Botany 87:83–90.

Myers, J. H., D. Simberloff, A. M. Kuris, and J. R. Carey. 2000. Eradication revisited: dealing with exotic species. Trends in Ecology and Evolution 15:316–320.

Myers, N., and A. H. Knoll. 2001. The biotic crisis and the future of evolution. Proceedings National Academy Sciences USA 98:5389–5392.

Myers, R. A., S. A. Levin, R. Lande, F. C. James, W. W. Murdoch, and R. T. Paine. 2004. Hatcheries and endangered salmon. Science 303:1980.

Nachman, M. W., H. E. Hoekstra, and S. L. D'Agostino. 2003. The genetic basis of adaptive melanism in pocket mice. Proceedings of the National Academy of Sciences, USA 100:5268.

Nachman, M. W., and J. B. Searle. 1995. Why is the house mouse karyotype so variable? Trends in Ecology and Evolution 10:397–402.

Nagorsen, D. W. 1990. The Mammals of British Columbia: a Taxonomic Catalogue. Memoir No. 4, Royal British Columbia Museum, Victoria, BC.

Nagy, E. S., and K. J. Rice. 1997. Local adaptation in two subspecies of an annual plant: implications for migration and gene flow. Evolution 51:1079–1089.

Nakajima, M., N. Kanda, and Y. Fujio. 1991. fluctuation of gene frequency in sub-populations originated from one guppy population. Nippon Suisan Gakkaishi: Bulletin of the Japanese Society of Scientific Fisheries 57:2223–2227.

Nauta, M. J., and F. J. Weissing. 1996. Constraints on allele size at microsatellite loci: implications for genetic differentiation. Genetics 143:1021–1032.

Nei, M. 1965. Variation and covariation of gene frequencies in subdivided populations. Evolution 19:256–258.

Nei, M. 1972. Genetic distance between populations. American Naturalist 106:283–292.

Nei, M. 1977. F-statistics and analysis of gene diversity in subdivided populations. Annals of Human Genetics 41:225–233.

Nei, M. 1978. Estimation of average heterozygosity and genetic distance from a small number of individuals. Genetics 89:583–590.

Nei, M. 1987. Molecular Evolutionary Genetics. Columbia University Press, New York.

Nei, M., and W.-H. Li. 1973. Linkage disequilibrium in subdivided populations. Genetics 75:213–219.

Nei, M., T. Maruyama, and R. Chakraborty. 1975. The bottleneck effect and genetic variability in populations. Evolution 29:1–10.

Neigel, J. E. 2002. Is F_{ST} obsolete? Conservation Genetics 3:167–173.

Neigel, J. E., and J. C. Avise. 1986. Phylogenetic relationships of mitochondrial DNA under various demographic models of speciation. Pp. 515–534 *in* Nevo, E., and S. Karlin, eds. Evolutionary Processes and Theory. Academic Press, New York.

Neigel, J. E., and J. C. Avise. 1993. Application of a random walk model to geographic distributions of animal mitochondrial DNA variation. Genetics 135:1209–1220.

Nelson, D. L., and 7 others. 1989. *Alu* polymerase chain reaction: a method for rapid isolation of human-specific sequences from complex DNA sources. Proceedings of the National Academy of Sciences, USA 86:6686–6690.

Nettancourt, D. De. 1977. Incompatibility in Angiosperms. Springer-Verlag, New York.

Nevo, E., A. Bailes, and R. Ben-Shlomo. 1984. The evolutionary significance of genetic diversity: ecological, demographic and life history correlates. Pp. 13–213 *in* S. Levin, ed. Lecture Notes in Biomathematics. Vol. 53: Evolutionary Dynamics of Genetic Diversity, (G. S. Mani, ed.). Springer-Verlag, Berlin.

Newman, D., and D. Pilson. 1997. Increased probability of extinction due to decreased genetic effective population size: experimental populations of *Clarkia pulchella*. Evolution 51:354–362.

Newman, D., and D. A. Tallmon. 2001. Experimental evidence for beneficial fitness effects of gene flow in recently isolated populations. Conservation Biology 15:1054–1063.

Nichols, R. 2001. Gene trees and species trees are not the same. Trends in Ecology and Evolution 16:358–364.

Nichols, R. A., M. W. Bruford, and J. J. Groombridge. 2001. Sustaining genetic variation in a small population: evidence from the Mauritius kestrel. Molecular Ecology 10:593–602.

Nichols, R. A., and K. L. M. Freeman. 2004. Using molecular markers with high mutation rates to obtain estimates of relative population size and to distinguish the effects of gene flow and mutation: a demonstration using data from endemic Mauritian skinks. Molecular Ecology 13:775–787.

Norman, J. A., C. Moritz, and C. J. Limpus. 1994. Mitochondrial DNA control region polymorphisms: genetic markers for ecological studies of marine turtles. Molecular Ecology 3:363–374.

Nunney, L. 1997. The effective size of a hierarchically structured population. Evolution 53:1–10.

Nunney, L. 2000. The limits to knowledge in conservation genetics – the value of effective population size. Evolutionary Biology 32:179–194.

Nunney, L., and K. A. Campbell. 1993. Assessing minimum viable population size – demography meets population genetics. Trends in Ecology and Evolution 8:234–239.

O'Brien, S. J., and 7 others. 1984. Giant panda paternity. Science 332:1127–1128.

O'Brien, S. J., and E. Mayr. 1991. Bureaucratic mischief: recognizing endangered species and subspecies. Science 251:1187–1188.

O'Brien, S. J., and 9 others. 1999. The promise of comparative genomics in mammals. Science 286:458–481.

O'Donald, P., and J. W. F. Davis. 1959. The genetics of the colour phases of the Arctic skua. Heredity 13:481–486.

O'Donald, P., and J. W. F. Davis. 1975. Demography and selection in a population of Arctic skuas. Heredity 35:75–83.

Ohta, T. 1971. Associative overdominance caused by linked detrimental mutations. Genetic Resource 18:277–286.

Ohta, T., and M. Kimura. 1969. Linkage disequilibrium due to random drift. Genetic Resource 13:47–55.

Ohta, T., and M. Kimura. 1973. A model of mutation appropriate to estimate the number of electrophoretcially detectable alleles in a finite population. Genetic Resource 22:201–204.

Olsen, E. M., and 6 others. 2004a. Maturation trends indicative of rapid evolution preceded the collapse of northern cod. Nature 428:932–935.

Olsen, J. B., P. Bentzen, M. A. Banks, J. B. Shaklee, and S. Young. 2000. Microsatellites reveal population identity of individual pink salmon to allow supportive breeding of a population at risk of extinction. Transactions of the American Fisheries Society 129:232–242.

Olsen, J. B., C. Habicht, J. Reynolds, and J. E. Seeb. 2004b. Moderately and highly polymorphic microsatellites provide discordant estimates of population divergence in sockeye salmon, *Oncorhynchus nerka*. Environmental Biology of Fishes 69:261–273.

Oostermeijer, J. G. B., S. H. Luijten, and J. C. M. den Nijs. 2003. Integrating demographic and genetic approaches in plant conservation. Biological Conservation 113:389–398.

Otis, D. L., K. P. Burnham, G. C. White, and D. R. Anderson. 1978. Statistical inference for capture data on closed animal populations. Wildlife Monographs 62:1–135.

Ouborg, N. J., Y. Piquot, and J. M. Vangroenendael. 1999. Population genetics, molecular markers and the study of dispersal in plants. Journal of Ecology 87:551–568.

Ovenden, J. R., and R. W. G. White. 1990. Mitochondrial and allozyme genetics of incipient speciation in a landlocked population of *Galaxias truttaceus*. Genetics 124:701–716.

Packer, L., and R. Owen. 2001. Population genetic aspects of pollinator decline. Conservation Ecology 5:4 (http://www.ecologyandsociety.org/vol5/iss1/art4/).

Paetkau, D. 1999. Using genetics to identify intraspecific conservation units: a critique of current methods. Conservation Biology 13:1507–1509.

Paetkau, D., W. Calvert, I. Stirling, and C. Strobeck. 1995. Microsatellite analysis of population structure in Canadian polar bears. Molecular Ecology 4:347–354.

Paetkau, D., R. Slade, M. Burden, and A. Estoup. 2004. Genetic assignment methods for the direct, real-time estimation of migration rate: a simulation-based exploration of accuracy and power. Molecular Ecology 13:55–65.

Paetkau, D., L. P. Waits, P. L. Clarkson, L. Craighead, and C. Strobeck. 1997. An empirical evaluation of genetic distance statistics using microsatellite data from bear (Ursidae) populations. Genetics 147:1943–1957.

Paetkau, D., L. P. Waits, P. L. Clarkson, L. Craighead, E. Vyse, R. Ward, and C. Strobeck. 1998. Variation in genetic diversity across the range of North American brown bears. Conservation Biology 12:418–429.

Pahl, M. 2003. U.S. "animal detectives" fight crime in forensics lab. National Geographic News, http://news.nationalgeographic.com/news/2003/04/0402_030402_tvwildlifecrimes.html (2 April).

Painter, T. S. 1933. A new method for the study of chromosome rearrangements and the plotting of chromosome maps. Science 78:585–586.

Palmer, A. R., and C. Strobeck. 1986. Fluctuating asymmetry: measurement, analysis, patterns. Annual Review of Ecology and Systematics 17:391.

Palmer, A. R., and C. Strobeck. 1997. Fluctuating asymmetry and developmental stability: heritability of observable variation vs. heritability of inferred cause. Journal of Evolutionary Biology 10:39–49.

Palsbøll, P. J., and 18 others. 1997. Genetic tagging of humpback whales. Nature 388:767–769.

Palumbi, S. R., and F. Cipriano. 1998. Species identification using genetic tools: the value of nuclear and mitochondrial gene sequences in whale conservation. Journal of Heredity 89:459–464.

Pamilo, P., and S. Pálsson. 1998. Associative overdominance, heterozygosity and fitness. Heredity 81:381–389.

Parker, I. M., J. Rodriguez, and M. E. Loik. 2003. An evolutionary approach to understanding the biology of invasions: local adaptation and general-purpose genotypes in the weed *Verbascum thapsus*. Conservation Biology 17:59–72.

Parker, M. A. 1992. Outbreeding depression in a selfing annual. Evolution 46:837–841.

Parris, M. J. 2004. Hybrid response to pathogen infection in interspecific crosses between two amphibian species (Anura: Ranidae). Evolutionary Ecology Research 6:457–471.

Patarnello, T., and B. Battaglia. 1992. Glucosephosphate isomerase and fitness – effects of temperature on genotype dependent mortality and enzyme activity in 2 species of the genus *Gammarus* (Crustacea, Amphipoda). Evolution 46:1568–1573.

Paterson, A. H., K. F. Schertz, Y.-R. Lin, S.-C. Liu, and Y.-L. Chang. 1995. The weediness of wild plants: molecular analysis of genes influencing dispersal and persistence of johnsongrass, *Sorghum halepense* (L.). Proceedings of the National Academy of Sciences, USA 92:6127–6131.

Paterson, S., K. Wilson, and J. M. Pemberton. 1998. Major histocompatibility complex variation associated with juvenile survival and parasite resistance in a large unmanaged ungulate populations (*Ovis aries L.*). Proceedings of the National Academy of Sciences, USA 95:3714–3719.

Peakall, R., D. Ebert, L. J. Scott, P. F. Meagher, and C. A. Offord. 2003. Comparative genetic study confirms exceptionally low genetic variation in the ancient and endangered relictual conifer, *Wollemia nobilis* (Araucariaceae). Molecular Ecology 12:2331–2343.

Pearce-Kelly, P., R. Jones, D. Clarke, C. Walker, P. Atkin, and A. A. Cunningham. 1998. The captive rearing of threatened Orthoptera: a comparison of the conservation potential and practical considerations of two species' breeding programmes at the Zoological Society of London. Journal of Insect Conservation 2:201–210.

Pella, J. J., and G. B. Milner. 1987. Use of genetic analysis in stock composition analysis. Pp. 247–276 *in* Utter, F., and N. Ryman, eds. Population Genetics and Fisheries Management. University of Washington Press, Seattle, WA.

Pemberton, J. 2004. Measuring inbreeding depression in the wild: the old ways are the best. Trends in Ecology and Evolution 19:613–615.

Petit, R. J., J. Duminil, S. Fineschi, A. Hampe, D. Salvini, and G. G. Vendramin. 2005. Comparative organization of chloroplast, mitochondrial and nuclear diversity in plant populations. Molecular Ecology 18:689–701.

Petit, R. J., A. Elmousadik, and O. Pons. 1998. Identifying populations for conservation on the basis of genetic markers. Conservation Biology 12:844–855.

Philipp, D. P. 1991. Genetic implications of introducing florida largemouth bass, *Micropterus salmoides floridanus*. Canadian Journal of Fisheries and Aquatic Sciences 48:58–65.

Phillips, B. L., G. P. Brown, and R. Shine. 2003. Assessing the potential impact of cane toads on Australian snakes. Conservation Biology 17:1738–1747.

Phillips, B. L., and R. Shine. 2004. Adapting to an invasive species: toxic cane toads induce morphological change in Australian snakes. Proceedings of the National Academy of Sciences, USA 101:17150–17155.

Phillips, P. C. 2005. Testing hypotheses regarding the genetics of adaptation. Genetica 123:15–24.

Pichler, F. B. 2002. Genetic Assessment of Population Boundaries and Gene Exchange in Hector's Dolphin. New Zealand Department of Conservation Science Internal Series No. 44, Auckland, New Zealand.

Pichler, F. B., and C. S. Baker. 2000. Loss of genetic diversity in the endemic Hector's dolphin due to fisheries-related mortality. Proceedings of the Royal Society of London B 267:97–102.

Pigliucci, M., and C. J. Murren. 2003. Perspective: genetic assimilation and a possible evolutionary paradox: can macroevolution sometimes be so fast as to pass us by? Evolution 57:1455–1464.

Pimentel, D., L. Lach, R. Zuniga, and D. Morrison. 2000. Environmental and economic costs of nonindigenous species in the United States. BioScience 50:53–65.

Pimm, S. L., H. L. Hones, and J. M. Diamond. 1988. On the risk of extinction. American Naturalis 132:757–785.

Pimm, S. L., and 31 others. 2001. Environment – can we defy nature's end? Science 293:2207–2208.

Piry, S., A. Alapetite, J.-M. Cornuet, D. Paetkau, L. Baudouin, and A. Estoup. 2004. GENECLASS2: a software for genetic assignment and first-generation migrant detection. Journal of Heredity 95:536–539.

Polovina, J. J., G. H. Balazs, E. A. Howell, D. M. Parker, M. P. Seki, and P. H. Dutton. 2004. Forage and migration habitat of loggerhead (*Caretta caretta*) and olive ridley (*Lepidochelys olivaceae*) sea turtles in the central North Pacific Ocean. Fisheries Oceanography 13:1–16.

Porter, C. A., and J. W. Sites. 1987. Evolution of *Sceloporus grammicus* complex (Sauria: Iguanidae) in central Mexico. II. Studies of nondisjunction and the occurrence of spontaneous chromosomal mutations. Genetica 75:131–144.

Powell, J. R. 1994. Molecular techniques in population genetics: a brief history. Pp. 131–156 *in* Schierwater, B., B. Streit, G. P. Wagner, and R. DeSalle, eds. Molecular Ecology and Evolution: Approaches and Applications. Burkhauser Verlag, Basel, Switzerland.

Price, D. K., G. E. Collier, and C. F. Thompson. 1989. Multiple parentage in broods of house wrens: genetic evidence. Journal of Heredity 80:1–5.

Primmer, C. R., M. T. Koskinen, and J. Piironen. 2000. The one that did not get away: individual assignment using microsatellite data detects a case of fishing competition fraud. Proceeding of the Royal Society of London B 267:1699–1704.

Pritchard, J. K., M. Stephens, and P. Donnelly. 2000. Inference of population structure using multilocus genotype data. Genetics 155:945–959.

Pritchard, P. C. H. 1999. Status of the black turtle. Conservation Biology 13:1000–1003.

Prout, T. 1973. Appendix to: Population genetics of marine pelecypods. III. Epistasis between functionally related isoenzymes of *Mytilus edulis*. Genetics 73:493–496.

Provan, J., W. Powell, and P. M. Hollingsworth. 2001. Chloroplast microsatellites: new tools for studies in plant ecology and evolution. Trends in Ecology and Evolution 16:142–147.

Provine, W. B. 1986. Sewall Wright and Evolutionary Biology. University of Chicago Press, Chicago.

Provine, W. B. 2001. The Origins of Theoretical Population Genetics with a New Afterword. University Chicago Press, Chicago.

Ptacek, M. B., H. C. Gerhardt, and R. D. Sage. 1994. Speciation by polyploidy in treefrogs: multiple origins of the tetraploid, *Hyla versicolor*. Evolution 48:898–908.

Punnett, R. C. 1904. Merism and sex in "*Spinax niger*". Biometrica 3:313–362.

Pyle, P. 1997. Identification Guide to North American Birds – Part 1. Slate Creek Press, Bolinas, CA.

Queller, D. C., and K. F. Goodnight. 1989. Estimating relatedness using genetic markers. Evolution 43:258–275.

Ralls, K., and J. Ballou. 1983. Extinction: lessons from zoos. Pp. 164–184 *in* Schonewald-Cox, C., S. Chambers, B. MacBryde, and L. Thomas, eds. Genetics and Conservation. Benjamin/Cummings, Menlo Park, CA.

Ralls, K., J. D. Ballou, B. A. Rideout, and R. Frankham. 2000. Genetic management of chondrodystrophy in California condors. Animal Conservation 3:145–153.

Ralls, K., J. D. Ballou, and A. Templeton. 1988. Estimates of lethal equivalents and the cost of inbreeding in mammals. Conservation Biology 2:185–193.

Ralls, K., S. R. Beissinger, and J. F. Cochrane. 2002. Guidelines for using population viability analysis in endangered-species management. Pp. 521–550 *in* Beissinger, S. R., and D. R. McCullough, eds. Population Viability Analysis. University of Chicago Press, Chicago.

Ralls, K., K. Brugger, and J. Ballou. 1979. Inbreeding and juvenile mortality in small populations of ungulates. Science 206:1101–1103.

Rannala, B., and J. L. Mountain. 1997. Detecting immigration by using multilocus genotypes. Proceedings of the National Academy of Sciences, USA 94:9197–9201.

Rawson, P. D., and R. S. Burton. 2002. Functional coadaptation between cytochrome c and cytochrome c oxidase within allopatric populations of a marine copepod. Proceedings of the National Academy of Sciences, USA 99:12955.

Reed, D. H., and E. H. Bryant. 2000. Experimental tests of minimum viable population size. Animal Conservation 3:7–14.

Reich, D. E., and 10 others. 2001. Linkage disequilibrium in the human genome. Nature 411:199–204.

Reilly, P. 2001. Legal and public policy issues in DNA forensics. Nature Reviews Genetics 2:313–317.

Reinartz, J. A., and D. H. Les. 1994. Bottleneck-induced dissolution of self-incompatibility and breeding system consequences in *Aster furcatus* (Asteraceae). American Journal of Botany 81:446–455.

Reisenbichler, R. R., and S. P. Rubin. 1999. Genetic changes from artificial propagation of Pacific salmon affect the productivity and viability of supplemented populations. ICES Journal of Marine Science 56:459–466.

Remington, D. L., and D. M. O'Malley. 2000. Evaluation of major genetic loci contributing to inbreeding depression for survival and early growth in a selfed family of *Pinus taeda*. Evolution 54:1580–1589.

Rennie, J. 2000. Cloning and conservation. Scientific American 283:1.

Rhymer, J. M., and D. Simberloff. 1996. Extinction by hybridization and introgression. Annual Review of Ecology and Systematics 27:83–109.

Rice, W. R. 1989. Analyzing tables of statistical tests. Evolution 43:223–225.

Rice, W. R. 1996. Evolution of the Y sex chromosome in animals. Bioscience 46:331–343.

Rich, W. H. 1939. Local populations and migration in relation to the conservation of Pacific salmon in the western states and Alaska. Contributions No. 1. Fish Commission of Oregon, Salem, OR.

Richards, A. J. 1986. Plant Breeding Systems. Allen & Unwin, London.

Richards, C. M. 2000. Genetic and demographic influences on population persistence: gene flow and genetic rescue in *Silene alba*. Pp. 271–291 in Young, A., and G. Clarke, eds. Genetics, Demography and Viability of Fragmented Populations. Cambridge University Press, Cambridge, UK.

Ricker, W. E. 1972. Hereditary and environmental factors affecting certain salmonid populations. Pp. 27–160 in Simon, R. C., and P. A. Larkin, eds. The Stock Concept in Pacific Salmon. University of British Columbia, Vancouver.

Ricker, W. E. 1981. Changes in the average body size and average age of Pacific salmon. Canadian Journal of Fisheries and Aquatic Sciences 38:1636–1656.

Ricklefs, R. E., and Miller, G. L. 2000. Ecology, 4th edn. W. H. Freeman & Co., New York.

Riechert, S. E., F. D. Singer, and T. C. Jones. 2001. High gene flow levels lead to gamete wastage in a desert spider system. Genetica 112–113:297–319.

Rieman, B., and J. Clayton. 1997. Wildlife and native fish: issues of forest health and conservation of sensitive species. Fisheries 22(11):6–15.

Rieman, B. R, D. Lee, J. McIntyre, K. Overton, and R. Thurow. 1993. Consideration of extinction risks for salmonids. Fish Habitat Relationships Technical Bulletin (US Forest Service) 14:1–12.

Rieseberg, L. H. 1997. Hybrid origins of plant species. Annual Review of Ecology and Systematics 28:359–389.

Rieseberg, L. H. 2001. Chromosomal rearrangements and speciation. Trends in Ecology and Evolution 16:351–358.

Rieseberg, L. H., M. A. Archer, and R. K. Wayne. 1999. Transgressive segregation, adaptation and speciation. Heredity 83:363–372.

Rieseberg, L. H., and J. M. Burke. 2001. A genic view of species integration. Journal of Evolutionary Biology 14:883–886.

Rieseberg, L. H., and D. Gerber. 1995. Hybridization in the Catalina Island mountain mahogany (*Cercocarpus traskiae*): RAPD evidence. Conservation Biology 9:199–203.

Rieseberg, L. H., and 9 others. 2003. Major ecological transitions in wild sunflowers facilitated by hybridization. Science 301:1211–1216.

Rising, J. D., and G. F. Shields. 1980. Chromosomal and morphological correlates in two new world sparrows (Emberizidae). Evolution 34:654–662.

Ritland, K. 1996. Inferring the genetic basis of inbreeding depression in plants. Genome 39:1–8.

Ritland, K. 2000. Detecting inheritance with inferred relatedness in nature. Pp. 187–199 in Mousseau, T. A., B. Sinervo, and J. A. Endler, eds. Adaptive Genetic Variation in the Wild. Oxford University Press, New York, NY.

Ritzema-Bos, J. 1894. Untersuchungen uber die Folgen der Zucht in engster Blutverwandtschaft. Biologisches Centralblatt 14:75–81.

Robert, J. 2000. Dossier Traffic Animal: la Mafia a l'assault de la nature. Terre Sauvage 155:35–50.

Robertson, A. 1952. The effect of inbreeding on the variation due to recessive alleles. Genetics 37:189–207.

Robertson, A. 1960. A theory of limits in artificial selection. Proceedings of the Royal Society of London B 153:234–249.

Robertson, A. 1962. Selection for heterozygotes in small populations. Genetics 47:1291–1300.

Robertson, A. 1964. The effect of non-random mating within inbred lines on the rate of inbreeding. Genetic Resource 5:164–167.

Robertson, A. 1965. The interpretation of genotypic ratios in domestic animal populations. Animal Production 7:319–324.

Robertson, A. 1967. The nature of quantitative genetic variation. Pp. 265–280 *in* Brink, R. A. and E. D. Styles, eds. Heritage from Mendel. University of Wisconsin Press, Madison, WI.

Robertson, B. C., and N. J. Gemmell. 2004. Defining eradication units in pest control programmes. Journal of Ecology 41:1042–1048.

Robertson, B. C., E. O. Minot, and D. M. Lambert. 2000. Microsatellite primers for the kakapo (*Strigops habroptilus*) and their utility in other parrots. Conservation Genetics 1:93–95.

Robichaux, R. H., J. Canfield, F. Warshauer, M. Bruegmann, and E. A. Friar. 1998. Restoring Mauna Kea's crown jewel. Endangered Species Bulletin 23:21–25.

Robinson, T. J., and F. F. B. Elder. 1993. Cytogenetics – its role in wildlife management and the genetic conservation of mammals. Biological Conservation 63:47–51.

Roca, A. L., N. Georgiadis, J. PeconSlattery, and S. J. O'Brien. 2001. Genetic evidence for two species of elephant in Africa. Science 293:1473–1477.

Roff, D. A. 1997. Evolutionary Quantitative Genetics. Chapman & Hall, New York.

Roff, D. 2003. Evolutionary danger for rainforest species. Science 301:58–59.

Roff, D. A., and P. Bentzen. 1989. The statistical analysis of mitochondrial DNA polymorphisms: chi-square and the problem of small samples. Molecular Biology and Evolution 6:539–545.

Roff, D. A., and T. A. Mousseau. 1987. Quantitative genetics and fitness: lessons from Drosophila. Heredity 58:103–118.

Rohde, D. L. T., S. Olson, and J. T. Chang. 2004. Modeling the recent human ancestry of all living humans. Nature 431:562–566.

Roman, J., and B. W. Bowen. 2000. The mock turtle syndrome: genetic identification of turtle meat purchased in the south-eastern United States of America. Animal Conservation 3:61–65.

Roman, J., and S. R. Palumbi. 2003. Whales before whaling in the North Atlantic. Science 301:508–510.

Rosenberg, N. A., and M. Nordborg. 2002. Genealogical trees, coalescent theory and the analysis of genetic polymorphisms. Nature Reviews Genetics 3:380–390.

Ross, H. A., G. M. Lento, M. L. Dalebout, et al. 2003. DNA Surveillance: web-based molecular identification of whales, dolphins, and porpoises. Journal of Heredity 94:111–114.

Roy, M. S., E. Geffen, D. Smith, E. A. Ostrander, and R. K. Wayne. 1994. Patterns of differentiation and hybridization in North American wolflike canids, revealed by analysis of microsatellite loci. Molecular Biology and Evolution 11:553–570.

Rozhnov, V. V. 1993. Extinction of the European mink: ecological catastrophe or a natural process? Lutreola 1:10–16.

Ruesink, J. L., I. M. Parker, M. J. Groom, and P. M. Kareiva. 1995. Reducing the risks of nonindigenous species introductions – guilty until proven innocent. BioScience 45:465–477.

Ruiz-Pesini, E., and 10 others. 2000. Human mtDNA haplogroups associated with high or reduced spermatozoa motility. American Journal of Human Genetics 67:682–696.

Ryder, O. A. 1986. Species conservation and systematics: the dilemma of subspecies. Trends in Ecol. Evol. 1:9–10.

Ryder, O. A. 1987. Conservation action for gazelles: an urgent need. Trends in Ecology and Evolution 2:143–144.

Ryder, O. A., and K. Benirschke. 1997. The potential use of "cloning" in the conservation effort. Zoo Biology 16:295–300.

Ryder, O. A., and L. G. Chemnick. 1993. Chromosomal and mitochondrial DNA variation in orang utans. Journal of Heredity 84:405–409.

Ryder, O. A., A. T. Kumamoto, B. S. Durrant, and K. Benirschke. 1989. Chromosomal divergence and reproductive isolation in diks-diks. Pp. 208–225 *in* Otte, D., and J. A. Endler, eds. Speciation and Consequences. Sinauer, Sunderland, MA.

Ryder, O. A., A. McLaren, S. Brenner, Y. P. Zhang, and K. Benirschke. 2000. Ecology – DNA banks for endangered animal species. Science 288:275–277.

Ryman, N., F. W. Allendorf, and G. Ståhl. 1979. Reproductive isolation with little genetic divergence in sympatric populations of brown trout (*Salmo trutta*). Genetics 92:247–262.

Ryman, N., R. Baccus, C. Reuterwall, and M. H. Smith. 1981. Effective population size, generation interval, and potential loss of genetic variability in game species under different hunting regimes. Oikos 36:257–266.

Ryman, N., P. E. Jorde, and L. Laikre. 1995. Supportive breeding and variance effective population size. Conservation Biology 9:1619–1628.

Ryman, N., and L. Laikre. 1991. Effects of supportive breeding on the genetically effective population size. Conservation Biology 5:325–329.

Saccheri, I. J., and P. M. Brakefield. 2002. Rapid spread of immigrant genomes into inbred populations. Proceedings of the Royal Society of London B 269:1073–1078.

Saccheri, I., M. Kuussaari, M. Kankare, P. Vikman, W. Fortelius, and I. Hanski. 1998. Inbreeding and extinction in a butterfly metapopulation. Nature 392:491–494.

Sage, R. D., D. Heyneman, K.-C. Lim, and A. C. Wilson. 1986. Wormy mice in a hybrid zone. Nature 324:60–63.

Saito, Y., K. Sahara, and K. Mori. 2000. Inbreeding depression by recessive deleterious genes affecting female fecundity of a haplo-diploid mite. Journal of Evolutionary Biology 13:668–678.

Sakai, A. K., and 13 others. 2001. The population biology of invasive species. Annual Review of Ecology and Systematics 32:305–332.

Salemi, M., and Vandamme, A.-M. 2003. The Phylogenetic Handbook: a Practical Approach to DNA and Protein Phylogeny. Cambridge University Press, Cambridge, UK.

Sarrazin, F., and R. Barbault. 1996. Reintroduction: challenges and lessons for basic ecology. Trends in Ecology and Evolution 11:474–478.

Savolainen, P., L. Arvestad, and J. Lundeberg. 2000. mtDNA tandem repeats in domestic dogs and wolves: mutation mechanism studied by analysis of the sequence of imperfect repeats. Molecular Biology and Evolution 17:474–488.

Schaal, B. B., and K. M. Olsen. 2000. Gene genealogies and population variation in plants. Proceedings of the National Academy of Sciences, USA 97:7024–7029.

Schlager, G., and M. M. Dickie. 1971. Natural mutation rates in the house mouse: estimates for five specific loci and dominant mutations. Mutation Research 11:89–96.

Schlötterer, C. 1998. Microsatellites. Pp. 237–262 *in* Hoelzel, A. R., ed. Molecular Genetic Analysis of Populations. Oxford University Press, New York.

Schlötterer, C. 2004. The evolution of molecular markers – just a matter of fashion? Nature Reviews Genetics 5:63–69.

Schnable, P. S., and R. P. Wise. 1998. The molecular basis of cytoplasmic male sterility and fertility restoration. Trends in Plant Science 3:175–180.

Schonewald-Cox, C. M., S. M. Chambers, B. MacBryde, and W. L. Thomas. 1983. Genetics and Conservation. Benjamin/Cummings, Menlo Park, CA.

Schwartz, M. K., L. S. Mills, K. S. McKelvey, L. F. Ruggiero, and F. W. Allendorf. 2002. DNA reveals high dispersal synchronizing the population dynamics of Canada lynx. Nature 415:520–522.

Schwartz, M. K., K. L. Pilgrim, K. S. McKelvey, E. L. Lindquist, J. J. Claar, S. Loch, and L. F. Ruggiero. 2004. Hybridization between Canada lynx and bobcats: genetic results and management implications. Conservation Genetics 5:349–355.

Schwartz, M. K., D. A. Tallmon, and G. Luikart. 1998. Review of DNA-based census and effective population size estimators. Animal Conservation 1:293–299.

Seal, U. S., S. A. Ellis, T. J. Foose, and A. P. Byers. 1993. Conservation assessment and management plans (CAMPs) and global action plans (GCAPs). Conservation Breeding Specialist Group Newsletter 4(2):5–10.

Searle, J. B. 1986. Meiotic studies of Robertsonian heterozygotes from natural populations of the common shrew. Cytogenetics and Cell Genetics 41:154–62.

Seeb, L. W., J. E. Seeb, G. H. Kruse, and R. G. Weck. 1989. Genetic structure of red king crab populations in Alaskan facilitates enforcement of fishing regulations. Pp. 491–502 in Proceedings of the International Symposium on King and Tanner Crabs, Anchorage, Alaska, USA. Alaska Sea Grant College Program, Fairbanks.

Seeb, L. W., and 6 others. 2000. Genetic diversity of sockeye salmon of Cook Inlet, Alaska, and its application to management of populations affected by the Exxon Valdez oil spill. Transactions of the American Fisheries Society 129:1223–1249.

Seehausen, O. 2004. Hybridization and adaptive radiation. Trends in Ecology and Evolution 19:198–207.

Seehausen, O., J. J. M. Vanalphen, and F. Witte. 1997. Cichlid fish diversity threatened by eutrophication that curbs sexual selection. Science 277:1808–1811.

Seuanez, H. N. 1986. Chromosomal and molecular characterization of the primates: its relevance in the sustaining of the primate populations. Pp. 887–910 in Benirschke, K., ed. Primates: the Road to Self Sustaining Populations. Springer-Verlag, New York.

Shaffer, M. L. 1978. Determining minimum viable population sizes: a case study of the grizzly bear (*Ursus arctos* L.). PhD Dissertation, Duke University, Durham, NC.

Shaffer, M. L. 1981. Minimum population sizes for species conservation. BioScience 31:131–134.

Shaffer, M. 1987. Minimum viable populations: coping with uncertainty. Pp. 69–86 in Soulé, M. E., ed. Viable Populations for Conservation. Sinauer, Sunderland, MA.

Shaffer, M., L. H. Watchman, W. J. Snape III, and I. K. Latchis. 2002. Population viability analysis and conservation policy. Pp. 123–142 in Beissinger, S. R., and D. R. McCullough, eds. Population Viability Analysis. University of Chicago Press, Chicago.

Shaw, F. H., C. J. Geyer, and R. G. Shaw. 2002. A comprehensive model of mutations affecting fitness and inferences for *Arabidopsis thaliana*. Evolution 56:453–463.

Shields, G. F. 1982. Comparative avian cytogenetics: a review. Condor 84:45–58.

Shields, W. M. 1993. The natural and unnatural history of inbreeding and outbreeding. Pp. 143–169 in Thorhill, N. W., ed. the Nature History of Inbreeding and Outbreeding. University of Chicago Press, Chicago.

Shivji, M., S. Clarke, M. Pank, L. Natanson, N. Kohler, and M. Stanhope. 2002. Genetic identification of pelagic shark body parts for conservation and trade monitoring. Conservation Biology 16:1036–1047.

Siemann, E., and W. E. Rogers. 2001. Genetic differences in growth of an invasive tree species. Ecology Letters 4:514–518.

Silva, P. J. N. 2002. HWpower, Version 1.0.

Simberloff, D. 1988. The contribution of population and community biology to conservation science. Annual Review of Ecology and Systematics 19:437–511.

Simberloff, D. 2001. Biological invasions – how are they affecting us, and what can we do about them? Western North American Naturalist 61:308–315.

Simon, J.-P., Y. Bergeron, and D. Gagnon. 1986. Isozyme uniformity in red pine (*Pinus resinosa*) in the Abitibi Region, Quebec. Canadian Journal of Forest Research 16:1133–1135.

Simpson, G. G. 1961. Principles of Animal Taxonomy. Columbia University Press, New York.

Singh, R. C., R. C. Lewontin, and A. Felton. 1976. Genetic heterogeneity within electrophoretic "alleles" of xanthine dehydrogenase in *Drosophila pseudoobscura*. Genetics 84:609–626.

Sissenwine, M. P. 1984. Why do fish populations vary? Pp. 59–94 in May, R. M., ed. Dahlem Konferenzen. Springer Verlag, Berlin.

Slate, J., T. Marshall, and J. Pemberton. 2000. A retrospective assessment of the accuracy of the paternity inference program CERVUS. Molecular Ecology 9:801–808.

Slatkin, M. 1977. Gene flow and genetic drift in species subject to frequent local extinctions. American Naturalist 12:253–262.

Slatkin, M. 1985. Rare alleles as indicators of gene flow. Evolution 39:53–65.

Slatkin, M. 1987. Gene flow and the geographic structure of natural populations. Science 236:787–792.

Slatkin, M. 1995. A measure of population subdivision based on microsatellite allele frequencies. Genetics 139:457–462.

Slatkin, M., and N. H. Barton. 1989. A comparison of three indirect methods for estimating average levels of gene flow. Evolution 43:1349–1368.

Smith, C. T., J. E. Seeb, P. Schwenke, and L. W. Seeb. 2005a. Use of the 5′-nuclease reaction for single nucleotide polymorphism genotyping in chinook salmon. Transactions of the American Fisheries Society 134:207–217.

Smith, G. R. 1992. Introgression in fishes – significance for paleontology, cladistics, and evolutionary rates. Systematic Biology 41:41–57.

Smith, J. J., K. Kump, J. Walker, D. M. Parichy, and S. R. Voss. 2005b. A comprehensive EST linkage map for tiger salamander and Mexican axolotl: enabling gene mapping and comparative genomics in *Ambystoma*. Genetics 171:1161–1171.

Smith, T. B., and Wayne, R. K. eds. 1996. Molecular Genetic Approaches in Conservation. Oxford University Press, Oxford.

Snyder, G. 1990. The Practice of the Wild. North Point Press, San Francisco.

Snyder, N. F. R., S. R. Derrickson, S. R. Beissinger, J. W. Wiley, T. B. Smith, W. D. Toone, and B. Miller. 1996. Limitations of captive breeding in endangered species recovery. Conservation Biology 10:338–348.

Soltis, D. E., and P. S. Soltis. 1999. Polyploidy: recurrent formation and genome evolution. Trends in Ecology and Evolution 14:348–352.

Sorensen, F. C. 1969. Embryonic genetic load in coastal Douglas fir, *Pseudotsuga menziesii* var. *menziesii*. American Naturalist 103:389–398.

Sorensen, F. C. 1999. Relationship between self-fertility, allocation of growth, and inbreeding depression in three coniferous species. Evolution 53:417–425.

Soulé, M. E. 1980. Thresholds for survival: maintaining fitness and evolutionary potential. Pp. 151–170 in Soulé, M. E. and B. M. Wilcox, eds. Conservation Biology: an Evolutionary–Ecological Perspective. Sinauer, Sunderland, MA.

Soulé, M. E. 1987. Introduction. Pp. 1–10 *in* Soulé, M. E., ed. Viable Populations for Conservation. Sinauer, Sunderland, MA.

Soulé, M. E., M. Gilpin, W. Conway, and T. Foose. 1986. The millennium ark: how long a voyage, how many staterooms, how many passengers? Zoo Biology 5:101–113.

Soulé, M. E., and L. S. Mills. 1998. No need to isolate genetics. Science 282:1658–1659.

Soulé, M. E., and L. S. Mills. 1992. Conservation genetics and conservation biology: a troubled marriage. Pp. 55–65 *in* Sandlund, O. T., K. Hindar, and A. H. D. Brown, eds. Species and Ecosystem Conservation. Scandinavian University Press, Oslo.

Soulé, M. E., and B. M. Wilcox. 1980. Conservation Biology: an Evolutionary–Ecological Perspective. Sinauer, Sunderland, MA.

Spencer, C. N., B. R. McClelland, and J. A. Stanford. 1991. Shrimp stocking, salmon collapse, and eagle displacement: cascading interactions in the food web of a large aquatic ecosystem. BioScience 41:14–21.

Spencer, H. G., and R. W. Marks. 1993. The evolutionary construction of molecular polymorphisms. New Zealand Journal of Botany 31:249–256.

Spitze, K. 1993. Population structure in *Daphnia obtusa* – quantitative genetic and allozymic variation. Genetics 135:367–374.

Spong, G. L., L. Hellborg, and S. Creel. 2000. Sex ratio of leopards taken in trophy hunting: genetic data from Tanzania. Conservation Genetics 1:169–171.

Sprague, G. F. 1978. Introductory remarks to the session on the history of hybrid corn. Pp. 11–12 *in* Walden, D. B., ed. Maize Breeding and Genetics. Wiley, New York.

Spruell, P., and 6 others. Submitted. The controversy about salmon hatcheries: objective assessment of available data. Fisheries.

Spruell, P., M. L. Bartron, N. Kanda, and F. W. Allendorf. 2001. Detection of hybrids between bull trout (*Salvelinus confluentus*) and brook trout (*Salvelinus fontinalis*) using PCR primers complementary to interspersed nuclear elements. Copeia 2001:1093–1099.

Spruell, P., A. R. Hemmingsen, P. J. Howell, N. Kanda, and F. W. Allendorf. 2003. Conservation genetics of bull trout: geographic distribution of variation at microsatellite loci. Conservation Genetics 4:17–29.

Spruell, P., and 9 others. 1999a. Inheritance of nuclear DNA markers in gynogenetic haploid pink salmon (*Oncorhynchus gorbuscha*). Journal of Heredity 90:289–296.

Spruell, P., B. E. Rieman, K. L. Knudsen, F. M. Utter, and F. W. Allendorf. 1999b. Genetic population structure within streams: microsatellite analysis of bull trout populations. Ecology of Freshwater Fish 8:114–121.

Stebbins, G. L. 1950. Variation and Evolution in Plants. Columbia University Press, New York.

Stebbins, G. L. 1959. The role of hybridization in evolution. Proceedings of the American Philosophical Society 103:231–251.

Steinberg, E. K., K. R. Lindner, J. Gallea, A. Maxwell, J. Meng, and F. W. Allendorf. 2002. Rates and patterns of microsatellite mutations in pink salmon. Molecular Biology and Evolution 19:1198–1202.

Stephens, J. C., and 27 others. 2001. Haplotype variation and linkage disequilibrium in 313 human genes. Science 293:489–493.

Stevens, N. M. 1908. The chromosomes in *Diabrotica vittata*, *Diabrotica soror*, and *Diabrotica 12-punctata*. Journal of Experimental Zoology 5:453–470.

Stockwell, C. A., and M. V. Ashley. 2004. Diversity – rapid adaptation and conservation. Conservation Biology 18:272–273.

Stockwell, C. A., A. P. Hendry, and M. T. Kinnison. 2003. Contemporary evolution meets conservation biology. Trends in Ecology and Evolution 18:94–101.

Stockwell, C. A., M. T. Kinnison, and A. P. Hendry. 2006. Restoring evolutionary potential: contemporary evolution during ecological restoration. In press *in* D. Falk, M. Palmer, and J. Zedler, eds. Foundations of Restoration Ecology. Island Press, New York.

Stone, G. 2000. Phylogeography, hybridization and speciation. Trends in Ecology and Evolution 15:354–355.

Stone, G. N., and P. Sunnucks. 1993. Genetic consequences of an invasion through a patchy environment – the cynipid gallwasp *Andricus quercuscalicis* (Hymenoptera: Cynipidae). Molecular Ecology 2:251–268.

Storfer, A. 1996. Quantitative genetics: a promising approach for the assessment of genetic variation in endangered species. Trends in Ecology and Evolution 11:343–348.

Storfer, A. 1999. Gene flow and endangered species translocations: a topic revisited. Biological Conservation 87:173–180.

Storz, J. F. 2002. Contrasting patterns of divergence in quantitative traits and neutral DNA markers: analysis of clinal variation. Molecular Ecology 11:2537–2551.

Streiner, D. L. 1996. Maintaining standards: differences between the standard deviation and standard error, and when to use each. Canadian Journal of Psychiatry 41:498–502.

Strickberger, M. W. 2000. Evolution, 3rd edn. Jones & Bartlett Publishere, Sudbury, MA.

Sturtevant, A. H. 1923. Inheritance of shell coiling in Limnaea. Science 58:269–270.

Sturtevant, A. H., and Beadle, G. W. 1939. An Introduction to Genetics. Dover, New York.

Sunnucks, P. 2000. Efficient genetic markers for population biology. Trends in Ecology and Evolution 15:199–203.

Sutherland, B., D. Stewart, E. R. Kenchington, and E. Zouros. 1998. The fate of paternal mitochondrial DNA in developing female mussels, *Mytilus edulis*: implications for the mechanism of doubly uniparental inheritance of mitochondrial DNA. Genetics 148:341–347.

Sutherland, W. J. 1996. Why census? Pp. 1–10 *in* Sutherland, W. J., ed. Ecological Census Techniques, a Handbook. Cambridge University Press, Cambridge , UK.

Syvanen, A.-C. 2001. Accessing genetic variation: a genotyping single nucelotide polymorphisms. Nature Reviews Genetics 2:930–942.

Taberlet, P., L. Gielly, G. Pautou, and J. Bouvet. 1991. Universal primers for amplification of three non-coding regions of chloroplast DNA. Plant Molecular Biology 17:1105–1109.

Taberlet, P., and 7 others. 1996. Reliable genotyping of samples with very low DNA quantities using PCR. Nucleic Acids Research 24:3189–3194.

Taberlet, P., and 8 others. 1997. Noninvasive genetic tracking of the endangered Pyrenean brown bear population. Molecular Ecology 6:869–876.

Taberlet, P., L. P. Waits, and G. Luikart. 1999. Noninvasive genetic sampling: look before you leap. Trends in Ecology and Evolution 14:323–327.

Tallmon, D. A., W. C. Funk, W. W. Dunlap, and F. W. Allendorf. 2000. Genetic differentiation among long-toed salamander (*Ambystoma macrodactylum*) populations. Copeia 2000:27–35.

Tallmon, D. A., G. Luikart, and M. A. Beaumont. 2004. Comparative evaluation of a new effective population size estimator based on approximate Bayesian summary statistics. Genetics 167:977–988.

Tarr, C. L., S. Conant, and R. C. Fleischer. 1998. Founder events and variation at microsatellite loci in an insular passerine bird, the Laysan finch (*Telespiza cantans*). Molecular Ecology 7:719–731.

Tave, D. 1984. Quantitative genetics of vertebrae number and position of dorsal spines in the velvet belly shark, *Etmopterus spinax*. Copeia 1984:794–797.

Taylor, A. C., W. B. Sherwin, and R. K. Wayne. 1994. Genetic variation of microsatellite loci in a bottlenecked species: the northern hairy-nosed wombat *Lasiorhinus kreffti*. Molecular Ecology 3:277–290.

Taylor, B. L., and A. E. Dizon. 1999. First policy then science: why a management unit based solely on genetic criteria cannot work. Molecular Ecology 8:S11–S16.

Taylor, E. B., C. J. Foote, and C. C. Wood. 1996. Molecular genetic evidence for parallel life-history evolution within a Pacific salmon (sockeye salmon and kokanee, *Oncorhynchus nerka*). Evolution 50:401–416.

Tear, T. H., J. M. Scott, P. H. Hayward, and B. Griffith. 1993. Status and prospects for success of the Endangered Species Act: a look at recovery plans. Science 262:976–977.

Templeton, A. R. 1982. Adaptation and the integration of evolutionary forces. Pp. 00–00 *in* Milkman, R., ed. Perspectives on Evolution. Sinauer, Sunderland, MA.

Templeton, A. R. 1991. Off-site breeding of animals and implications for plant conservation strategies. Pp. 182–194 *in* Falk, D. A., and K. E. Holsinger, eds. Genetics and Conservation of Rare Plants. Oxford University Press, New York.

Templeton, A. R. 1998. Species and speciation – geography, population structure, ecology, and gene trees. Pp. 32–43 *in* Howard, D. J., and S. H. Berlocher, eds. Endless Forms. Oxford University Press, New York.

Templeton, A. R. 2002. The Speke's gazelle breeding program as an illustration of the importance of multilocus genetic diversity in conservation biology: response to Kalinowski et al. Conservation Biology 16:1151–1155.

Templeton, A. R., and B. Read. 1984. Factors eliminating inbreeding depression in a captive herd of Speke's gazelle. Zoo Biology 3:177–199.

Templeton, A. R., and B. Read. 1994. Inbreeding: one word, several meanings, much confusion. Pp. 91–105 *in* Loeschcke, V., J. Tomiuk, and S. K. Jain, eds. Conservation Genetics. Birkhauser Verlag, Basel, Switzerland.

Templeton, A. R., and B. Read. 1998. Elimination of inbreeding depression from a captive population of Speke's gazelle: validity of the original statistical analysis and confirmation by permutation testing. Zoo Biology 17:77–94.

Templeton, A. R., R. J. Robertson, J. Brisson, and J. Strasburg. 2001. Disrupting evolutionary processes: the effect of habitat fragmentation on collared lizards in the Missouri Ozarks. Proceedings of the National Academy of Sciences, USA 98:5426–5432.

Thelen, G. C., and F. W. Allendorf. 2001. Heterozygosity-fitness correlations in rainbow trout: effects of allozyme loci or associative overdominance? Evolution 55:1180–1187.

Theodorou, K., and D. Couvet. 2004. Introduction of captive breeders to the wild: harmful or beneficial? Conservation Genetics 5:1–12.

Thomas, D. D., C. A. Donnelly, R. J. Wood, and L. S. Alphey. 2000. Insect population control using a dominant, repressible, lethal genetic system. Science 287:2474–2476.

Thomas, L. 1974. Lives of a Cell. Notes of a Biology Watcher. Viking Penguin USA, New York.

Thompson, G. G. 1991. Determining minimum viable populations under the Endangered Species Act. NOAA Technical Memoir NMFS-F/NWC-198. US Department of Commerce.

Thomson, K. S. 1991. Living Fossil – the Story on the Coelacanth. W. W. Norton, New York.

Thornton, I. W. B. 1978. White tiger genetics – further evidence. Journal of Zoology 185:389–394.

Thrall, P. H., C. M. Richards, D. E. McCauley, and J. Antonovics. 1998. Metapopulation collapse: the consequences of limited gene-flow in spatially structured populations. Pp. 83–104 in Bascompte, J., and R. V. Sole, eds. Modeling Spatiotemporal Dynamics in Ecology. Springer-Verlag, Berlin.

Throneycroft, H. B. 1975. A cytogenetic study of the white-throated sparrow, *Zonotrichia albicollis* (Gmelin). Evolution 29:611–621.

Tishkoff, S. A., and 16 others. 2001. Haplotype diversity and linkage disequilibrium at human G6PD: recent origin of alleles that confer malarial resistance. Science 293:455–462.

Triggs, S. J., R. G. Powlesland, and C. H. Daugherty. 1989. Genetic variation and conservation of kakapo *Strigops habroptilus* (Psittaciformes). Conservation Biology 3:92–96.

Trimble, H. C., and C. E. Keeler. 1938. The inheritance of "high uric acid excretion" in dogs. Journal of Heredity 29:280–289.

Tsutsui, N. D., A. V. Suarez, D. A. Holway, and T. J. Case. 2000. Reduced genetic variation and the success of an invasive species. Proceedings of the National Academy of Sciences, USA 97:5948–5953.

Turner, J. R. G. 1985. Fisher's evolutionary faith and the challenge of mimicry. Oxford Surveys in Evolutionary Biology 2:159–96.

Turner, T. F., J. C. Trexler, J. L. Harris, and J. L. Haynes. 2000. Nested cladistic analysis indicates population fragmentation shapes genetic diversity in a freshwater mussel. Genetics 154:777–785.

Tuttle, E. M. 2003. Alternative reproductive strategies in the white-throated sparrow: behavioral and genetic evidence. Behavioral Ecology 14:425–432.

Unger, F. 1852. Versuch einer Geschichte der Pflanzenwelt. Braumuller, Wein.

US National Science Board Committee on International Science's Task Force on Global Biodiversity. 1989. Loss of Biological Diversity: a Global Crisis Requiring International Solutions. National Science Board Report 89–171, Washington, DC.

USFWS. 1983. Endangered and threatened species listing and recovery priority guidelines. Federal Register 48(184):43098–43105.

USFWS. 1998. Recovery Plan for *Kokia cookei*. US Fish and Wildlife Service, Portland, OR.

USFWS and NOAA. 1996a. Endangered and threatened wildlife and plants; proposed policy and proposed rule on the treatment of intercrosses and intercross progeny (the issue of "hybridization"); request for public comment. Federal Register 61(26):4710–4713.

USFWS and NOAA. 1996b. Policy Regarding the Recognition of District Vertebrate Population; Notice. Federal Register 61(26):4721–4725.

USGS. 2002. White abalone restoration. US Geological Survey WERC Fact Sheet.

Utter, F. M. 2005. Farewell to allozymes (?). Journal of Irreproducible Results 49:35.

Utter, F. M., P. Aebersold, and G. Winans. 1987. Interpreting genetic variation detected by electrophoresis. Pp. 21–46 *in* Ryman, N., and F. Utter, eds. Population Genetics and Fishery Management. University of Washington Press, Seattle, WA.

Valdes, A. M., M. Slatkin, and N. B. Freimer. 1993. Allele frequencies at microsatellite loci: the stepwise mutation model revisited. Genetics 133:737–749.

Van Aarde, R. J., and A. van Dyk. 1986. Inheritance of the king coat color pattern in cheetahs *Acinonyx jubatus*. Journal of Zoology 209:573–578.

Van Noordwijk, A. J., and W. Scharloo. 1981. Inbreeding in an island population of the great tit. Evolution 35:674–688.

Van Oosterhout, C., and P. M. Brakefield. 1999. Quantitative genetic variation in *Bicyclus anynana* metapopulations. Netherlands Journal of Zoology 49:67–80.

Van Valen, L. 1973. A new evolutionary law. Evolutionary Theory 1:1–30.

Van Valen, L. 1976. Ecological species, multispecies, and oaks. Taxon 25:233–239.

VanCamp, L. F., and Henny, C. J. 1975. The screech owl: its life history and population ecology in northern Ohio. North American Fauna No. 71. US Fish and Wildlife Service, Washington, DC.

Vane-Wright, R. I., C. J. Humphries, and P. H. Williams. 1991. What to protect – systematics and the agony of choice. Biological Conservation 55:235–254.

Vergeer, P., E. Sonderen, and N. J. Ouborg. 2004. Introduction strategies put to the test: local adaptation versus heterosis. Conservation Biology 18:812–821.

Vieira, C. P., and D. Charlesworth. 2002. Molecular variation at the self-incompatibility locus in natural populations of the genera *Antirrhinum* and *Misopates*. Heredity 88:172–181.

Vigouroux, Y., M. McMullen, C. T. Hittinger, et al. 2002. Identifying genes of agronomic importance in maize by screening microsatellites for evidence of selection during domestication. Proceedings of the National Academy of Sciences, USA 99:9650–9655.

Vithayasai, C. 1973. Exact critical values of the Hardy–Weinberg test statistic for two alleles. Communications in Statistics 1:229–242.

Vogel, F., and Motulsky, A. G. 1986. Human Genetics: Problems and Approaches, 2nd edn. Springer-Verlag, Berlin.

Voipio, P. 1950. Evolution at the population level with special reference to game animals and practical game management. Papers on Game Research (Helsinki) 5:1–175.

Vollmer, S. V., and S. R. Palumbi. 2002. Hybridization and the evolution of reef coral diversity. Science 296:2023–2025.

Vonnegut, K. 1985. Galápagos. Delacorte Press, New York.

Vos, P., and 10 others. 1995. AFLP: a new technique for DNA fingerprinting. Nucleic Acids Research 23:4407–4414.

Voss, S. R. 1995. Genetic basis of paedomorphosis in the axolotl, *Ambystoma mexicanum*: a test of the single-gene hypothesis. Journal of Heredity 86:441–447.

Voss, S. R., K. L. Prudic, J. C. Oliver, and H. B. Shaffer. 2003. Candidate gene analysis of metamorphic timing in ambystomatid salamanders. Molecular Ecology 12:1217–1223.

Voss, S. R., and H. B. Shaffer. 1997. Adaptive evolution via a major gene effect: paedomorphosis in the Mexican axolotl. Proceedings of the National Academy of Sciences, USA 94:14185–14189.

Voss, S. R., and J. J. Smith. 2005. Evolution of salamander life cycles: a major effect quantitative trait locus contributes to discreet and continuous variation for metamorphic timing. Genetics 170:275–281.

Vrijenhoek, R. C. 1989. Genotypic diversity and coexistence among sexual and clonal forms of *Poeciliopsis*. Pp. 386–400 *in* Otte, D., and J. Endler, eds. Speciation and its Consequences. Sinauer, Sunderland, MA.

Vrijenhoek, R. C., and P. L. Leberg. 1991. Let's not throw the baby out with the bathwater – a comment on management for MHC diversity in captive populations. Conservation Biology 5:252–254.

Vrijenhoek, R. C., and S. Lerman. 1982. Heterozygosity and developmental stability under sexual and asexual breeding systems. Evolution 36:768–776.

Vrijenhoek, R. C., E. Pfeiler, and J. Wetherington. 1992. Balancing selection in a desert stream-dwelling fish, *Poeciliopsis monacha*. Evolution 46:1642–1647.

Waddington, C. H. 1961. Genetic assimilation. Advances in Genetics 10:257–290.

Wade, M. J., S. M. Shuster, and L. Stevens. 1996. Inbreeding: its effect on response to selection for pupal weight and the heritable variance in fitness in the flour beetle, *Tribolium castaneum*. Evolution 50:723–733.

Wahlund, S. 1928. Zusammensetzung von Populationen und Korrelationserscheinungen vom Standpunkt der Vererbungslehre aus betrachtet. Hereditas 11:65–106.

Waits, J. L., and P. L. Leberg. 2000. Biases associated with population estimation using molecular tagging. Animal Conservation 3:191–199.

Waits, L. P., G. Luikart, and P. Taberlet. 2001. Estimating the probability of identity among genotypes in natural populations: cautions and guidelines. Molecular Ecology 10:249–256.

Waits, L., P. Taberlet, J. E. Swenson, F. Sandegren, and R. Franzen. 2000. Nuclear DNA microsatellite analysis of genetic diversity and gene flow in the Scandinavian brown bear (*Ursus arctos*). Molecular Ecology 9:421–431.

Waits, L. P., S. L. Talbot, R. H. Ward, and G. F. Shields. 1998. Mitochondrial DNA phylogeography of the North American brown bear and implications for conservation. Conservation Biology 12:408–417.

Walker, B., and W. Steffen. 1997. An overview of the implications of global change for natural and managed terrestrial ecosystems. Conservation Ecology 1:http://www.consecol.org/vol1/iss2/art2.

Walker, N. F., P. E. Hulme, and A. R. Hoelzel. 2003. Population genetics of an invasive species, *Heracleum mantegazzianum*: implications for the role of life history, demographics and independent introductions. Molecular Ecology 12:1747–1756.

Wallace, A. R. 1892. Island Life. Macmillan & Co., London.

Wallace, A. R. 1923. Darwinism: an Exposition of the Theory of Natural Selection with some of its Applications. Macmillan & Co., London.

Wallace, B. 1968. Topics in Population Genetics. Norton, New York.

Wallace, B., and Th. Dobzhansky. 1959. Radiation, Genes, and Man. Holt, New York.

Walter, R., and B. K. Epperson. 2001. Geographic pattern of genetic variation in *Pinus resinosa*: area of greatest diversity is not the origin of postglacial populations. Molecular Ecology 10:103–111.

Wang, J. L. 2002. An estimator for pairwise relatedness using molecular markers. Genetics 160:1203–1215.

Wang, J. L., A. Caballero, P. D. Keightley, and W. G. Hill. 1998. Bottleneck effect on genetic variance: a theoretical investigation of the role of dominance. Genetics 150:435–447.

Wang, J. L., and N. Ryman. 2001. Genetic effects of multiple generations of supportive breeding. Conservation Biology 15:1619–1631.

Wang, W., and H. Lan. 2000. Rapid and parallel chromosomal number reductions in muntjac deer inferred from mitochondrial DNA phylogeny. Molecular Biology and Evolution 17:1326–1333.

Waples, R. S. 1987. A multispecies approach to the analysis of gene flow in marine shore fishes. Evolution 41:385–400.

Waples, R. S. 1989. A generalized approach for estimating effective population size from temporal changes in allele frequency. Genetics 121:379–391.

Waples, R. S. 1990. Conservation genetics of Pacific salmon: III. Estimating effective population size. Journal of Heredity 81:277–289.

Waples, R. S. 1991. Pacific salmon, *Oncorhynchus* spp., and the definition of "species" under the Endangered Species Act. Marine Fisheries Review 53:11–22.

Waples, R. S. 1995. Evolutionarily significant units and the conservation of biological diversity under the Endangered Species Act. American Fisheries Society Symposium 17:8–27.

Waples, R. S. 1998. Evolutionarily significant units, distinct population segments, and the endangered species act: reply to Pennock and Dimmick. Conservation Biology 12:718–721.

Waples, R. S. 1999. Dispelling some myths about hatcheries. Fisheries 24:12–21.

Waples, R. S. 2002. Definition and estimation of effective population size in the conservation of endangered species. Pp. 147–168 *in* Beissinger, S. R., and D. R. McCullough, eds. Population Viability Analysis. University of Chicago Press, Chicago.

Waples, R. S. 2006. Distinct population segments. *In* Scott, J. M., D.D. Goble, and F.W. Davis, eds. The Endangered Species Act at Thirty: Conserving Biodiversity in Human-Dominated Landscapes. Island Press, Washington, DC.

Waples, R. S. In press. A bias correction for estimates of effective population size based on linkage disequilibrium at unlinked gene loci. Conservation Genetics.

Ward, R. D., D. O. F. Skibinski, and M. Woodwark. 1992. Protein heterozygosity, protein structure, and taxonomic differentiation. Evolutionary Biology 26:73–160.

Waser, P. M., and C. Strobeck. 1998. Genetic signatures of interpopulation dispersal. Trends in Ecology and Evolution 13:43–44.

Waters, J. M., L. H. Dijkstra, and G. P. Wallis. 2000. Biogeography of a southern hemisphere freshwater fish: how important is marine dispersal? Molecular Ecology 9:1815–1821.

Watson, A. 2000. New tools. A new breed of high-tech detectives. Science 289:850–854.

Watt, W. B. 1995. Allozymes in evolutionary genetics: beyond the twin pitfalls of "Neutralism" and "Selectionism". Revue Suisse de Zoologie 102:869–882.

Wayne, R. K., L. Forman, A. K. Newman, J. M. Simonson, and S. J. O'Brien. 1986. Genetic markers of zoo populations: morphological and electrophoretic assays. Zoo Biology 5:215–232.

Wayne, R. K., and P. A. Morin. 2004. Conservation genetics in the new molecular age. Frontiers Ecology Environment 2:89–97.

Weber, E., and B. Schmid. 1998. Latitudinal population differentiation in two species of *Solidago* (Asteraceae) introduced into Europe. American Journal of Botany 85:1110–1121.

Weinberg, W. 1908. Uber den Nachweis der Vererbung beim Menschen. Jahresh. Verein f. vaterl. Naturk in Wruttemberg 64:368–382.

Weins, J. A. 1977. On competiton and variable environments. American Scientist 65:591–597.

Weir, B. S. 1996. Genetic Data Analysis II. Sinauer, Sunderland, MA.

Weir, B. S., and W. G. Hill. 1980. Effect of mating structure on variation in linkage disequilibrium. Genetics 95:477–488.

Welsh, J., and M. McClelland. 1990. Fingerprinting genomes using PCR with arbitrary primers. Nucleic Acids Research 18:7213–7218.

Westemeier, R. L., and 8 others. 1998. Tracking the long-term decline and recovery of an isolated population. Science 282:1695–1698.

Wheeler, L., and 7 others. 2000. Database resources of the National Center for Biotechnology Information. Nucleic Acids 28:10–14.

White, B. A., and J. B. Shaklee. 1991. Need for replicated electrophoretic analyses in multiagency genetic stock identification (GSI) programs – examples from a pink salmon (*Oncorhynchus gorbuscha*) GSI fisheries study. Canadian Journal of Fisheries and Aquatic Sciences 48:1396–1407.

White, M. J. D. 1973. Animal Cytology and Evolution, 3rd edn. Cambridge University Press, Cambridge, UK.

White, M. J. D. 1978. Modes of Speciation. W. H. Freeman & Co., San Francisco.

Whitlock, M. C., P. K. Ingvarsson, and T. Hatfield. 2000. Local drift load and the heterosis of interconnected populations. Heredity 84:452–457.

Whitlock, M. C., and D.E. McCauley. 1999. Indirect measures of gene flow and migration: F_{ST} doesn't equal $1/(4Nm+1)$. Heredity 82:117–125.

Whitlock, M. C., and S. P. Otto. 1999. The panda and the phage: compensatory mutations and the persistence of small populations. Trends in Ecology and Evolution 14:295–296.

Whitlock, M. C., P. C. Phillips, F. B. G. Moore, and S. J. Tonsor. 1995. Multiple fitness peaks and epistasis. Annual Review of Ecology and Systematics 26:601–629.

Wiegand, K. M. 1935. A taxonomist's experience with hybrids in the wild. Science 81:161–166.

Wilding, C. S., R. K. Butlin, and J. Grahame. 2001. Differential gene exchange between parapatric morphs of *Littorina saxatilis* detected using AFLP markers. Journal of Evolutionary Biology 14:611–619.

Williams, J. G. K., A. R. Kubelik, J. Livak, J. A. Rafalski, and S. V. Tingey. 1990. DNA polymorphisms amplified by arbitrary primers are useful as genetic markers. Nucelic Acids Research 18:6531–6535.

Williamson, E. G., and M. Slatkin. 1999. Using maximum likelihood to estimate population size from temporal changes in allele frequencies. Genetics 152:755–761.

Willis, J. H. 1993. Partial self-fertilization and inbreeding depression in two populations of *Mimulus guttatus*. Heredity 71:145–154.

Willis, J. H. 1999. The role of genes of large effect on inbreeding depression in *Mimulus guttatus*. Evolution 53:1678–1691.

Willis, J. H., and H. A. Orr. 1993. Increased heritable variation following population bottlenecks: the role of dominance. Evolution 47:949–957.

Willis, K., and R. J. Wiese. 1997. Elimination of inbreeding depression from captive populations: Speke's gazelle revisited. Zoo Biology 16:9–16.

Wilson, A. C., G. L. Bush, S. M. Case, and M. C. King. 1975. Social structuring of mammalian populations and rate of chromosomal evolution. Proceedings of the National Academy of Sciences, USA 72:5061–5065.

Wilson, A. C., and 10 others. 1985. Mitochondrial and two perspectives on evolutionary genetics. Biological Journal of the Linnean Society 26:375–400.

Wilson, E. O. 1984. Biophilia. Harvard University Press, Cambridge, MA.

Wilson, E. O. 2002. The Future of Life. Alfred A. Knopf, New York.

Wilson, F. 1965. Biological control and the genetics of colonizing species. Pp. 307–325 *in* Baker, H. G., and G. L. Stebbins, eds. The Genetics of Colonizing Species. Academic Press, New York.

Woinarski, J. C. Z., and A. Fisher. 1999. The Australian Endangered Species Protection Act 1992. Conservation Biology 13:959–962.

Wolf, D. E., N. Takebayashi, and L. H. Rieseberg. 2001. Predicting the risk of extinction through hybridization. Conservation Biology 15:1039–1053.

Wood, C. C., and C. J. Foote. 1990. Genetic differences in the early development and growth of sympatric sockeye salmon and kokanee (*Oncorhynchus nerka*), and their hybrids. Canadian Journal of Fisheries and Aquatic Sciences 47:2250–2260.

Wood, C. C., and C. J. Foote. 1996. Evidence for sympatric genetic divergence of anadromous and nonanadromous morphs of sockeye salmon (*Oncorhynchus nerka*). Evolution 50:1265–1279.

Wood, P. M. 2003. Will Canadian policies protect British Columbia's endangered pairs of sympatric sticklebacks? Fisheries 28(5):19–26.

Woodruff, D. S. 2001. Declines of biomes and the future of evolution. Proceedings of the National Academy Sciences, USA 98:5471–5476.

Woodruff, R. C., H. Huai, and J. N. Thompson Jr. 1996. Clusters of identical new mutations in the evolutionary landscape. Genetica 98:149–160.

Woodruff, R. C., and J. N. Thompson. 1992. Have premeiotic clusters of mutation been overlooked in evolutionary theory? Journal of Evolutionary Biology 5:457–464.

Woodworth, L. M., M. E. Montgomery, D. A. Briscoe, and R. Frankham. 2002. Rapid genetic deterioration in captive populations: causes and conservation implications. Conservation Genetics 3:277–288.

Wright, D. A., and C. M. Richards. 1983. Two sex-linked loci in the leopard frog, *Rana pipiens*. Genetics 103:249–261.

Wright, S. 1921. Systems of mating, I: the biometric relation between parent and offspring. Genetics 6:111–123.

Wright, S. 1922. Coefficients of inbreeding and relationship. American Naturalist 56:330–338.

Wright, S. 1923. Mendelian analysis of the pure breeds of livestock. I. The measurment of inbreeding and relationship. Journal of Heredity 14:339–348.

Wright, S. 1931. Evolution in Mendelian populations. Genetics 16:97–159.

Wright, S. 1939. Statistical Genetics in Relation to Evolution. Actualités scientifiques et industrielles 802. Herman & Cie, Paris. (Republished in Provine 1986.)

Wright, S. 1940. Breeding structure of populations in relation to speciation. American Naturalist 74:232–248.

Wright, S. 1943. Isolation by distance. Genetics 28:114–138.

Wright, S. 1945. Tempo and mode in evolution: a critical review. Ecology 26:415–419.

Wright, S. 1951. The genetical structure of populations. Annals of Eugenics 15:323–354.

Wright, S. 1960. On the number of self-incompatibility alleles maintained in equilibrium by a given mutation rate in a population of given size: a re-examination. Biometrics 16:61–85.

Wright, S. 1965a. The distribution of self-incompatibility alleles in populations. Evolution 18:609–619.

Wright, S. 1965b. The interpretation of population structure by F-statistics with special regard to systems of mating. Evolution 19:355–420.

Wright, S. 1969. Evolution and the Genetics of Populations. Vol. 2. The Theory of Gene Frequencies. University of Chicago Press, Chicago.

Wu, C.-I. 2001. The genic view of the process of speciation. Journal of Evolutionary Biology 14:851–865.

Xu, X., and U. Arnason. 1996. The mitochondrial DNA molecule of Sumatran and a molecular proposal for two (Bornean and Sumatran) species of orangutan. Journal of Molecular Evolution 43:431–437.

Young, A. G., A. H. D. Brown, B. G. Murray, P. H. Thrall, and C. H. Miller. 2000a. Genetic erosion, restricted mating and reduced viability in fragmented populations of the endangered grassland herb *Rutidosis leptorrhynchoides*. Pp. 335–359 in Young, A. G., and G. M. Clarke, eds. Genetics, Demography and Viability of Fragmented Populations. Cambridge University Press, Cambridge, UK.

Young, A. G., C. Millar, E. A. Gregory, and A. Langston. 2000b. Sporophytic self-incompatibility in diploid and tetraploid races of *Rutidosis leptorrhynchoides*. Australian Journal of Botany 48:667–672.

Young, A. G., and B. G. Murray. 2000. Genetic bottlenecks and dysgenic gene flow into re-established populations of the grassland daisy, *Rutidosis leptorrhynchoides*. Australian Journal of Botany 48:409–416.

Yunis, J. J., and O. Prakash. 1982. The origin of man: a chromosomal pictorial legacy. Science 215:1525–1529.

Zakharov, V. M. 2001. Ontogeny and population: developmental stability and population variation. Russian Journal of Ecology 32:146–150.

Zapata, C. 2000. The D' measure of overall gametic disequilibrium between pairs of multiallelic loci. Evolution 54:1809–1812.

Zapata, C., and G. Alvarez. 1992. The detection of gametic disequlbrium between akkozyme loci in natural populations of Drosophila. Evolution 46:1900–1917.

Zardoya, R., A. Garrido-Pertierra, and J. M. Bautista. 1995. The complete nucleotide sequence of the mitochondrial DNA genome of the rainbow trout, *Oncorhynchus mykiss*. Journal of Molecular Evolution 41:942–951.

Zeder, M. A., and B. Hess. 2000. The initial domestication of goats (*Capra hircus*) in the Zagros mountains 10,000 years ago. Science 287:2254–2257.

Zhi, L., and 9 others. 1996. Genomic differentiation among natural populations of orang-utan (*Pongo pygmaeus*). Current Biology 6:1326–1336.

Zimmer, C. 2002. Darwin's avian muses continue to evolve. Science 296:633–635.

Index

Page numbers in *italic* refer to figures and / or tables, those in **bold** refer to guest boxes

corn snake (*Elaphe guttata*), 458
correlation, 549
cottonwood (*Populus deltoides*), 9
countergradient variation, 28–9, **29–30**
coyote (*Canis latrans*), 198, 337, 392, 509
cpDNA *see* DNA, chloroplast
crocodile (*Crocodylus*), 508
 cladogram, *389–90*
 phenogram, *389–90*
 phylogeny, *386*
crops, loss of primitive varieties, 12, **13**
cryopreservation, 458
cultivars, use in restoration programs, 476, 478, 479

D' coefficient, 239
dace
 finescale (*Phoxinus neogaeus*), 427
 northern redbelly (*Phoxinus eos*), 427
dalmatian, 139
Daphnia, 277
Darwin, Charles, 16, 139, 171, 305, 334
Darwin's finches, *156–7, 264*, 425
deer, 11, 140
 fallow (*Cervus dama*), 334
 muntjac (*Muntiacus* spp.), 40
 red (*Cervus elaphus*), 275
degrees of freedom, 100–1
demes, 197–232
 complete isolation, 204–5
 and conservation, 226–7
 cytoplasmic genes, 211–13
 differentiation for quantitative traits, 276–8,
 281–2
 F-statistics, 199–204
 gametic disequilibrium, 245–6
 gene flow among, 205–10, 214–18
 genetic drift, 206–10
 genetic variation between, 52–4, 81–2, 198, 201,
 218–20, 276–8, 281–2
 and hierarchical population structure, 219–20
 mutations, 294–5
 natural selection, 214–18
 sex-linked markers, 213–14
demographic rescue, 220
demographic stochasticity, 10, 342
demographic swamping, 428
demography
 criteria, 349–50
 and extinction, 10, 334–62
dendrograms, 394–8
desert spider, 424, *425*
developmental noise, 23
diagnostic loci, 435, 437

dik-dik (*Madoqua* spp.), 35
dinosaurs, 385
dioecy, 149
diploidy, 39
directional selection, 175–6, 186–7
 differential, 217
 divergent, 216
disassortative mating, 46
disease resistance, 355
dispersal
 definition, 199
 estimates of, 224
distinct population segment (DPS), 226, 381, 407,
 408
DNA
 arrays, 80–1
 "barcoding", 504
 chloroplast, 65, 67, 159, 504
 chromosomal, 36–7
 deletion, 288
 "fingerprinting", 75–6, 505, 509–13
 insertion, 288
 inversion, 288
 markers, 504–5
 mitochondrial, 64–5, 67
 effect of bottlenecks, *160–2*
 in genetic identification, 504, 513
 genetic variation, 159–62
 introgression, 426
 lineage sorting, 162–3
 mutations, 288, 355–7
 nonrandom associations with nuclear loci,
 239
 variation between demes, 211–13
 multilocus techniques, 74–7
 mutations, 288–90
 recombination, 288
 sequencing, 78, 80
 single locus techniques, 69–74
 substitution, 288
 variations in populations, 63–90
DNA Surveillance program, 505, *506–7*
Dobzhansky, Theodosius, 35, 543
Dobzhansky-Muller incompatibilities, 432
dog (*Canis familiaris*), 4, 139
dolphin
 Hector's (*Cephalorhynchus hectori*), **227–9**
 Maui's (*Cephalorhynchus hectori maui*), *229*
domestication of animals, 4
dominance, 176
donkey (*Equus asinus*), 40, 429
Douglas fir (*Pseudotsuga menziesii*), 320, *321*
DPS, 226, 381, 407, *408*